Contents

Preface vii

Acknowledgements ix

1 Environmental Geology: Definition, Scope and Tools 1

2 The Geology of Resource Management 27

3 Economic Mineral Resources 37

4 Construction Resources: Geomaterials 87

5 Water Resources 129

6 Aesthetic and Scientific Geological Resources 161

7 Engineering Geology 195

8 Engineering Geology in Extreme Environments 245

9 Waste and Pollution Management 277

10 The Geology of Natural Hazards 341

11 Exogenic Hazards 351

12 Endogenic Hazards 421

13 Environmental Geology: an Urban Concept 457

Index 489

Environmental Geology

Geology and the Human Environment

MATTHEW R. BENNETT

and

PETER DOYLE

School of Earth & Environmental Sciences,
University of Greenwich, UK

JOHN WILEY & SONS
Chichester • New York • Weinheim • Brisbane • Singapore • Toronto

National 01243 779777
International (+44) 1243 779777
e-mail (for orders and customer service enquiries):
cs-books@wiley.co.uk.
Visit our Home Page on http://www.wiley.co.uk
or http://www.wiley.com

Reprinted April 1999

Other Wiley Editorial Offices

John Wiley & Sons, Inc., 605 Third Avenue,
New York, NY 10158-0012, USA

WILEY-VCH Verlag GmbH, Pappelallee 3,
D-69469 Weinheim, Germany

Jacaranda Wiley Ltd, 33 Park Road, Milton,
Queensland 4064, Australia

John Wiley & Sons (Asia) Pte Ltd, 2 Clementi Loop #02-01,
Jin Xing Distripark, Singapore 129809

John Wiley & Sons (Canada) Ltd, 22 Worcester Road,
Rexdale, Ontario M9W 1L1, Canada

Library of Congress Cataloging-in-Publication Data

Bennett, Matthew (Matthew R.)
Environmental geology : geology and the human environment / by Matthew R. Bennett and Peter Doyle.
p. cm.
Includes bibliographical references (p. –) and index.
ISBN 0-471-97459-5 (pbk.)
1. Environmental geology. I. Doyle, Peter. II. Title.
QE38.B46 1997
550—dc21
97–15135
CIP

British Library Cataloguing in Publication Data

A catalogue record for this book is available from the British Library

ISBN 0-471-97459-5

Typeset in 10/12pt Palatino from authors' disks by Mayhew Typesetting, Rhayader, Powys
Printed and bound in Great Britain by Bookcraft (Bath) Limited, Midsomer Norton, Somerset

This book is printed on acid-free paper responsibly manufactured from sustainable forestation, for which at least two trees are planted for each one used for paper production.

Preface

In the past few years interest in the environment has reached a peak as popular opinion has become aware of the extent of the human impact on natural systems. A proliferation of degrees has followed this wave of 'environmentalism'; their focus has been on natural areas, endangered species and the damage caused by human exploitation of the biological and physical resources of our planet. Environmental geology – the interaction of humans with the geological environment – is becoming increasingly popular as a degree option, but is not a new subject. In fact, it is a meld of three related Earth science disciplines – economic geology, engineering geology and applied geomorphology – each of which has been given a new focus through the need for greater environmental management. The true challenge of environmental geology does not lie in rural areas or in green issues, but in the urban environment. By the year 2000 over 3.5 billion people, 50% of the world's population, will live in urban areas covering just 1% of the Earth's surface. It is here that human interaction with the geological environment is at its most intense; it is here that many of the practical challenges lie. Urban growth fuels the demand for mineral and water resources, tests our skill as engineering geologists, produces vast volumes of waste which must be managed, and increases human vulnerability to natural hazards. Environmental geology is a practical subject, and environmental geologists have a crucial role to play in managing this human interaction with the geological environment. It is with this background that we have tried to produce a textbook which emphasises the practical contribution that geologists can make in managing our interaction with the physical environment.

Matthew R. Bennett
and Peter Doyle
Chatham Maritime, 1997

Acknowledgements

The writing of any text inevitably draws upon the accumulated wisdom of innumerable colleagues. This book is written for undergraduates, and is consequently broad in approach, with little space for the explanation of detail. We apologise if this appears to neglect the valuable work of others; we alone are responsible for the style, content and any errors or omissions. We are indebted to many people for their help, advice, encouragement and forbearance in the production of this book. In particular, Julie Doyle helped both of us with her wise words and encouragement. Drafts of the text were reviewed by Alistair Baxter, Andy Bussell, Neil Glasser, Murray Gray, Jonathon Larwood, Alan McKirdy, Duncan Pirrie, Colin Prosser and Ted Rose. Their comments have done much to improve the clarity and accuracy of our message. We would like to thank all those who supplied photographs for the book, particularly Andrew Bennett, who deserves special mention. Linda Murr, at the University of Greenwich, provided valuable secretarial assistance during times of crisis. Many of the illustrations within this book are not original, although all have been modified to some extent, and where this is the case, the source used for the new drawing is indicated in the caption. To all these people and the many more not mentioned we would like simply to say 'Thank you'.

1
Environmental Geology: Definition, Scope and Tools

1.1 ENVIRONMENTAL GEOLOGY: A DEFINITION

Environmental geology is the interaction of humans with the geological environment. The geological environment includes not only the **physical constituents** of the Earth – its rocks, sediments, soils and fluids – but also the **surface** of the Earth, its landforms, and in particular the processes which operate to change it through time. This environment is both a resource and a hazard to human development, and is essential to life. It provides us with water, industrial minerals, building materials and fuel. It constrains the location and architecture of our urban settlements and transport infrastructure. It provides us with the means for the effective disposal of our waste products. However, although the geological environment provides us with the essential elements for human endeavour, it also generates some of the most potent hazards to our existence in the form of earthquakes, volcanoes and floods.

Environmental geology may be considered as a subset of environmental science. **Environmental science** is the study of the interaction of humans with all aspects of their environment, its physical/geological, atmospheric and biological components. Environmental science is our basic understanding of the Earth. From this starting point it considers both the debits (impacts) and credits (benefits) of our coexistence with other species and environments. Put simply, environmental science gives us the scientific basis with which to maximise human success, while minimising its adverse impacts on other elements of the environment. Environmental geology therefore has as its central philosophy the concept of environmental management, working with natural geological systems to sustain development, but not at an unacceptable environmental cost. It has four main components:

1. Managing geological resources, such as fossil fuels, industrial minerals and water: this is a process which involves not only mineral exploration and exploitation, but also limitation and mitigation of the environmental damage caused by this resource use.
2. Understanding and adapting to the constraints on engineering and construction imposed by the geological environment, a subject of particular importance in regions of climatic extreme.
3. Appropriate use of the geological environment for waste disposal in order to minimise problems of contamination and pollution.
4. Recognition of natural hazards and mitigation of their human impacts.

These areas represent the scope of environmental geology, the human interaction with the geological environment. In the next section we discuss the concept that this interaction is at its most acute in the urban environment.

1.2 ENVIRONMENTAL GEOLOGY: A MODEL

To some, environmental geology may appear to be simply the documentation of the environmental impacts of the geological exploitation of mineral resources, the geological disposal of waste, major construction and engineering projects, or natural hazards. The emphasis is placed on documenting environmental damage, identifying its cause and providing for its mitigation. This approach, although valid, tends to cast all geological exploitation as potentially damaging, leading to conflict with society's need for resources and for defence against natural hazards.

An alternative view is that environmental geology is about management of a natural system, a concept which has recently been subsumed into the popular term 'sustainable development'. While this term often means different things to different people, it is here defined as: the management of natural resources to support continued social and economic development in such a way that renewable resources are not depleted, and the impact of the extraction and use of non-renewable resources is minimised. Put simply, this provides for development, but not at an unacceptable cost to the natural environment. The key to sustainable development is effective environmental management; for example, in balancing the need for the exploitation of geological resources with the impact that this may cause. This is the definition and approach which is followed throughout this book. We examine not only the environmental consequences of human interaction with the geological environment, but also its effective management in order to provide future mineral resources, a safe built environment and appropriate waste disposal strategies, and to mitigate the impact of natural hazards. Environmental geology is therefore about managing human interaction with the geological environment, an interaction which is at its most challenging within, or close to, the urban environment.

By the end of the century over 3.5 billion people, approximately 50% of the world's population, will live in urban areas which will cover just 1% of the

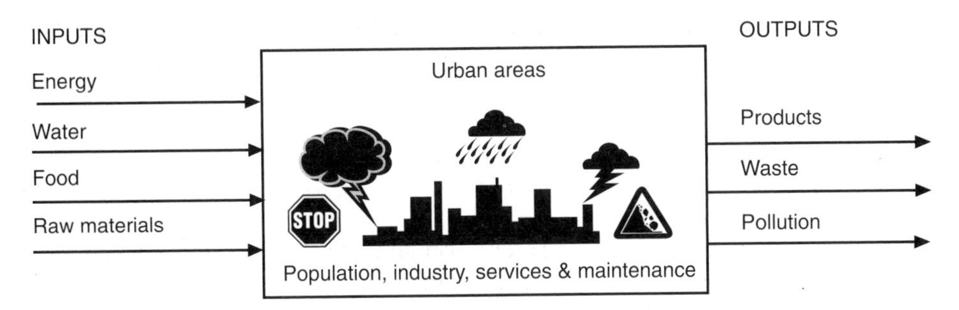

Figure 1.1 *The urban machine: a balance of inputs, outputs and maintenance*

Earth's surface. It is in these areas where human interaction with the geological environment will be at its most intense; it is here that environmental geology has its main focus. Urban growth fuels the demand for mineral and water resources, tests our skills as engineering geologists, produces vast volumes of waste which must be managed, and increases our vulnerability to natural hazards. These are the component subjects of environmental geology discussed within this book, and help define its scope.

The urban environment can be likened to a machine which consumes inputs from the natural environment and produces outputs (Figure 1.1). Inputs include: (1) water, derived either locally or from distant reservoirs and rivers; (2) raw materials, in the form of mineral resources for industry and construction, traditionally extracted from the immediate hinterland, but with increasing transport infrastructure, now derived globally; (3) food resources, both locally produced and via import; and (4) energy, as the end product of the use of physical resources such as coal, gas and uranium. Output from this urban machine includes: (1) the products of industry and commerce; (2) waste, in the form of worn-out materials, by-products from industry and day-to-day wastes from domestic, industrial and commercial sources; and (3) pollution caused by poor waste management strategies which overload the ability of the natural atmospheric, land and water systems to recycle and redistribute waste gases, solids and liquids. The machine requires constant maintenance in the form of infrastructure upgrading and rebuilding; its foundations rely upon the stability of the geological setting; and its security is threatened by natural hazards both from within the Earth and from its surface processes.

Figure 1.2 shows a schematic model of an urban centre and its resource hinterland as a machine. In this model the city draws much of its basic bulk mineral resource, such as aggregates, water and industrial minerals, from its immediate area. Historically, the growth of most major urban centres has been driven by the availability of such resource hinterlands, although with sophisticated transport links this may be increasingly removed from the city (Figure 1.2). These distant links with the urban environment have their own problems, and distant and remote mineral workings will remain economic only while the unit price of the minerals extracted is greater than the cost of exploitation and transport. Environmental debits may be high, but urban demand fuels their

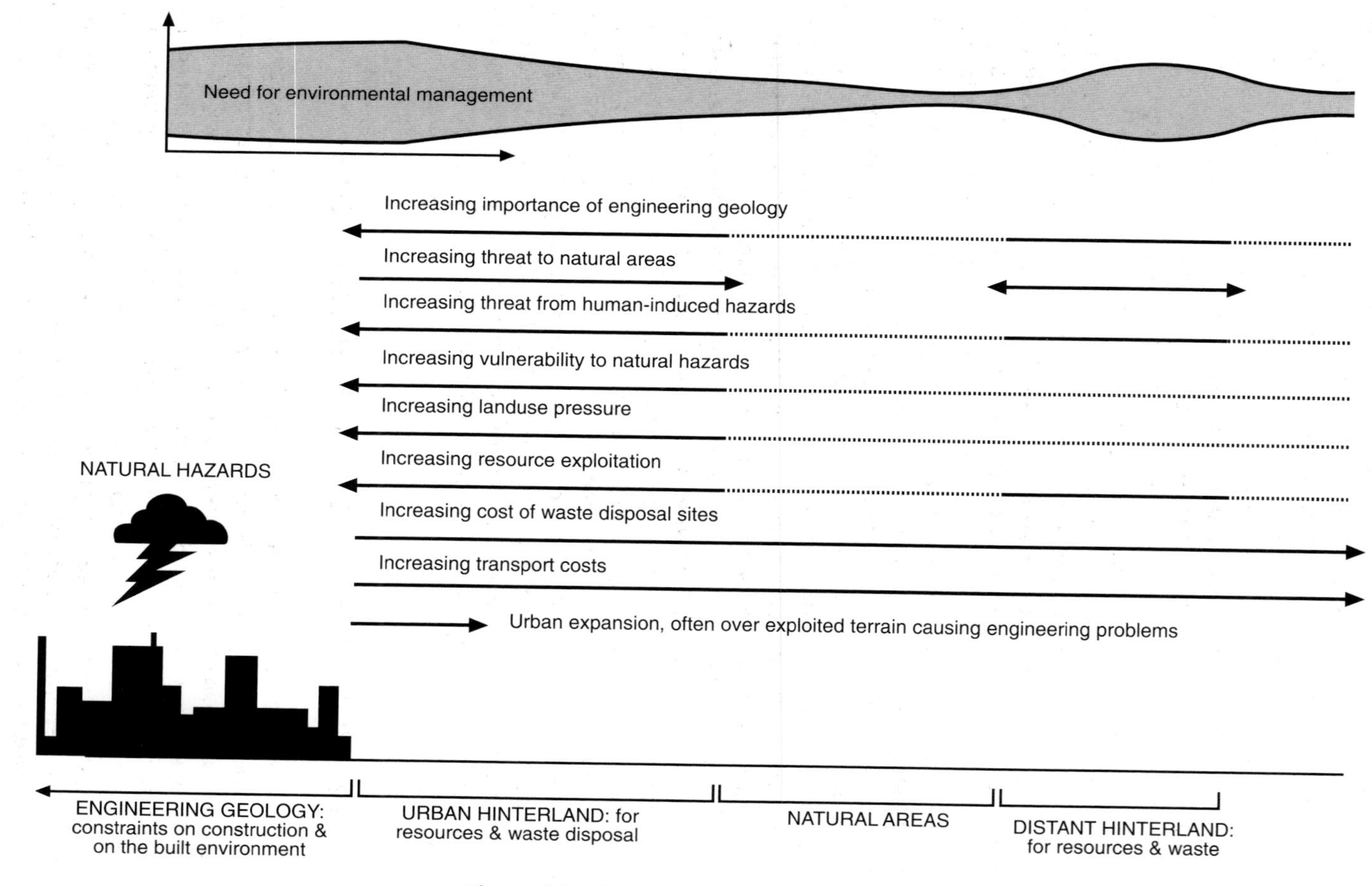

Figure 1.2 *Environmental geology: a model*

continued success, and they remain a far-removed part of the urban machine. The immediate urban hinterland is most important in its supply of cheap bulk minerals such as aggregates, which cannot support high transport costs. Equally, the majority of waste produced by an urban area must be disposed of close to the city to minimise disposal costs, usually as landfill in disused or exhausted quarries. London provides an excellent example, as it is ringed by sand and gravel pits, most of which have been infilled with waste. Consequently, the need for effective environmental management of the resource hinterland is greatest close to the edge of the city, particularly since this area often contains agricultural, aesthetic, recreational and scientific resources which will subsequently be subject to land-use pressure as the city grows. Geological ground conditions, the subject of engineering geology, constrain and influence both construction and urban development within the city (Figure 1.2). Equally, the environment itself may threaten the urban machine through the action of natural hazards. Geological hazards are natural processes which only become a problem when they interact with human infrastructure. The concentration of people in urban areas increases human vulnerability to even the most minor hazard.

The model presented demonstrates that while environmental geology is not restricted to the urban environment, particularly since the adverse impacts of mineral extraction and waste disposal can occur beyond the city limits (Figure 1.2), many of the components, such as engineering geology, waste management and hazard mitigation, are focused on it for the most part.

In the first part of this book we look at the management of geological resources: mineral resources, construction resources, water resources, and the aesthetic and scientific resources provided by the landscape. This is followed by two chapters which examine engineering geology: the constraints imposed on the urban environment by the rocks and soils which occur beneath it. We then consider the management of waste and pollution generated by the built environment. Three chapters follow which consider the impact and management of natural hazards. Finally, the application and integration of these various aspects of environmental geology are illustrated with respect to urban areas in the concluding chapter.

1.3 THE HISTORY OF ENVIRONMENTAL GEOLOGY

Humans have long been interested in the exploitation of the physical resources of the Earth, and in taming the landscape for purposes of agriculture, habitation and commerce. Traditionally these subjects have been the preserve of applied Earth scientists who have focused their attention on limited aspects of resource exploitation, civil engineering and hazard mitigation. Environmental geology has been born from the need for interaction of the three main fields of applied Earth science (Figure 1.3): applied geomorphology, economic geology and engineering geology. **Applied geomorphology** has its origins in physical geography and is primarily concerned with the hazards and limits to human development imposed by the processes which shape the Earth's surface. The focus of **economic geology**

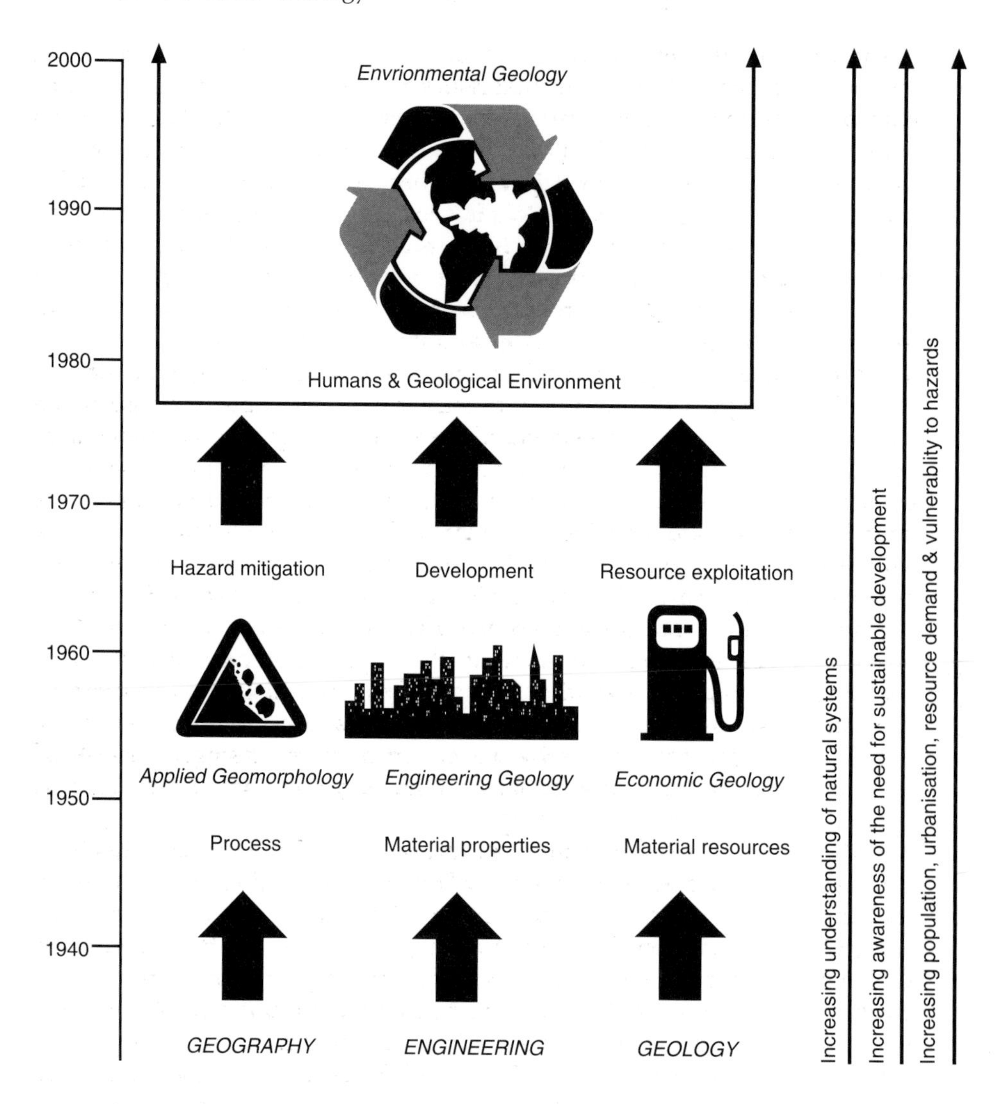

Figure 1.3 *Environmental geology as the product of applied geomorphology, engineering geology and economic geology*

is the economic exploitation of Earth's materials. **Engineering geology** is an essential part of civil engineering which requires information on the strength and behaviour of the Earth's surface and its materials during and after construction. These subject areas are broadly complementary, and assembled as a coherent whole in environmental geology they provide a focus for research and training. The need for this approach has become increasingly apparent during the last 30 years (Figure 1.3). The increasing interaction between these three fields of applied research, and the greater environmental focus present within them, is the product of three main trends.

1. **Sustainable development**. In the last 30 years there has been an increasing political awareness of environmental issues and in particular the need for sustainable development. The term was popularised by the United Nations Earth Summit in Rio de Janeiro in 1992, when 170 world leaders agreed a global agenda to help promote development within what is naturally sustainable by the Earth. In geological terms this has led to a shift in focus from resource exploitation to managing the impacts caused by such exploitation and disposing safely of the wastes produced by it. In practice this usually means seeking compromise between the interests of development/exploitation and the conservation of the natural environment and its regulatory system.
2. **Conflict versus management of natural process**. In dealing with natural hazards, two types of response can be identified: a structural response, and a process-based response. The **structural response** usually involves civil engineering work to combat and contain problems; a typical example is the use of sea walls to prevent coastal erosion. In contrast, the **process-based response** is founded on an understanding of the natural system and attempts to work with it to combat a given problem. For example, this might involve enhancing natural defences to coastal erosion by beach nourishment schemes. Experience has shown that the structural response may simply transfer a problem elsewhere, often exacerbating it in doing so. This is particularly the case in coastal erosion problems, where erosion may simply be transferred down-coast by the construction of a sea wall or similar structural defence. Fortunately, the application of process-based solutions is becoming increasingly possible as our understanding of natural surface processes has improved dramatically, particularly during the last 20 years.
3. **A shift from reactive to proactive involvement**. Applied Earth scientists have traditionally had a reactive role: a hazard or resource requirement is identified and an engineering solution or new exploration frontier is sought. However, in dealing with geomorphological hazards, prevention is often better than cure. For example, by promoting development away from vulnerable areas, costly mitigation work in the future can be avoided. Over the last 20 years the need for the reactive response has diminished. There has been a progressive shift towards proactive work within applied Earth science as the importance of its contribution to environmental and resource management has become better appreciated by decision-makers (Figure 1.3). Typically, the information required by such decision-makers as land-use planners is very broad in scope, requiring information on resource distribution, hazards and ground conditions. The supply of this information requires a greater integration between the three branches of applied Earth science and consequently has led to the first real and effective integration between applied geomorphology, engineering geology and economic geology in order to deliver the appropriate data.

There is a need, therefore, to take an integrated view of applied geology in its modern context: human interaction with the geological environment. The subject of environmental geology allows this and encompasses all three of the traditional applied Earth science subjects. Its scope understandably varies according to the perspective and training of the observer. For example, traditional geomorphologists might give more emphasis to hazard and construction than a traditional geologist, who might emphasise the need for mineral resource exploration and exploitation. Given time, such artificial boundaries will disappear as the modern subject of environmental geology continues to develop.

1.4 ENVIRONMENTAL GEOLOGY AND COMMERCIAL REALITY

The activities of environmental geologists are primarily driven by the requirements of business, and of local, regional and national governments. There are two types of research and development activity within environmental geology (Figure 1.4): (1) reactive research/action; and (2) proactive research/action. Reactive research/action occurs in response to some crisis or other urgent demand. The flooding of a town centre or the threat of accelerated coastal erosion are typical examples. Finding solutions to these pressing problems is clearly an important task for environmental geologists. In contrast, proactive research/action involves forward planning to influence environmental management so that perceived or real problems are avoided. Clearly, in practice it is usually better and cheaper to avoid a problem than to find an urgent solution to some crisis, and proactive environmental geology seeks to achieve this.

In most cases, research in environmental geology, whether proactive or reactive, is undertaken by contract, and therefore the commercial relationship between client and consultant is central to its practice. This relationship can take a variety of forms, but is summarised in generic terms in Table 1.1.

Initially, the client recognises a problem or the need for geological information and makes the decision to acquire it. After this initial decision has been taken, which is often the most difficult stage, a project brief or outline is formulated which identifies the nature of the problem, the constraints on its solution and the type of output or product required. This brief is then put out to tender. Tenders may be internally derived from within a company or organisation, but it is often more common to expect external, competitive tendering. Usually, a range of commercial consultants, academics or government scientists will tender for the project. Each potential consultant will examine the project brief and will attempt to answer the following questions: How can/should the project be tackled? How much work is involved? Can it be done in the required time? What resources will be required? How much will it cost, in terms of person-hours, materials and expenses? On this basis a tender will be prepared for the client which outlines the professional credentials of the consultant, the approach to be followed and the cost. The emphasis is on finding the quickest, most effective, most efficient and therefore cheapest way of tackling the problem. Planning is essential: mistakes made at this stage may be costly. For example, if the bid is too low and work proves more expensive, then the consultant will make a loss since the client will usually pay only in accordance with the original bid. When all the bids have been submitted, the client will usually make a decision on the basis of

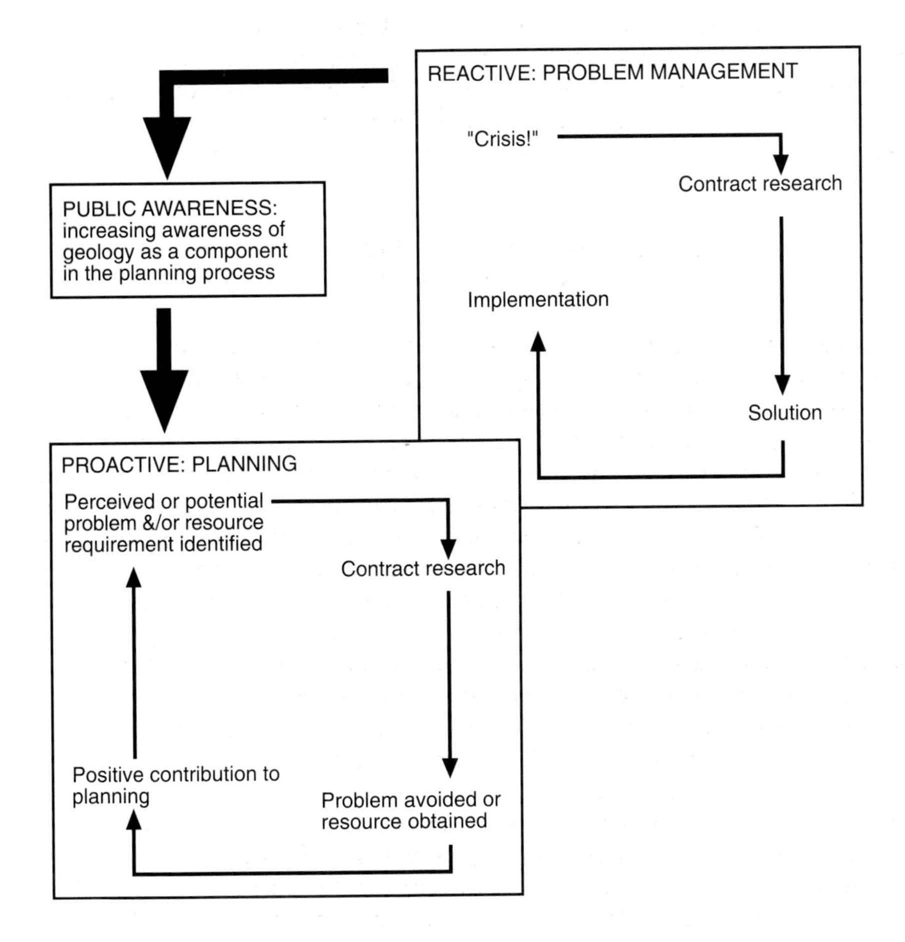

Figure 1.4 *Proactive versus reactive research/action in environmental geology. In the last 30 years there has been a progressive shift from reactive to proactive activity*

cost, quality of approach and the credibility of the consultant. Once contracted, the consultant, together with the client, then refines the project brief to identify the key objective and to establish a control structure, such as a programme of meetings, with which to monitor/report progress. As the work nears completion a draft final report and/or a set of maps is produced which will be subject to rigorous internal audit before its contents are discussed with the client. On this basis, a final version is produced and submitted to the client on the required date. Payment will only be forthcoming after a careful audit in relation to the original brief. For an environmental geologist, producing good quality work is one of the best ways of ensuring future contracts.

The details of this process may vary with the nature of the project and the type of clients involved, but several aspects are consistent: (1) the research needs to be tackled in the most cost-efficient manner (remembering that staff time is

Table 1.1 *The consultant – client interaction in commercial environmental geology*

Client	Consultant	Activities
Stage 1		*Recognition of the problem or information requirement*
Stage 2		*Decision to bring in a consultant*
Stage 3		*Project brief formulated*
Stage 4		*Project put out to tender, usually by invitation based on reputation and contacts*
	Stage 5	Examine the project brief. Can it be tackled in the required time? How? How much work is involved and of what type? How much will it cost?
	Stage 6	Realistic price put on the work and tender submitted with an outline of approach and costing. If the costing is wrong then any profit may be lost; if it is too high then it will not be competitive
Stage 7		*Lowest suitable tender accepted and the contract allocated. Cost and ability to do the job are assessed*
	Stage 8	Consultant starts work. Further definition of the project brief through discussion with client
	Stage 9	Strategy for the project drawn up. Staff appointed or elements of the project subcontracted
	Stage 10	Desk survey, review of maps and published/unpublished literature
	Stage 11	Field survey
	Stage 12	Simulations, modelling and data analysis
	Stage 13	Report stage. Presentation and interpretation of data, and recommendations
	Stage 14	Internal quality control. The quality of the work is important: a reputation for professional work is essential if future contracts are to be won
Stage 15		*Is the work acceptable? Does it meet the requirements set out in the brief? If so the consultant is paid*

an important cost); (2) time is a critical factor – results are usually required quickly; (3) the research is tightly focused on a specific objective; and (4) the product is often used by the non-specialist (i.e. the client) and therefore must be presented concisely, unambiguously and in a format which is intelligible to the non-expert.

The emphasis on time and commercial imperatives is central to the difference between pure 'blue-sky' and applied research methodologies. This difference can be illustrated by considering the two models in Figure 1.5. In this figure, the difference between the applied and pure approaches is represented by two boxes – a black box and a transparent box – both seeking to establish a relationship between beach sediment depletion and cliff erosion. The black box, representing applied research, does not show the detailed processes but identifies a strong empirical (i.e. statistical) and therefore predictive relationship between beach depletion and erosion. The detailed processes which link cause and effect are not resolved. The transparent box, representing pure 'blue-sky' research, shows the intricate detail of process relationships between beach depletion and coastal erosion. In both cases, the end result of the research may be the same – the

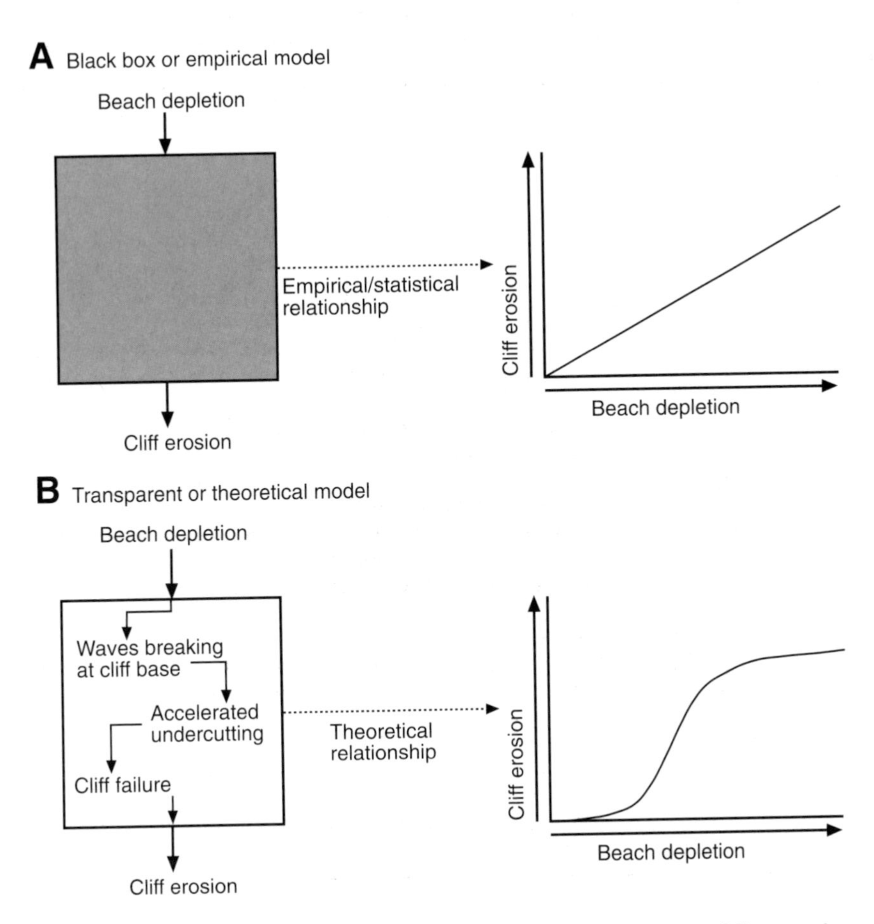

Figure 1.5 *Black box versus transparent box models, illustrating the difference between applied and pure research. Both show a link between beach depletion and coastal erosion, but the detailed basis for this link is only resolved in the transparent model and is only required for academic purposes*

determination of cause and effect – but the need for a complete understanding of the problem is overshadowed by the demands of cost and time for applied research.

Commercial reality is clearly the most important impetus to advancing applied research. In many cases it can have a symbiotic relationship with its pure or 'blue-sky' cousin, highlighting key questions for the advancement of knowledge which cannot be addressed under the constraints of time and cost, and which can only be resolved in universities and research institutes. Applied research is end-product focused, working to specific briefs within set frameworks of time and cost. The themes of cost and time have a recurrent influence on the tools, techniques and strategies adopted by environmental geologists, which are discussed in the following section.

1.5 THE TOOLS OF THE ENVIRONMENTAL GEOLOGIST

The tools used by environmental geologists can be grouped into: (1) desk surveys; (2) field-based documentation or mapping; (3) terrain analysis; (4) environmental monitoring; and (5) presentation tools. Each of these is discussed briefly below.

1.5.1 Desk Surveys

A large amount of information on a region, site or specific problem can be obtained by recourse to archival data sources. These provide a rapid means of obtaining data at a relatively low cost. Typical data sources include: published and unpublished maps; published papers, books and reports; unpublished papers and reports; local and national government records; company records; and statistical records such as meteorological data. The availability of such data depends on the nature of the region or country and the type of problem being tackled.

In addition to archival data, remotely sensed images of all types, from traditional air photographs to satellite images, are becoming commercially available for most parts of the world and contain valuable data for both regional and site-specific projects. Air photographs allow rapid field mapping of land use, geomorphology, geological structure and lithological outcrop. More importantly, temporal variation within the environment can be monitored effectively through sequential air photographs, both those flown specifically for a given project and those available within air photo libraries. The increasing sophistication of image-processing systems now allows digital terrain models to be generated from stereo air photographs, digitally scanned into the computer, allowing effective terrain visualisation and quantification of landform attributes such as slope steepness. Satellite images which contain a much wider range of spectral data can provide information on such things as vegetation, land use, geomorphology, lithology, geological structure, surface water content and pollution plumes both on land and in water bodies. The increasing availability of sequential satellite images has provided new scope in environmental monitoring. Remote sensing is probably, therefore, one of the most important tools available to the environmental geologist.

1.5.2 Field-based Documentation or Mapping

A very wide range of field-based tasks is undertaken by environmental geologists, reflecting the breadth of problems which they tackle. The following list gives some of the key points which must be considered in collecting these data, and which reflect the principles of commercial reality.

1. **Scoping the task**. To be cost efficient only those data pertinent to the particular problem or task in hand must be collected in the field. Equally, all the

information required must be collected during the field visit: it may be impossible or costly to send an environmental geologist back into the field to obtain that missing item. It is important not to duplicate data already obtained from published or unpublished sources. A list of variables relevant to the problem under investigation must therefore be drawn up at the start of the project, either on the basis of past experience or with reference to published literature.

2. **Types of field task**. Three main types of fieldwork can be identified: (1) field checking of an existing map or data source, derived from desk survey or by remote sensing, often referred to as **ground truthing**; (2) the collection of additional or supplementary data to enhance an existing source, which might involve field sampling and testing of, for example, geochemical data, physical properties of rock units or lithological details; and (3) the collection of completely new field data. The type of job being tackled and the quality of the data obtained during the desk survey will determine the relative balance between these three tasks.
3. **Spatial scale**. The spatial scale of a project may vary from point-specific (usually a few metres) to regional (usually many kilometres) in size. Site-specific tasks are usually tackled via detailed site plans, field notes, and annotated sketches and photographs, while problems which involve a larger area are usually map-based, increasingly derived at a regional scale from remote-sensing techniques. In map-based projects a systematic approach is required for data collection. Many geological map-based tasks depend on the identification of land parcels with distinct boundaries, such as the exposure of a particular lithology. However, other data may be better collected with reference to a systematic grid of sampling points. Decisions about data collection must be made with reference to the methods being employed in its subsequent analysis; it is important to collect the data in the correct format.
4. **Observational versus instrumental data**. Observational data are usually compiled systematically on maps, along transects, or on detailed logs and sketches. Typical basic data types include geological outcrop maps, structural maps, hydrogeology maps, geomorphological maps and land-use maps. Geological outcrop maps record the spatial distribution of lithological units, along with varying amounts of structural information, while a structural map records the principal faults, folds and other tectonic features of an area. Hydrogeology maps contain a variety of data items including the extent of hydrological units (see Section 5.1), depth to water table and the principal groundwater flow paths. Geomorphological maps contain details on the morphology of the land surface or the processes which operate on it. They are usually produced by annotating a topographic map, air photograph or detailed site plan with symbols to represent either breaks of slope (morphological map), specific landforms (morphogenetic map) or processes. Land-use maps may be constructed to record a variety of different data and range from broad land-use classification maps, in which land parcels are assigned to such classes as arable, pastoral, industrial or residential land, to detailed land-use maps recording variation in different vegetation communities. In addition to

Table 1.2 *Some examples of environmental geology maps, with the principal name commonly used for them [Modified from: McCall & Marker (1989)* Earth science mapping for planning, development and conservation, *Graham & Trotman, Table 1, p. 10]*

Name	Description
Geopotential maps	Maps demonstrating the resource and development potential of land
Engineering geology maps	Maps which record ground conditions, and rock/soil properties useful in the design of engineering works
Thematic maps	Maps devoted to specific specialist topics
Element maps	Thematic maps showing observational or factual data on a single theme. Examples of these include solid geology maps, surficial geology maps, structural geology maps, geomorphological maps, land-use maps and soil maps
Derived maps	Interpretation maps based on one or more element maps, synthesising several types of information relevant to a single issue
Potential maps	Derived maps which demonstrate potential uses of land, or the potential for processes to occur
Constraint maps	Thematic maps which indicate limitations on the use of land (e.g. agricultural land capability maps in the UK)
Hazard maps	Thematic maps which show the known extent and types of hazards in an area
Risk maps	Maps which attempt to quantify the likelihood of a damaging event of a given type and size occurring
Vulnerability maps	Thematic maps which assess the vulnerability of a population or environment to a particular hazard
Resource maps	Thematic maps indicating the nature, extent and quality of resources, on and under the ground

these basic data maps, a plethora of more specialist maps may be constructed, examples of which are listed in Table 1.2.

Instrumental data are either laboratory-based or field-based and are usually compiled graphically, although plotting of relative values on maps is also common. Laboratory data involve sampling an area according to a specific strategy in order to obtain material for such things as geochemical or geotechnical testing. Field-based instrumental data are obtained, for example, from geophysical techniques, the use of boreholes and trial pits, or in-situ water quality monitoring.

1.5.3 Terrain Analysis

In its broadest sense, terrain analysis is the evaluation or assessment of the physical landscape and its materials in the light of a specific objective. There are two types of terrain analysis: (1) parametric terrain analysis; and (2) physiographic terrain analysis.

Parametric terrain analysis involves the derivation of a secondary map, or series of maps, which summarises some aspect of terrain relevant to a specific task. It is the basic tool by which all hazard maps, ground classification maps, site-location maps and planning zone maps are produced, and is therefore an essential tool in environmental geology. There are two broad approaches to parametric terrain analysis.

1. **Interrogative approach**. Here, individual land parcels or sites are scored against a list of predetermined parameters (Figure 1.6A). Areas may be divided into grids (pixels), slope facets or some other arbitrary unit, and each area is then classified according to the predetermined scheme. Usually it will involve the operator visiting each parcel of land.
2. **Cartographic summation**. Here, a series of single-attribute maps (e.g. soil type, geomorphology, geology, land value) are combined to produce a derived map. The method by which the maps are combined may be based on a variety of different strategies. For example, a '**sieve map**' may be produced by simply overlaying a series of single-attribute maps, each showing the areas unsuitable for a given activity, such as landfill. The land parcels which 'fall through' each map are identified as possible locations and are marked on the derived or product map (Figure 1.6B). Alternatively, the maps may be overlaid and a series of classes devised on the basis of the range of possible attribute combinations. The resultant product map shows the distribution of each possible attribute combination (Figure 1.6C). More commonly, however, some form of scoring system is used to derive the product map. This may be based on intuitive knowledge of a process, or on the basis of a statistical model (Figure 1.6D). For example, statistical comparison of a series of single-attribute maps with a landslide inventory map can be used to identify those combinations of factors most commonly associated with the occurrence of landslides. Such statistical data can then be used predictively to produce a hazard map from the single-attribute maps. Most hazard maps involve some form of intuitive or probabilistic scoring system. Geographical Information Systems (GIS) have played a major role in recent years in developing sophisticated analytical techniques of this sort.

Physiographic terrain analysis uses the physical properties of the land surface to break it down into a natural sequence of units known as **landsystems**. The idea of this type of analysis is to recognise recurrent or repetitive patterns within the landscape, which are referred to as landsystems. Each landsystem is then calibrated by field observations at typical locations to determine its suitability for a given objective such as civil engineering schemes or resource exploitation. Having established and calibrated a landsystem at one specific site, it can then be used predictively to map or identify areas likely to have similar characteristics elsewhere. In this way, engineering problems can be anticipated or possible mineral prospects identified. An example of a landsystem, applicable in this case to glacial terrain, is given in Figure 1.7 and consists of a block diagram showing typical terrain elements, along with specific characteristics typical of subglacially modified regions. Physiographic terrain analysis has been assisted by the increasing availability of satellite remote-sensing data. The digital data obtained

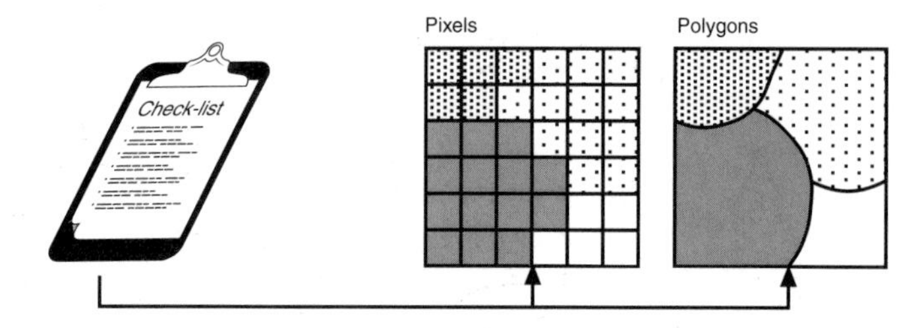
A The interrogative approach
Check-list
Pixels
Polygons

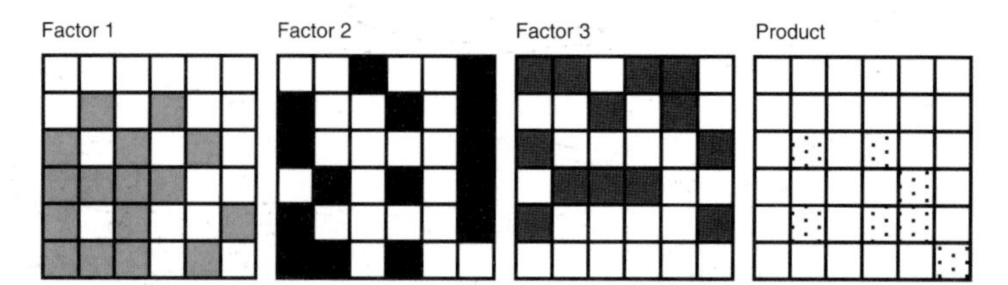
B Sieve maps
Factor 1
Factor 2
Factor 3
Product

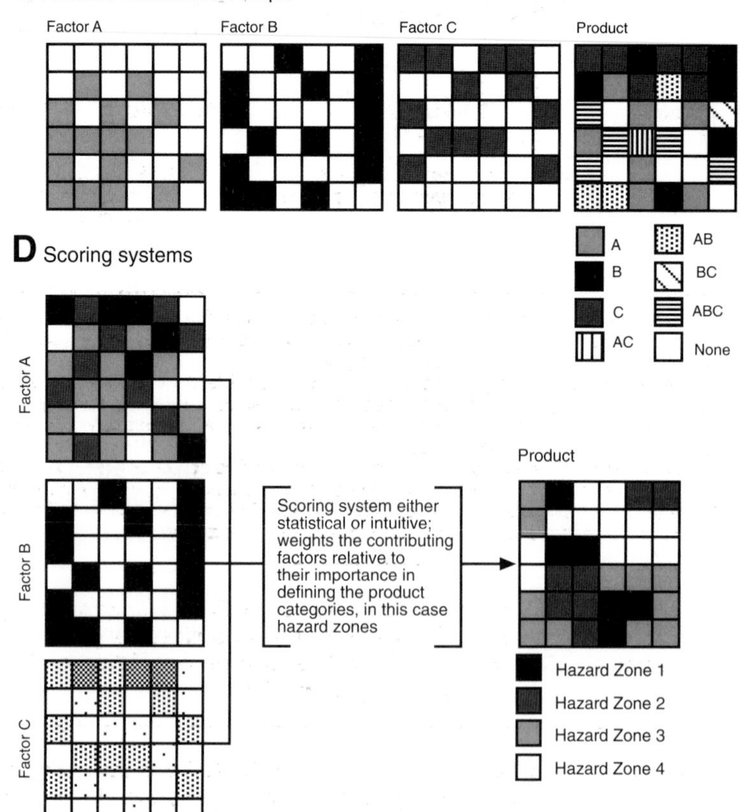
C Attribute combination maps
Factor A
Factor B
Factor C
Product
A
B
C
AC
AB
BC
ABC
None
D Scoring systems
Factor A
Factor B
Factor C
Scoring system either statistical or intuitive; weights the contributing factors relative to their importance in defining the product categories, in this case hazard zones
Product
Hazard Zone 1
Hazard Zone 2
Hazard Zone 3
Hazard Zone 4

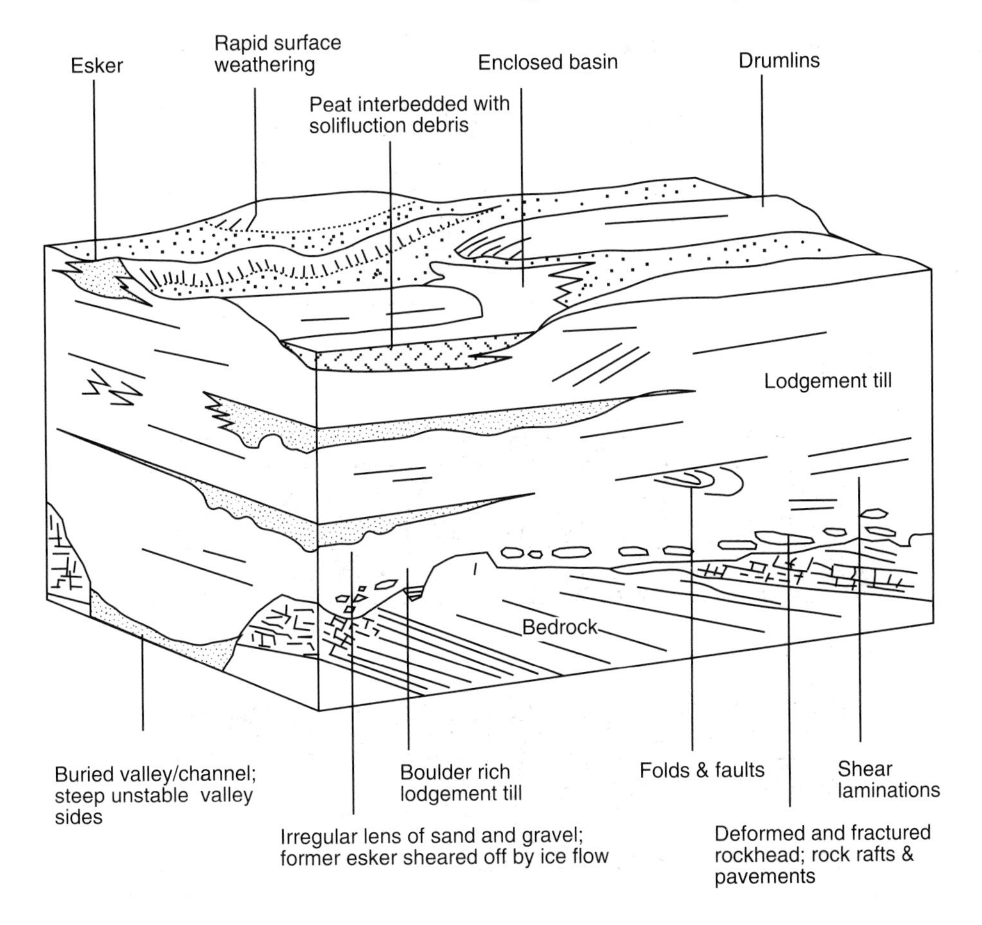

MORPHOLOGY: streamlined mounds (drumlins) elongated in the direction of former glacier flow. Sinuous, sharp crested ridges (eskers). Intermound depressions contain peat filled hollows and irregular bodies of sand and gravel

SEDIMENTS: lodgement till—a fine grained diamicton with variable clast size, sheared fabric, fissures and shear planes, locally folded, may contain lenses of sand and gravel originally deposited in subglacial drainage channels. Peat and irregular sand and gravel deposits occur on the surface

ENGINEERING PROBLEMS: the following characteristics need to be considered: (1) fractured and displaced rockhead; (2) oversteepened, buried channels; (3) rock pavements; (4) shear planes and fissures; (5) soft zones caused by lenses of sand and gravel; and (6) prone to rapid surface weathering and loss of strength

Figure 1.7 *An example of a physiographic landsystem, in this case for an area subjected to subglacial processes. The emphasis in this landsystem is on engineering geology [Modified from: Eyles & Dearman (1981)* Bulletin of the International Association of Engineering Geology, *24, Fig. 2, p. 176]*

Figure 1.6 *(opposite) Types of parametric terrain analysis.* **A.** *The interrogative approach, which uses either a grid of pixels or mapping boundaries (polygons).* **B.** *Sieve maps.* **C.** *Attribute combination maps.* **D.** *Product maps based on statistical or intuitive scoring systems* from such data sources allow the statistical as opposed to visual identification of landsystems. Physiographic terrain analysis has been used extensively in some countries as the basis for national resource surveys.

1.5.4 Environmental Monitoring

Environmental monitoring is time-dependent, which is a considerable problem given that there is not usually sufficient time with which to observe phenomena under commercial constraints, or even within the human life-span. This is a considerable problem in predictive environmental geology. Commonly posed questions, particularly about geomorphological processes, are: How fast is it eroding? When will the river next flood? How will sedimentation change with time? Although these are difficult to answer in the short term, it is necessary for the environmental geologist to find a solution and some monitoring must be carried out, although predictive modelling using computer-based analysis of a range of parameters is of increasing importance. Two broad types of environmental monitoring can be identified: (1) direct methods, and (2) indirect methods.

Direct methods involve the observation of a process or landform for a period of time. The length of time required to obtain satisfactory results depends on the rate of change being measured. The accuracy of the measuring equipment necessary is determined by the magnitude of the change being monitored: the larger the change, the less accurate the equipment needs to be. The rate of change must always be greater than the error margin of the equipment used. This type of approach involves repetitive surveys, the use of fixed-point photography or the use of instrumentation. The application of this type of approach is usually limited by the time available in which to obtain the data.

Indirect methods provide information relatively quickly and are therefore more commonly used in environmental geology. There are five broad types of indirect or secondary method.

1. **Inference from morphology or sediment volume**. The amount of sediment or size of landform produced by a process is usually indicative of its magnitude. For example, if it is possible to estimate the volume of soil deposited by a storm in drains and on pavements, then an estimate of the rate of soil erosion can be obtained as long as the source area for the deposits can be constrained. Essentially, the more sediment there is, the greater the amount of soil erosion.
2. **Prediction from a known location to an unknown location**. Here, empirical relationships developed at one site, perhaps an experimental plot, are used to predict the rate of change at another.
3. **Physical and numerical simulation**. In the past, scale models of harbour works, estuaries and rivers have been made to examine problems of sedimentation in response to engineering work. Today, however, these have been largely replaced by computer models, which are increasingly sophisticated tools in environmental monitoring and prediction (Box 1.1). For

BOX 1.1: COMPUTER MODELS AND THE NEXT ICE AGE

Computer modelling is playing an increasing role in monitoring and predicting environmental change. This has been assisted not only by increasing computer power, but also by our ability to design realistic environmental models. Computer modelling has also played a role in the design of nuclear repositories for high-level waste. One of the applications of this modelling is an assessment of the likely impact of the growth of a future ice sheet. Although the present is optimistically called the Postglacial Period, glaciers are likely to return in the future to northern Europe as they have done at regular intervals during the last two million years of the Cenozoic Ice Age. Due to the long-lived nature of high-grade nuclear waste, the potential impact of a future ice sheet over the waste installation must be considered. Ice sheets not only erode and deposit sediment but also have a dramatic impact on groundwater flow beneath them. Predicting the geometry of future ice sheets, the pattern of erosion and deposition within them as well as the likely reorientation of groundwater flow is relevant to the design of nuclear waste facilities in former glacial areas such as Sellafield in Britain. The diagram below shows the geometry of a future European ice sheet predicted by a computer model.

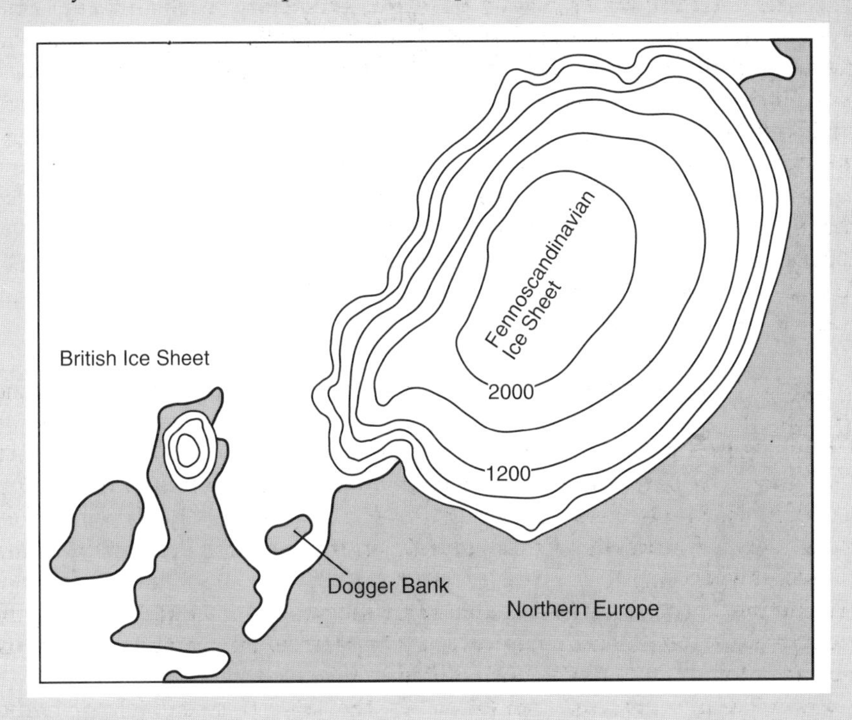

Source: Boulton, G.S. 1992. The spectrum of Quaternary global change – understanding the past and anticipating the future. *Geoscientist*, **2**, 10–12. [Diagrams modified from: Boulton (1992) *Geoscientist*, **2**, Fig. 4, p. 11]

example, concern was expressed about the effect of the construction of a second bridge on tidal currents and sedimentation within the Severn Estuary in Britain and its possible impact on conservation interests and navigation within the estuary. A computer model of the tidal flows within the estuary was produced to predict the likely impact of the crossing and to identify potential problem areas. This type of modelling exercise is becoming routine in environmental research.

4. **Space–time transformation**. Space–time transformation (ergodic reasoning) involves the use of spatial data to indicate a temporal trend. For example, in a population of landforms of mixed age, there should be representatives of most of the stages in its evolution. By measuring all the examples present they can therefore be arranged into a temporal model of landform evolution which is useful in the prediction of long-term environmental change (Box 1.2).
5. **Historical data**. The use of historical data is an important proxy for accurately monitored environmental data. Data may be of four main types: (A) graphical, including old maps, old air photographs, paintings and oblique photographs; (B) written reports, such as newspaper articles (Box 1.3), diaries and other accounts of events; (C) oral evidence from eye-witnesses; and (D) statistical series data, such as meteorological, tidal or river records. Historical data are used very widely within environmental geology where good records exist, although care is required in their interpretation. Errors can be introduced into historical data during the data collection, during a subsequent stage of copying/editing, or during the storage and treatment of documents. In using historical data it is usual to look for both internal and external corroboration of change or event. Internal corroboration essentially involves establishing the reliability of the evidence. This is done with reference to a series of checks, as outlined in Table 1.3. External corroboration involves looking for a second, independent piece of historical data to support the first.

Direct methods such as dating and geological evidence may be used to establish the frequency of an event within a time series. Such data might be the recurrence of certain stratigraphical horizons with a sequence which indicate floods, fires and other events.

1.5.5 Data Presentation

Adequate, concise and factual data presentation is an essential part of the role of environmental geologists. For the most part they are working for the non-specialist and it is therefore essential that the information is clear, concise and in a format which can be readily interpreted by the layperson. In summarising their results, environmental geologists have to consider the following points.

1. **The nature of the brief**. It is important to establish whether specific recommendations are required, or simply the evidence with which decision-makers can make an informed choice. The end product and its presentation must address the needs of the user as posed in the project brief.

BOX 1.2: THE TRANSFORMATION OF SPACE AND TIME: RIVER CLIFF EVOLUTION ON THE MISSISSIPPI

Brunsden & Kesel (1973) use space–time transformation to study the evolution of river cliffs progressively abandoned by the Mississippi migrating across its floodplain near Port Hudson, Louisiana. The section of river bank studied was progressively abandoned from 1722 as a meander loop migrated downstream. This can be determined from historical maps, but these maps do not provide detail on the morphological changes which have occurred to the river banks since they were abandoned. A series of cross-sectional profiles was taken along banks abandoned at different times between 1722 and 1971. These fall into three categories: (1) those at which the river is no longer eroding the base of the banks and basal retreat is of the order of 0.2 m a^{-1}; (2) those at which the river is some distance away from the base of the slope and only causes scour at high water levels, with basal retreat of the order of 0.3 m a^{-1}; and (3) those at which the river is actively undercutting the bank at rates of up to 23 m a^{-1}. By placing these profiles, collected in space, into their relative age order, a temporal model for the evolution of these river cliffs can be obtained (see diagram). Basal erosion by the river maintains the slopes in a steep state; however, as this ceases, mass movement debris starts to accumulate at the base of the slope. The frequency of mass failure of the river cliff declines due to accumulation of debris at the base of the slope. In this way a stable slope profile, in equilibrium with the subaerial processes acting upon it, evolves over time. This process appears to take of order of the 90 years on the Mississippi according to the model of Brundsen & Kesel (1973). This work illustrates the potential of space–time transformations (ergodic reasoning) in environmental monitoring.

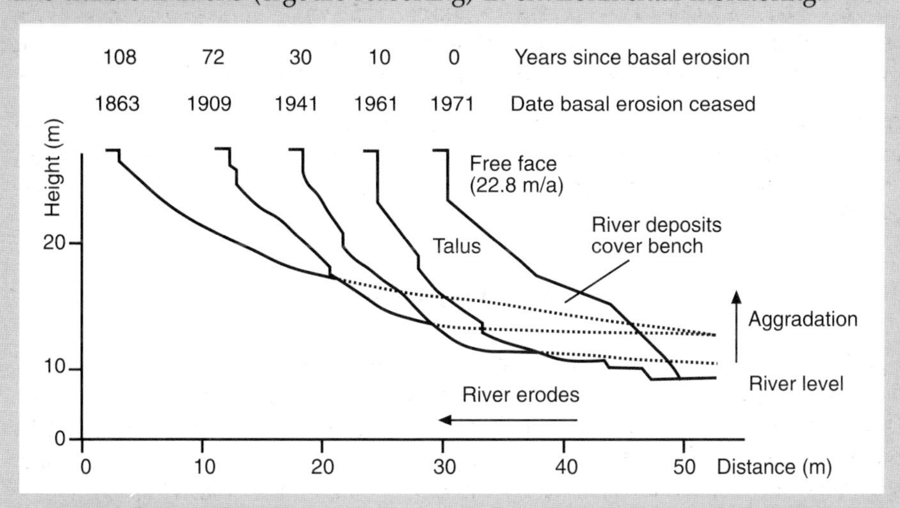

Source: Brunsden, D. & Kesel, R.H. 1973. Slope development on a Mississippi River bluff in historic time. *Journal of Geology*, **81**, 576–597. [Diagram modified from: Brunsden & Kesel (1973) *Journal of Geology*, **81**, Fig. 14, p. 595]

BOX 1.3: THE APPLICATION OF HISTORICAL DATA: NEWSPAPERS AND SNOWFALL

Pearson (1976) provides a good example of how newspaper records can be used to reconstruct environmental events and conditions. He was interested in the pattern of climate variability during the eighteenth and nineteenth centuries – a period of time before organised meteorological records. Pearson used newspaper reports obtained from the regular Edinburgh paper *The Caledonian Mercury*, supplemented, where missing, by issues of the *Edinburgh Evening Courant*, for the period from 1729 to 1830. This involved looking at over 15 000 newspapers for stories about snow. As the diagram below illustrates, the number of days with snowfall, deduced from the stories within the newspapers, shows a marked increase in the 1780s and 1790s. Pearson's problem was to decide where this was a real increase in snowfall, due to climatic variability, or the result of an increase in reporting. Throughout the later part of the eighteenth century, Edinburgh was developing as a commercial centre and consequently its postal links with London and other centres were of increasing importance, particularly when severed due to snow. Similarly, its inhabitants were travelling more easily and widely at this time. Consequently the increased reporting may simply reflect an increase in the inconvenience caused by snow due to a more mobile population. Pearson argued, however, that this is not the case since the incidence of snow reports declines towards the start of the nineteenth century.

This example illustrates both the potential of newspapers and also some of the problems in interpreting historical data, in particular in deciding whether trends reflect changes in the environment or changes in reporting.

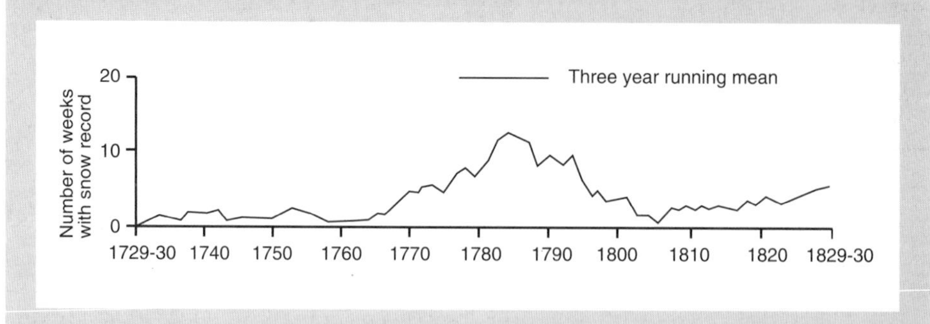

Source: Pearson, M.G. 1976. Snowstorms in Scotland 1729 to 1830. *Weather*, **31**, 390–393. [Diagram modified from: Pearson (1976) *Weather*, **31**, Fig. 3, p. 392]

2. **The nature of the audience**. The nature of the audience – that is, whether the people who commissioned and who are paying for the work are specialists or non-specialists – must be established and the report and all graphics must be produced to convey information at the appropriate level. If

Table 1.3 *Internal checks or corroboration of historical data of all types: graphical, written, and oral [Modified from: Hooke & Kain (1982)* Historical change in the physical environment. *Butterworths, Table 3.2, p. 73]*

Characteristics	Checks
Originality	Whether original – check author, data, source
Contemporaneity	Closeness in time of document to reality portrayed (e.g. was an eyewitness account written months later?)
Propinquity	Closeness in space of document to situation it describes (e.g. was the observer on the scene of an event or some distance from it?)
Generalisation	Level of detail and style
Transmission of information	Reliability of secondary reporting
Purpose	Aim of the original record (e.g. some early estate maps were designed to flatter landowner rather than to record their estates factually)
Observer	Who compiled the original record?
Intellectual and social milieu	Social, intellectual and technical attitudes and knowledge of the time
Methods and instruments	Type and design of instruments and methods
State of documents	Are they damaged or distorted?

Table 1.4 *Avalanche hazard zones in the Swiss Alps. This system of hazard zones illustrates the use of a clear, unambiguous zonal scheme in which the land-use planning significance of each zone is apparent [Modified from: Smith (1992)* Environmental hazards, *Routledge, Table 8.3, p. 177]*

Red Zone: High hazard. No buildings or winter parking lots allowed. Special bunkers needed for equipment. Areas affected by avalanches with a frequency of less than 30 years

Blue Zone: Potential hazard. Public buildings that encourage gatherings of people should not be erected. Private houses may be erected if they are strengthened to withstand impact forces. They may be closed during periods of hazard

White Zone: No hazard. Very rarely affected, if at all only by minor air blast. No building restrictions

the data and recommendations cannot be understood by the intended audience then they will not be used.

3. **Map format**. To be useful, any maps, such as hazard maps, and their implications need to be clear, intelligible and unambiguous. For example, avalanche hazard zones in the Swiss Alps are represented by just three colours and are expressed in terms of the planning constraints required (Table 1.4). The need for a clear presentation of geological data for non-specialists has led to the development of **thematic maps** (Table 1.2). The concept of thematic maps has been applied to land-use planning in many parts of Britain and North America. Typically, each sheet in a set of maps presents a different data theme in a simplified fashion relevant to the end user. For a particular planning area, a series of maps is produced, each of which summarises a variable relevant to planning, such as slope steepness,

ground conditions, areas of scientific or natural beauty, mineral resources, and so on. Much of the information in these map sets is already available in specialist or technical maps and databases, but the aim of the thematic set is to provide a single repository of data at a standard scale which can be interpreted by the layperson or informed decision-maker.

4. **Internal and external audit**. All information, summary maps and recommendations must be accurate, particularly where questions of public safety, and therefore liability, are at issue. In particular, recommendations should be shaped through consultation with currently employed quality standards for the country or state. In addition, a good finished product is the best way of ensuring future consultancy contracts. Final reports must therefore be subject to both internal as well as external quality audit. In order to ensure that this is possible, a clear statement of all methods and data sources used in the report is required. If there is any ambiguity about how information, maps or recommendations are obtained, then clearly their reliability will be doubted.

Essentially, environmental geologists and their products act as the interface between the non-specialist decision-maker and the technical detail of the subject. Six key words summarise the desired characteristics in a piece of commercial work, which should be: intelligible, concise, clear, justified, reasoned and relevant.

1.6 SUMMARY OF KEY POINTS

- Environmental geology is the interaction of humans with the geological environment, which encompasses all the rocks, fluids, sediments and soils of the solid Earth as well as its surface landscape and the processes which change this surface over time.
- Environmental geology can be divided into three main areas: (1) the geology of resource management, including exploration, exploitation and their consequences; (2) the geology of the built environment, including both the constraints of ground conditions on development and the problems associated with the disposal of waste generated by this development; and (3) the geology of natural hazards which threaten human life and infrastructure.
- The modern subject of environmental geology embraces all three of the traditional applied Earth science subjects: applied geomorphology, economic geology and engineering geology. The increasing interaction of these three disciplines coupled with greater environmental awareness has created environmental geology.
- Environmental geology is an applied subject driven for the most part by the commercial reality of the marketplace, since most research in this field is financed commercially.
- The tools used by environmental geologists can be grouped into: (1) desk surveys; (2) field-based documentation or maps; (3) terrain analysis; (4) environmental monitoring; and (5) presentation.

1.7 SUGGESTED READING

The definition and scope of environmental geology are covered in most standard texts on the subject; in particular, Coates (1981), Keller (1992), Lumsden (1992) and Woodcock (1994) provide good overviews. Some of the tools used in environmental geology are reviewed in Hooke & Kain (1982), McCall & Marker (1989), Cooke & Doornkamp (1990), Barnes (1995) and Tucker (1995). The volume edited by Culshaw *et al.* (1987) contains a wide selection of papers on geology, the urban environment and planning, with the section on land evaluation and site assessment providing a useful review of mapping techniques in environmental geology.

Barnes, J. 1995. *Basic geological mapping*, third edition. John Wiley & Sons, Chichester.
Coates, D.R. 1981. *Environmental geology*. John Wiley & Sons, New York.
Cooke, R.U. & Doornkamp, J.C. 1990. *Geomorphology in environmental management*. Oxford University Press, Oxford.
Culshaw, M.G., Bell, F.G., Cripps, J.C. & O'Hara, M. (Eds) 1987. *Planning and engineering geology*. Geological Society, London.
Hooke, J.M. & Kain, R.J.P. 1982. *Historical change in the physical environment*. Butterworths, London.
Keller, E.A. 1992. *Environmental geology*, sixth edition. Macmillan, New York.
Lumsden, G.I. (Ed.) 1992. *Geology and the environment in western Europe*. Oxford University Press, Oxford.
McCall, J. & Marker, B. (Eds) 1989. *Earth science mapping for planning, development and conservation*. Graham & Trotman, London.
Tucker, M. 1995. *Sedimentary rocks in the field*, second edition. John Wiley & Sons, Chichester.
Woodcock, N. 1994. *Geology and environment in Britain and Ireland*. UCL Press, London.

2
The Geology of Resource Management

The Earth is a repository for a wide range of materials which have a diverse application in all walks of human life. The water we drink, the houses we build and the metals we use in our cars and homes are all derived from the Earth's geological resource. In the following four chapters we explore the resource potential of the geological environment: its mineral resources; its construction resources; its water resources; and finally, the aesthetic, scientific and cultural resources provided by the geological environment. In this chapter we define the nature of a resource, consider the economics of resources, and review the aims of effective resource management.

2.1 WHAT IS A GEOLOGICAL RESOURCE?

In a geological context, **resources** can be defined as naturally occurring solids, liquids or gases known or thought to exist in or on the Earth's crust in concentrations which make extraction economically feasible either at present or at some time in the future. The key point in this definition is that it includes not only those deposits or rocks known to exist but also those which are thought to exist through the extrapolation of our current geological knowledge. A **reserve** is a subset of a resource, and is that portion of an identified resource which can be extracted economically using current technology (Figure 2.1). This type of definition works well for economic minerals, but other geological resources, such as the aesthetic resource provided by the landscape, cannot be quantified easily. This type of resource is based upon **perception** and is related to human values rather than economic requirements. Attempting to constrain such resources depends on the recognition of geographical areas which are of greater or lesser aesthetic, cultural and scientific significance.

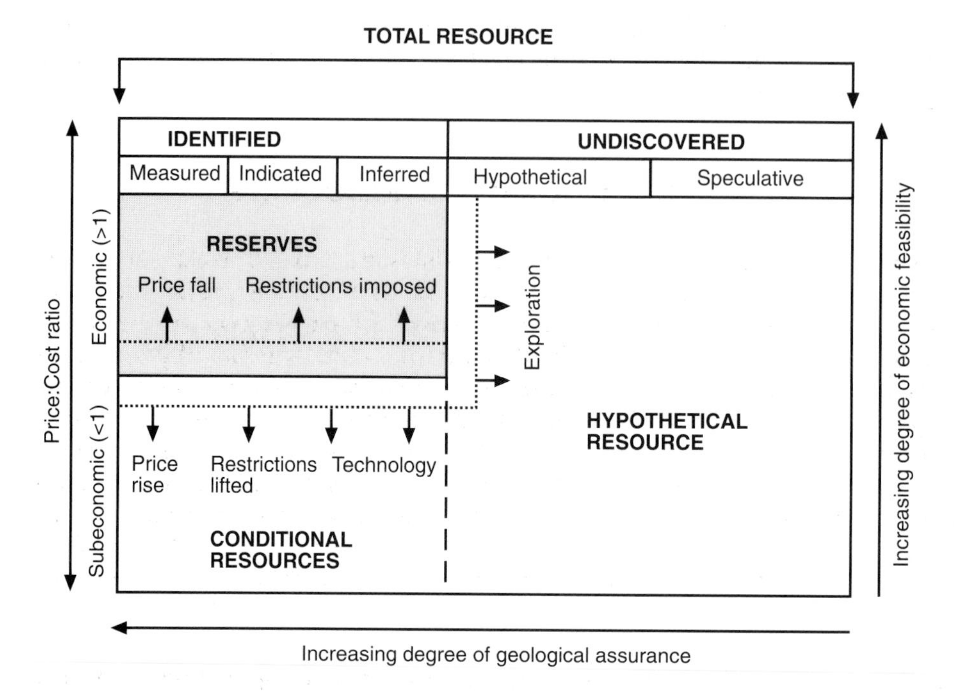

Figure 2.1 *The definition of resources and reserves. The size of a reserve relative to the total resource varies with such things as exploration, commodity price, technology and political regulations*

In the short term, resource management must focus on reserves – those deposits which are currently economically viable. Environmental considerations, which are of increasing importance in resource management, have the potential to reduce available reserves by altering the economics of their exploitation. In contrast, long-term planning must focus on the total resource. Long-term commercial and public planning is based on the probability of new geological resources being discovered or, alternatively, that technological developments will allow extraction from currently uneconomic deposits. The scope of our total geological resource is therefore constantly changing with both geological and technological development. Components within this total geological resource can be classified on the basis of both the economic feasibility of exploitation and the geological assurance that the resource actually exists (Figure 2.1).

The size of the available reserves relative to the total resource (Figure 2.1) can be changed by such factors as: (1) commodity price or value, for example an increase in price may make uneconomic deposits economic and vice versa; (2) exploration, which may increase the proportion of known resource; (3) technological developments, which may improve the efficiency of mineral extraction thereby reducing the unit cost of production; and (4) changes in regulatory requirements, for example, environmental measures that may increase the cost of quarry restoration or mine waste disposal. One of the most important factors in determining the size of reserves and whether a given mineral deposit is economic is cost versus the price of production, which is intimately linked with supply and demand.

2.2 THE ECONOMICS OF RESOURCES

Geology provides a distinct series of commodities, the price of which varies from precious gemstones through fossil fuels and water to bulk aggregates, depending on the value placed by society upon them and their ubiquity. Aesthetic and cultural resources are notoriously difficult to value economically but their conservation may exact a high cost. In most situations, geological resources are only exploited when it is economical to do so; that is, when the cost of production is less than the price obtained by selling the commodity in question. Exceptions exist when, for political or strategic reasons, some geological reserves are exploited. For example, the initial development of Norwegian and Russian mining settlements on the high-arctic island of Spitsbergen owed more to Cold War politics than to economics. In general, the cost of production can be split into two components.

1. **Capital costs**. These are associated with the original exploration to identify the reserve, the cost of the mining rights, planning or political costs incurred, and the cost of mining and processing infrastructure. They may also include investment in transport infrastructure where no suitable transport is available to move the product to the marketplace. They may also include the cost of quarry/mine aftercare and restoration.
2. **Operational costs**. These are the costs of producing a unit volume of marketable product. They are primarily controlled by the quality and extent of the reserve, and the cost of removing it from the ground and extracting the required product from the waste rock. In the case of gas, oil or water only the product is removed, but in most situations minerals have to be extracted from the host, or unwanted rock (overburden) has to be removed to allow access to the deposit. Political considerations such as taxation may also affect the price of a unit of product. In addition, economies of scale may affect the unit cost; generally the bigger an operation the greater the saving per unit of output.

The price obtained for a mineral, and therefore the economics of its extraction, is closely linked with the concepts of supply and demand. Supply can be defined as the quantity of a commodity that suppliers are prepared to sell at a given price. In contrast, demand is the quantity of a commodity a consumer is prepared to buy at a given price. Supply, demand and product price are completely interwoven in a free-market economy. Figure 2.2A shows a hypothetical supply and demand curve for a particular commodity. If the unit price of a product falls, then the incentive to supply that product will decrease, since the supplier will get less money for it. Equally, if the unit price of a product rises

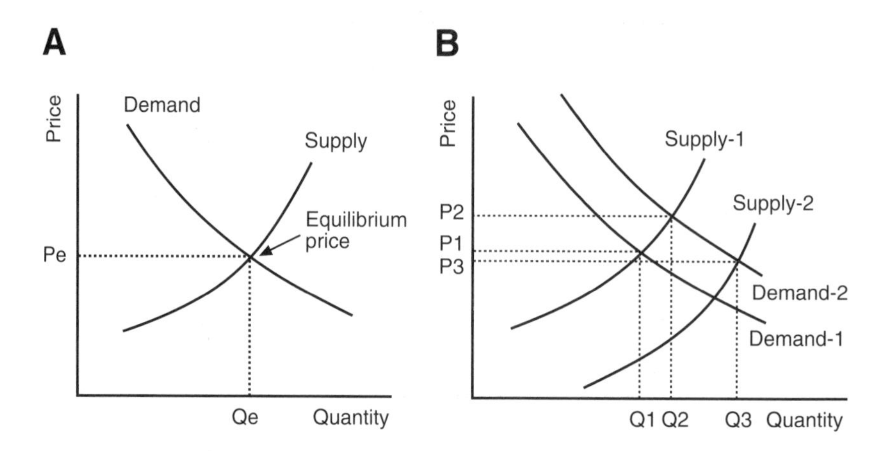

Figure 2.2 *Supply and demand curves.* ***A.*** *Simple supply and demand curves showing their intersection to give the equilibrium price.* ***B.*** *Changes in the equilibrium price due to changes in supply and demand curves. The equilibrium price shifts from P1 to P2 as the supply curve changes and then to P3 in response to a change in the demand curve*

then the incentive to supply it will increase. Demand for a product usually falls with increasing price and rises as its unit cost falls. The steepness of both the supply and demand curves depends on what is termed elasticity of supply and demand. If demand is inelastic the curve will be steep; that is, people require the product whatever the price and changes in price therefore have little effect on demand. For example, water is essential to life and increasing its price is unlikely to change demand, at least in the short term. If demand is elastic, then the curve will be very shallow. A slight increase in price causes a large reduction in demand. Supply may also be elastic or inelastic depending on its sensitivity to changes in price.

The intersection of supply and demand curves gives the **equilibrium price** for a given product, as here the price per quantity that the suppliers are willing to pay equals the quantity the consumers demand (Figure 2.2A). A commodity will tend to move towards this equilibrium price, although it may never get there owing to the constantly changing nature of the marketplace. This picture is complicated by changes in demand and supply. A change in demand may be caused by such factors as increasing consumer spending power, change in consumer priorities, or the availability of cheaper substitutes. Equally, changes in supply may occur, allowing more resource to be delivered at the same unit price; this may be due to increased output due to technological advances or the discovery of new reserves. Such changes will cause changes in the equilibrium price (Figure 2.2B). Despite this, supply and demand tend to operate for many commodities to maintain the price at a similar level over a number of years.

Copper production provides a good example of these processes in action. Figure 2.3 shows that copper has maintained a relatively constant price during the twentieth century. This price stability reflects several feedback systems in

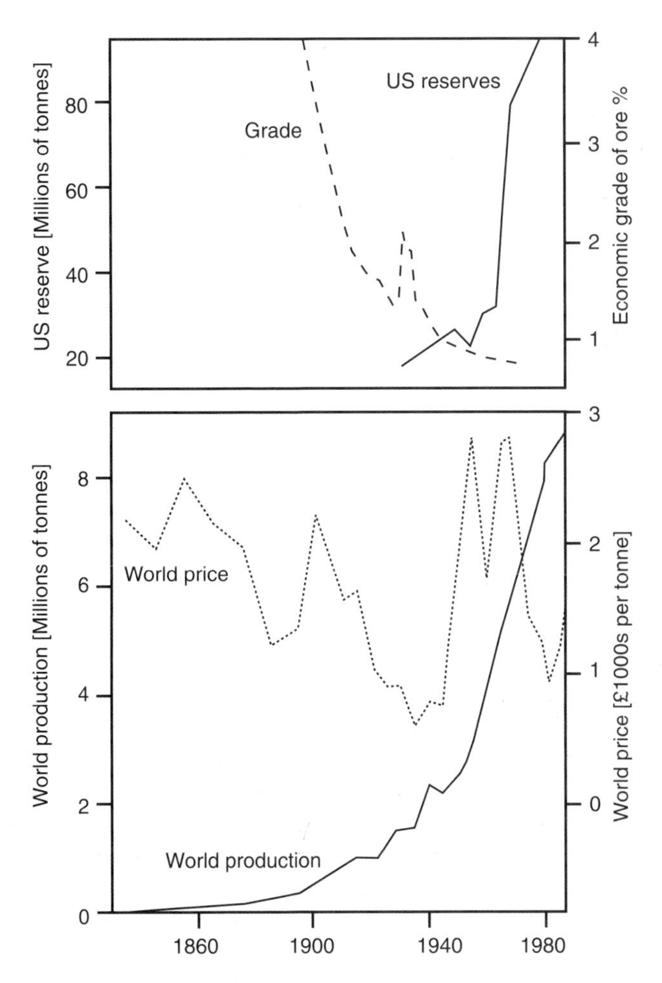

Figures 2.3 *Variation in production, price and US copper reserves [Modified from: Woodcock (1994)* Geology and environment in Britain and Ireland. *UCL Press, Fig. 5.4, p. 36]*

supply and demand. For example, an increase in demand increases the commodity price which in turn increases the incentive of suppliers to produce copper; this leads to exploration and the introduction of new supplies. This price increase is regulated by the fact that increased prices cause consumers to look for alternatives, such as the use of aluminium in electrical appliances. Both the introduction of new suppliers and substitution of cheaper products cause a surplus of supply, reducing the price of copper. In this way copper prices are maintained at a relatively stable level and are only subject to large fluctuations during major world events, such as the oil crisis of the early 1970s, the Wall Street crash of 1930 and the first and second world wars (Figure 2.3). Figure 2.3 also illustrates another important trend. Copper production has risen throughout the last hundred years and logically should have depleted copper reserves.

However, reserves appear to have increased, a paradox explained by the steady decrease in grade (concentration) of copper ore that it is economic to mine. With each fall in grade, a new fraction of resource is added to the reserve – in the case of copper, at a greater rate than it is consumed.

Not all commodities are subject to free-market economy. The price of oil and aluminium has been fixed in the past by cartels of producers to maintain prices at artificially high values. The most famous is the Organisation of Petroleum Exporting Countries (OPEC) which was formed in the 1960s by 13 Middle Eastern oil-producing countries which accounted for 60% of all oil production and over 90% of oil export. The artificially high prices set increased the value of other oil reserves, such as the North Sea, and by 1991 OPEC's share of world export had fallen to 38%.

Finally, the value given to cultural, scientific and aesthetic resources will rarely be associated with the needs of the market other than the fact that it may be more acceptable to conserve sites in those areas where economic demands do not necessitate their destruction.

2.3 THE AIM OF EFFECTIVE RESOURCE MANAGEMENT

There are two issues concerning geological resource management: (1) the sustainability of the resource; and (2) the environmental impact associated with its exploitation.

Resources may either be sustainable, that is renewable through natural process, or non-sustainable, that is finite. The majority of traditional geological resources are non-sustainable. There is only a finite amount of mineral resource on Earth. Groundwater provides an exception, since natural recharge of water-bearing strata is part of the hydrological cycle. Sustainable management of groundwater resources involves ensuring that artificial abstraction is balanced by natural recharge or artificial recharge. Mineral resources cannot be managed sustainably; once used they are gone, since the geological processes necessary to replace them may take millions of years. Management of these resources focuses on determining the total amount of resource and reserve available. This is particularly true of mineral resources. There are two alternative perspectives on the future availability of mineral resources: (1) the Ricardian perspective; and (2) the Malthusian perspective.

The **Ricardian paradigm** states that as reserves are consumed, growth in demand and technological evolution will allow the exploitation of increasingly low-grade (i.e. low concentration of the desired metal or commodity) resources; this may be at an increasing price to society, but reserves will continue to be available. The assumption here is that resources have a unimodal probability distribution, with society currently exploiting only the high-grade tail (Figure 2.4A). Low-grade resources are available at increasing cost provided that society is prepared to pay for them.

In contrast, the **Malthusian paradigm** states that economically viable resources are finite and will be consumed at an exponential rate. Figure 2.4B

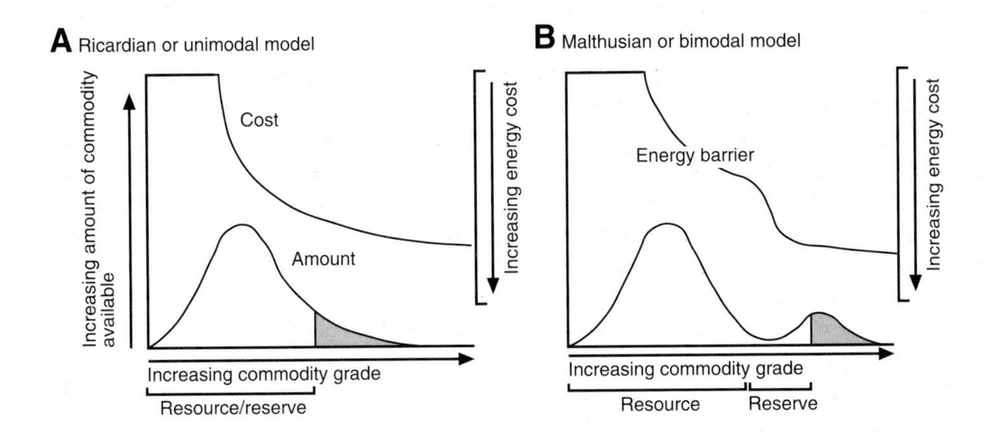

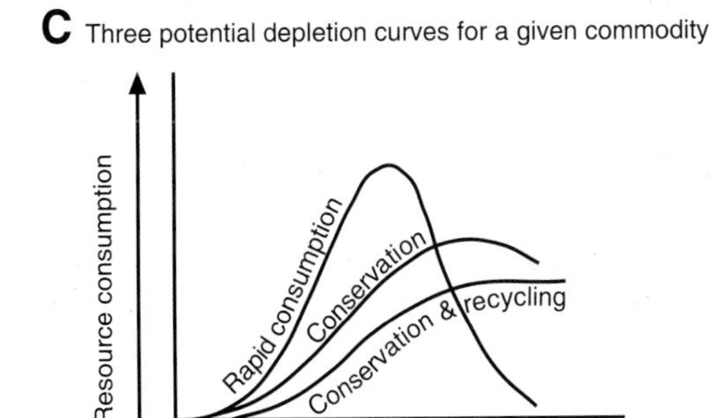

Figure 2.4 *Two alternative paradigms for the future of mineral resources.* ***A.*** *The Ricardian paradigm.* ***B.*** *The Malthusian paradigm.* ***C.*** *The effect of recycling and and mineral conservation on resource consumption [A & B Modified from: Woodcock (1994)* Geology and environment in Britain and Ireland. *UCL Press, Fig. 5.5, p. 37]*

shows a bimodal distribution and is believed to be typical of some metal ores. Most of the metal is dispersed in low-grade mineral deposits and only a small fraction is concentrated into economically viable reserves. Once these reserves are consumed, there is a huge technological and practical barrier to the exploitation of the more disseminated, low-grade deposits. There is also the cost to the environment, which would be considerable in extracting such low-grade minerals, since vast amounts of material would need to be excavated to obtain small amounts of mineral. The reserve is therefore finite.

The choice of model depends largely on the type of mineral concerned. A unimodal Ricardian view works well for some materials such as limestone, aggregates or iron ore, but is less well suited to many metal ores, which probably have a bimodal distribution. The choice of model also depends on one's faith in technology either to provide alternatives, or to overcome the problems of economic mineral extraction. The use of technological alternatives is well illustrated by the optical fibres now used to carry information formerly sent via copper wires. It has been estimated that in Germany the use of this technology has saved over 200 000 tonnes of copper, which is approximately 18% of German production. In more practical terms, recycling and reuse of minerals, particularly metals, may reduce the demand for natural minerals, effectively preventing resource depletion when combined with resource conservation (Figure 2.4C).

The second issue of mineral resource management is concerned with the environmental impact of mineral extraction. Society needs mineral resources, which are often won at a cost to the natural environment. In the past, geologists have been associated only with exploration and exploitation of these resources. However, with increasing environmental awareness the role of the geologist now encompasses management so as to minimise environmental impact and destruction of other resources, such as agricultural productivity and landscape aesthetics. However concerned about the environment we are as individuals, damage cannot be avoided. It can, however, be minimised and managed. This is increasingly the role of the environmental geologist, in devising working practices which reduce environmental impact. This is probably one of the most important contributions that geologists can make to the sustainable management of our environment as a whole.

2.4 SUMMARY OF KEY POINTS

- A geological resource is any naturally occurring solid, liquid or gas known or thought to exist in or on the Earth's crust in concentrations which make extraction economically feasible either at present or in the future. Aesthetic and cultural associations of landscape and geology are also a valuable resource.
- A reserve is a subset of a resource: it is that portion which can be extracted economically using current technology. The size of a reserve relative to the resource as a whole varies with the price a commodity commands and the technology available for extraction. As technology improves it may become economically feasible to extract more of a mineral resource, thereby increasing the reserve.
- The economics of resources and their exploitation are controlled primarily by supply and demand. Supply can be defined as the quantity of a commodity that suppliers are prepared to sell at a given price. In contrast, demand is the quantity of a commodity a consumer is prepared to buy at a given price. For some commodities the interaction of supply and demand may help maintain a stable or constant commodity price.
- Resources may be sustainable (renewable) or non-sustainable (finite). With the exception of water, most geological resources are non-sustainable. The future availability of mineral resources can be viewed either optimistically via a Ricardian paradigm, in which technology holds the key to the exploitation of increasingly disseminated mineral deposits, or alternatively

via a more pessimistic Malthusian paradigm, in which mineral resources will simply run out. Both these visions may be modified by the conservation of mineral reserves, by the use of substitute products, and by recycling.

2.5 SUGGESTED READING

The best introductions to physical resources and their economics are provided by Kesler (1994) and Sheldon (1995). Pearce & Turner (1990) provide a more advanced text on mineral economics while McLaren & Skinner (1987) contains several chapters relevant to resource assessment. Woodcock (1994) discusses Malthusian versus Ricardian paradigms and the Malthusian view is contained within the famous work of Meadows *et al.* (1972).

Kesler, S.E. 1994. *Mineral resources, economics and the environment*. Macmillan, New York.
McLaren, D.J. & Skinner, B.J. 1987. *Resources and world development*. John Wiley & Sons, Chichester.
Meadows, D.H., Meadows, D.L., Randers, J. & Behrens, W.W. 1972. *The limits to growth*. Universe, New York.
Pearce, D.W. & Turner, R.K. 1990. *Economics of natural resources and the environment*. Harvester Wheatsheaf, New York.
Sheldon, P. 1995. *Physical resources: an introduction*. The Open University, Milton Keynes.
Woodcock, N. 1994. *Geology and environment in Britain and Ireland*. UCL Press, London.

3
Economic Mineral Resources

Geology is all around us; its products touch every facet of modern life in the developed world. It is well known that geology provides fuel and construction materials, but its numerous by-products are also found in such diverse things as toothpaste, floor cleaner, dog biscuits, and the paper on which the book is written. For example, consider the car in Figure 3.1. Geology provides the steel out of which the body is formed, the chrome for the bumpers, the sand out of which the glass is made, the hydrocarbon-based plastic and rubber of the interior and wheels, as well as the fuel to make it go. The applications and uses of minerals are vast and essential to modern life.

Economic minerals are defined here in their broadest sense as any geological material which is of commercial value to human society. It is a definition which encompasses fossil fuels, construction materials, industrial minerals, metals and gemstones. In practice, although these materials are very different from one another, the generic principles of exploration, exploitation and environmental impact are very similar and, together with the formation of these resources, form the subject of this chapter.

3.1 TYPES AND FORMATION OF ECONOMIC MINERAL RESOURCES

A simple classification of economic minerals is given in Figure 3.2. It identifies three main mineral groups: mineral fuels, industrial and metal minerals, and construction minerals also known as geomaterials. In practice there is often overlap between these categories, since mineral fuels such as petroleum give rise to plastics and other materials which can be used in construction. Equally, some valued construction materials, such as limestones, may be used as industrial minerals in a crushed form. However, construction materials and industrial or non-radioactive metal minerals cannot be used as fuels. In this section we briefly

Figure 3.1 Photograph of a Desoto car, La Plata, Argentina

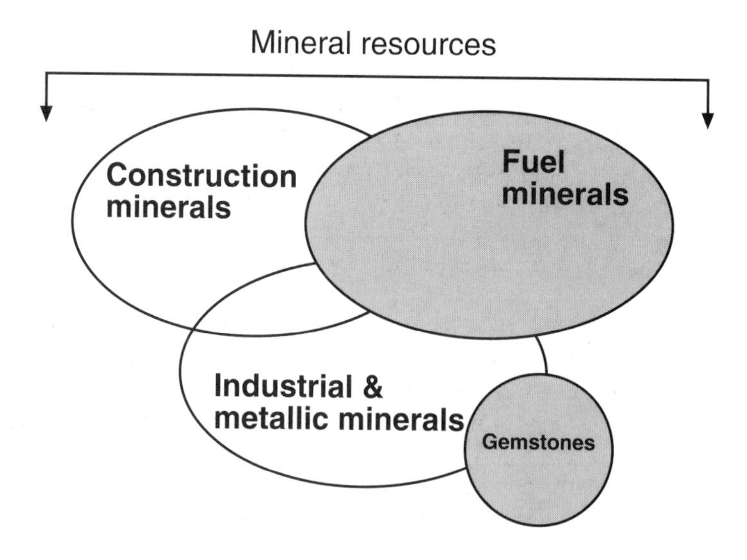

Figure 3.2 The three main types of economic mineral discussed in this chapter

review the formation of the principal mineral groups before going on to discuss the generic principles of their exploration, exploitation and extraction.

3.1.1 Mineral Fuels

There are three main mineral fuels: (1) coal and peat; (2) petroleum; and (3) uranium. Coal and peat are traditional mineral fuels which, at their simplest, are extracted, transported and burned to produce energy without recourse to processing. Both materials are products of the accumulation of organic matter, mostly land plants, in a humid environment. **Peat** is composed of partially decomposed and compressed plant matter and forms extensive surface deposits today in many mid- to high-latitude countries; it is, for example, the traditional fuel source in rural parts of Scotland and Ireland. **Coal** is derived from peat, but has been buried and matured, and is therefore mostly a geologically older deposit. There are two main requirements for coal and peat formation: (1) land plants, which restricts the age of coal and peat formation to post-Silurian; and (2) an anaerobic depositional environment such as swamps and bogs in which organic decomposition occurs in the absence of free oxygen. A typical example of such an environment today is the Okie Fenokie Swamp in the southern United States, a modern-day coal forest in a swampy deltaic setting. Since coal requires compaction for its maturation, two other factors are important in the formation of economic deposits: (1) subsidence of the peat swamps in order to accumulate sufficient organic matter; and (2) burial under a great thickness of sediment. In the coalfields of northern Europe and Pennsylvania these conditions developed most favourably during the Carboniferous, when extensive anaerobic swamps and peat deposits were formed in association with a succession of prograding deltas within slowly subsiding basins. These swamps were periodically covered by new prograding delta lobes and a cyclic stratigraphy developed in which units of organic matter (coal) are sandwiched between layers of deltaic sandstone. The degree of biochemical decomposition and thermal maturation of the buried peat deposits determines the **rank** or maturity of the coal produced. The term rank refers to the ratio of carbon to oxygen and hydrogen present, which in turn controls the heat content or calorific value of the coal (Figure 3.3).

Petroleum is used here to cover a broad spectrum of natural hydrocarbons, including crude oil, natural gas, and solid hydrocarbons such as asphalt. The formation of hydrocarbons is the product of four components: (1) the nature of the source; (2) maturation; (3) migration to a reservoir; and (4) the provision of a suitable trap and associated cap-rock seal.

Petroleum source rocks need to have high organic content, usually expressed as the total organic carbon (TOC) content. This may be in a concentrated form such as coal or more usually in a dispersed form such as that found commonly in mudrocks. In mudrocks, organic matter is derived from the land and carried out to sea, or may be derived directly from marine plants and micro-organisms. Both types of material accumulate as a steady rain of matter which settles out from

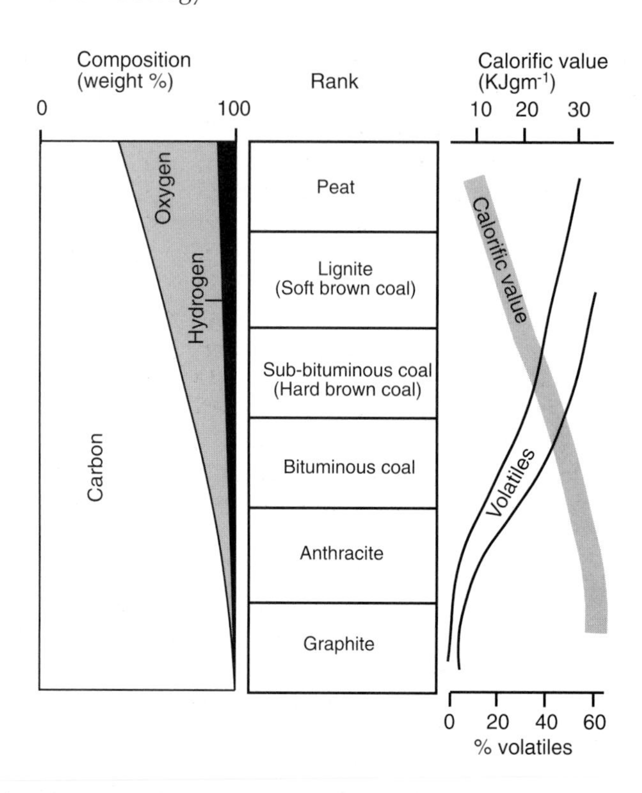

Figure 3.3 *The change in the characteristics of coal with increasing rank or maturation [Modified from: Woodcock (1994)* Geology and environment in Britain and Ireland. *UCL Press, Fig. 10.5, p. 73]*

suspension during low-energy conditions, to be incorporated into the accumulating muds. The process is assisted if the conditions are anaerobic, which prevents the decomposition of the organic matter. Typical oil **source rocks** are often referred to as black shales because of the abundant organic matter they contain. On burial, the organic matter is converted primarily to kerogen. The type of organic matter incorporated determines the likely hydrocarbon product. For example, if the rock is rich in fossil land plants the kerogen produced is known as Type III which produces mainly gas. In contrast, if the organic matter consists primarily of fossil marine algae, then kerogen Type I is produced which gives light oils. Kerogen Type II is the product of algae, spores and plant cuticle and this gives rise to heavy oils and some gas. Initial decay of the organic matter may produce biogenic gas which is normally lost to the atmosphere, although it may be frozen as gas hydrate in the Arctic or on the beds of deep seas.

Maturation of organic matter occurs as the source rock is heated in the temperature range from 50 to 200°C. With increasing temperature, oil is released followed by wet gas, which yields a liquid component when condensed, and dry gas. These hydrocarbon products then migrate from the source rock until they are trapped by an impermeable horizon usually known as a **cap-rock** or seal. Primary migration occurs when the oil and gas leave the source rock due to increases in pore pressure caused either by compaction or by the production of the hydrocarbons themselves. Once produced, hydrocarbons will migrate away from the source rock until they reach a potential reservoir. Secondary migration occurs when the hydrocarbons migrate from the source to the reservoir and beyond, if not constrained by a trap and seal. Typical **reservoir rocks** are porous and permeable and good examples include sandstones and limestones. Secondary migration is driven by buoyancy and will therefore always be directed towards the surface; it continues until an impermeable barrier, the cap-rock, is reached, at which point the oil and gas accumulate, the lighter gas lying above the denser oil (Figure 3.4A). Oil and gas migration has to be constrained beneath the cap-rock to prevent loss through seepage to the surface, and this is a function of many factors which combine to form a **trap**. There are two broad types of trap: (1) stratigraphical traps which reflect primary depositional geometry of the rocks; and (2) structural traps which result from tectonic activity. Examples of commonly occurring traps are shown in Figure 3.4B.

Once recovered, crude oil can be fractionated to give a wide range of different hydrocarbon products with a very wide range of applications, although 90% of oil and gas is burnt as fuel. Hydrocarbon products include: liquid petroleum gas (8% of crude oil), which is used for cooking, heating and as fuel; naphtha (6% of crude oil) used in the production of detergents, plastics, textiles, agrochemicals, rubber and cosmetics; petrol (gasoline; 29% of crude oil) used as fuel for cars; kerosene (8% of crude oil) used in heating, lighting and as aircraft fuel; diesel (gas oil; 27% of crude oil) used in diesel engines; fuel oil (16% of crude oil), which is used in generating electricity, in heating and as ship fuel; and bitumen (6% of crude oil), used on roads and chemically in the same way as naphtha. Unconventional petroleum products include heavy and extra-heavy oils, bitumen, oil shales and methane clathrate. **Heavy oils** have a viscosity far greater than conventional crude oil, which makes their extraction difficult. Steam injection into the reservoir rocks may reduce their viscosity and allow extraction. **Bitumen** coats the pores of some reservoir rocks and acts as a cement. These rocks are sometimes referred to as tar sands. Injection of steam may help mobilise the bitumen for recovery from wells, but more commonly tar sands are quarried where they occur near the surface. Oil shales are source rocks which have not been subject to maturation and have therefore not yielded free hydrocarbons. Oil shales can be mined, and heated at over 450°C to liberate the hydrocarbons as shale oil, although experimental plants have attempted to ignite the oil shale in the ground to liberate the hydrocarbons without the need for extensive mining. Some oil shales can also be burnt directly. **Methane clathrate** or gas hydrate is a frozen form of methane and occurs within frozen ground and in marine sediments on continental shelves. The high water pressure in deep oceans and seas promotes freezing of this gas as pockets within sea-floor sediment. The primary source for this gas is the biogenic decay of organic matter within the sediment. It may provide a potential source of gas in the future, but is also of concern as global warming may cause its release from frozen arctic regions helping to further accelerate the process (Box 8.1).

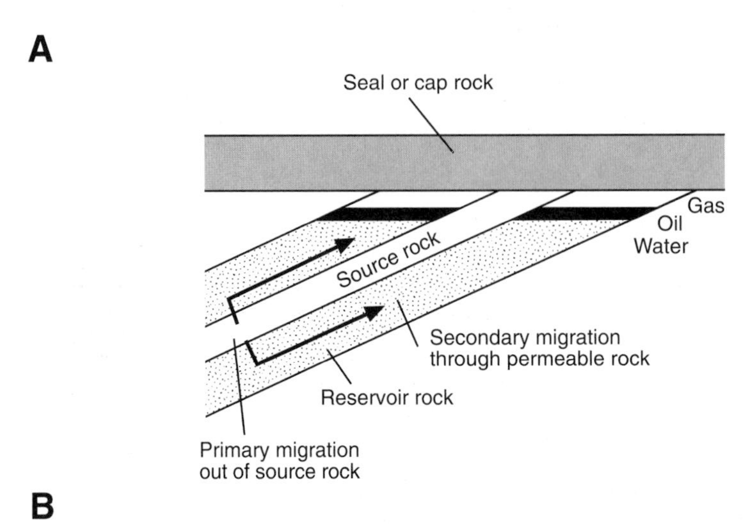

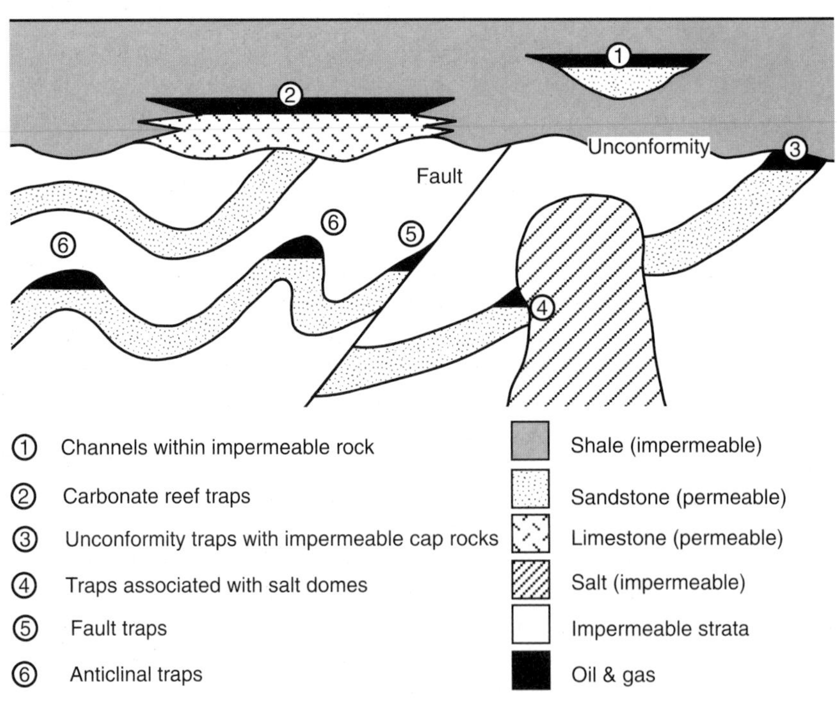

Figure 3.4 *The geology of oil and gas.* ***A.*** *Oil and gas migration.* ***B.*** *Examples of some oil and gas traps*

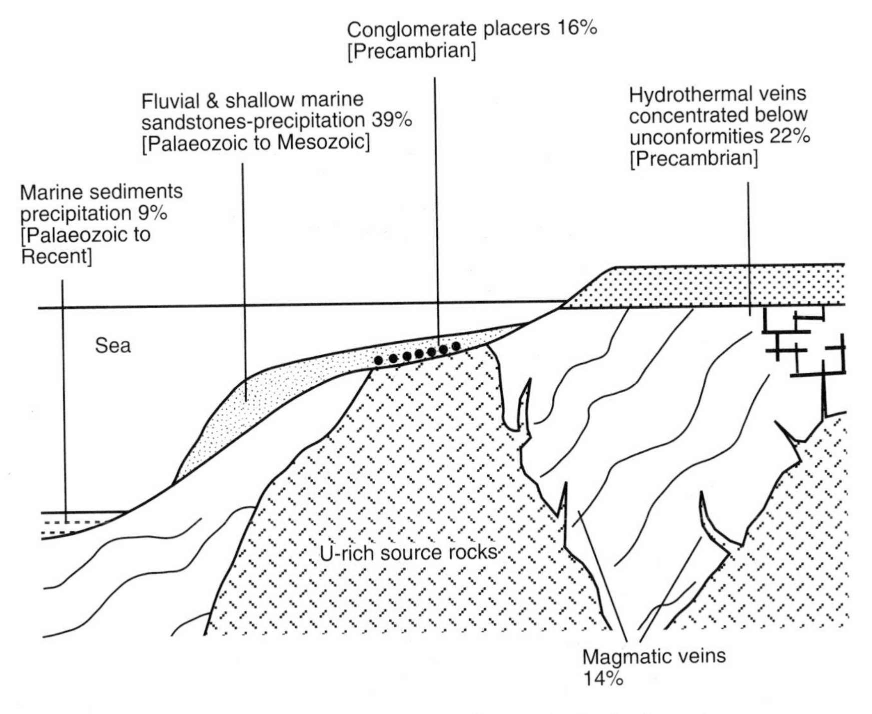

Figure 3.5 *The location, origin and age of principal uranium ores*

Uranium is a mineral fuel not by virtue of combustion, but through its radioactivity. One kilogram of the uranium nuclide ^{235}U releases three million times more energy than the combustion of 1 kg of coal. It is this energy potential which stimulated the development of the nuclear power industry during the 1960s and 1970s. Most nuclear reactors use the heat produced when ^{235}U breaks down under bombardment from neutrons. This process involves a chain reaction. Initially each ^{235}U forms the highly unstable isotope ^{236}U by the capture of a neutron. The decay of this new isotope is rapid and liberates neutrons which in turn bombard more ^{235}U perpetuating the reaction. The radioactive decay of ^{236}U gives isotopes of iodine and yttrium. This chain reaction would run out of control and explode if not for the neutron modulators which absorb some neutrons and are placed between the uranium fuel rods in the reactor core. Typical modulators are graphite or heavy water. The heat generated by the reactor is used to convert water to steam and drive turbines which generate electricity.

The principal locations in which uranium ore occurs are shown in Figure 3.5 and can be divided into two broad categories: (1) veins, either magmatic or hydrothermal in origin, which are often concentrated beneath impermeable unconformities; and (2) placer deposits. Uranium ore is processed by milling and solvent extraction to concentrate the product from 0.025%, a typical ore concentration, to over 85% before it is used to produce either uranium metal or uranium hexafluoride, both of which can be used in reactors.

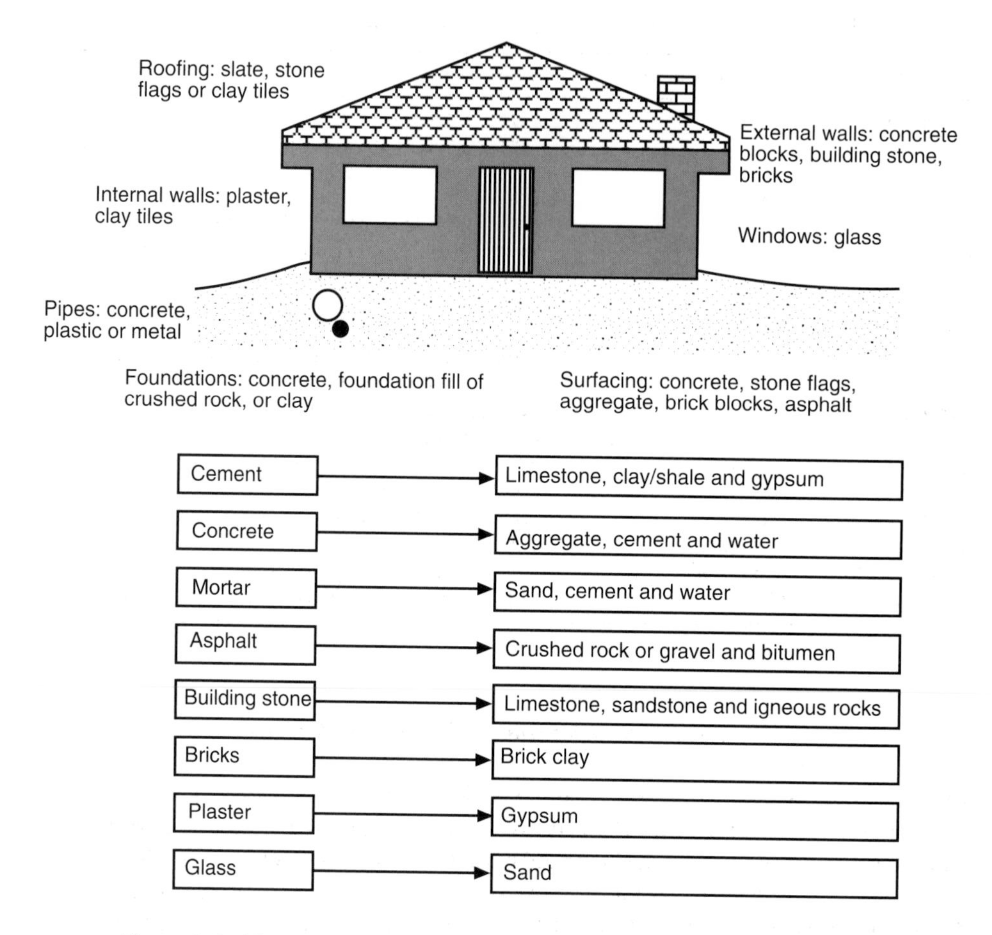

Figure 3.6 *The main construction minerals (geomaterials) and their uses*

3.1.2 Construction Minerals

Minerals, aggregates and stones which are used in building are known collectively as construction minerals, and these represent the most widely used of all geological materials. In this case the term mineral is used in an industrial sense, referring to any geological material of economic value, rather than the strict geological definition of minerals as the components of rocks. The extraction and processing of construction minerals give the **geomaterials** from which the fabric of our urban environment is built. A very diverse range of minerals are used as geomaterials, including: limestones, sandstones and crystalline rocks used as building and armourstones; sands and gravels for aggregates; clays for bricks; limestones for cement; and sand and gypsum used for glass and plaster. Geomaterials are discussed in more detail in Chapter 4, and their application in buildings is illustrated in Figure 3.6.

Table 3.1 *Some important industrial minerals, their mineralogy and use*

Product	Principal minerals	Main uses
Limestone	Calcite	Filler in animal feed, polymers, paint, paper and pharmaceuticals; agriculture; flux in iron, steel and glass production; chemicals
Salt	Halite	Chemicals
Silica sand	Quartz	Glass, foundry moulds, abrasives
Dolomite	Dolomite	Agriculture; iron, steel and glass production, abrasives
Gypsum	Gypsum	Filler for textiles
Chalk	Calcite	Filler in animal feed, polymers, paint, paper and pharmaceuticals; agriculture; flux in iron, steel and glass production; chemicals
Fire clay	Kaolinite	Refractory bricks, pipes and tiles
Potash	Sylvite, langbeinite	Fertiliser, chemicals
China clay	Kaolinite	Filler and coating for paper; porcelain
Phosphate	Apatite	Fertiliser
Barytes	Barite	Filler for paper; drilling mud
Ball clay	Kaolinite	Ceramics
Fuller's earth	Montmorillonite	Filler for paints; drilling mud
Fluorspar	Fluorite	Flux in steel production; aluminium smelting

3.1.3 Industrial, Metallic and Precious Minerals

Industrial minerals are any minerals valued for their chemical or physical properties and which are used in the manufacture of a wide range of products and materials (Table 3.1). **Metallic minerals**, known as **ores**, are exploited for their metal content which is used in the primary manufacture of metal products. **Precious minerals** are by definition rare and hard to come by, and are valued for their aesthetic beauty as well as for their application in specialist tasks. For example, diamonds are not only aesthetically pleasing, but are also vital components of some rock drill-bits.

Industrial, metallic and precious minerals form in a variety of different environmental settings and through the action of a wide range of processes. These can be broadly grouped into one of two categories: as either endogenic or exogenic. **Endogenic processes** are those which operate within the Earth's crust and are primarily associated with igneous rocks, particularly the metamorphism and hydrothermal activity associated with their intrusion. In contrast, **exogenic processes** are those which operate on the Earth's surface to concentrate minerals, either physically or chemically.

Endogenic mineralisation may occur through: (1) differential settling within a magma chamber; (2) separation of immiscible liquids in magmas; (3) metamorphism; (4) hydrothermal precipitation; and (5) hydrothermal leaching. Early fractionation, crystallisation and settling of some heavy minerals within magma bodies may give concentrations of economic value. For example, concentrations of magnetite, ilmenite and chromite can form in large basic intrusions. In other cases, immiscibility may develop in basic magmas, with the separation of a nickel sulphide liquid as dense droplets which settle out and coalesce within the magma body. On crystallisation, pods of pyrite (iron sulphide), chalcopyrite (copper sulphide) and occasionally gold and platinum may form economic concentrations. Metamorphism associated with igneous bodies may cause mineral deposits to be concentrated in country rock. Acidic magmas are most effective since they contain abundant water which can transport metals from the magma into the metamorphic aureole where they can be precipitated, particularly if the country rock is a carbonate. A wide variety of minerals are created which may be concentrated into economically viable amounts; for example, magnetite, haematite, cassiterite, pyrite, chalcopyrite, galena, sphalerite and molybdenite form in this way. The intrusion of igneous rocks also results in hydrothermal precipitation where abundant water is present. Superheated water travels through joints and fissures in the surrounding country rocks carrying dissolved compounds which are then precipitated and concentrated as the water cools to form mineral veins. Equally, the hot magma may cause convective circulation of water contained in the country rocks which leaches out metals and concentrates them in veins and fissures as the water cools. The passage of cooler hydrothermal water may also chemically leach certain rocks and minerals leaving valuable concentrations of secondary minerals such as kaolinite (china clay) as a residue.

The circulation of superheated groundwater on the ocean floor, associated with volcanic activity such as that found along mid-ocean ridges, may also give rise to mineral deposits. The circulation of seawater within rocks surrounding the basalt magma chambers at mid-ocean ridges gives rise to spectacular hot springs referred to as black smokers. Water leaves the vents at temperatures as high as 350°C and is prevented from boiling only by the high water pressures. As the vent water mixes with the cooler sea water, metals and sulphur are precipitated and may form economically valuable deposits. Although today's black smokers cannot be mined owing to the considerable water depths involved, palaeo-examples exist within the stratigraphical record and form economic mineral deposits.

Economic minerals may also be concentrated by the circulation of basinal brines. During sedimentation, water may be incorporated into sediments as formation water. This water may become saline either by evaporation during deposition or by dissolution of evaporites and other salts within the sedimentary sequence. Saline fluids can be very effective in dissolving and transporting metals. Movement of basinal water is largely a product of gravity and tectonic deformation within the basin (basin evolution). Water flow caused in this way may concentrate dissolved metals either within the original basin or beyond its limits. Some types of lead and zinc ores may be formed in this way, and this family of ore types is generally referred to as Mississippi Valley type deposits.

Exogenic processes involve: (1) transport, concentration and deposition of heavy minerals; (2) surface weathering and leaching; (3) groundwater leaching and precipitation; (4) evaporation of surface water; and (5) chemical precipitation within water bodies. The erosion of primary rocks in which the desired

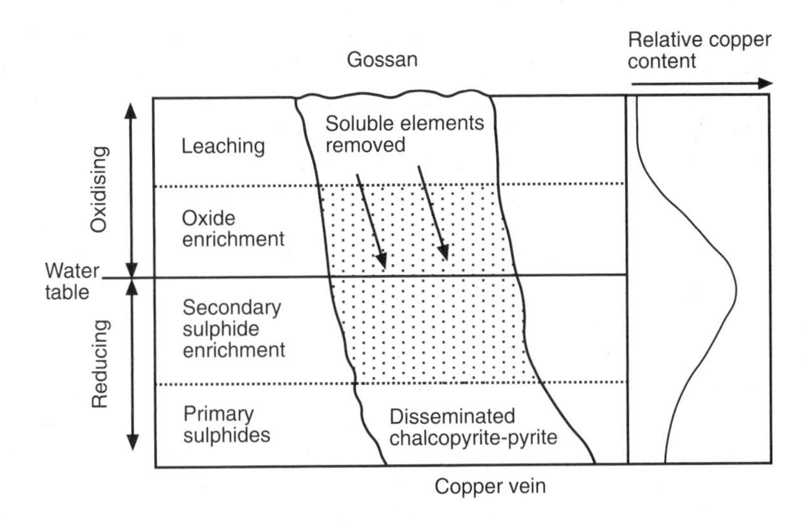

Figure 3.7 *Schematic diagram showing the copper concentration (gossan) formed by weathering of a copper sulphide vein [Modified from: Moon & Whately (1995) In: Evans (Ed.)* Introduction to mineral exploration. *Blackwell Science, Fig. 5.6, p. 72]*

mineral is disseminated widely, followed by sediment transport may provide the opportunity for heavy minerals to separate out preferentially and become concentrated on deposition. This type of deposit is known as a **placer**. Similar surface weathering may lead to mineral concentration either in the leached residue (e.g. bauxite) or due to precipitation of the removed mineral (e.g. copper). Bauxite may be produced by deep tropical weathering of soils and rocks which are iron-poor to produce iron- and aluminium-rich deposits. Copper deposits may be produced by weathering, oxidation and leaching of the surface outcrop of a copper deposit, initially causing copper depletion at the surface. However, below the water table, oxidising conditions cease and reduction occurs, and as the copper is not soluble in such conditions it is precipitated to form an enriched copper zone (Figure 3.7).

Evaporation of shallow saline pools and seawater may give rise to thick mineral deposits; for example, the salt deposits of the Cheshire Basin in Britain formed by the evaporation of shallow seas. Chemical precipitation may also occur in seawater. If the sea is well oxygenated, iron and manganese may be precipitated to produce manganese nodules. These nodules are forming today in economically viable concentrations on the ocean floor, although exploitation is extremely problematic given the water depths involved. Some nodules can also carry high concentrations of other economically important metals. In anoxic conditions, sulphides rich in copper, lead and zinc may be precipitated and are frequently associated with marine mudstones in the geological record. Figure 3.8 provides a diagrammatic summary of some of the mineral-forming processes discussed in this section.

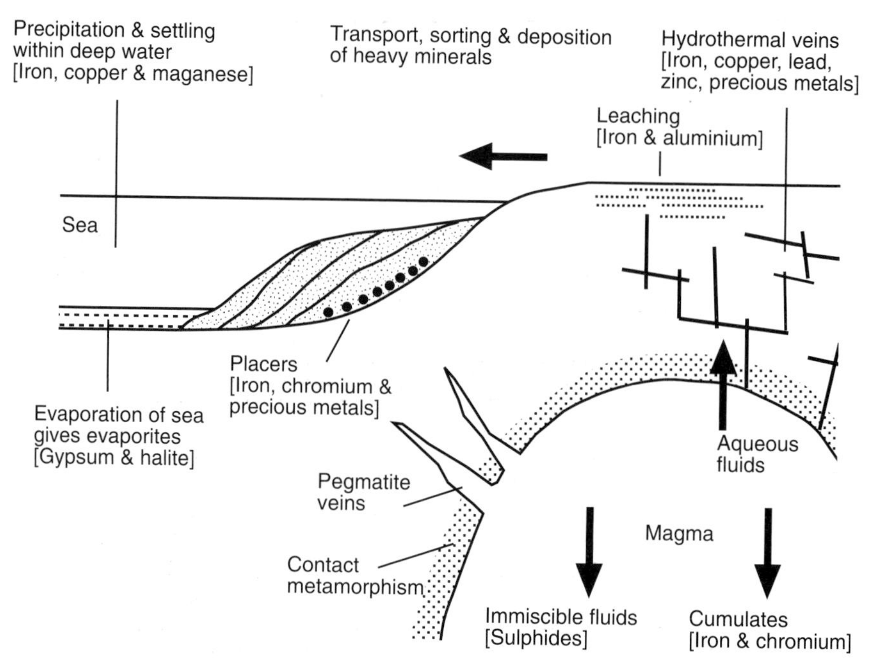

Figure 3.8 *Schematic diagram of common industrial and metal-forming mineral environments*

3.2 METHODS OF MINERAL EXPLORATION

The subject of mineral exploration covers a vast range of information. The details of exploration will vary from one mineral to the next and in this section we cannot do justice to this wealth of detailed information. However, the generic stages of mineral exploration are the same whatever commodity is sought, and the stages involved in a typical mineral exploration programme are discussed below.

Mineral exploration is a high-risk and high-cost exercise which is vital to the continued prosperity of a mineral company. There are two main types of organisation involved in mineral evaluation: (1) the private company, ranging from the large multinational to the small venture capital company; and (2) state organisations, whether as part of nationalised industry or as geological surveys. The cost of exploration, and therefore the associated risks involved in failing to secure return on investment, increases as one moves through the various stages from prospecting, to pre-development to operational mines and quarries (Figure 3.9). Generically there are seven main stages in this process, which are outlined below.

1. **Conceptual planning**. The aim here is to anticipate market demand for a particular commodity and to initiate exploration so that future demands can be met. The location of the market may also be important, especially

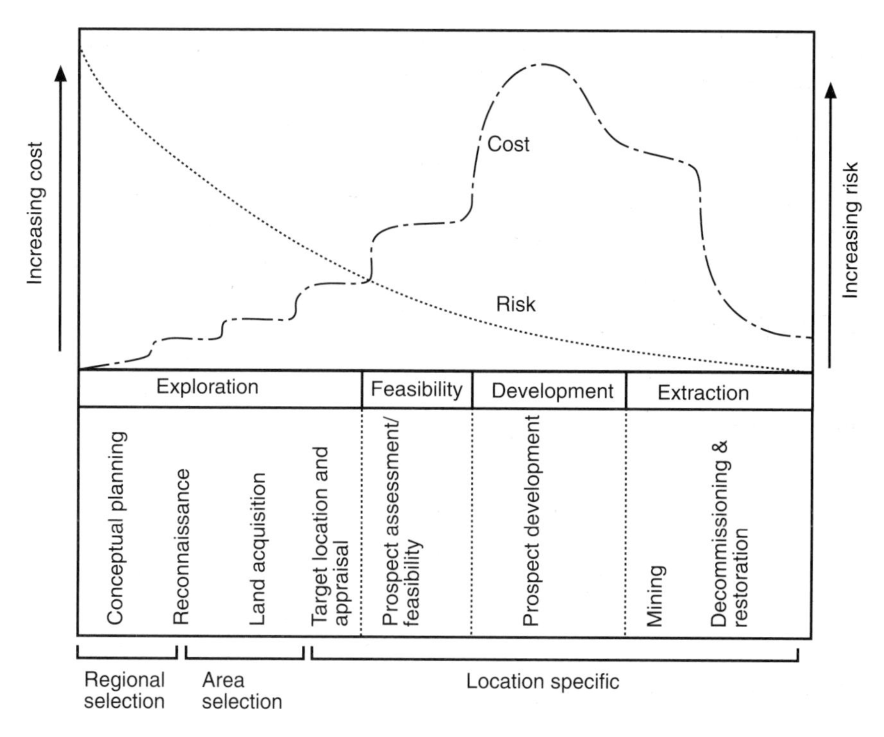

Figure 3.9 *Principal stages in a mineral exploration project [Modified from: Moon & Whately (1995) In: Evans (Ed.)* Introduction to mineral exploration. *Blackwell Science, Fig. 4.1, p. 45]*

for those commodities, such as aggregates, which become unprofitable if they have to be transported too far. This may control the choice of the search area. Political and strategic considerations may also play an important role in determining the choice and location of potential exploration targets. The opening up of new search areas due to the removal of political restrictions or changes in technology must also be anticipated. As a consequence, such planning is highly dependent on national and international economic, political and social fluctuations, as well as on the economic strategy of the individual exploration company or organisation. Exploration is costly and is often linked to the financial health of a commercial company. In the last 20 years the 'boom and bust' nature of many western economies has led to a cyclic pattern of exploration. During periods of economic growth, exploration proceeds rapidly and the number of exploration geologists employed is increased, while during economic recession exploration by smaller companies may cease altogether and exploration geologists are made redundant. Developments in geological theory and knowledge may also play a role at this stage in changing the proportion of the available mineral resources of different regions. The end products at this stage are: (1) a decision to explore for a specific commodity within a defined region, the size of which will vary depending on the commodity; (2) a definition of the broad search parameters, for example, what grade of mineral is required and how much; (3) a timescale for that exploration; and (4) a budget or cost envelope for the project.

2. **Geological models**. The next stage is to establish an exploration model. Put crudely, this should provide an image of the exploration target: it is no good looking for a needle in a haystack unless one knows what a needle looks like! An exploration model links the occurrence of a mineral to its geological setting and at its simplest is a list of search parameters. It is based on a combination of geological experience (i.e. where minerals have been found before) and geological theory. An exploration model can have a variety of forms, and may be either empirical or generic. Empirical models are based on a synthesis of the geological descriptions of a particular type of mineral-bearing site. The logic is that if you can find the same characteristics at another site, then you should find the mineral required. The alternative approach is to develop generic models based on an understanding of how a particular mineral is formed. In this case the search is for geological locations at which a similar range of conditions may have developed. The approach can be illustrated through the search for gravel resources. In regions such as northern Britain, most natural gravel reserves were deposited by glaciers during the Cenozoic Ice Age. By mapping and interpreting these landforms it is possible to identify target areas which may be rich in gravel. This approach was used in the search for sand and gravel on the Llyn Peninsula of North Wales. A model of the internal composition of different glacial landforms was established from a review of the literature (Table 3.2). This model lists the types of sand and gravel deposit likely to be found in a particular landform. From this it is clear that outwash fans and kames may be gravel-rich. The glacial geomorphology of the search area on the Llyn Peninsula was then mapped and these landforms where identified as potential gravel reserves and subjected to further investigation. In this case, the model allowed valuable exploration resources to be focused into those areas most likely to contain the required mineral. A more complex example involves the search for uranium, which is commonly found in hydrothermal veins beneath unconformities (Figure 3.5). These deposits result from: (1) a pre-concentration of uranium and associated elements in basement sedimentary rocks; (2) the concentration of these elements during weathering/erosion of the basement prior to deposition of the overlying sediment which forms the unconformity; and (3) the mobilisation of the uranium by oxidising fluids and its precipitation in reducing environments at the junction between fault zones and the overlying unconformity. Table 3.3 shows a simplified exploration model for such a mineral target. The model is an important and vital component of the exploration process, as it not only provides search parameters, but may also help determine the exploration strategies used in subsequent stages of the process (Box 3.1).

There are a number of problems with the use of exploration models which have been likened to a series of religious cults: (1) the 'cult of fad or fashion'

Table 3.2 *Exploration model for sand and gravel deposits within an area of recent glaciation. The characteristic composition of different landforms is determined from a review of the literature and potential as mineral targets assessed [Modified from: Crimes* et al. *(1992)* Engineering Geology, **32**, *Table 2, p. 141]*

Landform	Characteristics	Probable composition	Suitability
Till sheet (ground moraine)	Glacial sediment deposited beneath a glacier; low relief, undulating plane; may contain small elliptical hills (drumlins)	Clay with disseminated clasts	*
Cross-valley recessional moraine	Accumulation of debris at a glacier snout during a stillstand and readvance; arcuate or straight ridges, perpendicular to the former ice flow direction	Very variable, may contain several different types of till and glaciofluvial sands/gravels	*
Eskers	Linear, sinuous accumulations of sand and gravel deposited by subglacial rivers	Usually rich in well sorted aggregate, often with mineral fines content	*****
Kames	Mounds, hummocks and discontinuous terraces	Usually contain sand and gravel, but can contain silt and clay; commercial potential limited by size	**
Kame terraces	Terrace formed by glaciofluvial deposition between a glacier margin and the valley side may contain kettle holes on the outside edge of the terrace	Usually dominated by sand and gravel, although silt and clay content may be high	***
Dead ice topography	Complex assemblage of eskers, kames and irregular mounds produced by glaciofluvial deposition over buried ice	Very complex sedimentologically; large silt and clay contents; requires careful investigation to determine potential	**
Outwash fans	Extensive flat or fan-shaped proglacial accumulations of glaciofluvial sediment	Contain large sand gravel sheets, although variable in grain size and continuity	****

Suitability: * poor to ***** excellent

in which certain types of model may be more fashionable than others, leading to neglect of proven models; (2) the 'cult of the panacea', in which one model is held to explain all occurrences; (3) the 'cult of the classicists', in which all new models developed by academic research are rejected as unproven; (4) the 'cult of the corporate iconoclast', in which only those

Table 3.3 *Simplified example of an exploration model for uranium beneath an impermeable unconformity [Modified from: Moon & Whately (1995) In: Evans (Ed.)* Introduction to mineral exploration. *Blackwell Science, Table 4.4, p. 50]*

Element	Description
Commodity	Vein-like uranium
Description	Uranium mineralisation in fractures and breccia fills in metapelites, metapsammites and quartz arenites below, across and above an unconformity separating early and middle Proterozoic rocks
Rock types	Regionally metamorphosed carbonaceous pelites, psammites and carbonates. Younger unmetamorphosed quartz arenites
Texture	Metamorphic foliation and later brecciation
Age range	Early middle Proterozoic affected by Proterozoic regional metamorphism
Depositional environment	Host rocks are shelf deposits and overlying continental sandstone
Tectonic setting	Intracratonic sedimentary basins on the flanks of Archaean domes. Tectonically stable since the Proterozoic
Associated deposits	Gold- and nickel-rich deposits may occur
Mineralogy	Pitchblende + uraninite ± coffinite ± pyrite ± galena ± sphalerite ± arsenopyrite ± niccolite. Chlorite + quartz + calcite + dolomite + haematite + siderite + sericite. Late veins contain native gold, uranium and tellurides. Latest quartz–calcite veins contain pyrite, chalcopyrite and bituminous matter
Ore controls	Fracture porosity controlled ore distribution in metamorphics. Unconformity acted as disruption in fluid flow but not necessarily ore locus
Examples	Rabbit Lake, Saskatchewan, Canada Cluff Lake, Saskatchewan, Canada

models generated within one's own company are considered to be any good; and (5) the 'cult of the specialist', in which only the specialist components of a model are tested in the field. Clearly, models need to be used with caution to ensure that valuable mineral targets are not overlooked and that attention is not focused in the wrong place. However, without a list of search parameters – which essentially is all that an exploration model is – exploration cannot proceed.

3. **Reconnaissance and detailed planning**. Having established the search region and the exploration model or search parameters, the next stage is to find mineral targets worthy of detailed exploration. This involves a combination of desk work and field reconnaissance. An initial desk survey will involve assembling all the published and unpublished data available on the region and both collating and compiling the relevant information. The availability of data depends primarily on the availability of geological base-line surveys and the level of academic or commercial interest which has been shown within an area in the past. The use of remote sensing and airborne geophysics (e.g. seismic surveys, magnetic, electromagnetic, gravity

BOX 3.1: THE SEARCH FOR SAND AND GRAVEL RESERVES IN BRITAIN

The majority of onshore sand and gravel resources in Britain are of glacial or periglacial origin, formed during the Cenozoic Ice Age. There are two commonly used exploration strategies: (1) geomorphological mapping; (2) systematic borehole and geophysical exploration. The geomorphological mapping approach involves mapping the landforms present and grouping them into landsystems on the basis of their likely sand and gravel content. This grouping is based on published literature and knowledge of landform genesis (Table 3.2). Only those areas containing landforms likely to be mineral-rich are targeted for detailed borehole and geophysical investigation. This reduces exploration cost by targeting limited or reduced resources. The process is also relatively quick, with over 5 km^2 per day being mapped on average. However, this approach is only possible where the mineral is associated with identifiable surface landforms; if the deposits have no landform expression the approach does not work.

The alternative approach is to lay out a grid of boreholes and follow a systematic geophysical approach. This is more expensive and uses valuable resources in areas with little or no economic minerals. It is also slower, with approximately 1 km^2 per day being covered.

The approach chosen depends on the nature of the sand and gravel resource in a given country. In Britian one can identify three main geographical areas with different types of sand and gravel resource. A different exploration strategy is appropriate in each region, as outlined below.

Type One deposits occur within the limits of the last ice sheet (Devensian); landforms are pronounced and are associated with most gravel deposits. In these areas the geomorphological approach works well.

Type Two deposits occur beyond the Devensian Ice Sheet but within the maximum extent of glacial ice (Anglian Ice Sheet) in Britain during the Quaternary. The glacial landforms within this area are older and more subdued due to the action of periglacial processes during the last glacial (Devensian) cycle. More importantly, the Anglian Ice Sheet advanced over large areas of fluvial sands and gravels deposited by a pre-Anglian River Thames, which flowed through East Anglia. These sand and gravel reserves have no landform expression, being buried beneath glacial till. In this area the geomorphological approach fails to pick out the hidden reserves and a more traditional borehole survey is required.

Type Three areas occur beyond the glacial limit. Here the sand/gravel reserves are more fragmentary and have variable landform expression. In this area neither approach is particularly efficient.

continues overleaf

BOX 3.1: *(continued)*

A conceptual model of the glacial history of an area such as Britain is therefore an important starting point in the exploration of sand and gravel reserves, and illustrates the potential role of exploration models in determining the exploration strategy used.

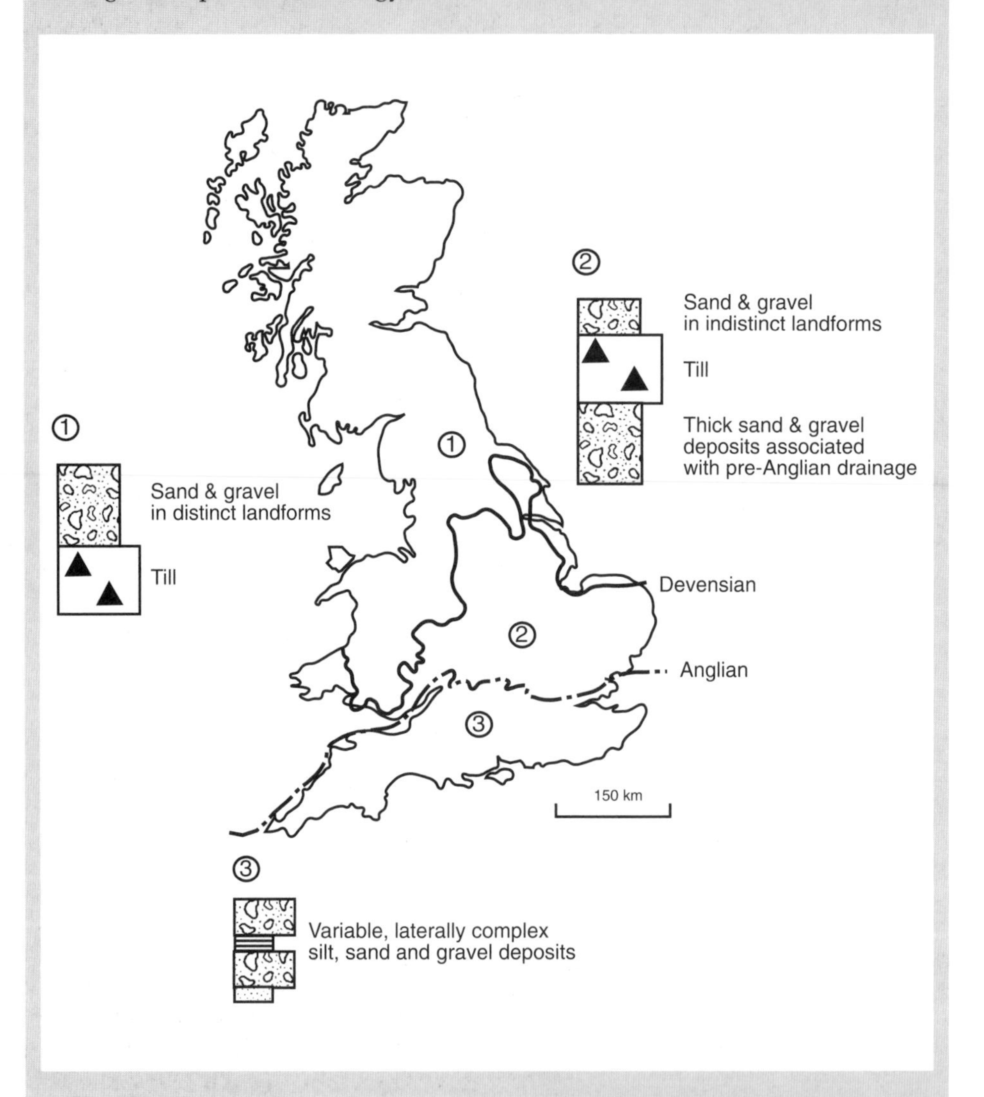

Source: Merritt, J.W. 1992. A critical review of the methods used in the appraisal of onshore sand and gravel resources in Britain. *Engineering Geology*, **32**, 1–9.

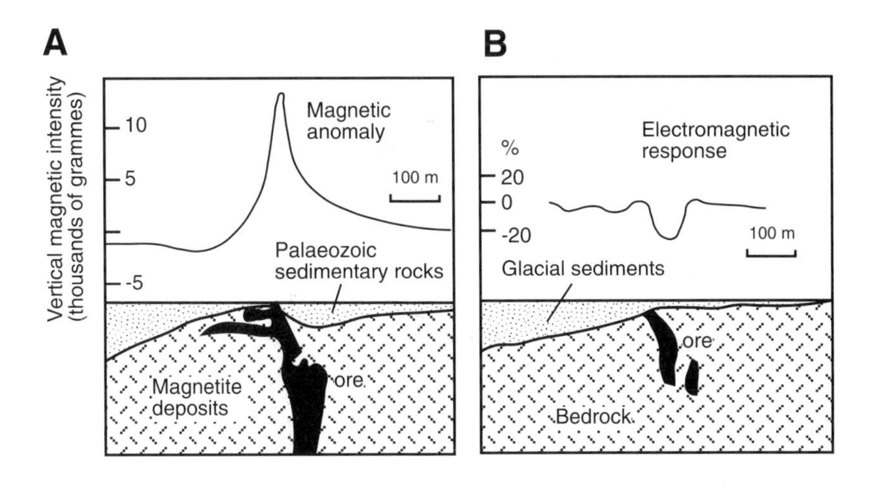

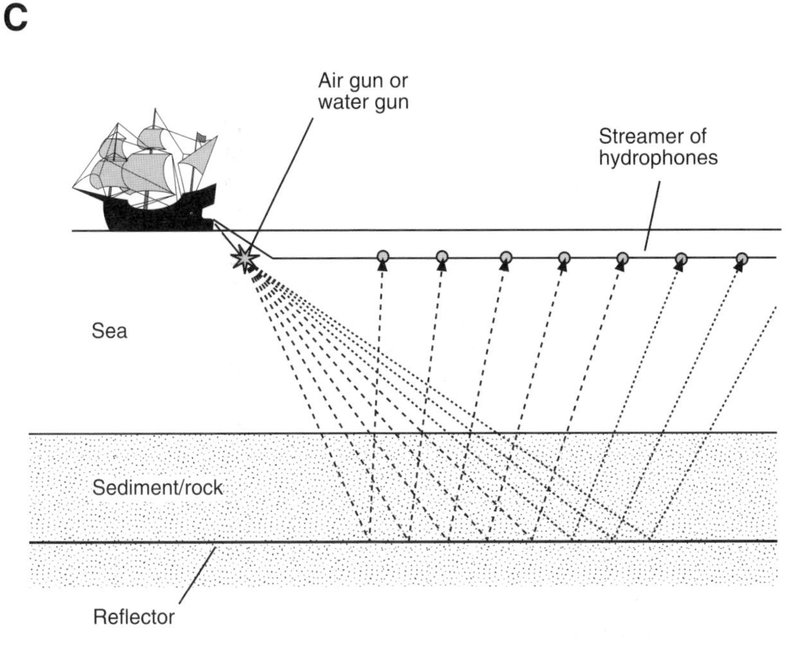

Figure 3.10 *The use of geophysics in mineral exploration.* ***A.*** *Magnetic anomaly of an iron ore body at Iron Mountain, Missouri, USA.* ***B.*** *Electromagnetic anomaly over a nickel ore body in Ontario, Canada.* ***C.*** *The principles of seismic data collection [A & B Modified from: Kesler (1994)* Mineral resources, economics and the environment, *Macmillan, Fig. 4.10, p. 64]*

and gamma-ray measurements) is of particular importance here (Figure 3.10). Seismic surveys are a vital part of the exploration for oil and gas. Sound waves generated by a water or air gun penetrate the subsurface, are reflected by different geological horizons and are recorded by instruments on the surface (geophones on land and hydrophones on water; Figure 3.10).

Table 3.4 *Colour of some economic minerals in surface outcrops*

Mineral or metal	Outcrop colour
Iron sulphides	Yellows, browns, reds
Manganese	Blacks
Arsenic	Greens, greenish yellows
Cobalt	Black, pink
Lead	White, yellow
Silver	Waxy green, yellow
Zinc	White
Uranium	Bright green, yellow

This allows a picture of the subsurface geology to be established, which is essential in the identification of potential oil and gas reservoirs and traps. Remote sensing also allows geological reconnaissance mapping where data are poor or unavailable. The desk study phase may be followed up by field reconnaissance within areas identified as being of interest. The aim at this stage is primarily to confirm interest in specific areas but not to evaluate the detailed geology of a region. The end product of this stage of exploration is a number of identified land parcels of varying size in which further exploration effort is to be concentrated.

4. **Land acquisition**. The aim here is to acquire the rights, preferably exclusive, to explore and potentially to extract a deposit from the land parcels identified in the previous stage. Normally a company will obtain the rights to explore the property for a specific period of time and the option to convert this into the rights to mine. This is a complex legal problem which varies from country to country depending on the owner's mineral rights and national legislation. In general there are two extremes: (1) all the mineral rights are owned by the state, which can grant access without regard to the occupier; and (2) all the mineral and surface rights are owned privately. In most cases a mixture of the two may occur. For example, in Britain all gold and oil reserves are owned by the state, whereas onshore aggregates and other minerals are owned privately. The end product at this stage is to obtain the rights to explore some or all of the identified target areas.
5. **Target location and appraisal**. The aim here is to locate all the potential mineral reserves within the target area and define the extent of the mineral bodies present. It involves a wide range of intensive fieldwork and laboratory tasks, which may include field mapping, float mapping, geochemical exploration, geophysical surveying, the investigation of abandoned mines, geobotany, and trenching or drilling. Detailed field mapping may help define the surface outcrop and geometry of mineral bodies. Mineral deposits may be recognisable on the surface by distinctive colours or coloured weathering products, some of which are listed in Table 3.4. In areas of poor exposure or with thick surficial deposits, such as in glacial terrain, float mapping may be used, in which mineralised boulders or rock fragments are traced back to their source. This method has been used extensively in

Finland where the general public are encouraged to post, free of charge, unusual rocks to the geological survey. The survey receives over 10 000 boulders/samples each year through the post, of which 5% lead to field investigation, and 0.25% actually lead to detailed target appraisal. In glacial terrain this involves determining dispersal plumes (Box 3.2) and a knowledge of ice flow direction is vital. Sediment sampling from streams may also be important in tracing mineralisation. Geochemical exploration also has an important role in float mapping. The detection of geochemical anomalies within soils, weathering horizons and glacial tills is particularly important. For example, Figure 3.11 shows an empirical model of a mineral dispersal plume in a basal till deposit. The mobility of different minerals is an important consideration in tracing geochemical anomalies: the less mobile they are within soil or groundwater, the closer they will be associated with the actual outcrop. Detailed geophysical surveys are important not only in the recognition of mineral bodies but also in determining the geological structure of a region; again, this is of particular importance in oil and gas exploration (Figure 3.10). Such surveys are the principal exploration tools for oil exploration, where the aim is to identify potential oil and gas traps.

The investigation of abandoned mineral workings may form an important part of an exploration programme in an area of former mining activity. Not only do old workings provide subsurface information, but it may also be cheaper to reopen them than to start again.

Geobotany may also provide information on the location and geometry of mineral bodies. When certain minerals are concentrated in soils they may be toxic to plants, causing problems of disease and growth retardation. Equally, certain plants may be tolerant of such conditions and may therefore thrive in such areas, particularly where there is reduced competition. For example, the small mauve flower *Beccium homblei*, common in the Zambian Copper Belt, usually occurs in areas with a soil copper content of between 50 and 1600 parts per million, which is toxic to most other plants, and it has been widely used as an indicator of copper targets in this region.

When specific mineral targets have been identified, a drilling programme may be required. Drilling is used to confirm the subsurface structure, define the extent and volume of any deposits present, establish their continuity and recover data on subsurface conditions. Ideally the pattern of drilling needs to be as systematic as possible to ensure adequate recovery of the data required. An example of a drilling programme on a gold deposit is shown in Figure 3.12. Data may be recovered from boreholes directly using core samples or remotely using well-logging techniques, which are particularly importantly in oil exploration. **Wireline** or **well logging** involves lowering a device, known as a sonde, down the borehole or well and taking regular readings from a range of instruments as it descends. In this way a log of various physical rock attributes is obtained for the length of the borehole. The sonde may contain a variety of instruments, depending upon the data required and the sophistication of the enterprise. Typical wire logs include calliper logs, resistivity logs, gamma-ray logs, neutron logs and sonic/acoustic logs.

BOX 3.2: ERRATIC TRACING AS AN AID IN MINERAL EXPLORATION

In glacial terrain the ground surface is often covered by a thick layer of glacial till, preventing direct observation of the outcrop geology and any associated mineral deposits. Glacially transported boulders (erratics) can sometimes be used, if ice flow directions are known, to pinpoint mineral deposits beneath the till surface. This has been used widely in Scandinavia in the search for metal minerals.

The principle is well illustrated by an exercise carried out by Shakesby (1978) in the Central Lowlands of Scotland. Here a distinctive igneous rock, essexite, crops out over a very limited area. The outcrop was eroded by the

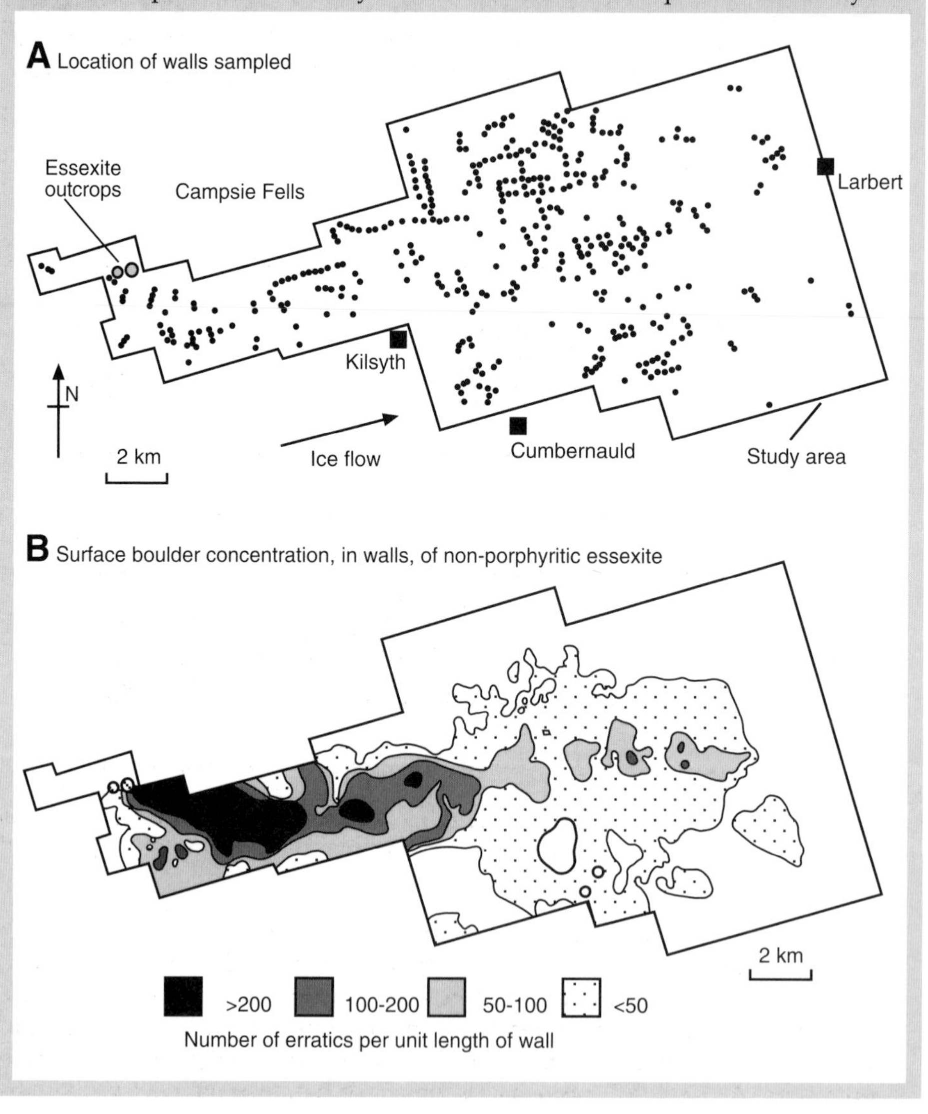

passage of the last Scottish Ice Sheet and essexite boulders have been dispersed in the downflow direction. Many of these boulders were deposited in the surface layers of the till and have been used by farmers to build field boundary walls. Shakesby mapped the geometry of the dispersal plume by recording the number of essexite boulders per unit length of dry-stone wall. In this way he was able to establish the dispersal plume, the apex of which is located on the essexite outcrop, as shown in the diagram below. The work demonstrates how glacial erratics deposited over a wide area can be used to locate a very localised rock outcrop. This principle has been used widely in mineral exploration within glacial terrain.

Source: Shakesby, R.A. 1978. The dispersal of glacial erratics from Lennoxtown, Stirlingshire. *Scottish Journal of Geology*, **14**, 81–86. [Diagrams modified from: Shakesby (1978) *Scottish Journal of Geology*, **14**, Fig. 1, p. 82]

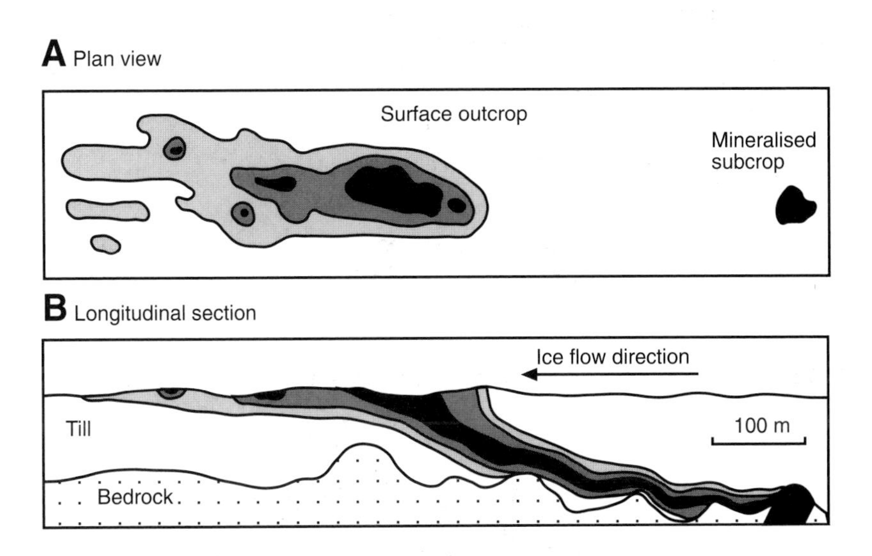

Figure 3.11 *Schematic model of a geochemical anomaly associated with a layer of lodgement till overlying a mineral body. This type of model has been validated by field evidence and the rising plume anomaly probably reflects subglacial deformation of the till layer during deposition [Modified from: Gray (1992)* Quaternary Proceedings, **2**, *Fig. 1, p. 2]*

Calliper logs measure the width of the borehole, which is related to rock hardness and fracture density, and consequently provides lithological data. Resistivity logs measure the electrical resistance within the rock of the borehole wall. This is influenced by the volume and nature of the fluid within the rock pores, thereby providing information on porosity and pore fluid content. If a rock is porous and contains brine, then its resistivity will be

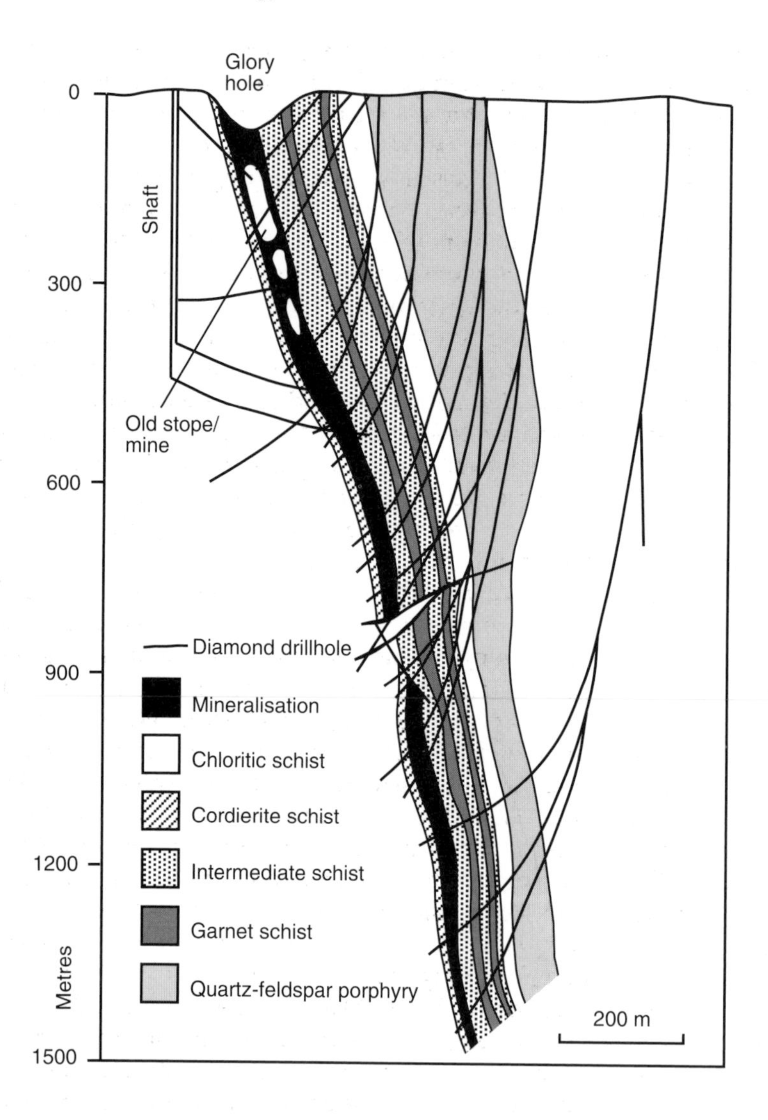

Figure 3.12 *Typical drilling programme for a previously mined deep-vein gold deposit, Big Bell, Western Australia [Modified from: Moon & Whately (1995) In: Evans (Ed.)* Introduction to mineral exploration. *Blackwell Science, Fig. 5.13, p. 79]*

low; if the same rock contained hydrocarbons, then its resistivity would be much higher. Gamma-ray logs record the rock's natural radioactivity and may provide information about lithological composition. Neutron logs are produced by bombarding the rock in the well wall with neutrons emitted from a radioactive source. As these neutrons collide with the rock and pore fluids they lose energy, particularly if they collide with hydrogen atoms. This loss of energy is monitored and provides information about the presence of water and hydrocarbons in a formation. Also, most importantly, it usually enables oil to be distinguished from gas. Finally, sonic/acoustic logs measure the velocity of sound waves within the rock and provide vital information in the interpretation of seimic data, which is also sound-based.

Drilling is expensive and the design of drilling programmes must involve careful consideration of their precise purpose in order to ensure that this will be achieved with the minimum number of holes. Equally, the decision to stop drilling is also of importance and usually occurs when: (1) no mineralisation has been encountered; (2) mineralisation has been intersected, but is not of sufficient economic grade or width; (3) mineralisation has been established but its continuity is poor; (4) a body of potential economic grade mineral has been found; or (5) the drilling budget is exhausted.

Ideally, the end product of this stage in the exploration process is an identified target or site at which mineral deposits of sufficient grade and volume occur to justify the expense of extraction. In practice, however, all potential targets may turn out to be uneconomic, putting the exploration programme back to the reconnaissance stage.

6. **Site assessment/feasibility studies**. The aim here is to establish the feasibility and economic viability of a proposed mine, quarry or oil/gas field. This is determined by a very wide range of factors, which can be broadly grouped into geology, economics and socio-political factors. Geological factors include establishing the grade (**assay**) and tonnage of mineral present. For most minerals this is based on detailed sampling and geochemical examination of the ore. These ore samples are then used statistically to predict the grade (i.e. the ratio of the required material to the waste) of the mineral likely to be produced from the deposit. The amount of mineral or tonnage obtained requires a detailed knowledge of the geometry of the mineral body and any variation in grade present. For oil and gas prospects, detailed information on the geometry of the oil reservoir and trap and its content is critical in evaluating its economic feasibility. Next, the methods of both mineral extraction and processing need to be established, based on a detailed knowledge of ore chemistry, metallurgy, the geometry of the ore body, and the rock mass strength of the surrounding rocks. From this information the methods of mineral extraction and processing can be determined, providing information on the likely cost of exploitation. In the context of oil and gas, the production strategy is determined by such things as the porosity/permeability of the reservoir and its geological content, as well as surface conditions. This last factor is particularly important in offshore locations where water depth and the nature of the marine environment may affect the cost and feasibility of production.

   It then becomes simply a question of economics. From a knowledge of the current commodity price and its likely long-term value it should be possible to establish an estimate of the likely cash flow of the mine; that is, the balance between revenue and expenditure. Expenditure includes extraction costs, processing costs (including energy costs), transport costs, sales costs, capital costs, interest payments on capital loans needed to establish the mine or oil/gas field, and taxes. Revenue will depend on estimates of the long-term price of a commodity, determined by market demand. If the accounts do not balance then the prospect is not economically viable. If the prospect is to be viable then the next stage is to consider the socio-political factors. Firstly, it is necessary to establish the stability of the host country: is there any threat to the long-term productivity of the site from government intervention or regulation? This can profoundly influence a company's ability to borrow money in order to finance a new mine. In many countries there is a risk to investment from war or civil strife. In more general terms, different countries have different tax requirements and regulations, all of which may affect the economics of a mine in that country. The environmental impact of the proposed mineral extraction must also be considered to ensure that it is both socially and politically acceptable in the host country, which may ultimately determine whether planning permission for the site or operation will be given. The majority of the prospects analysed at this stage will prove uneconomic. For example, a Canadian mining company examined the feasibility of over 1000 prospects during a 40 year period, of which only 18 were brought into production and only seven actually made a profit.

If all these considerations suggest that a prospect is economically and politically viable, and therefore likely to go ahead, a feasibility report is drawn up, usually by external consultants. In the case of mines, this will include detailed engineering plans for all mine infrastructure. This is the final stage prior to commissioning the prospect, and it is on the basis of this document that investors and banks will be asked to put up the necessary capital to establish the venture. A report of this type may therefore be audited several times during the process of obtaining the necessary venture capital.

7. **Prospect development.** Having obtained the necessary financial support for the project, planning permission or any extraction licence required must be sought, and if successful, the next stage is to let the contracts for construction of the basic site infrastructure.

Only after this long and often expensive process is there a return on the money invested in exploration. The large number of stages at which this process can go astray demonstrates just what a high-risk exercise commercial mineral exploration can be.

3.3 METHODS OF MINERAL EXTRACTION

Methods of extracting mineral wealth from the ground can be summarised as: (1) fluid removal using wells; (2) underground mining; and (3) surface mining or quarrying.

1. **Fluid removal.** The removal of fluids via wells is usual practice for most oil and gas operations, and is also employed for a variety of soluble minerals

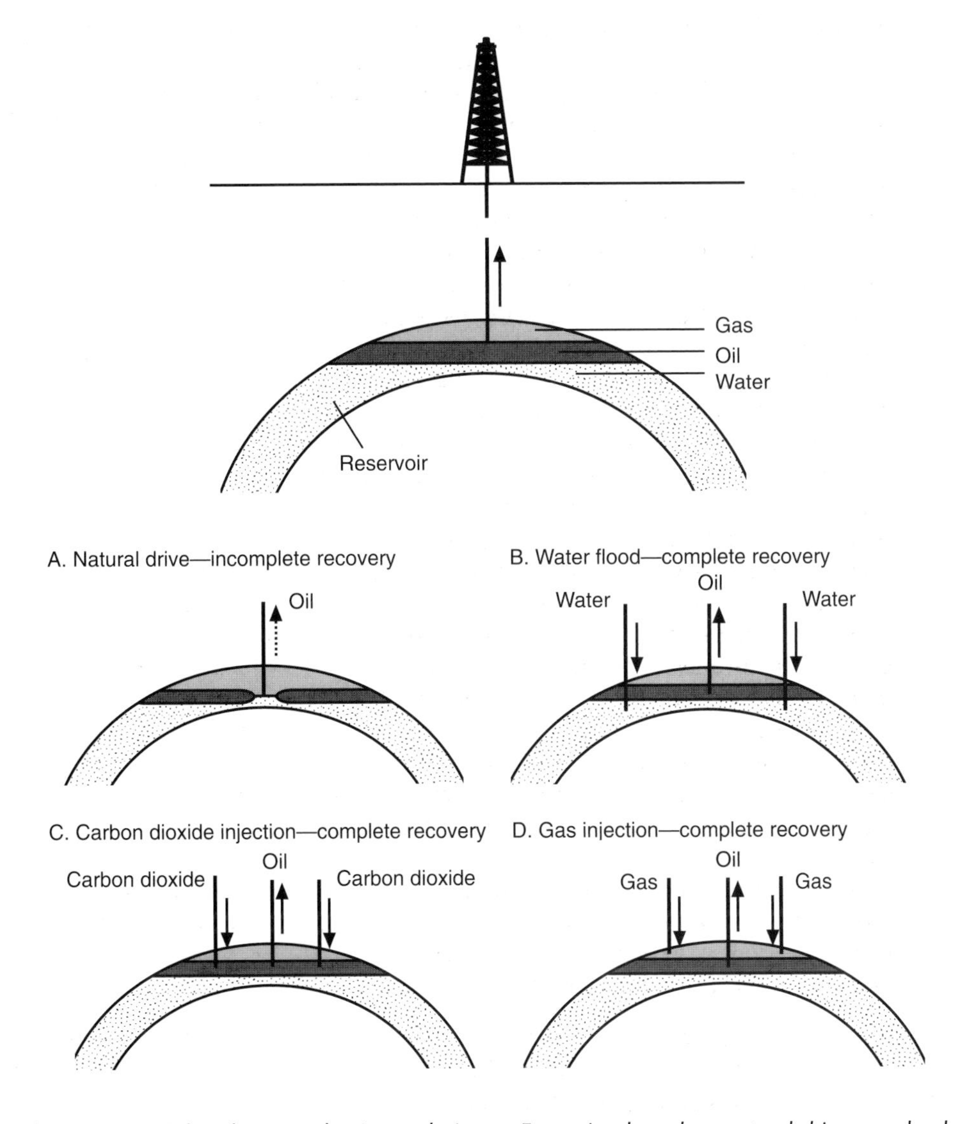

Figure 3.13 *Oil and gas production techniques. Extraction based on natural drive may lead to incomplete recovery of the oil body; the alternative involves pumping either water, carbon dioxide or natural gas into the reservoir*

such as salt. Oil recovery commonly uses either single drill holes or an array of boreholes, often in non-vertical directions, a technique referred to as deviated drilling. The number of drill holes sunk and their orientation is controlled by the geometry, porosity and permeability of the oil reservoir. An important consideration is the method of pressure removal, either by injection of gas from above or water from below; if natural drives are used this can lead to incomplete recovery of the oil (Figure 3.13). Oil recovery usually proceeds in two principal stages (primary and secondary recovery), although a third (enhanced recovery) is increasingly being used in order to maximise yield. During primary recovery, 30–40% of the oil is removed, while during secondary recovery a further 40–50% is recovered through the injection of gas or water into the reservoir from additional boreholes (Figure 3.13). Enhanced recovery is achieved through the addition of steam or chemical or bacterial additives to the oil in order to reduce its viscosity. Oil production usually brings to the surface a variable amount of salty toxic water. Once separated from the oil, this may either be pumped back into the well or alternatively disposed of by evaporation or deep groundwater recharge outside the oil-field.

Pumping of dissolved salts from subsurface evaporite beds is also widely used, particularly in the Cheshire Saltfield of Britain. Brine is pumped from the salt-rich beds and is either replaced by the injection of fresh water (**controlled brining**) or is replaced naturally by groundwater (**wild brining**). Injection of groundwater is preferable since some control on the form of subsurface solution cavities can be exercised. Wild brining is less predictable and subsurface cavities may eventually coalesce, causing subsidence problems.

2. **Underground mining**. Methods of underground mining depend largely upon the geometry of the exploration target. Five principal methods of underground mining are identified: (1) longwall mining; (2) pillar-and-stall working; (3) bell pits; (4) adit mines; and (5) deep cavern mining. Modern underground coal mining is now usually undertaken using **longwall mining**. The coal is removed using a mechanical cutter from a continuous face. As the coal seam is worked, the roof above the cutter is supported by hydraulic jacks, while that beyond the cutter is allowed to collapse onto the former floor of the coal seam, known as the **goaf** (Figure 3.14A). This causes ground subsidence, the level of which depends on the depth of the seam and its thickness. Subsidence is usually less than the thickness of the seam since overlying rock volume is increased by opening of joints and fractures. The extent of subsidence, the subsidence wave, moves forward as the coal face is worked and can therefore be anticipated, and affected buildings can be strengthened where necessary. This type of technique is particularly suited to laterally continuous coal seams with gentle dips.

   Where mineral deposits are less continuous, more traditional methods, using **pillar-and-stall working,** are often used (Figure 3.14B). Only part of the mineral deposit is removed, leaving the rest to provide support for the roof. Between 50 and 85% of the mineral is removed, and the pattern of rock support is calculated carefully to leave a regular pattern of pillars. In the past, however, the pattern was often more irregular and the size of support was based on tradition, experience and intuition. Pillar-and-stall working has been used as a traditional method of mining in Britain for a wide range of different commodities, including coal, ironstone, gypsum, clays and some building stones. Another traditional method of mining is the **bell pit** (Figure 3.14C). It is only really suited for shallow seams in horizontal strata, and is rarely used today.

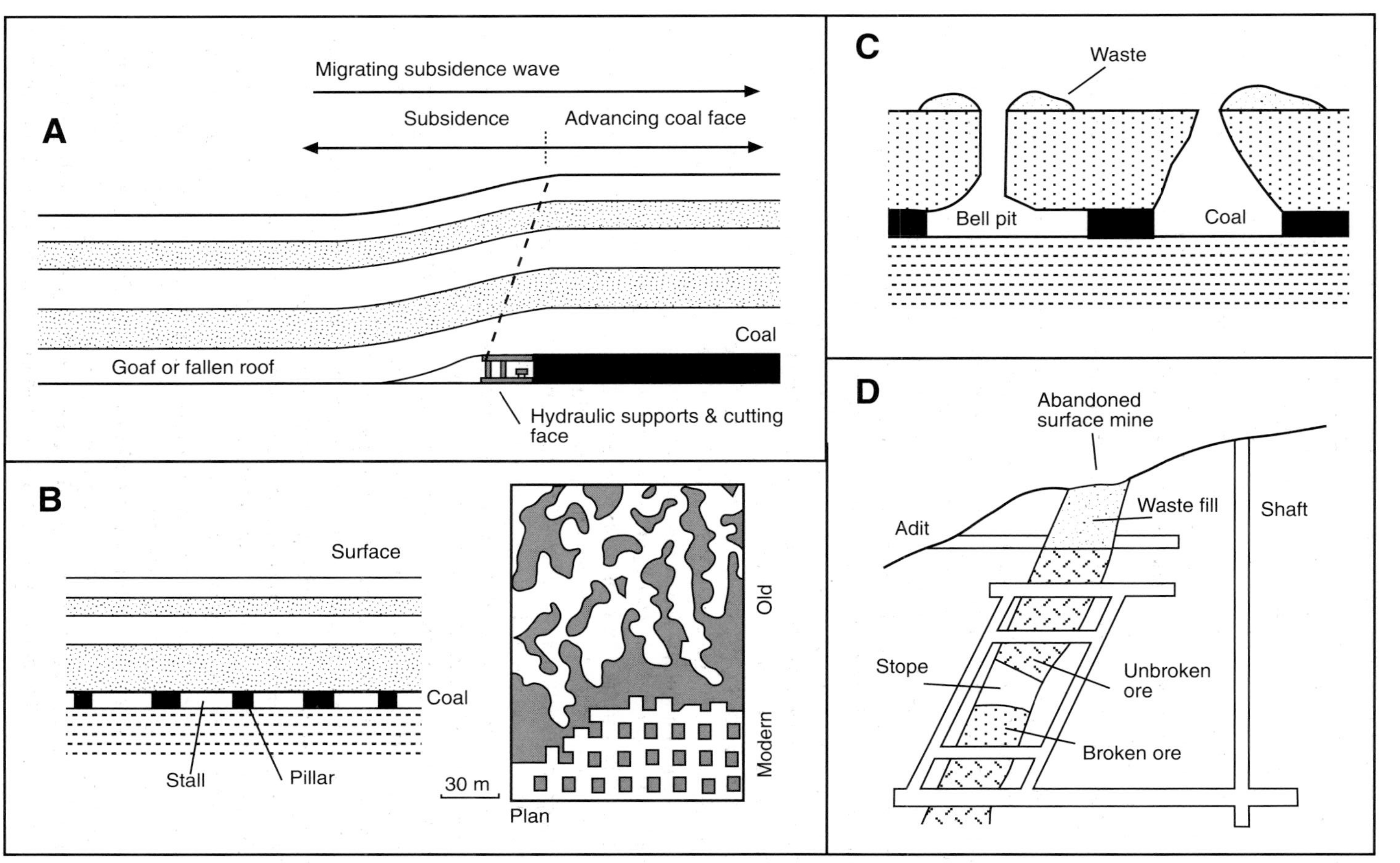

Figure 3.14 *Types of mine.* ***A.*** *Longwall coal mine.* ***B.*** *Pillar and stall working.* ***C.*** *Bell pit.* ***D.*** *Adits and deep cave mines [B Modifed from: Waltham (1994)* Foundations of engineering geology. *Blackie, p. 60]*

Adit mines are horizontal tunnels which are driven into the hillside either along mineral deposits or, more usually, to intersect inclined mineral-bearing units which are then worked from either above or below the main adit. **Deep cavern mining** is used to remove steeply dipping or irregular minerals. A vertical shaft is sunk and is connected to a series of horizontal levels which cut the main mineral body (Figure 3.14D). Ore is removed from the body to form a void or **stope**. As the stope is expanded upwards, the floor is raised either by using broken ore or by importing waste rock produced by surface processing of the ore.

3. **Quarrying**. Surface working or quarrying is used to obtain a wide variety of minerals, particularly bulk minerals such as construction stone and aggregates, as well as gold and copper deposits and, increasingly, coal. One of the critical factors determining the economics of such operations is the amount of **overburden**, the unwanted soil and rock, that is above the desired mineral. This is determined by the depth of mineral below the surface and the geometry of the mineral body (Figure 3.15A). The economics are usually expressed as a ratio of overburden to mineral; that is, the number of units of overburden that must be removed for every unit of mineral. There are five basic methods of surface working: (1) deep open-cut excavations; (2) shallow open-cut excavations; (3) area strip mining; (4) contour strip mining; and (5) dredging.

   Deep and **shallow open-cut excavations** are formed as open quarries (hard rocks) or pits (unconsolidated sediments) of varying size. The overburden is removed and the mineral is extracted to leave an open excavation equal in size to the volume of material removed. Problems of flooding, where the excavation lies below the water table or is cut in impermeable rock, are of particular concern. An alternative is **area strip mining** in which an excavation is backfilled as it is cut. The backfilled areas are restored as work proceeds (Figure 3.15B). This method has the advantage of minimising the disrupted area and has been used extensively in Britain to recover near-surface coal and ironstone deposits. On steep hillsides with horizontal or near-horizontal mineral-bearing units, **contour strip mining** is possible (Figure 3.15C). A contour-parallel trench or excavation is cut to recover the mineral. The width of this trench is determined by the stripping or overburden ratio; that is, the point where so much overburden has to be removed that the operation is rendered uneconomic. It is an approach which has been used widely in the Appalachian coalfields of the USA. Finally, surface removal of alluvial placer sediments by dredging is common in some regions. This technique is used extensively off the British coast to supply aggregates, and is used for economic concentrations of metallic minerals in the Far East.

A wide range of mining systems and techniques may be used to exploit a given mineral depending upon the geometry of the mineral body and the nature of the mineral. Geological structure has a vital role in determining the geometry of a mineral deposit, and its control on the mining system employed is illustrated by the different approaches used in slate extraction in North Wales (Figure 3.16)

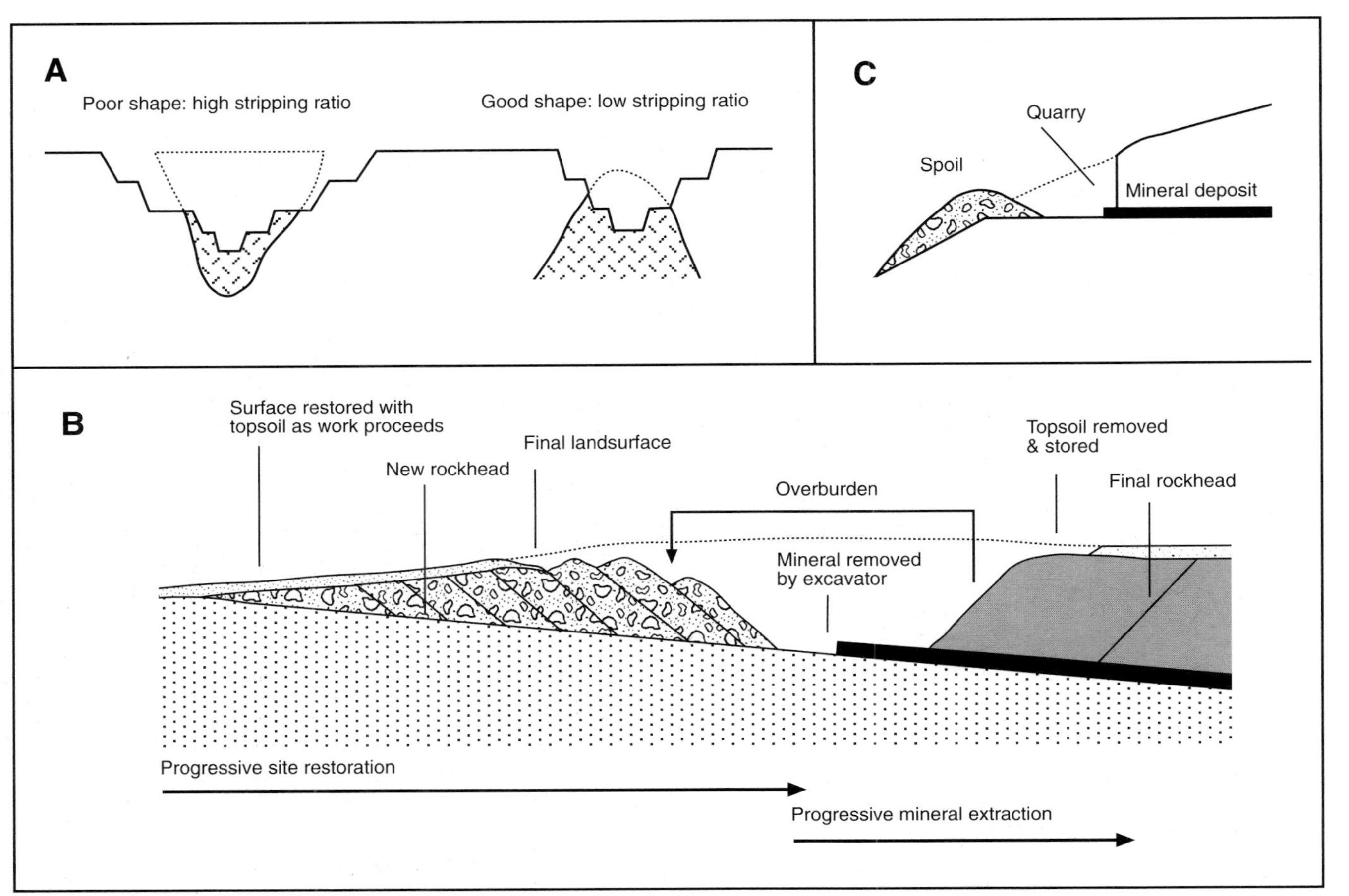

Figure 3.15 *Types of surface mining or quarrying.* ***A.*** *The geometry of a mineral deposit as a control on the amount of overburden which has to be removed.* ***B.*** *Strip mining/quarrying.* ***C.*** *Contour mining/quarrying*

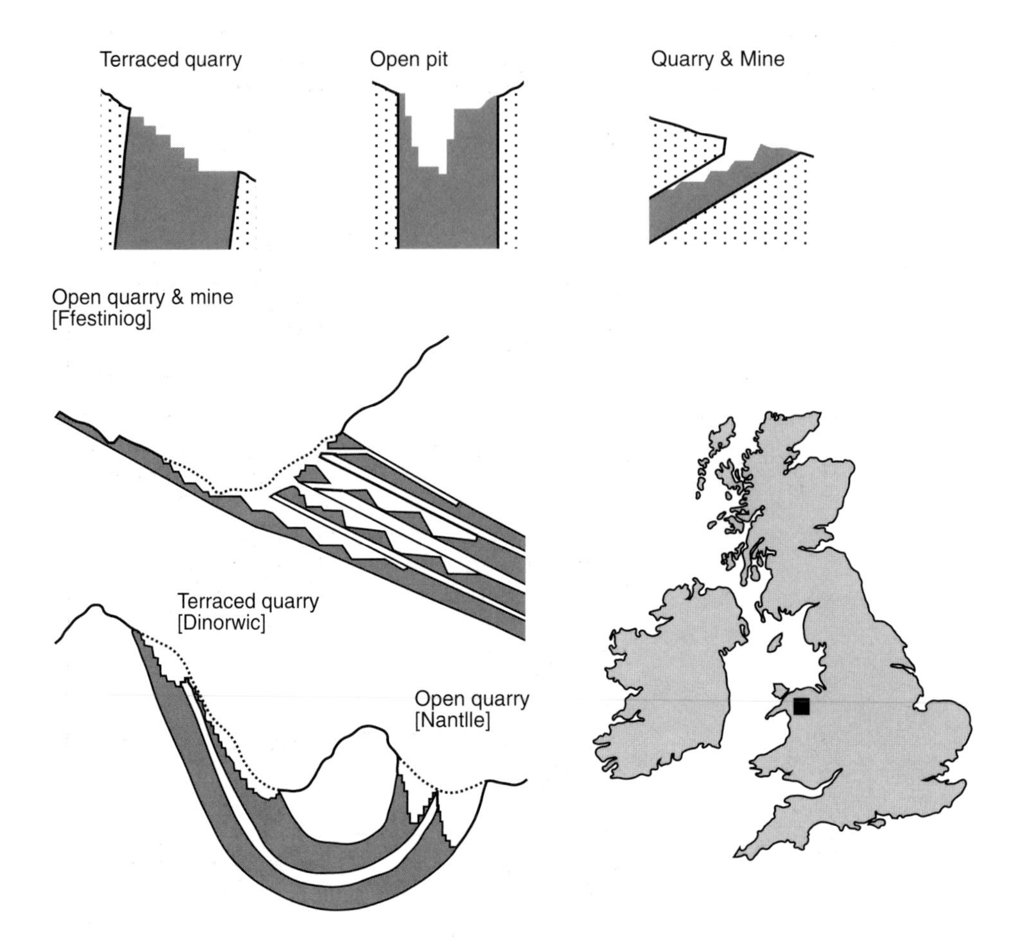

Figure 3.16 *The type of mine or quarry used to extract Welsh slate varies with the geometry of the mineral body*

In most cases the removal of the mineral from the ground is only the first stage in the process of mineral extraction. Some minerals, such as fuel minerals and construction minerals, occur in sufficient concentrations to be transported directly from a mine or quarry to the marketplace. However, most metal ores and industrial minerals require processing before they can be transported from the mine site, as the concentration of the required metal or compound may be relatively small. Most processing involves separating the valuable mineral from the waste rock or **gangue**. Methods by which this is achieved are varied, but in the case of most non-ferrous metals include either flotation (e.g. copper, lead, zinc) or leaching (e.g. bauxite, uranium, gold). **Flotation** works on the principle that once crushed and milled to a fine grain size, it is possible to confer, using a variety of reagents, a hydrophobicity to metal particles. This means that the particles will naturally separate from water, allowing them to become attached

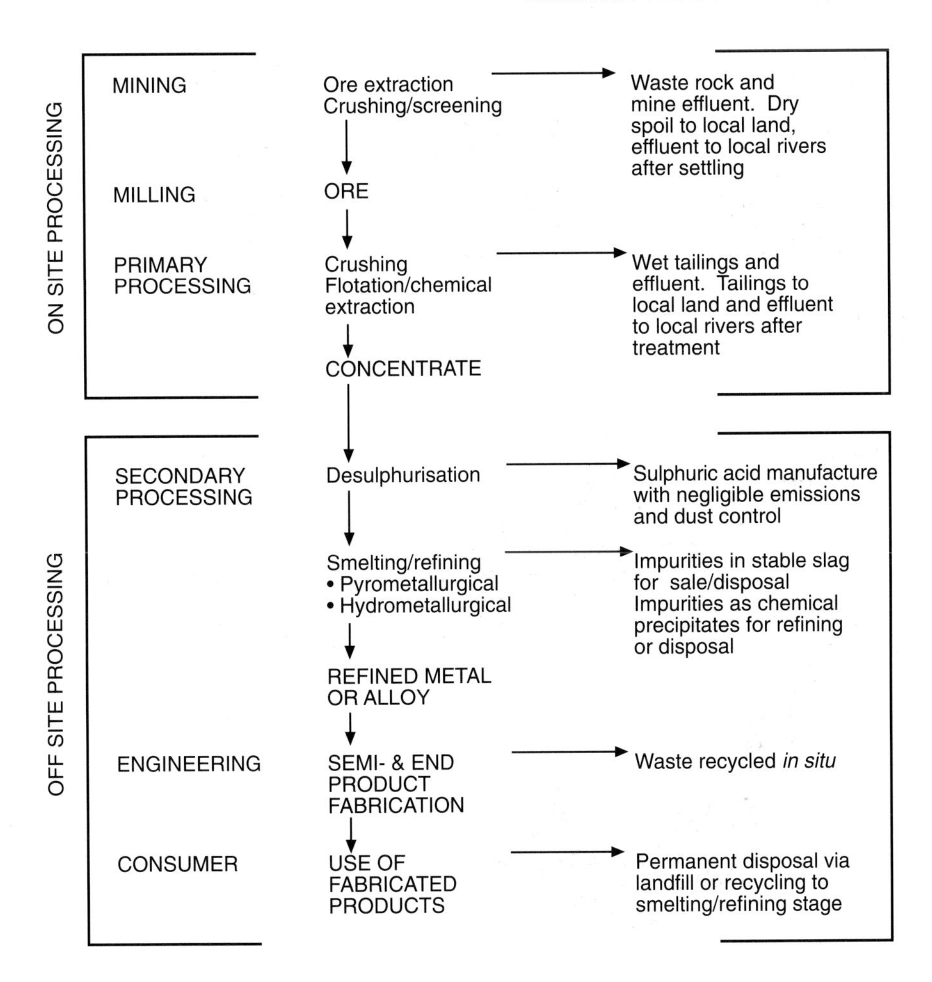

Figure 3.17 *Generic stages in the processing of non-ferrous metal minerals [Modified from: Barbour (1994) In: Hester & Harrison (Eds)* Mining and its environmental impact. *Royal Society of Chemistry, Fig. 1, p. 3]*

to air bubbles and therefore float as a surface froth which can be skimmed off, while the gangue settles out. The concentrate produced in this way is then sufficiently pure to stand the cost of transport to smelting plants. The process is illustrated generically for non-ferrous minerals in Figure 3.17.

Leaching involves the addition of a chemical reagent which selectively dissolves the required mineral. For example, caustic soda is used to dissolve hydrated aluminium oxide from bauxite in what is known as the **Bayer process**, while sodium cyanide is used for the leaching of gold. The concentration of the mineral at the mine site is important since it produces waste rock known as **spoil** and **tailings**, which have to be disposed of at or near to the mine site, adding to its environmental impact, particularly where hazardous chemical substances such as sodium cyanide have been employed in the process.

3.4 ENVIRONMENTAL IMPACTS OF MINERAL EXTRACTION

All mineral extraction will have some degree of environmental impact. In most developed countries mineral extraction is now closely regulated and environmental impacts are increasingly being controlled. Modern operations are bound by current environmental legislation which is becoming stringent in the developed world. However, in most countries there is a legacy of older workings, many of which have been abandoned, that were not subject to the same level of environmental control. Consequently there is a mixture of modern impacts coupled with historical legacies. In developing countries, where for economic or other reasons environmental control is less stringent, the impact of current mineral extraction may be more marked.

There are a variety of different types of environmental impact depending on the mining/quarrying operation involved. The key impacts commonly encountered both during and after the cessation of mineral extraction include: (1) mining subsidence; (2) problems associated with the disposal and management of mine spoil and tailings; (3) environmental impact of mine operations such as blasting, land loss, noise, traffic problems and problems of water quality; and (4) quarry/mine restoration. Each of these management problems is dealt with in turn.

3.4.1 Mining Subsidence

Mining subsidence is a direct result of underground removal of material and it may occur in a controlled or an uncontrolled fashion. Longwall mining involves controlled or planned subsidence as part of the mining activity. In contrast, the collapse of pillar-and-stall working may occur long after a mine has been abandoned (see Section 13.4.1). Subsidence may cause the lowering of the ground surface or alternatively, if the mine was close to the surface, a circular depression or crown hole may form (Figure 3.18A). The failure of multiple pillars may give rise to graben-like structures associated with regional subsidence. The nature of the collapse is influenced by the properties of the material involved and the depth of the water table, as shown in Figure 3.18B. Subsidence affects a larger area than simply that immediately above the mineral deposit or coal seam removed, and the width of the area of subsidence, known as the **subsidence bowl**, is a function of the depth of the extracted deposit (Figure 3.18C). The critical parameters which determine the size and shape of the subsidence bowl are the depth of working, the width of extracted mineral, and the extracted thickness. In general, the effects

---

Figure 3.18 *(opposite) Subsidence associated with underground mining.* ***A&B.*** *Types of ground subsidence and the effects of groundwater on the type of ground response.* ***C.*** *The pattern of compression, extension, ground tilting and subsidence associated with underground mining [Modified from: Waltham (1994)* Foundations of engineering geology. *Blackie, pp. 61 & 62]*

A

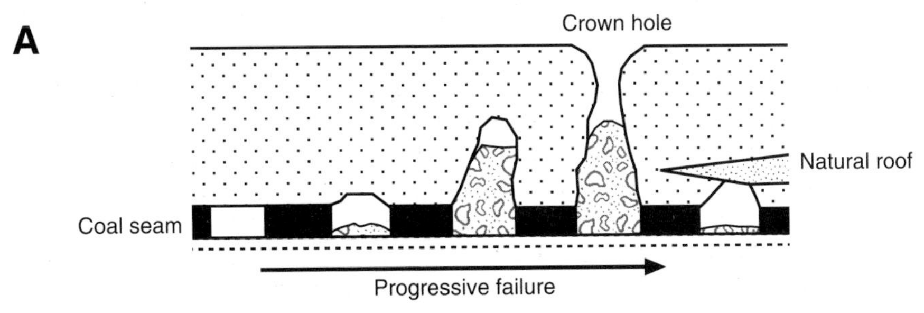
Crown hole
Natural roof
Coal seam
Progressive failure

B

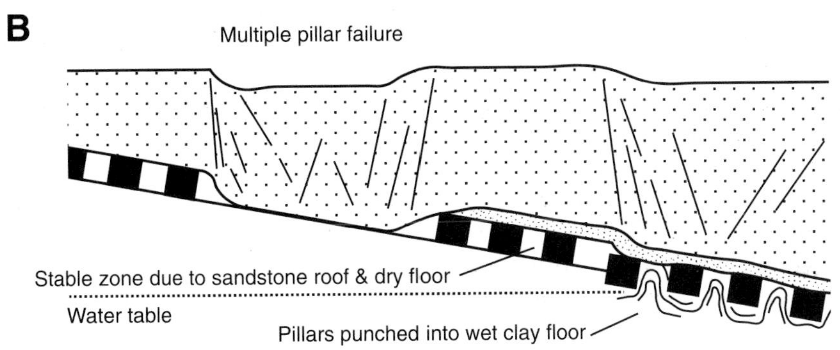
Multiple pillar failure
Stable zone due to sandstone roof & dry floor
Water table
Pillars punched into wet clay floor

C

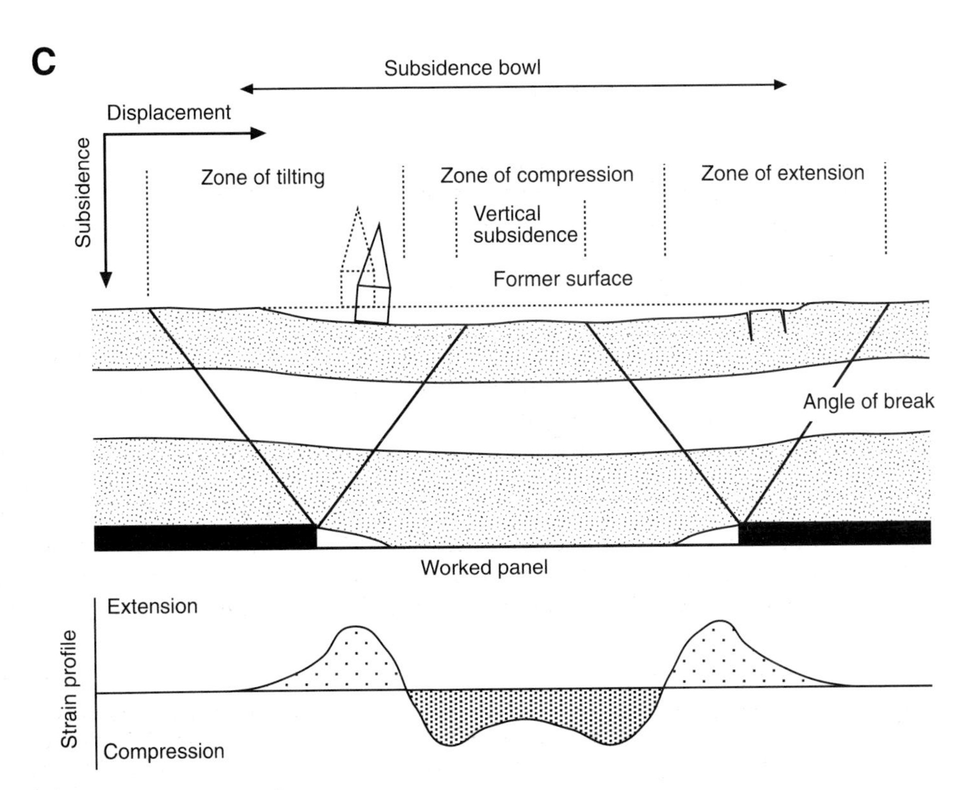
Subsidence bowl
Displacement
Subsidence
Zone of tilting
Zone of compression
Zone of extension
Vertical subsidence
Former surface
Angle of break
Worked panel
Extension
Strain profile
Compression

Table 3.5 *Classification of the level of damage due to subsidence [Modified from: Waltham (1994)* Foundations of engineering geology. *Blackie, p. 63]*

Class	Distortion (mm)	Typical features
Very slight	<30	Barely noticeable hair cracks in plaster
Slight	30–60	Slight internal fractures, doors and windows may stick
Appreciable	60–125	Slight external fractures, service pipes may fracture
Severe	125–200	Floors slope and walls lean, door frames distorted
Very severe	>200	Severe floor slopes and wall bulge, floor and roof beams lose bearing, needs partial or complete rebuild

of ground subsidence tend to decrease with the depth of the mineral unit and may be minimal if mine depths exceed 50 m, unless the extracted deposit or seam is very thick. Most property damage occurs in the zone of maximum tilting (Figure 3.18C). The passage of a subsidence wave, caused by the advance of a longwall mine, is also associated with ground strain, first extension and then compression, which is particularly important in damaging subsurface infrastructure such as pipes (Figure 3.18C). Micro-earthquakes may also occur, causing minor vibration damage.

The amount of strain, tilting and subsidence can be calculated precisely from a knowledge of the dimensions of extracted mineral and empirical curves developed for the geology of a specific area. Certain geological conditions, however, make prediction very difficult: these include fractures, faults and steep dips. Fractures, such as master joints within competent rock, may localise all subsidence adjustment causing very high strain rates between stable zones. If buildings are located over such fractures, damage may be very severe. Steep dips may also cause distortion of the subsidence bowl.

Classification of structural damage is usually in relation to the amount of strain and distortion (Table 3.5). Subsidence may be reduced by stowing waste in abandoned mines. The types of foundation designed to accommodate subsidence are shown in Figure 3.19 and include deep pile, grouted and raft foundations.

Subsidence is caused not only by the extraction of minerals such as coal and metal ores, but may also occur due to brining and the removal of such products as gas and oil. Wild brining of evaporites in the Cheshire Basin in England has caused the formation of long, linear subsidence hollows following subsurface drainage routes. Similarly, extraction of oil and gas may cause subsidence in oilfields (Figure 3.20). Oil and gas extraction reduces the pore pressure, thereby causing the reservoir to compact in response to the overburden pressure. This compaction may cause surface subsidence, particularly where the reservoir is unconsolidated. Surface subsidence of greater than 6 m has been recorded for the Wilmington and Lagunillas Oilfields in California and Venezuela respectively; in both areas there are relatively unconsolidated Tertiary sedimentary reservoirs. Beneath the Ekofisk Oilfield in the North Sea, compaction of the chalk reservoir is causing sea-floor subsidence of 0.3 to 0.7 m a^{-1}.

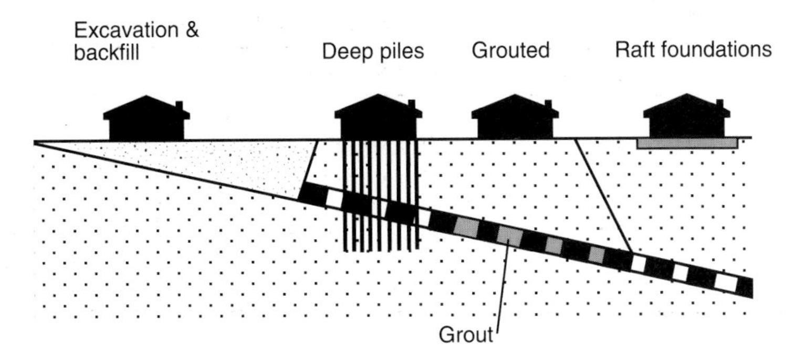

Figure 3.19 *Foundation design in areas with a known subsidence problem [Modified from: Waltham (1994)* Foundations of engineering geology. *Blackie, p. 61]*

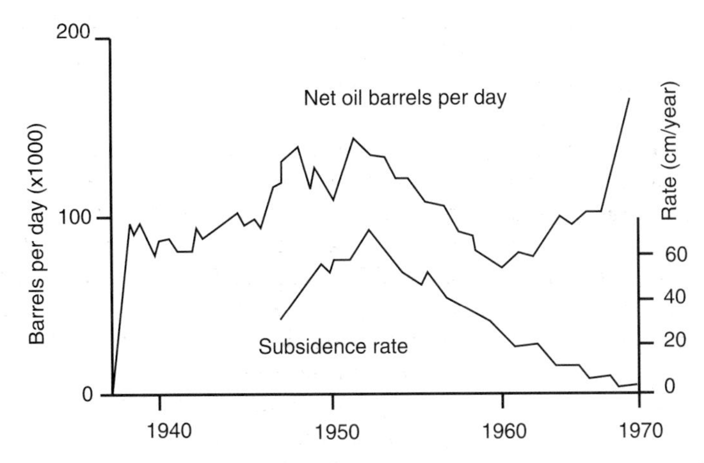

Figure 3.20 *Subsidence caused by oil extraction in the Wilmington Oilfield at Long Beach, California [Modified from: Cooke & Doornkamp (1990)* Geomorphology and environmental management. *Oxford University Press, Fig. 5.9, p. 134]*

3.4.2 Mine Waste

Mine wastes are produced at two stages during mineral extraction: firstly during mining, when unwanted rock (**spoil**) may be produced; and secondly during mineral processing or concentration (**beneficiation**), which produces a further set of waste products (**tailings**). The nature of mine wastes may be variable. Rock spoil usually has low mineral concentrations, is often coarse with a variable grain size and is usually dry waste. In contrast, tailings produced by mineral processing may be more mineral-rich and tend to be of a more uniform and finer grain size as a result of milling. Tailings are also typically transported as a fluid which may be enriched in the chemicals used during the separation process. The relative proportion of dry spoil to wet tailings produced by a particular mining operation depends on the mineral being extracted. In coal mining, for example, 80% is usually dry spoil, while for non-ferrous metals this percentage is much lower. Greater problems are usually encountered with the disposal of wet tailings, particularly from non-ferrous metal extraction due to high residual concentrations of toxic metals, although modern processing, which is more efficient at extracting the target mineral, has reduced this problem.

Mine wastes need to be disposed of close to the mine site, although wet tailings can be pumped or moved under gravity to more suitable locations. In most cases, dry spoil is disposed of in slag heaps or waste tips close to the mine entrance. The importance of this point is illustrated by estimates of the cost of spoil disposal at Asfordby in Belvoir Coalfield in northern England. Spoil production was predicted to exceed 600 000 tonnes per year, at a cost of £2.10 per tonne for disposal adjacent to the mine, but at up to £11.34 per tonne for disposal at a more distant site. The disposal of mine wastes is an even greater problem for low-grade metal ores, where typically only a few grams per tonne of economic metals are present. Operations may involve the mining of over 100 000 tonnes per day and produce potentially enormous chemical waste problems associated with the processing and mineral extraction of this volume of rock.

Careful choice of location for waste tips is essential to prevent problems of instability. This issue is particularly emotive in Britain since 112 school children were killed at Aberfan in South Wales in 1966 when 110 000 m^3 of debris moved 610 m down a slope of 13° from a spoil tip. This tip failure was caused by the siting of the tip above a natural spring which caused the colliery waste to become saturated. Two previous failures in 1944 and 1956 on adjacent spoil tips failed to generate effective action to deal with the problem. The aesthetic impact of slag heaps is also being increasingly considered in the choice of location.

The storage/disposal of wet tailings is usually done behind some form of check dam, where settling can occur. Where possible, disposal in narrow deep valleys is best, minimising the amount of land lost as well as limiting the aesthetic impact. The safety of retaining dams must be considered carefully owing to the potential hazard of failure, not only in generating fluid mud flows but also in terms of potential contamination that release could cause. Containment of tailing fluids is an essential requirement to avoid uncontrolled contamination of groundwater and surface water supply. In many cases mine water is recycled, either in mineral processing or to dampen dust on roadways and slag heaps. Despite its widespread reuse, some waste water derived from mining activity must ultimately be disposed of. The toxicity of tailing effluents is highly specific to a particular mine and the chemicals used in mineral processing. Containment and treatment may be essential for very toxic effluents. In many older mines, tailing effluents are simply stored and/or evaporated in ponds lined with an impermeable membrane or sealant. However, this creates a toxic 'hot spot' which may be difficult to manage subsequently during mine restoration. Alternatives include dilution and discharge to adjacent water bodies or the sea. This is not always environmentally sensitive and depends on the human value, if any, placed on the receiving water body. Treatment is frequently required before discharge. This may involve the removal of heavy metals such as arsenic, adjustment of water pH and removal of unwanted salts and nitrates. Improving water turbidity may also be an important component of treatment.

Exposed wastes are subject to erosion by both wind and, more importantly, water. Wind erosion may be minimised by dampening waste heaps and roadways with waste water. However, water erosion is difficult to control and is particularly pronounced on tailings from mineral processing, which tend to be of a uniform and fine grain size and consequently are very easily eroded owing to the lack of vegetation cover. There are two problems: (1) runoff from mine dumps will be extremely turbid, that is, rich in suspended sediment, and this may cause problems for adjacent water bodies; and (2) runoff may be acidic, particularly where either pyrite-rich deposits or high-sulphur coal is being worked, since oxidation of these deposits gives sulphuric acid which releases toxic metals from primary sulphides and can leach more from wastes. Where an adverse environmental impact will occur, it is necessary to intercept this runoff and manage it. Management may involve dilution, deep groundwater injection, or treatment via settling or with lime to neutralise the acid present.

The management of rock waste and mineral wastes after the cessation of mine working can also be a particular problem. Today, most environmental regulation requires some form of permanent management and/or restoration of waste areas. Waste rock or dry spoil are usually found in slag heaps and mounds. A variety of disposal strategies may be used, which include: (1) use as fill in civil engineering projects; (2) use as aggregate; (3) disposal as inert landfill; (4) infill of quarry or mine voids; and (5) recontouring and vegetation. The commonest solution is to reshape the waste tips and vegetate them. Where the size fractions of the waste are very large, cover material may also be required. In most cases the areas involved are simply too large to make it economic to cover the area with topsoil. Where restored mine sites are subsequently used for building, problems of differential compaction and settling may be encountered. The problem of restoration is much greater for the finer tailings produced during mineral processing, particularly from the extraction of non-ferrous metals. During mine operation these tailings are usually covered by mineral effluent and the first stage in restoration is to dewater these tailings. This may adversely affect the toxicity of the tailings, due to oxidation of pyrite and other sulphide-rich components which creates sulphuric acids. Older tailings are commonly much more toxic than those produced today owing to increased efficiency of mineral extraction (Figure 3.21A & Table 3.6). Restoration of mine tailings is often difficult because of adverse physical and chemical properties. The uniform particle size, typically in the clay or fine sand fraction, may give tailings a high bulk density, poor permeability, and may make them susceptible to erosion. Chemically, tailings may contain toxic levels of the original target metals, plus elevated levels of non-target metals such as arsenic and cadmium. Oxidation of pyrite and other sulphur-rich components may produce extremely acidic conditions. Mine tailings may also be highly saline and poor in essential plant nutrients such as nitrogen and phosphorus. Finally, mine tailings contain little or no organic matter. The net result is that they are extremely hostile and sterile to

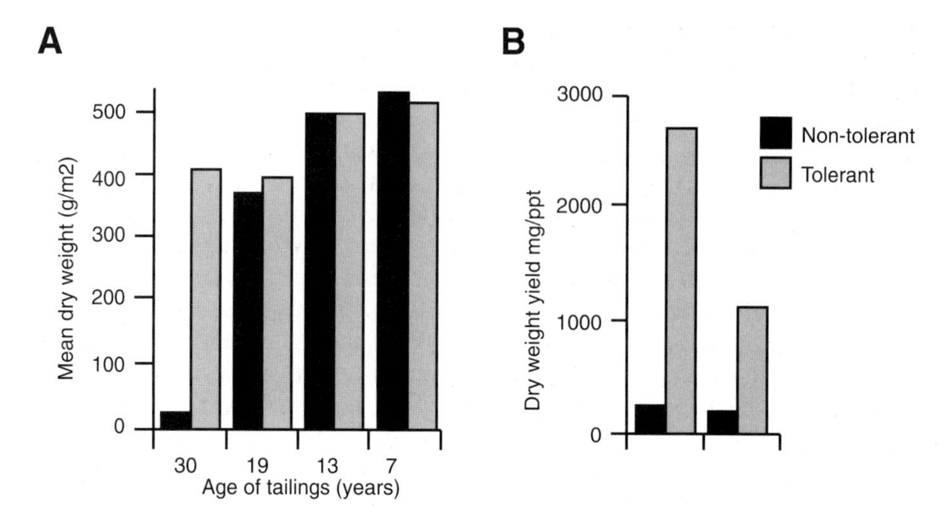

Figure 3.21 *Plant growth on mine tailings.* ***A.*** *Growth of lead/zinc-tolerant and non-tolerant* Festuca rubra *on tailings of different ages in Derbyshire, Britain. The increasing efficiency of mineral extraction has reduced the toxicity of the mine tailings.* ***B.*** *Growth of lead/zinc-tolerant (*Festuca rubra, *left, and* Agrostis stolonifera, *right) and non-tolerant plant populations on lead and zinc mine waste after seven months' growth, with the addition of slow release fertiliser [Modified from: Johnson* et al. *(1994) In: Hester & Harrison (Eds)* Mining and its environmental impact. *Royal Society of Chemistry, Figs 1 & 2, pp. 40 & 41]*

Table 3.6 *Base metal concentrations in some abandoned mine waste dumps in Britain (in mg kg^{-1}). [Modified from: Johnson* et al. *(1994) In: Hester & Harrison (Eds)* Mining and its environmental impact. *Royal Society of Chemistry, Table 1, p. 33]*

Region	No. of sites surveyed	Cu	Pb	Zn
SW England	16	65–6140	48–2070	26–1090
N Pennines	8	20–140	605–13 000	470–28 000
S Pennines	17	23–97	10 800–76 500	12 700–42 000
Lake District	7	77–3800	2070–7630	4690–7370
N Wales	19	30–5750	6400–76 000	11 300–12 700

most plants. Despite this, one of the key management options is to encourage vegetation of tailing deposits. The aims of revegetation are: (1) to achieve the long-term stability of the tailing surface in order to minimise erosion and thereby stop tailings moving into the water system; (2) to reduce leaching, and therefore the entry of toxic elements into groundwater, by increasing surface water loss by evapotranspiration; (3) to develop a vegetation ecosystem in harmony with the surrounding natural area; and (4) to provide some positive value in an aesthetic, economic or nature conservation context.

There are three approaches to revegetation of both fine tailings and coarser spoil dumps: (1) the **ameliorative approach**, in which the surface is conditioned to neutralise adverse chemical components and treated with fertiliser and/or mixed with a cover layer before planting with vegetation suitable for the desired afteruse (e.g. agriculture, recreation or conservation); (2) the **adaptive approach**, in which tolerant species are chosen which can withstand the adverse conditions; and (3) the **agricultural and forestry approach**, which involves restoration to agriculture or forestry by using a thick layer of topsoil or overburden. This last method is practical where the tailings are less toxic (e.g. ironstones and bauxite), but is generally too expensive a practice for more toxic tailings owing to the amount of soil cover required. The potential impact of adapted species is shown in Figure 3.21B, although their use is restricted by their availability within a given region and the length of time required to propagate tolerant species where no natural examples exist. Tolerant species may also concentrate harmful quantities of certain metals in their leaf tissue, and grazing or other types of land use may require restriction. In some cases perimeter belts of prickly plants are used to discourage grazing. Some widely used revegetation schemes are outlined in Table 3.7. The success of a particular scheme depends on careful supervision to ensure adequate plant cover and long-term soil development, and to monitor the uptake of metals within the plants.

3.4.3 Mining Operations

There may be a wide variety of problems associated with the actual operation of mines or quarries. For example, blasting in open quarries may cause problems of vibration in adjacent areas and has been linked to increased rockfall and landslip activity in some areas. In practice, however, establishing this link, and therefore liability, is often difficult. Other operational problems include dust and noise, which may be associated directly with the quarry and also with the increased traffic it generates. As far as possible, mine traffic is concentrated on agreed routes, which are sometimes upgraded at the expense of the quarry/mine company. Problems of noise and visual intrusion are combated using earthworks such as linear berms and mounds made from the overburden or topsoil removed from a site. The use of vegetation belts may also help to ameliorate these problems. Dust control, especially in dry regions, is also an essential component of sensitive mine/quarry management. This may be achieved by dampening all roadways, spoil heaps and areas of loose waste, or through the use of strategically placed shelter belts. The paving of roadways as soon as is practicable is also of importance in reducing dust problems. The emission of methane from coal mines and other deep excavations can in some cases cause a hazard, and such emission is increasingly being controlled because of its impact as a greenhouse gas and its potential to cause explosions (Box 9.2).

Quarries and mines which occur below the water table will need to be pumped dry to allow operation, which may cause local disruption of the water

Table 3.7 *Different approaches to revegetation [Modified from: Johnson* et al. *(1994) In: Hester & Harrison (Eds)* Mining and its environmental impact. *Royal Society of Chemistry, Table 2, p. 37]*

Waste characteristics	Revegetation technique	Problems
Low metal toxicity. No major acidity or alkalinity problems	*Amelioration and direct seeding with agricultural or amenity grasses and legumes.* Apply lime if pH <6. Add organic matter if physical amelioration required. Add nutrients	Probable commitment to long-term maintenance. Grazing may need restricting due to movement of toxic metals into vegetation
Low metal toxicity, but severe climatic limitation. No major acidity or alkalinity problems	*Amelioration and direct seeding with native species.* Seed or transplant native species adapted to climatic conditions. Add lime, organic matter and fertiliser as required	Irrigation often necessary during establishment in arid climates. Selection of native flora critical. May be labour-intensive
Medium to high metal toxicity. High salinity	*A. Amelioration and direct seeding with tolerant ecotypes.* Sow with metal- and/or salt- and/or acid-tolerant seed. Add lime, organic matter and fertiliser as necessary	Possible commitment to regular application of fertiliser. Grazing inadvisable. Few commercially available species
	B. Surface treatment and seeding with agricultural or amenity grasses and legumes. Amelioration with 10 to 50 cm of innocuous mineral water. Apply lime and fertiliser as required	Expensive waste layer must be thick enough to be effective
Extreme toxicity. Very high toxic metal content. Intense salinity or acidity	*Barrier layer.* Surface treatment with 30 to 100 cm of innocuous barrier material and surface covering with suitable rooting medium. Apply lime, organic matter and fertiliser as required	Susceptibility to drought according to the nature and depth of the surface covering. High cost and potential limitation of availability of barrier material. Integrity of barrier layer may be affected by root penetration

table. For example, in the East Durham and Nottinghamshire Coalfields of England, over 36 billion and 14 billion litres of water, respectively, are pumped from mine systems each year. Water pumped from quarries and mines needs careful disposal to avoid contamination of adjacent water bodies or groundwater. Mining increases the surface area of a mineral deposit and also increases the flow of groundwater, which increases the rate of leaching of mineral deposits. This frequently has an adverse effect on groundwater quality. Most mine water has a brown coloration due to the presence of dissolved iron. Another particular problem is an increase in acidity due to the presence of pyrite and other sulphides which oxidise to give sulphuric acid. Iron pyrites is common in coal and associated mudstones. In Britain up to 10% of a coal seam may consist of iron pyrites. This may increase the acidity of groundwater to such an extent that its neutralisation is essential to protect mine machinery. Oxidation will cease once a mine floods after extraction stops, but this may simply bring into solution all the available sulphuric acid produced to date. A dramatic example of this was the cessation of pumping at the metal mine at Wheal Jane in southwest England, which resulted in a dramatic red plume of contaminated water being discharged into the Fal Estuary. A variety of toxic metals may become concentrated in mine water, which can also contain large amounts of dissolved salts. Finally, in soft rock quarry sites, water pumped from the pit may be extremely turbid. Turbidity may cause both aesthetic and ecological problems in adjacent rivers.

Release of untreated mine water into natural water bodies may have a variety of adverse impacts including: (1) depletion of number and diversity of free-swimming and benthic organisms; (2) loss of spawning gravel for fish; and (3) direct fish mortality. In the majority of cases, mine discharge is dealt with simply by dilution within the receiving water body. Particularly acidic or turbid water may be treated with lime or settled before release.

The problem of water discharge from mines is not simply restricted to active mines and quarries, but may continue once they are abandoned and flooded. In Britain there are over 10 000 recorded disused coal mine workings, all of which could potentially have an adverse impact on water quality. Around 400 km of river is affected adversely by discharge from abandoned mines in England and Wales. In regions subjected to intensive non-ferrous mineral mining, ground and surface waters may contain dangerous levels of lead, copper, zinc and in some cases arsenic, all of which need treatment if adverse affects are to be avoided.

3.4.4 Quarry/Mine Restoration

Restoration of mines and quarries is an essential part of the aftercare plan of all mineral workings. Increasingly, aftercare plans must be set out before a mine or quarry obtains planning consent. Numerous abandoned workings exist, however, which require some aftercare to bring them up to modern environmental standards.

Mine restoration involves tackling three problems: (1) ensuring the safety of all underground workings and shafts; (2) the removal of pit head infrastructure; and (3) the disposal or restoration of mine waste. The restoration of mine wastes has already been discussed. At cessation of mining, a decision must be made as to whether the underground mine infrastructure is to be abandoned and allowed to collapse, or whether it is to be maintained ('moth-balled'). Maintenance of shafts and principal underground infrastructure may be undertaken for two reasons: (1) mining operations may be restarted in the future, or an option to do so is to be retained; and (2) collapse of mine shafts might have an adverse effect on the hydrogeology or ventilation of adjacent mines. Maintenance usually involves the up-keep of the principal shafts and haul-ways as well as continued groundwater pumping. In Britain, for example, concern has been expressed about the closure of many mines and the associated loss of access to known coal reserves. Effectively these reserves have been sterilised, since the development of new access shafts and tunnel systems would be prohibitively expensive. In cases like these, there may be strategic reasons for maintaining mine access.

Where mines are to be abandoned, shafts must be stabilised and capped. Detailed plans of all underground workings must be kept to ensure that any future developer is aware of the engineering problems present on a site. Mine shafts may be partially filled and capped with concrete slabs, a process which may cost in excess of £10 000 per shaft.

Quarries may be restored in a variety of ways and there are essentially two situations: (1) a void in the ground; and (2) a topographic loss. In the latter case a topographic feature has been removed by quarrying to create new surface contours. Restoration here involves simply grading and consolidating the surface before restoring the topsoil, which is usually removed and stockpiled prior to quarrying. This is common in situations where aggregate has been removed from glacial landforms such as eskers or kames. Reshaping of landform remnants, perhaps to mimic the original, may in some cases form part of such a restoration scheme.

In the case of voids or conventional quarries the options are numerous, but stem from a choice between either maintaining the void or infilling it. Methods of infill may include the use of the quarry as a landfill site either for inert waste, in which case little engineering of the void space is required, or for the disposal of domestic, industrial or toxic waste, in which case a containment system must be introduced into the void. In either case, the original contours of the land surface are reinstated either for agriculture or some other land use. Quarry infill is often in conflict with the interests of geological conservation, since quarries provide vital rock exposure and are essential for research and teaching. Attempts have been made in Britain, through the use of conservation voids, to achieve compromise between the needs of geologists to maintain exposure and the wishes of mineral companies to infill quarries completely, thereby maximising valuable landfill space (Box 6.5).

Not only are original land contours restored, but attempts may also be made to reinstate natural ecosystems following quarry infill. For example, at a number of chalk quarries in southern Britain the original chalk grassland, a threatened natural habitat in Britain, has been removed prior to quarrying and stored on an adjacent slope. In some cases soil berms have been built around the rescued turf in order to encourage rabbits to inhabit the site, graze the grass and thereby maintain it in its natural ecological state. When the quarries are infilled with waste, the surface is covered with chalk rubble and the grass is returned to its original location. By transplanting and replanting the natural vegetation it is hoped that the natural ecosystem will not, in the long term, be adversely affected by the quarrying operation.

Where infilled quarries are used for subsequent construction, problems of differential consolidation may be experienced. Partial infill of a quarry may also be an option where insufficient fill is available or where there are other factors involved, such as the need to maintain geological exposure for conservation (Box 6.5).

In those situations where restoration involves maintaining the void, afteruses may be wide-ranging. In such situations the stability of the final faces must be considered carefully. Where soft rocks are involved, slopes are battered back to a suitable angle and drained to provide stability. In hard rock quarries a stability analysis of the various faces is required in which the rock mass strength (see Section 7.1.3) of each part of a quarry face is determined, along with its susceptibility to failure (Figure 3.22). Having identified problem faces, a new face geometry is designed and produced by blasting, ripping or in some cases by direct excavation (see Section 7.3). In very exceptional cases, faces may be stabilised by the use of **rock bolts**, '**dental masonry**' infilling cavities, and wire mesh to retain loose material. Where stable faces are not achievable, **catch fences** may be built at the base of a face, designed to prevent both excessive development of scree slopes and human access. However, in such cases afteruse of the quarry floor is severely restricted.

Typical quarry floor afteruses include: (1) industrial, commercial and domestic construction; (2) recreation; (3) restoration to agriculture; and (4) restoration for nature conservation. Quarry floors may provide ideal construction sites, particularly where it is desirable for a structure or industrial complex to remain out of sight for aesthetic reasons. In recent years, major out-of-town shopping centres have been sited in quarry complexes in Britain. One such complex, at Thurrock in Essex, is ideally situated adjacent to the M25 London orbital motorway. In Edinburgh the supermarket giant J.S. Sainsbury has constructed a large store in the Carboniferous Craigleath Quarry, incorporating this local stone into its construction. Finally, Salthill Quarry in Clitheroe, northern England, has been used for industrial units, the untreated faces being conserved for their scientific interest through the use of perimeter fences designed to protect buildings from rockfall.

Care may be required in foundation design if the quarry floor contains unconsolidated fill, and blasting may have decreased the rock mass strength of near-surface layers by increasing the fracture density. Construction on quarry floors is only possible where the water table is below the quarry floor.

In situations where the cessation of quarry pumping leads quickly to quarry flooding, restoration to recreation is often chosen as an alternative. Water sports and parkland can be created from flooded quarries where sufficient space exists between the water edge and the quarry walls. Flooded quarries with near-vertical walls may, however, be extremely dangerous. In this case restoration options may be limited and preservation of site security may be imperative. The potential of these water bodies as small holding reservoirs for domestic water supply, particularly in urban areas, has yet to be fully explored, but provides a potential afteruse.

Restoration to nature conservation is an increasingly attractive option for many quarry companies attempting to improve the 'green' image of the extractive industry. Blasting techniques have been developed to create more natural quarry faces which can be colonised more effectively (Box 3.3). Careful landscape design in the quarry floor can be used to create water bodies and a variety of habitats.

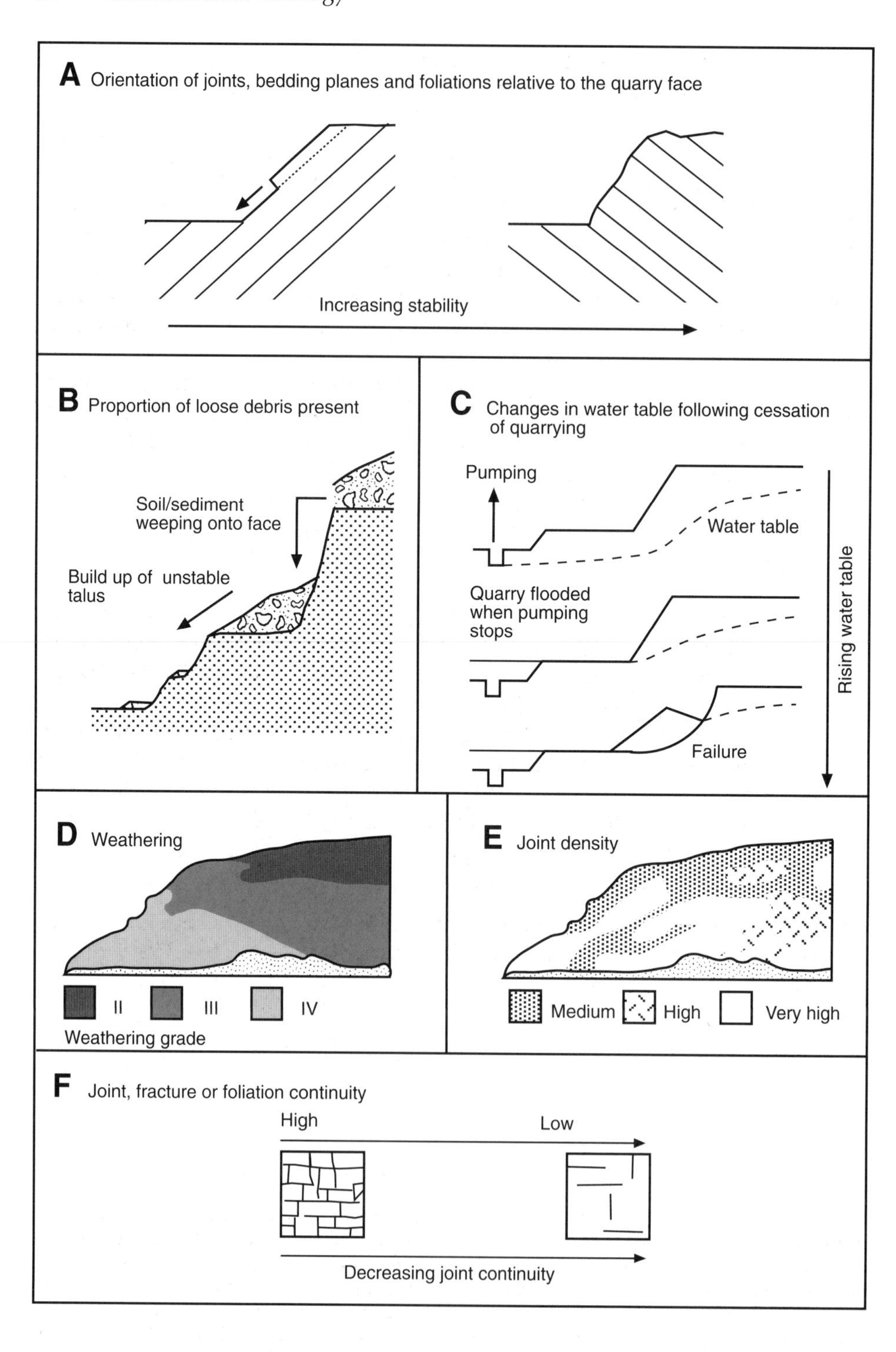
A Orientation of joints, bedding planes and foliations relative to the quarry face
Increasing stability
B Proportion of loose debris present
Soil/sediment weeping onto face
Build up of unstable talus
C Changes in water table following cessation of quarrying
Pumping
Water table
Quarry flooded when pumping stops
Failure
Rising water table
D Weathering
II
III
IV
Weathering grade
E Joint density
Medium
High
Very high
F Joint, fracture or foliation continuity
High
Low
Decreasing joint continuity

BOX 3.3: RECREATING NATURAL CLIFFS FROM QUARRIES: RESTORATION BLASTING

In Britain over 90 million tonnes of limestone are produced each year, 25% of which is extracted from the Peak District of Derbyshire and Staffordshire, both areas of outstanding natural beauty. Reconciling the conflict between the economic need for limestone and the visual intrusion of quarries is particularly challenging. One solution has been to develop methods of quarry restoration which mimic the natural rock outcrops and the geomorphology of the region, by the creation of natural cliff exposures from the artificial quarry walls and faces abandoned by quarrying.

Modern blasting produces regular and relatively uniform rock faces which are not only unnatural, but also extremely hazardous. Recent research has developed techniques for creating natural cliffs through restoration blasting. The objective is to use controlled blasting to produce skeletal rock landforms which can be colonised by vegetation and, more importantly, continue to evolve in a natural way. Four objectives for restoration blasting have been identified:

- reduction of the face height by the construction of scree blast piles to mask the regular sequence of scorch marks associated with the blast holes, cover the quarry face to varying heights to reduce cliff regularity, and vary the degree of block size in the scree to add texture;
- indent the crest line or quarry top with a series of semicircular cut-backs to mimic the presence of buttresses and recesses in the quarry face;
- produce 'ragged' rock headwalls to replace the smooth blasted faces;
- stabilise the scree slope created by leaving or inserting rock stumps and coarser material at the foot to prevent the scree encroaching onto the quarry floor or onto lower slopes.

In short, the aim is to recreate a natural cliff from an artificial one, as shown in the diagram below. This approach illustrates one way of restoring a quarry to create a new landscape resource.

continues overleaf

Figure 3.22 *(opposite) Some of the factors which determine the stability of a quarry face and which must be considered when evaluating a site.* **A.** *Orientation of bedding surfaces, joints or foiliations with respect to the angle of the face determine stability; outward dipping surfaces tend to lead to slab failure.* **B&C.** *The presence of loose debris on ledges and cliff tops may provide a hazard, as may variations in the water table.* **D&E.** *Variation in weathering and joint density may cause variation in stability across a face.* **F.** *The continuity of joints, fractures and foliations also helps to determine face stability*

BOX 3.3: *(continued)*

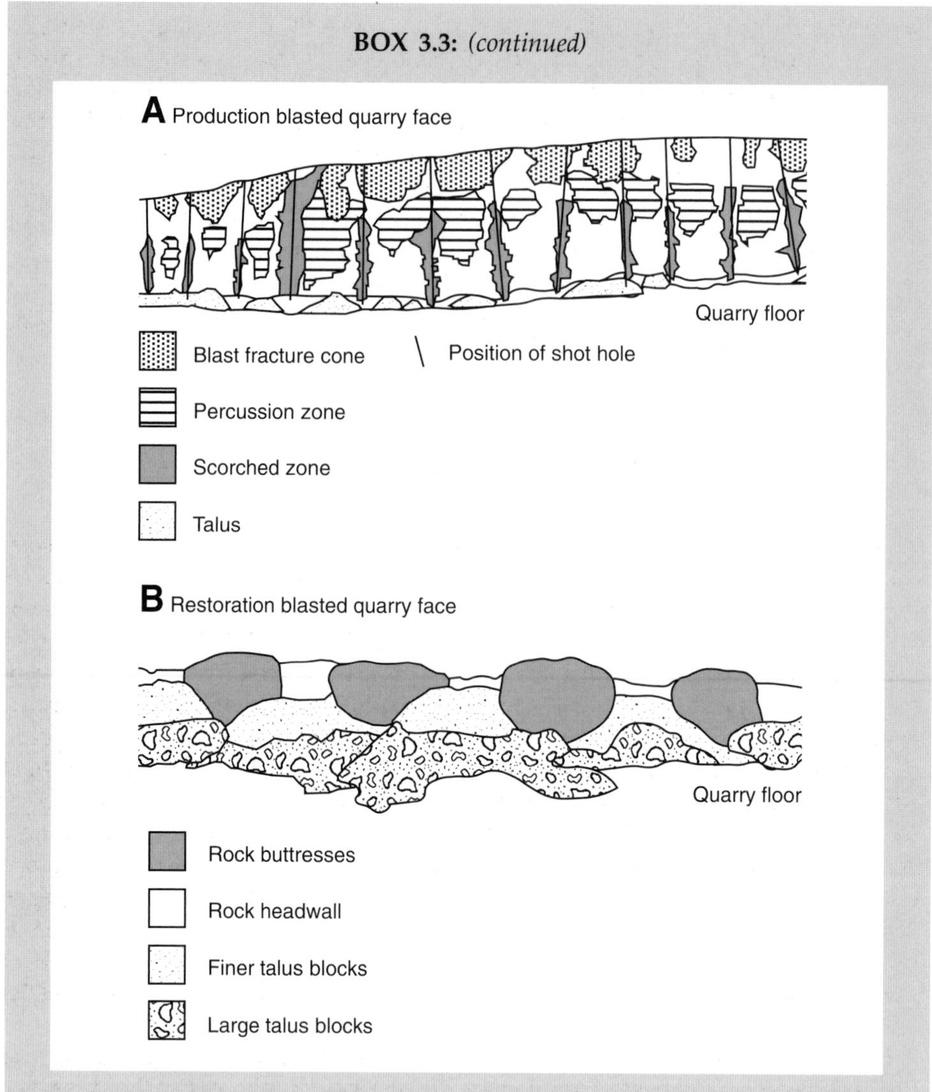

Source: Gagen, P. & Gunn, J. 1988. In: Hooke, J.M. (Ed.) *Geomorphology in environmental planning*. John Wiley & Sons, Chichester, 121–142. [Diagrams modified from: Gagen & Gunn (1988) In: Hooke (Ed.) *Geomorphology in environmental planning*, John Wiley & Sons, Fig. 8.7, p. 137]

3.5 SUMMARY OF KEY POINTS

- The three main mineral groups, which form in a variety of geological environments, are mineral fuels, industrial and metal minerals, and construction minerals.

- Mineral exploration is a high-risk enterprise which involves seven stages: (1) conceptual planning; (2) selection of geological models or search parameters; (3) reconnaissance and planning; (4) land acquisition; (5) target location and appraisal; (6) mine assessment and feasibility; and (7) mine development. The cost associated with each stage escalates as the risk associated with failing to obtain a return on that investment increases.
- A wide variety of mineral extraction methods exist, including fluid removal, underground mining, and surface mining or quarrying.
- All mining activity is associated with some environmental impact prior to, during and after mineral extraction. The aim of the environmental geologist is to minimise and manage this impact

3.6 SUGGESTED READING

There is a wide range of specialist texts on mineral exploration, most of which focus on specific mineral types. General information about specific minerals and exploration can be obtained from the books by Ward (1984), Evans (1985), North (1985), Shackleton (1986), Barnes (1988) and Kesler (1994). Information on construction minerals can be found in Chapter 4 but they are dealt with in an industrial context by Manning (1995). Basic field exploration techniques are covered by Peters (1987) and Chaussier & Morer (1987). Float mapping in glacial terrain is covered in the work of Grip (1953), Miller (1984) and in the book edited by Kujansuu & Saarnisto (1990). Exploration geochemistry is covered in the book by Levinson (1980). Annels (1991) covers methods for evaluating a chosen mineral target. Crimes *et al.* (1992, 1995) provide an interesting case study on some of the exploration methods used in the search for aggregate reserves in the UK. Similarly, the volume edited by Glasson & Rattigan (1990) contains useful case histories of mineral exploration. A general overview of mineral resources and restoration is provided in the volume edited by Lumsden (1994) and the paper by McCall & Marker (1996).

Hester & Harrison (1994) provide an excellent collection of edited papers on the environmental impact of mining and its control; in particular, the papers by Barbour (1994) and Johnson *et al.* (1994) review mining impacts and the problems of vegetating mine tailings, respectively. Additional information on this subject can be obtained from Culshaw *et al.* (1987) and Bell (1996). The problem of mining subsidence is covered by Gray & Bruhn (1984), Waltham (1989, 1994) and Bell (1992). Mayuga & Allen (1969) report the subsidence in the Wilmington Oilfield in California.

Annels, A.E. 1991. *Mineral deposit evaluation: a practical approach*. Chapman & Hall, London.

Barbour, A.K. 1994. Mining non-ferrous metals. In: Hester, R.E. & Harrison, R.M. (Eds) *Mining and its environmental impacts*. Royal Society of Chemistry, Cambridge, 1–15.

Barnes, J.W. 1988. *Ores and minerals: introducing economic geology*. Open University Press, Milton Keynes.

Bell, F.G. 1992. Salt mining and associated subsidence in mid-Cheshire, England, and its influence on planning. *Bulletin of the Association of Engineering Geologists*, **29**, 371–386.
Bell, F.G. 1996. Dereliction: colliery spoil heaps and their rehabilitation. *Environmental & Engineering Geoscience*, **2**, 85–96.
Chaussier, J.B. & Morer, J. 1987. *Mineral prospecting manual*. Elsevier, Amsterdam.
Crimes, T.P., Chester, D.K. & Thomas, G.S.P. 1992. Exploration of sand and gravel resources by geomorphological analysis in the glacial sediments of the Eastern Llyn Peninsula, Gwynedd, North Wales. *Engineering Geology*, **32**, 137–156.
Crimes, T.P., Chester, D.K., Hunt, N.C., Lucas, G.R., Mussett, A.E., Thomas, G.S.R. & Thompson, A. 1995. Techniques used in aggregate resource analyses of four areas in the UK. *Quarterly Journal of Engineering Geology*, **27**, 165–192.
Culshaw, M.G., Bell, F.G., Cripps, J.C. & O'Hara, M. (Eds) 1987. *Planning and engineering geology*. Geological Society, London.
Evans, A.M. (Ed.) 1985. *Introduction to mineral exploration*. Blackwell Science, Oxford.
Glasson, K.R. & Rattigan, J.H. (Eds) 1990. *Geological aspects of the discovery of some important mineral deposits in Australia*. Australasian Institute of Mining and Metallurgy, Melbourne.
Gray, R.E. & Bruhn, R.W. 1984. Coal mine subsidence: eastern United States. *Geological Society of America Reviews in Engineering Geology*, **6**, 123–149.
Grip, E. 1953. Tracing of glacial boulders as an aid to ore prospecting in Sweden. *Economic Geology*, **48**, 715–725.
Hester, R.E. & Harrison, R.M. (Eds) 1994. *Mining and its environmental impacts*. Royal Society of Chemistry, Cambridge.
Johnson, M.S., Cooke, J.A. & Stevenson, J.K.W. 1994. Revegetation of metalliferous wastes and land after metal mining. In: Hester, R.E. & Harrison, R.M. (Eds) *Mining and its environmental impacts*. Royal Society of Chemistry, Cambridge, 31–48.
Kesler, S.E. 1994. *Mineral resources, economics and the environment*. Macmillan, New York.
Kujansuu, R. & Saarnisto, M. (Eds) 1990. *Glacial indicator tracing*. Balkema, Rotterdam.
Levinson, A.A. 1980. *Introduction to exploration geochemistry*. Applied Publishing, Wilmette.
Lumsden, G.I. (Ed.) 1994. *Geology and the environment in Western Europe*. Oxford University Press, Oxford.
Manning, D.A.C. 1995. *Introduction to industrial minerals*. Chapman & Hall, London.
Mayuga, M.N. & Allen, D.R. 1969. Subsidence in the Wilmington oil field Long Beach, California, USA. *Publication of the Institute of Scientific Hydrology*, **88**, 66–79.
McCall, G.J.H. & Marker, B.R. 1996. Mineral resources. In: McCall, G.J.H., De Mulder, E.F.J. & Marker, B.R. (Eds) *Urban geoscience*. A.A. Balkema, Rotterdam, 13–34.
Miller, J.K. 1984. Model for clastic indicator trains in till. In: *Prospecting in areas of glaciated terrain*. Institution of Mining and Metallurgy, London, 69–77.
North, F.K. 1985. *Petroleum geology*. Unwin Hyman, Winchester.
Peters, W.C. 1987. *Exploration and mining geology*, second edition. John Wiley & Sons, New York.
Shackleton, W.G. 1986. *Economic and applied geology*. Croom Helm, London.
Waltham, A.C. 1989. *Ground subsidence*. Blackie, Glasgow.
Waltham, A.C. 1994. *Foundations of engineering geology*. Blackie, Glasgow.
Ward, C.R. 1984. *Coal geology and coal technology*. Blackwell Scientific Publications, Oxford.

4
Construction Resources: Geomaterials

In this chapter we examine the application of geological materials in construction. **Geomaterials** are those construction materials which are recognisable as being fundamentally geological in origin (see Section 3.1.2). This definition includes all stone- and aggregate-based materials used in construction, but excludes such things as metals which have undergone an extensive and complex process of smelting in order to liberate the metals from their ores. There are six basic types of geomaterials: (1) **construction stone**, comprising quarried blocks of natural stone which are either left in a rough state or worked to produce a finished building stone; (2) **aggregates**, encompassing both naturally occurring coarse clastic sediments, such as glacial or river gravels and alluvium, and quarried stone which is subsequently crushed, both of which are used in many guises in the creation of construction materials; (3) **cement** and **concrete**, which are admixtures of ground limestone and clay with aggregates; (4) **structural clay**, which is fired to make bricks; (5) **gypsum**, which is calcined to form plaster; and (6) **glass sand**, which is used to make glass in combination with certain other elements. Each of these is discussed below.

4.1 CONSTRUCTION STONE

Stone which is actively quarried or mined to provide materials for buildings, or for major engineering projects such as coastal defences and roads, is known as construction stone. Construction stone is more often than not used in a raw, unprocessed state, although blocks may be sized, shaped or crushed to form suitable materials for major construction works. In general, complex processing is mostly limited to architectural stone, in which accurate cutting, cleaving, shaping and dressing is necessary. Three main types of construction stone can be identified: (1) **dimension stone**, which is shaped and dressed into regular sizes and shapes as architectural stone; (2) **decorative stone**, which is similar to dimension stone in that it is cut and shaped, but differs in that it is intended only for decoration or non-load-bearing use, although it may still have to be durable; and (3) **armourstone** and **rip-rap**, which are irregular but durable stone blocks or coarse aggregates intended to provide protection in sea walls, breakwaters and similar structures.

4.1.1 Dimension Stone

Stone which has been actively quarried, and then **dressed** or formed either by hand or by machine into regularly shaped blocks for use in the construction of buildings, is known as dimension stone (Figure 4.1). Dimension stone may be classified according to use and rock type. Typical uses include: **masonry stone**, cut as regular blocks and intended for load-bearing use, forming the fabric of walls and foundation works; **stone cladding**, cut as relatively thin, regular sheets which have primarily a decorative function and which are fixed to the exterior of buildings constructed from other materials; **flooring stone** which is machine-cut into tiles, or split along bedding or cleavage surfaces to form a decorative and/or utility floor covering; **roofing**, employing lithologies which have a well developed cleavage or bedding which is naturally split and then machine- or hand-trimmed to form thin sheets of stone; and **pavement construction**, comprising paving flagstones, kerb-edging blocks and cobble stones, the majority of which are specially shaped for use in street and road construction. Typical rock types include: limestones and sandstones, employed for all uses; igneous rocks, usually loosely classed as 'granites' by the industry, but including a range of rock types from basalts to gabbros used for masonry, cladding, flooring and pavement construction; marble, employed mostly as decorative cladding and flooring, although this term is sometimes incorrectly used in industry to encompass some fine-grained micritic limestones; and slate, mostly used as roofing material, but also employed as cladding and flooring.

Historically, natural stone has been used as a construction material in architecture as diverse as grand public buildings and simple private dwellings. This is clearly reflected in the fact that in many cases the architectural character of a district is associated with the nature of the locally quarried stone. In Britain alone, the unpretentious vernacular architecture of private houses and other simple buildings of the pre-Industrial Revolution era is clearly influenced by the local availability of natural stone. This has left a heritage which mirrors the geology of Britain, from the granite cities of the north to the flint-knapped cottages of the southeast. In fact, the use of dimension stone has a long history and many of the most important surviving prehistoric and early historic buildings are constructed from stone. In general, the earliest buildings were constructed from naturally occurring, otherwise unshaped blocks, but later

Figure 4.1 Photographs showing typical uses of dimension stone in masonry. **A.** Rock-faced blocks. **B.** Cut and shaper pillars

examples include truly quarried and shaped stones. Famous examples include: Stonehenge in southwest England, an ancient Neolithic construction built from large sarsen stones which were obtained from the landscape, but which do show evidence of working in order to provide the necessary architectural function; the Pyramids and other Egyptian structures of between 3000 and 2000 BC, which were constructed from regularly shaped rectangular blocks of limestone, and later from granite, porphyry and other crystalline rocks; and finally the architecture of the Inca civilisations of the South American Andes who managed to achieve a high degree of sophistication in their masonry without complex tools, a feat which still defies explanation. Today, dimension stone is rarely employed as a load-bearing construction material in our cities; it is more commonly used as a cladding for steel-framed and concrete buildings, and in this function the use of dimension stone in our urban areas is undiminished.

Not all stone is of sufficient quality to allow it to be used as dimension stone. There are five main factors which are of importance in the selection of stone for dressing and working. These are: (1) structural strength; (2) durability; (3) appearance; (4) ease of working; and (5) availability. The environmental factors of quarrying the stone must also be addressed, as with all quarrying activities; however, these are often associated not simply with the rock type, but rather with a combination of factors which include location and methods of extraction.

Structural strength is defined as the ability of a stone to carry a load without failure, and the nature of the structural strength varies according to the proposed function of the stone. Large blocks intended to support buildings, for example, are required to have compressive strength as supplied by the weight of the building. Cladding materials, on the other hand, which often consist simply of sheets of dressed stone of between 40 and 100 mm in thickness and are attached to the structural frame of a building, do not have a load-bearing function and are therefore not required to have high levels of compressive strength. In general terms, it has been estimated that an average block of masonry stone will have at least as much compressive strength as concrete or bricks (Table 4.1). The structural strength of a given block of stone is a function of two factors associated with its geological properties, namely: (1) its mineralogy; and (2) its fabric. The mineralogy of a stone is significant simply because some minerals are inherently strong under loading or other extreme conditions, while others are inherently weak. Good examples would be: quartz, which is hard, does not possess a cleavage, is resistant to weathering, and is therefore inherently strong; and mica, which is soft, has a pervasive cleavage, is prone to chemical weathering and is therefore inherently weak. Quartzites, composed of in excess of 90% quartz grains, therefore make excellent load-bearing stones; schists, with their high percentage of micas, do not.

The fabric of a stone is also significant, particularly where there are prominent primary sedimentary bedding characteristics, discontinuities such as joints or fractures, or tectonically induced cleavage or other foliations. Such features represent planes of weakness within a stone, which can lead to failure if they are heavily loaded. In most cases, the compressive strength of a stone will be dependent on the direction in which a force is applied in relation to such natural

Table 4.1 *Compressive strengths of typical dimension stones compared with other construction materials [Modified from: Jefferson (1993)* Quarterly Journal of Engineering Geology, **26**, *Table 4, p. 310]*

Material	Compressive strength (MN m^{-2})
Basalt, dolerite, some quartzites	250
Microgranite, microdiorite, basalt, well cemented sandstones, quartzites and limestones	160–250
Sandstone, limestone, medium and coarse grained granite, granodiorite	60–160
Porous sandstone, limestone and mudstone	30–60
Tuff, chalk, very porous sandstone and siltstone	<30
Fired clay bricks	10–60
Concrete	typically 48

discontinuities. It is therefore necessary to minimise the effects of these defects in the fabric of the rock by laying the stone so that the dominant forces, such as the weight of the building, are normal to the planes of weakness, thereby preventing failure along these planes. Discontinuities can also weaken dimension stone where they enhance its permeability. For example, fractures and dominant bedding planes will promote the ingress of water even where the stone is normally considered to be impervious, as in crystalline rocks. Water penetration has many damaging effects, including the promotion of salt crystal growth in pore voids; this exerts internal pressures which can promote failure along natural planes of weakness within a given stone block. Despite these reservations, natural discontinuities can also allow the stone to be worked at source, particularly in the production of workable materials. Good examples of this are: slates, where splitting along cleavage planes can provide successively thinner sheets; flagstones, such as those from the British Coal Measures of Carboniferous age, usually referred to as 'Yorkstone', which are split parallel to bedding planes to provide paving slabs; and limestones, which may often be worked into crudely rectangular construction blocks through the exploitation of the bedding plane–joint intersections.

Durability is defined as the ability of a stone to withstand exposure to a given environment (Figure 4.2). Typical environments include: (1) the atmosphere, often of polluted industrial towns and cities, which is directly in contact with structural stone and cladding; (2) the subsurface, associated with stone built foundations; and (3) marine and estuarine waters, where stone has at least tidal contact with saline waters. Each of these environments can present problems for the durability of natural stone. For example, urban atmospheric conditions are often acidic, caused by the sulphurous emissions from local industries, and chemical attack of limestone monuments in cities is of great concern to those seeking to preserve them. Contact with damp soils or constant wetting and drying of stone walls can lead to the promotion of salt crystal growth in pore spaces which ultimately causes physical breakdown of the stone (see Section 8.2.2; Box 8.3). The ability of stone to withstand chemical and physical attack in

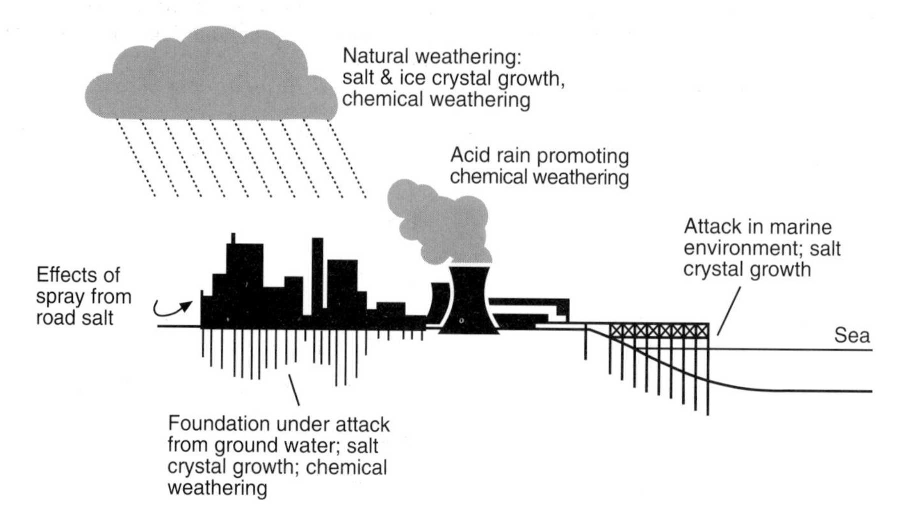

Figure 4.2 *Conceptual diagram demonstrating durability of building stones as a function of exposure to different environments*

such environments should clearly be an important consideration in selecting it as a building material.

The mechanism of stone weathering is complex and may often be the result of several different processes acting together. For example, physical disintegration may be achieved through secondary crystal growth, temperature changes, wetting and drying, and the effects of plant roots and other organic growth, while chemical decomposition may be the result of many different reactions, including acid solution (Table 4.2). However, despite the complexity of the processes involved in the degradation of building stones, industry standards for durability in most countries are primarily concerned with the resistance of stone to secondary crystal growth, whether by ice or salts, acid attack, and thermal and wetting/drying effects.

The ability of a given building stone to withstand weathering processes is a function of several factors, but is particularly related to porosity and original mineralogy. In crystalline rocks, primary porosity is not a factor, although differential thermal expansion of the constituent minerals can lead to the development of a secondary porosity. In igneous and metamorphic silicate-rich rocks, durability is directly linked to the susceptibility of certain minerals to chemical weathering, and this is particularly associated with the decay of some non-quartz silicates, such as micas and feldspars, into clay mineral assemblages. This can also lead to the development of secondary porosity in such rocks. In porous sedimentary rocks, chemical weathering of constituent minerals may also be important, as it is clear that limestones and calcite-cemented sandstones may be attacked by the acidic atmosphere of many industrial cities.

Although chemical weathering effects can be significant, physical weathering associated with secondary crystal growth in pore voids is considered to be the

Table 4.2 *Processes of atmospheric decay in natural building stone [Based on data in: Honeybourne (1990) In: Ashurst & Dimes (Eds)* Conservation of building and decorative stone, Volume 1. *Butterworth-Heinemann, 153–178]*

Process	Stones	Effect
Salt crystallisation	All types	Efflorescence: harmless but unsightly surface growth.
	Limestones & sandstones	Cryptoflorescense: growth within pore voids causing damage through expansion pressure.
Acid decay	Limestones & marbles	Solution followed by formation of calcium sulphate (gypsum) skin.
	Dolomite	Similar to limestones but with magnesium sulphate skin beneath gypsum causing blistering.
	Sandstones	Loss of calcite cement; growth of subsurface gypsum crust causing differential thermal expansion.
	Some slates	Deterioration through acid attack of calcite content.
	Igneous rocks	Largely unaffected unless already chemically weathered.
Frost damage	Limestones & sandstones	Separation of stone wafer from exploitation of exposed fracture; ice growth within pore voids causing expansion pressure.
	Igneous rocks	Largely unaffected unless already chemically weathered.
Heating & cooling	Marbles	Differential expansion of calcite crystal axes causes bending of slabs.
	Granites	Differential expansion of mica, quartz and feldspar causes microcracking.
	Limestones & calcite cemented sandstones	Differential expansion of gypsum crust leading to surface loss.
Wetting & drying	Porous rocks	Salt crystallisation; fatigue failure from expansion on wetting and contraction on drying.

most important factor in the assessment of durability in sedimentary rocks. Pore volume and size, and the ability and rate of fluid uptake, directly influence the activity of both ice crystal and mineral salt growth in pore voids, both of which can have a seriously detrimental effect. Under frost conditions, pore waters are liable to freeze, and the process of ice crystal growth exerts a pressure through expansion. Subsequent thawing of the ice crystals produces contraction, and together the pressures developed by the processes of expansion and contraction

Table 4.3 *Damaging salts in building stones and their common sources [Modified from: Honeybourne (1990) In: Ashurst & Dimes (Eds)* Conservation of building and decorative stone, Volume 1. *Butterworth-Heinemann, Table 7.1, p. 154]*

Salt	Common source
Sodium sulphate	Washing powder, soil, some bricks, some solid fuels, action of acidic atmosphere on sodium carbonate
Sodium carbonate	Washing powder, domestic cleaning aids, fresh concrete and cement-based mortars
Magnesium carbonate	Some bricks, rain wash from dolomite affected by acidic atmosphere
Potassium carbonate	Fresh concrete and cement-based mortars, fuel ashes and ash-based mortars
Potassium sulphate	Some bricks, action of acidic atmosphere on potassium carbonate
Sodium chloride	Seawater, road-salt and salt in general, soil
Potassium chloride	Soil
Calcium sulphate	Many types of bricks, limestone and dolomite affected by acidic atmosphere, gypsum-based wall plaster
Sodium nitrate	Soil, fertilisers
Potassium nitrate	Soil, fertilisers

may severely damage the integrity of the stone, in some cases leading to the total loss of structural strength. Similar pressures and results are achieved through the incremental growth of salt crystals in pore spaces, associated with successive wetting and drying of the stone. This may be the most significant physical degradation process in the urban environment (see Section 8.2.2). The growth of mineral salts may be a direct result of chemical weathering of minerals in the stone or mortar, or may occur through direct precipitation of minerals from the groundwaters (Table 4.3). In most cases, weathering by the precipitation of mineral salts is a phenomenon which is increased in the polluted atmosphere of major urban and industrial conurbations, as sulphurous and other gases in the atmosphere commonly promote chemical reactions within the materials themselves.

Accurate measurement and testing of durability in natural building materials is obviously desirable if the stone is to withstand the demands of its environment and function, and if it is to have a long working life. In most countries, the use of stone in buildings is subject to national standards, but standardisation is often difficult to achieve in non-uniform natural materials. However, an approximation of standardisation has been achieved through measurement and application of a series of tests. Typically, standards reflect measurement of porosity, and the testing of porous stone to observe its tolerance of frost action and salt crystallisation. Other tests are employed to determine the ability of a stone to withstand decay from acidic atmospheric conditions (Table 4.4). It is recognised that these tests are often a poor approximation of the behaviour of natural stone subjected to weathering, and that in many cases the only truly accurate method is the observation of stone in place on buildings over

Table 4.4 Summary of British Standard (BS) tests for stone materials

End use	Stone property tests required	General visual quality assessment
Dressed kerb stone (BS 435)	None required	Yes
Roofing slates (BS 680 & 5543)	Water absorption, wetting and drying, sulphuric acid immersion	Yes
Stone masonry (BS 5390)	Crushing strength, salt crystallisation, porosity, saturation coefficient, microporosity, wet and dry density, freeze–thaw, petrography (occasionally)	No
Stone cladding (BS 8298)	None required	Yes

an extended time interval (Box 4.1). However, the tests as they stand provide at least a basis for the sound study of natural building materials.

Porosity is usually measured by the comparison of a sample of rock when it is dry and after it has been water-saturated. Four aspects are currently measured, particularly in load-bearing stones (Figure 4.3): (1) porosity percentage; (2) saturation coefficient; (3) microporosity; and (4) capillarity. The **porosity percentage** is a simple measurement of the overall percentage of pores in a given stone, and is calculated through direct comparison of the weights of dry and saturated stone samples. In general, stones with a high percentage of pores, such as sandstones and limestones, are more susceptible to physical weathering than those with a low pore percentage, such as granites and marbles. The **saturation coefficient** of a given stone is the ratio of the volume of water which can be absorbed by the sample after total immersion over a 24 hour period to the total volume of the sample. This is intended to demonstrate the ability of a stone to absorb water, a process which is clearly a factor in promoting mineral salt and ice crystal growth in pore spaces. **Microporosity** is the volume percentage of pores in the sample which are less than 5 μm in diameter. It is expressed as the volume of water left in a fully saturated sample after it has been subjected to a vacuum. These extremely small pores are known to retain their interstitial water even after extensive drying and therefore a high degree of microporosity is an undesirable characteristic of a building stone. Finally, **capillarity** expresses the rate of uptake of water in a stone, and it is measured through comparison of a partially immersed sample with the sample after a given period of total immersion. Of these factors, the microporosity of the stone is considered to be an important influence on its durability, as a highly microporous stone will be unlikely to drain its interstitial water and therefore will potentially promote the growth of ice crystals. Similarly, capillarity reflects the ability of a stone to saturate from partial immersion and to fully dry out, and this is important as multiple wetting and drying of a stone will effectively promote the growth of mineral salts in the pore spaces. Where stones are perpetually damp, typically in foundations, salts may also be

BOX 4.1: THE RATE OF NATURAL WEATHERING: ST PAUL'S

The rate at which different stones weather when exposed to the urban atmosphere is difficult to determine precisely. In rare cases an indication of weathering rates can be obtained from historic buildings. This is true of St Paul's Cathedral in London.

A decision to add a balustrade around the base of the dome of the Cathedral was taken in October 1717 and construction was carried out the following year. The balustrade was constructed of Portland Stone, a bioclastic limestone. The blocks were hoisted into position by means of a pulley system, using holes drilled through each stone. These holes were later filled with scrap wood and other rubbish before being filled flush with lead. The lead plugs now stand 8 to 30 mm proud of the rock surface. Assuming that the lead has not been weathered in any way this would suggest that the limestone has weathered/eroded in the London atmosphere at a rate of 0.078 mm a^{-1}. This example illustrates how historic buildings can be used to gauge the weathering resistance of traditional building stones, such as Portland Stone.

Source: Sharp, A.D., Trudgill, S.T., Cooke, R.U., Price, C.A., Crabtree, R.W., Pickles, A.M. & Smith, D.I. 1982. Weathering of the balustrade on St. Paul's Cathedral, London. *Earth Surface Process and Landforms*, **7**, 387–389.

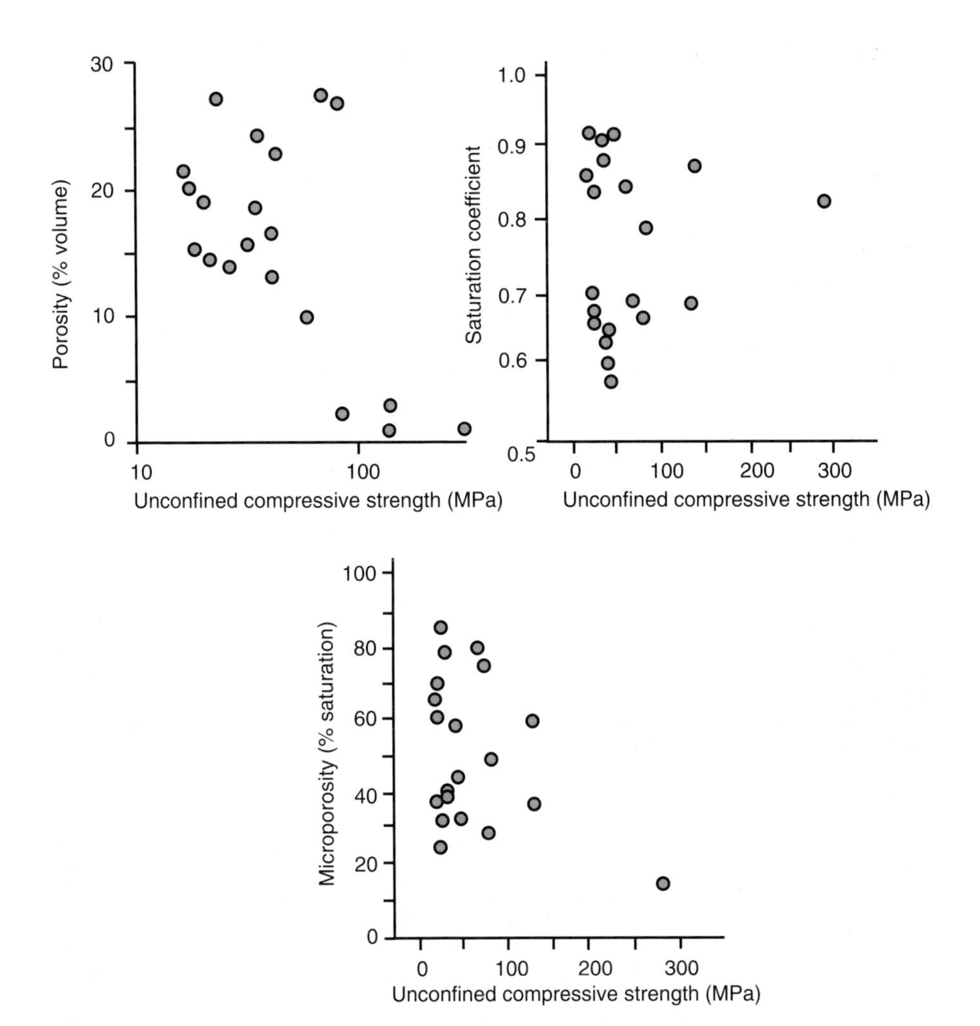

Figure 4.3 *The relationship between the compressive strength of building limestone and porosity, microporosity and the saturation coefficient [Modified from: Prentice (1990)* Geology of construction materials. *Chapman & Hall, Figs 2.5 & 2.6, pp. 40 & 41]*

concentrated at boundaries between a damp inner core of a building stone and the dry outer levels of the stone.

The capability of a given rock type to withstand physical weathering from freeze–thaw and salt crystallisation processes, and chemical weathering through acid attack on carbonates, is determined through experimentation and testing in a number of countries. The **crystallisation test** attempts to simulate the action of ice and salt crystal growth in pore voids through alternate soaking and drying of a sample in a salt solution. The test involves 15 cycles of soaking in a solution of sodium sulphate followed by drying in a humid oven. The resulting growth of sodium sulphate crystals allows examination of the effect of salt crystal growth in the pore spaces of a given stone, but is also intended as a method of emulating the action of ice crystal growth. The **acid immersion test** is intended to assess the ability of a stone to withstand the rigours of a polluted, acidic atmosphere. This test involves the immersion of a stone sample into sulphuric acid of varying molarity, normally 20% but sometimes 40% where long life is demanded of the stone. This test is probably not representative of normal conditions and may be of limited value, as all carbonates and many silicates will react, given the concentrated nature of the acid. The use of these and other tests has led to the development of building stone classifications based on viability of use in a wide variety of settings and climates. These vary from country to country, but are generally associated with basic recommendations for use and situation. However, many of the tests currently available are insufficient to fully assess durability, and the international stone industries across the world have yet to agree on the most appropriate way of testing the durability of natural stone under a variety of atmospheric and climatic conditions and settings.

Appearance of a dimension stone is obviously an important factor in its selection as a building stone. In Britain alone there are at least 150 domestically quarried and commercially available stones with varied appearance, strength and durability. In general, the most important aesthetic aspects of stone are **colour** and **texture**. The use of contrasting colours of natural stone has been a feature of architecture for centuries, and the re-emergence of stone cladding for decoration has seen the widespread use of colourful stones on many public and private buildings. Typical examples include the distinctively blue larvikite, a coarse grained igneous rock from Scandinavia famed for its iridescent feldspars, the rich reds of many commercially available granites, and the striking greens of serpentinites. Architecturally, there are two important factors to consider when selecting stones on the basis of colour: the inherent and unpredictable colour variations of natural stone; and the variation in colour with age, which is usually predictable and is most often caused by the weathering characteristics of the stone, particularly with exposure to different atmospheric conditions. Texture is also extremely important architecturally, as it provides an opportunity for a greater degree of artistic experimentation. For example, many architects enhance the appearance of their buildings by utilising in their design prominent features such as the tectonic fabric of metamorphic rocks, or the depositional structures and fossil content of sedimentary rocks.

Ease of working is important in the selection of stone, especially as it determines its suitability for specific purposes. Dimension stones, particularly those for masonry and cladding, clearly need to possess inherent qualities which allow them to be worked into precise and regular blocks. For example, well jointed and massively bedded rocks such as limestones and sandstones produce excellent dimension stones, as the spacing of the joint–bedding plane intersections allows for the quarrying of good sized blocks which are capable of being dressed. However, those sedimentary rocks which are restricted in lateral and vertical extent and which display a great deal of facies variation are not easily standardised and therefore cannot provide enough regularity to guarantee supply to the user. The often pervasively jointed but otherwise massive central cores of plutonic igneous rocks, such as granites, also provide extremely important sources of dimension stone, the joint spacing allowing the working of regular stone blocks in much the same way as the bedding–joint intersections of sedimentary rocks. Variation is also found in igneous bodies, but may be less significant than in sedimentary rocks. A term often used in Britain is **freestone**, which describes the best possible state for dimension stone – the ability to freely dress quarried stone in any direction in order to provide extremely regular blocks. This ability also determines whether a stone can be worked to provide the three-dimensional sculptural aspects which are characteristic of some of the finest architectural masterpieces in the world. Typical freestones include uniform, evenly grained and well cemented sandstones such as the red Penrith Sandstone of northern England, and oolitic limestones, such as the famous Portland Stone of southern England. Both stones can be worked in any direction to provide such varied architectural finishes as **ashlar**, blocks of stone with completely smooth faces, and **rock-faced**, blocks of stone with a smooth border but with an externally rough texture resembling unworked stone (Fig. 4.1A). Construction stones intended for roofing or flooring are also dependent on the way in which they may be worked. Slate, for example, is an extremely efficient roofing material which is effective largely because it is easily split into slabs and thin sheets along its well defined cleavage planes. **Flagstones**, by way of their natural splitting surfaces provided by bedding or cleavage, can be worked into slabs or flags of stone for use in flooring, pavement construction and roofing. Typical examples include fissile sandstones, such as the 'Yorkstones' used as paving in many British Victorian towns and cities, and slates, used extensively for roofing.

Availability of stone and costs of transport are obviously extremely influential in the choice of natural building stones for use in construction. For example, prior to the Industrial Revolution, most stone in Britain was quarried locally and the early, pre-Victorian fabric of many of the urban centres reflects this, as they are usually constructed from locally quarried stone. The quarries from which such stone was recovered were mostly small affairs, which may have been opened for a specific purpose if the building was particularly grand. This has caused problems of stone matching in the repair of ancient buildings which have suffered decay in the polluted atmosphere of the now industrial cities. Today, stone is derived from across the world and this may be seen in any high street, as the transport of materials by freight container and other means has opened up the availability of options. This is particularly true of stone utilised in cladding, which is cut as thin sheets, and many exotic marbles and other materials are shipped across the world from the main mediterranean production centres. Load-bearing masonry stone is, however, more often indigenous as it is more expensive to transport because of its bulk.

Periodically, stone used in buildings will require cleaning and/or restoration, particularly where there has been significant deterioration due to atmospheric conditions (Box 4.2). Stone cleaning and restoration techniques are highly technical and have considerable environmental impacts. **Stone cleaning** includes the chemical or physical removal of accumulated dirt and residue from building stones, particularly in the urban environment, but also in the rural setting. This includes the 'restoration' of naturally weathered stone to its original state. Typical techniques include: (1) water washing, involving directed water sprays followed by surface scrubbing intended to remove surface accumulations of dirt, which is commonly used on limestones; (2) chemical cleaning, using preparations of hydrofluoric acid which are highly reactive and dangerous, and are often used to clean the surfaces of sandstones, unpolished crystalline rocks and brickwork; and (3) air abrasion, either dry or wet, which involves the pressurised jetting of abrasive-charged air or water to remove the surface layer of the stone, generally applied to limestones and sandstone buildings.

Restoration encompasses a great variety of techniques from stone replacement through to stabilisation of weathering. This is particularly sensitive in ancient monuments and other heritage sites, as poorly thought out restoration techniques have, in the past, often marred the original beauty of a building. Restoration is usually required when there is: (1) excessive physical or chemical weathering of the natural stone by salt crystallisation and acid attack, when replacement with matched stone blocks is the usual remedy; (2) poor repairs with inappropriate stone or other materials which need replacement; (3) corroded metalwork, which causes expansion and spalling of stonework; (4) accidental damage by fire or other activities; and (5) structural failure, requiring reinforcing and restitching of the stonework.

4.1.2 Decorative Stone

Decorative stone can be defined as any natural stone, other than that used in the fabric of buildings, which is selected on aesthetic grounds for a wide variety of uses. The most common uses are: (1) sculptural stone; (2) gravestones and simple monuments; and (3) garden stone. Complex architectural monuments, such as the Albert Memorial in London (Box 4.2), are buildings in their own right and the stone employed for these uses therefore needs to meet the same conditions of structural strength and durability as in other buildings.

Sculptural stone needs to meet at least three requirements before it is selected, namely: (1) aesthetic beauty; (2) ease of working; and (3) durability. Of these, aesthetic beauty is perhaps the most important, but final selection is ultimately tempered by the ease with which a stone may be worked to create an aesthetically pleasing sculpture, and by the durability of the stone to ensure survival of sculptural detail in a wide range of environments. The most appropriate materials for sculpture remain marbles and micritic limestones. These satisfy at least two of the criteria: they are attractive, particularly the much admired marbles widely used in classical sculpture; and many have the properties of freestones, being equally workable in any direction. A classic sculptural stone is Carrara Marble from northern Italy, which has a uniform texture and fine white colour much admired by artists. Durability is more of a problem, especially where sculptures are intended for external display, as carbonate rocks are particularly susceptible to chemical weathering by acidic atmospheric conditions. For example, stone sculptures dating from AD 1170 in the industrialised

BOX 4.2: STONE CLEANING AND THE ALBERT MEMORIAL, LONDON

The Albert Memorial was built in the mid-nineteenth century as a record of the achievements of Prince Albert the Prince Consort in sponsoring the arts and sciences in Britain during the Victorian age. The memorial was designed by George Gilbert Scott and is a fine example of high Victorian gothic architecture. It is based on a medieval reliquary and is encrusted with jewels and mosaics and was originally gilded to provide a glittering spectacle which often jarred with the eye in later, more sombre days. In the late 1980s the memorial was in a sorry state: pieces had fallen from it, and continued to do so; there were some particularly poor repairs; the stonework, statuary and mosaics were all covered with the grime of a hundred years and more of poor atmospheric conditions. Restoration was imperative. The restoration of the memorial is a mammoth task and involves much attention to the steel framework and to the leadwork on the outside of the memorial. Attention has

continues overleaf

BOX 4.2: *(continued)*

also been focused on the restoration of the stonework. The memorial itself is particularly rich in British building stones; for example, paving slabs which show differential weathering, Portland Stone framework and granite columns (Robinson, 1987). It is the Portland Stone which has needed the most attention. The accumulated grime has been removed through the use of chemical poultices designed to draw out the dirt. After this has been completed, small, directional water jets are used with scrubbing brushes to clean the stone. In some places, small, badly weathered sections have been cut out and replaced with new stone. Finally, artists are employed in painting the detail in bright blues and reds of this gothic jewel in the centre of London.

Sources: Robinson, E. 1987. The geology of the Albert Memorial. *Proceedings of the Geologists' Association*, **98**, 19–37; Brooks, C. 1995. *The Albert Memorial.* English Heritage, London.

Rhine–Ruhr region of Germany have undergone accelerated decay since the early part of this century, leading to the overall loss of definition in the sculptures. These sculptures are composed of sandstones with a high percentage of calcite as both grains and cement (64%), and weathering has involved a combination of chemical weathering of the calcite attacking the cement, and physical weathering through the growth of mineral salts in pore spaces, both of which are promoted by the polluted atmosphere. Igneous rocks were particularly favoured by ancient civilisations, and their durability has ensured their survival into the modern world. For example, Egyptian obelisks hewn from red and pink granites some 3500 years ago and removed in the early part of the nineteenth century are to be found in the variable atmospheric conditions of central London, Paris, Rome, Istanbul and New York. These monuments are up to 15 m long, each one cut from a single piece of granite, carefully selected by the Egyptian masons who sank shafts into the granite body to test the joint spacing. Despite fears for their continuing integrity in some of the most polluted cities in the world, the impervious nature of the granite has ensured that their copious hieroglyphics are still clear after 35 centuries.

Gravestones are another type of decorative stone in which the decoration reflects a need for sombre remembrance. The attributes of good gravestones are in essence similar to those required of sculptural stone. Stone used for grave markers needs to reflect the following qualities: (1) a decorative quality which is consistent with its purpose, that is, not gaudy or inelegant; (2) durability; and (3) the ability to be slabbed and to retain the definition of inscriptions. Typical gravestone types include limestones, marbles, granites, gabbros and other igneous rocks, and, more rarely, sandstones. In Britain, Portland Stone, Dartmoor Granite, a range of imported gabbros, and slate are amongst the most important gravestone types. In particular, the uniform pale colour of Portland Stone is much admired, and its qualities as a freestone mean that it may be easily worked and inscribed. The qualities of this stone led to its adoption, almost worldwide, for the graves of the war dead of the British Commonwealth (Box 4.3). English Dartmoor Granite, particularly from the Merrivale Quarry, is commonly employed for its overall grey colour, while the so-called 'black granites' – actually gabbro – which are imported from a range of countries provide a black or otherwise dark and therefore appropriate colour. All of these materials are capable of being cut into slabs, dependent on the mode of working in the quarry, and are capable of retaining the required inscriptions. These materials are also durable, especially within the lifetime of a typical graveyard. Slate makes a much admired and attractive gravestone. Its colour is highly appropriate, and deeply incised inscriptions commonly retain their crispness, even with advancing age. Finally, marble is common in many British churchyards, although some authorities have suggested that its very whiteness jars in the more subtle colouring surrounding it. Marble is easily worked, but is prone to chemical weathering in acidic environments.

Garden stone is widely used and appreciated both as a rockery material and in the provision of minor decorative construction. Garden rockeries have a long history which extends back to at least the eighteenth century in Britain and the

BOX 4.3: THE COMMONWEALTH WAR GRAVES

The Commonwealth War Graves Commission was set up during the first world war in order to supervise the erection of cemeteries to British and Commonwealth war graves across the world. Similar organisations exist for other countries, but only Britain and the United States of America use natural stone to commemorate the war dead. British and Commonwealth cemeteries

were meant to resemble an English garden or churchyard as a remembrance of home. The stone selected for the gravestones was Portland Stone. Portland Stone combines ease of working with an attractive and fitting colour. Each of the gravestones inscribed with: the name and rank of the soldier, sailor airman/woman or nurse; a regimental badge; a religious symbol where appropriate; and, in most cases, a private epitaph. Slabs with a rounded top were chosen to provide a uniform feel to the cemeteries. The selection of Portland Stone to mark British and Commonwealth war dead across the world was based on: appropriateness of colour; uniformity of texture; availability of the resource; and ability to take a high degree of inscription. Durability is a problem, and decay from physical weathering has meant that in some settings the inscriptions are losing definition after standing for 70 years. These are now being replaced by Botticino Limestone from Italy which is less porous and is less prone to weathering from salt crystallisation effects, although problems of acid attack remain.

Source: Ward, G.K. & Gibson, E. 1995. *Courage remembered*. HMSO, London.

rest of Europe, while in oriental settings they have a much longer heritage. The most important factors which influence the selection of garden stone are: (1) decorative appeal in terms of texture, colour and shape; and (2) durability. Cost and availability are important factors, as for most domestic gardens exotic stones are prohibitive in cost. In oriental gardens, stone is of the utmost importance, and is selected on the basis of pattern and form rather than its pure function. Blocks and boulders of highly coloured or patterned stones alternate with the use of aggregate mixes that provide an essential background to the garden, which often has a sculptural quality. In Europe, the use of quarried stone in gardens to mimic the natural setting of alpine plants was developed in the early nineteenth century, coincident with the move towards the development of more naturalistic gardens. The most favoured stones are porous limestones and sandstones rather than impervious igneous rocks, as the weathering of these rocks helps provide valuable nutrients for the alpine plants. In Britain and Ireland, the use of karst weathered limestone pavement blocks in gardens has been widespread and has led to serious damage to a finite resource and unique landscape feature (Box 4.4). Such blocks are widely favoured for their weathered appearance, the characteristic solution features of the weathered surface providing refuge for plants and creating an impression of age not seen in freshly constructed rockeries.

4.1.3 Armourstone and Rip-rap

Armourstone and rip-rap are terms which refer to stone blocks of variable size used in the construction of breakwaters, dams, sea defences and other flood,

BOX 4.4: ROCK GARDENS AND THE THREAT TO LIMESTONE PAVEMENT

The delicate and sculpted form of limestone pavement has meant that it has long been prized by horticulturists for use in rock gardens. Limestone pavement is, however, a unique and finite part of Britain's physical landscape (Photograph A) and is under threat along with other forms of naturally weathered rock from exploitation by the horticultural market (Photograph B). Legal extraction is restricted, but this has not stopped its exploitation, since blocks of pavement can sell at between £90 and £120 per tonne. Limestone pavement is unique to Britain and is one of its major contributions to the world's natural heritage. In recognition of its importance it is protected under the Wildlife and Countryside Act (1981) in Britain through Limestone Pavement Orders as well as in a series of Sites of Special Scientific Interest. In this way, all significant areas of limestone pavement in Britain have been, or are being, designated and protected. Unauthorised destruction can incur fines of up to £5000. Despite this level of protection, limestone pavement is still under threat; the market demand far exceeds the legal consequences for the destruction of an area of limestone pavement. The only way to stop this destruction is to re-educate the horticultural market away from the use of limestone pavement. In recent years this has led to a reaction amongst conservationists against the use of natural stone in rock gardens. However, this is short-sighted, since the use of stone in gardens has the potential to improve the

appreciation of natural stone and thereby foster interest in its conservation; it is only the use of limestone pavement which must stop.

In summary, the use of limestone pavement involves the destruction of a unique and finite part of our landscape and should stop, but use of natural stone in gardens and landscaping is, however, ultimately beneficial to geological and landscape conservation and should be encouraged.

Source: Bennett, A.F., Bennett, M.R. & Doyle, P. 1995. Paving the way for conservation? *Geology Today*, **11**, 98–100. [Photographs: A.F. Bennett]

coastal erosion and mass movement mitigation measures which require extremely strong and durable materials (see Section 11.2; Figure 11.17). It has been estimated that 95% of the world's breakwaters have used natural stone blocks in their construction.

Armourstone is also sometimes known as **derrick stone**, a term which refers to the fact that it consists of stone blocks which usually weigh between 0.5 and 20 tonnes, and which often require a crane or derrick to lift them into place. Armourstone is typically used in the construction of coastal and harbour jetties, breakwaters, and in coastal berms (Figures 4.4 & 11.17). In these situations the wave energy is considerable, and is often capable of moving even large and weighty blocks of stone; consequently armourstone blocks are usually of large dimensions and are employed en masse in carefully constructed and engineered structures. **Rip-rap** is quarry rubble which is most often used in stream bank

Figure 4.4 *Photograph of typical armourstone blocks on the Devon coast, England*

erosion, road and rail embankments, levees and small-scale coastal erosion projects. Rip-rap normally comprises smaller scale stone blocks, ranging from coarse aggregates to small boulder size (up to approximately 500 mm in size), which are dumped from road and rail trucks and barges in order to provide the appropriate protection. In most cases, rip-rap is used as the central core to breakwaters, forming an initial dumped mass of material constituting around 50–80% by volume of the breakwater. In such cases, the rip-rap is capped by gravel-sized aggregates to act as a filter layer, which helps trap any fines, on to which coarse material is then laid. Finally, a course of armourstone blocks, carefully chosen for their uniform size, strength and durability, is laid over the rip-rap and aggregate layers. The stability of the breakwater is usually dependent on the weight and shape of the individual pieces of amourstone, the care of their emplacement, and the slope of the core.

The selection of appropriate armourstone is dependent primarily on the reaction of the stone to the aggressive environments into which it will be placed, as well as the availability of appropriate stone. In constant contact with the sea, breakwater stone has to withstand a number of differing environmental conditions, depending on its location in the structure (Figure 4.5). Armourstone blocks forming the lower slopes of breakwaters are usually below the low-water mark and as such are constantly submerged. These blocks are therefore protected both from subaerial weathering and mechanical attrition from crashing waves. In the intertidal zone, armourstone blocks are subjected to direct wave attack and regular tidal wetting and drying. In this situation, armourstone is

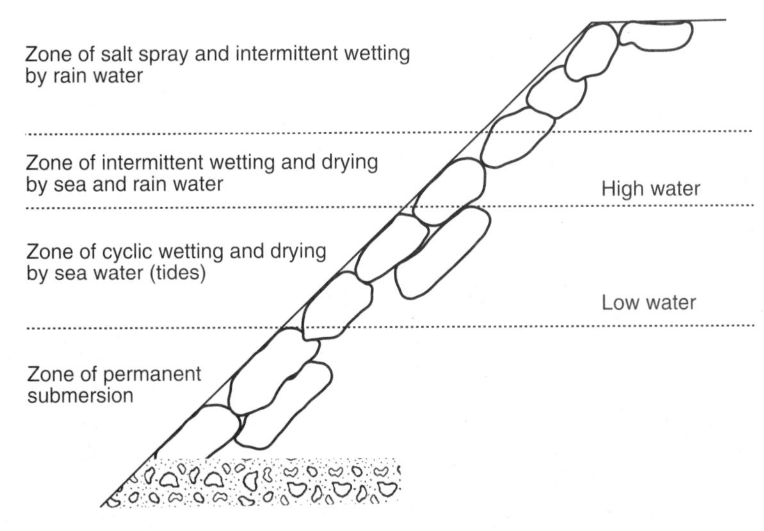

Figure 4.5 *The four main weathering zones of the coastal marine environment [Modified from: Fookes & Poole (1981)* Quarterly Journal of Engineering Geology, ***14****, Fig. 1, p. 98]*

most frequently damaged by physical attrition from the crashing waves. This involves two processes: (1) water load abrasion, caused by material held in suspension by the turbulent waves being driven against the blocks; and (2) rocking abrasion, caused by the effects of armourstone blocks rocking against each other and causing physical erosion. Finally, blocks forming the upper levels of a breakwater may never be submerged, but instead may be subjected to the effects of salt spray and intermittent wetting from rain waters, both of which accelerate weathering through propagation of salt crystal growth. In addition, bioerosion from a range of boring organisms, such as pholad and mytilid bivalves, sponges and crustacea, may lead to physical erosion of the armourstone, particularly in the intertidal and submerged zones.

In selecting stone for the construction of breakwaters, the following factors are of importance: (1) block size, which is controlled by the availability of natural stone with widely spaced fractures in the quarry; (2) uniformity of the fabric of the stone, particularly an absence of physical discontinuities and fractures; (3) durability, the resistance to weathering; (4) hardness, the resistance to abrasion; and (5) toughness, which is the resistance to fracture under impact. **Block size** is usually determined at source in the quarry, and is a product of natural joint spacing and intersection, and the methodology of extraction (Figure 4.6). The production of armourstone of 20 tonnes or more requires widely spaced joints and/or faults, usually several metres apart, in order to provide blocks of the desired size. Blasting also leads to the propagation of fractures, and in order to produce large blocks blast holes must be relatively widely spaced. In some cases, rough handling while loading or unloading blocks can also cause size diminution. **Uniformity** is related to block size, in that physical discontinuities or distinct facies changes can provide planes of weakness which are easily

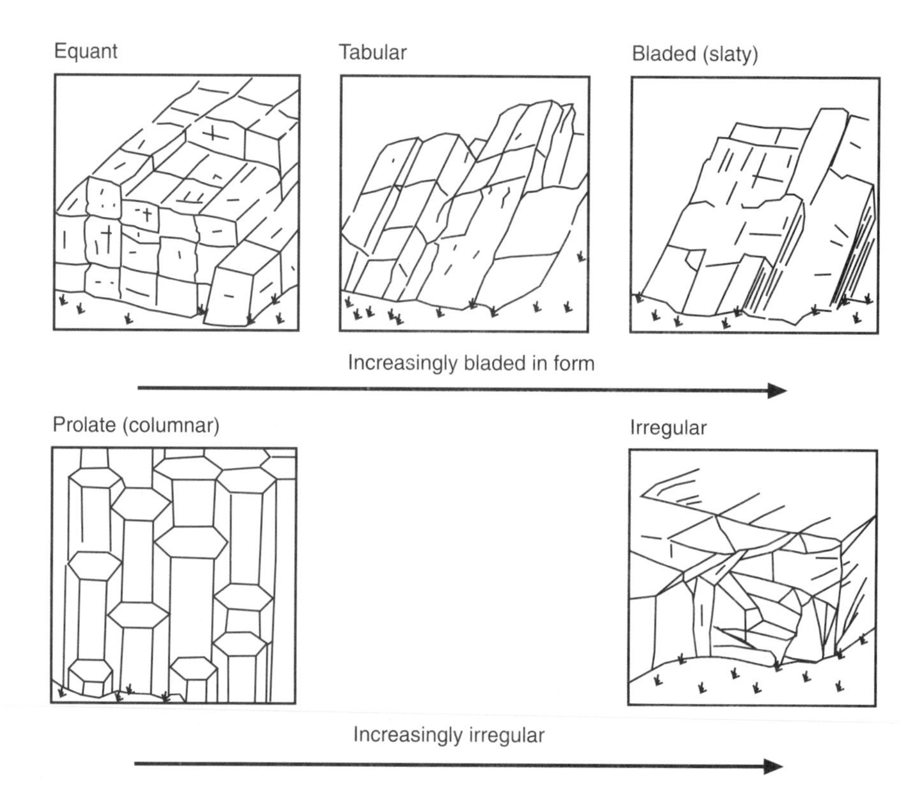

Figure 4.6 *The control of joints and bedding in the production of block size for armourstone [Modified from: Fookes & Poole (1981)* Quarterly Journal of Engineering Geology, ***14****, Fig. 26, p. 118]*

exploited in marine environments through both physical attrition and weathering. It is particularly important to examine crystalline rocks petrologically for signs of chemical degradation of feldspars and other minerals into clay mineral complexes which would swell in the marine environment and cause spalling and other weathering effects. **Durability** is of importance given the salt water environment and the continuous process of wetting and drying experienced by the upper, intermittently submerged levels of a typical breakwater structure. Measures of porosity and resistance to the salt crystallisation test, as employed in building stones, allow some assessment of the durability of armourstone. The physical strength of armourstones is expressed in two properties: **hardness** and **toughness**. Hardness is expressed in the ability of the stone to undergo physical abrasion, and the tests employed for roadstone and other aggregates are the most appropriate ones for measuring this property. Those rocks with a high degree of quartz are clearly going to be more resistant to abrasion that those composed mostly of calcite. Toughness equates with compressive strength and, as with building stones, it is those rocks with an impervious and interlocking crystal structure which are capable of resisting attack. The twin properties of hardness and toughness are clearly of extreme importance in the selection of stone for breakwaters, which are usually under intense attack from the attritional effects of waves.

4.2 AGGREGATES

Aggregates may be defined as particles of rock which, when brought together in a bound or unbound condition, form a construction material. As geomaterials they consist of particulate rocks which vary in size from sand to pebble and/or cobble grade. Aggregates are of extreme economic importance, and as such they represent the greatest volume of any geological material extracted in most countries (Figure 4.7). In Britain alone, the quarrying of aggregates makes up over 70% of the mineral productivity, three times greater than the production of coal. In the United States, the production of aggregates averages about 1.1 billion tonnes annually. The importance of aggregates lies with their widespread use in the construction industry, either alone or as the basis for other geomaterials such as concrete and cement. There are two basic aggregate types: natural and processed. **Natural aggregates** are mostly derived from unconsolidated sand and gravel deposits, which are often, but not exclusively, of Quaternary age. Such materials may be quarried from both subsurface or surface deposits, dug from river and beach deposits, and dredged from offshore sources. Natural aggregates usually undergo minimal processing to produce the end material, often a simple extraction process followed by separation into appropriate grades. The commonest natural aggregates consist of glacial, river and marine sands and gravels.

Processed aggregates are produced from stone which is quarried or mined and is then crushed, usually on site, to provide the grade required. Crushed aggregates are mostly coarse, consisting of rock particles which have a diameter of more than 5 mm, and are commonly produced from a variety of lithologies, including limestones, crystalline rocks such as granite or dolerite, and, less commonly, sandstones. For example, in the USA, 70% of crushed rock production is of limestone, while 20% is granite and basalt, and the remainder, sandstone and quartzite. In some cases, crushed aggregates are produced as a by-product of the quarrying operations for dimension stone. In many cases, such quarrying activities produce stone that is unsuitable by virtue of its inconsistent appearance, or the presence of pervasive joints or other features which prevent the production of large unflawed blocks for masonry. In addition to natural and processed aggregates, by-products of other industrial activities are also used as aggregates, typically blast furnace slag, broken rubble and pulverised fuel ash or **fly ash** from power stations. These are economically less important than aggregates sourced from primary geological materials, although they are sometimes used in the construction of roads. Fly ash is particularly common as it is often used to make simple construction blocks.

Classification of aggregates is important as it provides the potential user with the basic information needed to make a judgement on the suitability of an aggregate for a specific purpose. It is important that classification schemes are

Figure 4.7 *The production of aggregates from Quaternary sand and gravel deposits in southern England*

simple for wide accessibility, and use terms which can be precisely defined in order to be consistent with contract specifications. Typical classifications employ three important characteristics, namely type, physical characteristics and petrology. **Aggregate type** refers to whether the material comprises crushed rock, or natural sand and gravel derived from land or marine sources. **Physical characteristics** help to define the nature of the aggregate material, particularly with respect to shape, size and textural features, all of which are of value in the consideration of end use. Typically, the following characteristics are noted: size, shape, surface texture, colour, contamination by fines such as dust, silt or clay, and the presence of surface coatings. **Petrology** is also important in defining specifications for aggregate type, because the specifications for some uses can exclude certain rock types, and the performance of some rock types to aspects such as physical abrasion may be different. Petrological examination, at least at a basic level, is therefore fundamental. The use of all or part of this classification scheme is recommended by the appropriate national standards authority. For example, in Britain and the United States, some measure of standardisation is achieved through the classification of aggregates on the basis of basic type, origin, size and petrology.

The applicability of aggregates for their intended end use is governed by a number of testable qualities. These tests have been devised in order to provide estimates of suitability for a wide range of applications but, like the tests for building stones, their appropriateness is a matter of current debate, and is dependent on the end destination of the aggregate material. Typically, this destination is in one of two areas: as roadstone, discussed below, and as the major component of concrete, discussed in Section 4.3.

4.2.1 Roadstone

Roadstone can be defined as aggregates which are used in the construction of roads, either in a loose, unbound state, or bound with bitumen to form the familiar asphaltic or 'black top' roads. Road construction and maintenance in Britain alone consume well over 50 million tonnes of quarried aggregate, and there is no indication that this demand will lessen in the near future. The principles of aggregate-based road building were developed by the Romans, who not only constructed unbound roads but may also have employed a rudimentary bitumen binder. However, the concept of the modern, bitumen-bound road may be traced back to John Loudon McAdam who developed it in the early nineteenth century.

The majority of modern roads are constructed using a series of stacked aggregate layers of variable thickness and construction. These are both bound with bitumen and unbound, and there are usually four layers: the **sub-base**, forming a drainage layer of unbound coarse crushed rock aggregate; the **road-base** and **base course**, composed of bitumen-bound coarse aggregate; and finally, the **wearing course**, which is composed of asphalt with 30% aggregates topped with bitumen-coated chippings. In general, these layers must be able to withstand both the great weight imposed by heavy traffic, and the wear of the frictional contact with vehicle tyres, which is important in itself to provide grip for the cars and other vehicles using the road. Probably the most important layer is the wearing course, as it provides the basic contact with motor vehicles. Ideally, the wearing course should satisfy the following demands: (1) it should be hard wearing and strong; (2) it should present a relatively low tractive resistance in order to prevent both unnecessary slowing of traffic and surface noise, while preventing skidding; and (3) it should be easy to clean, excavate and reinstate. Meeting these requirements is largely a function of using the appropriate aggregate type, as some are stronger and more durable than others, and some are more likely to be polished to a state where they may create skidding accidents (Box 4.5). Clearly, appropriate aggregate selection for road construction is of great importance.

Aggregates used in the construction of roads need to meet several basic requirements: (1) aggregate **size**, which is often made deliberately variable, should be less than 38 mm diameter for the base course, and less than 19–25 mm diameter for the wearing course; (2) **shape** is particularly important as flattened aggregates with pervasive cleavage may lead to 'flakiness', the ability of the stone to chip or wear away along certain defined planes; (3) **strength**, the ability to withstand mechanical crushing; (4) **durability**, the resistance to normal weathering effects, usually measured using the sulphate crystallisation test as applied to dimension stone; and (5) **wear resistance** to abrasion from the frictional effect of vehicle tyres, and in particular polishing, a property which determines to some degree the skid resistance of the road surface (Box 4.5). In general, roads are constructed using non-porous, crushed igneous and other crystalline rocks which have a high compressive strength, possess a high percentage of free silica enabling them to resist abrasion, and can withstand frost action as well as acid and other forms of chemical attack. However, the suitability of crushed rock aggregates for roadstone construction is the subject of vigorous testing across the world, with particular emphasis on strength and wear resistance.

Strength is measured by using the application of two tests: the aggregate crushing value (ACV) and the aggregate impact value (AIV) (Figure 4.8). The **aggregate crushing value** measures the percentage of fines (i.e. particles of < 2.36 mm in diameter) produced from a standard sample of aggregate which has been subjected to a continuous load of 400 kN for 10 minutes. This is intended to simulate crushing by continuous heavy traffic, and the lower the value, the more resistant the aggregate is to this form of crushing. The **aggregate impact value** measures similar properties, but is assessed by measuring the fines produced after dropping a hammer of standard weight, from a standard height, on to a sample of aggregate. Typically, close-textured siliceous rocks such as hornfels and fine grained igneous rocks perform well in these tests, while coarser grained igneous rocks and sedimentary rocks perform less well (Table 4.5).

Wear resistance is assessed through the measurement of a further two properties: the aggregate abrasion value (AAV) and the polished stone value (PSV). The **aggregate abrasion value** is a measure of mechanical abrasion through comparison of the weight of an aggregate sample before and after it has

BOX 4.5: SKID RESISTANCE OF ROAD AGGREGATES

Skidding accidents on our roads are a major cause of death, bodily injury and property damage. Although accidents of this kind are often caused by a combination of factors, including local weather conditions, vehicle condition and driving ability, it is recognised that the condition of the road surface is also an important contributing factor. Although it is very difficult to equate specific road accidents with the nature of the road surface quality, it is recognised that the quality of wearing-course aggregates plays an important role in the maintenance of the skid resistance of the road surface. Since the 1950s the National Crushed Stone Association (NCSA) of the United States has been conducting controlled experiments in order to measure the skid resistance of indigenous aggregates. This has equated with the ability of the roadstone to resist polishing. Early experiments included subjecting a test track to simulated severe traffic action through the continuous passage, several thousand times, of a bus wheel loaded with a weight of up to 1000 kg. Measurement of polishing was then carried out using a standard testing device, known as the British portable tester (BPT), which expresses polishing as a single number quantifying the resistance to polishing: the higher the number, the greater the resistance.

Many of the aggregates used in road construction in the USA are composed of crushed carbonates. Renninger & Nichols (1978) carried out a series of tests to compare their skid resistance with non-carbonate rocks. Their work suggested that the acid solubility of carbonate aggregates was a major factor in the rapid development of a polished surface. However, they demonstrated

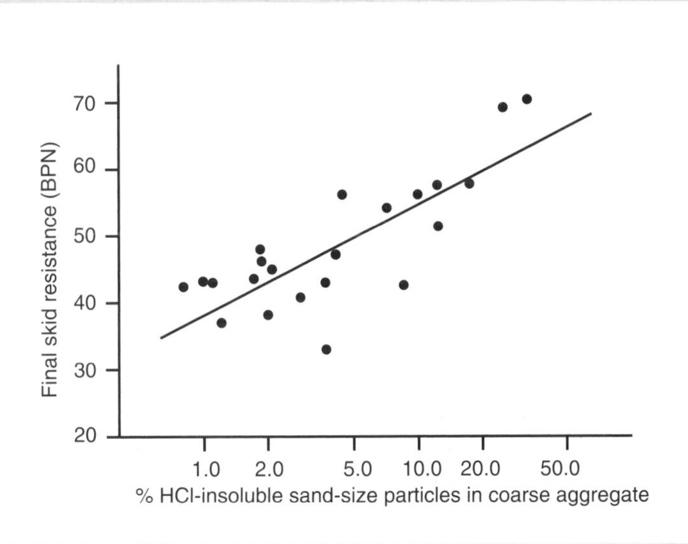

continues overleaf

BOX 4.5: *(continued)*

that carbonates were not necessarily a poor choice in road surface construction provided that they were mixed with at least 5% of sand-sized insoluble (i.e. non-carbonate) particles. As the plot on previous page indicates there is a direct relationship between the percentage of non-carbonate clastic material in an aggregate, and its skid resistance number. This work demonstrates that carbonates can be used with the same degree of safety as many 'hard' rocks in the design of skid-resistant road surfaces provided that they are mixed with small amounts of non-carbonate aggregates.

Source: Renninger, F.A. & Nichols, F.P. (1978) Aggregates and pavement skid resistance. In: Winkler, E.M. (Ed.) *Decay and preservation of stone*. Geological Society of America Engineering Geology Case Histories, 11, 25–29. [Diagram modified from: Renninger & Nichols (1978) In: Winkler (Ed.) *Decay and preservation of stone*. Geological Society of America Engineering Geology Case Histories, 11, Fig. 1, p. 26]

been abraded for a set time using a rotating lapping machine fed by sand. A similar test is the Los Angeles abrasion test which subjects loose aggregate in a rotating cylinder to abrasion from the action of steel balls. As with the strength tests, the lower the number, the greater the resistance to abrasion. In general, there is a correlation between this measure of abrasion and the indicators of strength, such that rocks with a low AAV number often have low crushing and impact values (Table 4.5). The **polished stone value** measures the resistance of an aggregate to the action of polishing from vehicle tyres. This is simulated by first applying a polish for a set interval of time through the action of a loaded rubber wheel fed with abrasive slurry. The friction given by the polished stone after this process is then measured in order to attain the polished stone value. The higher the value, the greater the resistance to polishing (Figure 4.9). This property is of great importance as it is thought that the ability of an aggregate to take on a high polish significantly increases the possibility of road accidents through skidding (Box 4.5), and thus minimum polished stone values are usually specified for a variety of situations in road construction schemes. For example, values in excess of 45 are suitable for lightly used roads, while heavily used roads and important junctions should have aggregate wearing courses with PSV numbers which exceed 65. Relatively few aggregates have a PSV of more than 65 (Figure 4.9). Limestones display the greatest propensity to take a polish, although this is dependent to a certain extent on the purity of the limestone (Box 4.5). The best rocks are coarse sandstones with a high grain density, a close texture and variation in clast types allowing the constituents to abrade at different rates, thus helping to prevent polishing. As an indication of this, uniform grained quartzites are also capable of being polished and therefore usually have relatively low PSV numbers (Figure 4.9).

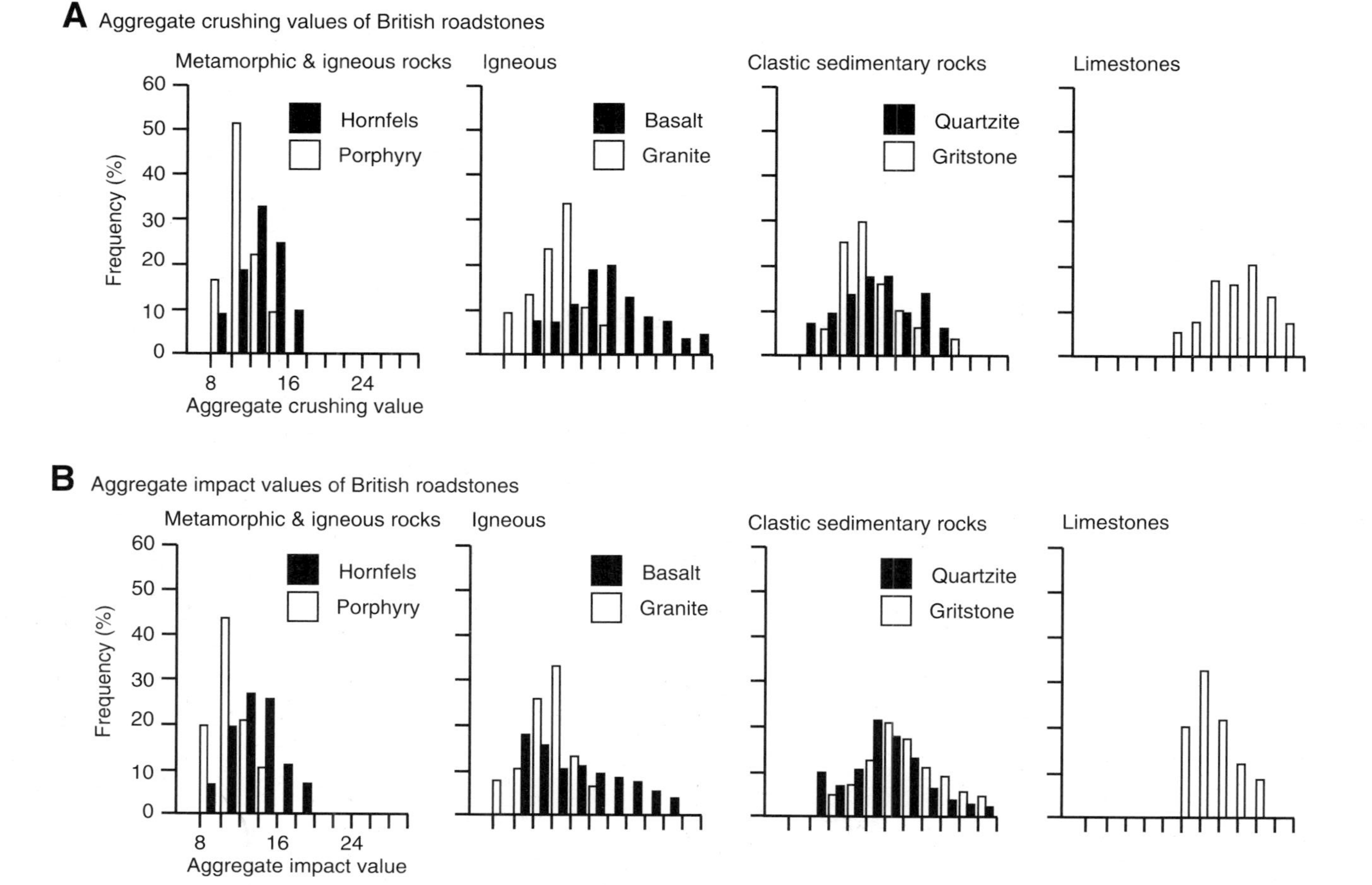

Figure 4.8 *Properties of British roadstones.* ***A.*** *Aggregate crushing values.* ***B.*** *Aggregate impact values [Modified from: Prentice (1990)* Geology of construction materials. *Chapman & Hall, Figs 3.2 & 3.3, pp. 74 & 75]*

Table 4.5 Test values for typical roadstone aggregates [Modified from: Manning (1995) Introduction to industrial minerals. Chapman & Hall, Table 2.1, p. 23]

	Strength		Wear resistance	
Aggregate	Impact value (AIV)	Crushing value (ACV)	Abrasion value (AAV)	Polished stone value (PSV)
Limestone	22	23	10.0	37–39
Pennant sandstone	20	18	7.1	67
Dolerite	8	11	3.2–3.6	50–56
Gabbro	17	17	4.1	54
Granite	20	20	3.0	52
Hornfels	9	10	1.4	55
Flint gravel	22	18	1.7	43

4.2.2 Railway Ballast

Crushed rock aggregates are commonly used for ballast in the construction of rail track, where it is banked up as the seating for railway sleepers. In view of the great weight of the rolling stock which such material is expected to withstand, the selection of aggregates for such a purpose is primarily based upon compressive strength and durability. Igneous and non-platy metamorphic crystalline rocks have the greatest compressive strengths, are impervious and resistant to weathering, and are therefore preferred. However, some limestones, such as the Carboniferous Limestone of Britain, demonstrate reasonable compressive strength and are widely used, particularly as the PSV of the material is unimportant in this case.

4.3 CEMENT AND CONCRETE

Cement is an extremely important industrial material, particularly when mixed with aggregate to form concrete, the most versatile and widely used construction material of all. Almost every nation produces cement, the prime ingredients of which are crushed limestone and clay; annually over 1.1 billion tonnes are produced, making it a multi-million pound industry which is of extreme importance to world economy.

4.3.1 Cement

Cement is a geomaterial which may be used on its own in the production of thin floor (**screed**) and wall (**render**) covering materials, mixed with sand to produce the **mortar** needed to construct brick buildings, and, most importantly, mixed with aggregate to produce **concrete**. Cement is made from the combination of crushed limestone and clay, which when fired in a kiln produces a substance

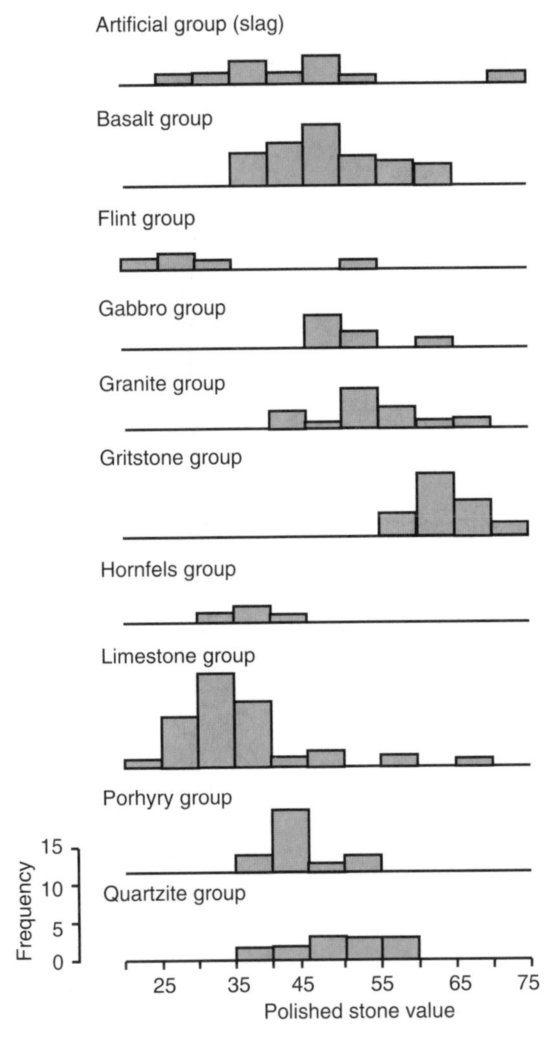

Figure 4.9 *Polished stone values (PSV) for different groups of rock type [Modified from: Prentice (1990)* Geology of construction materials. *Chapman & Hall, Fig. 3.6, p. 79]*

capable of setting under water. For this reason cement is more properly referred to by the title **hydraulic cement**, of which there are two types, pozzolan and Portland. **Pozzolan** has a long history which extends back to at least classical times when both Greeks and Romans discovered that volcanic ashes could be combined with slaked lime to produce a cement; the name is derived from the village of Pozzuoli on the slopes of Mount Vesuvius (see Section 13.4.1). The importance of pozzolan was superseded on the advent of Portland cement, which was patented in the early part of the nineteenth century.

Portland cement was so named because it was meant to resemble, when cured, the texture of the freestones from the Isle of Portland in southern England. This cement is by far the most common today and its use is world-wide; its success is associated with the simplicity of its ingredients, which combine crushed limestone with around 18–25% clay, providing a mixture of lime (CaO), silica (SiO_2) and alumina (Al_2O_3). Portland cement can therefore be readily produced wherever there is a juxtaposition of limestone and clay; in Britain, typical occurrences include the limestones and shales of the Carboniferous, and the Cretaceous chalks and Tertiary clays of southeastern England. Some limestones contain a relatively high degree of fine clastic material, and these rocks, called **cementstones**, may be crushed to provide the basic raw material. The limestone and clay are crushed and combined either in a wet or dry state, and then fired in a rotary kiln (**calcined**) to produce the cement. The use of water slows the process and raises fuel costs, and appears to serve no other purpose than to allow a thorough mixing of the primary materials. On heating to temperatures of up to 1500°C, the calcining firstly evaporates any water in the mixture, and then thermally decomposes both the clay minerals and calcite before fusing the minerals into a glassy state. This glassy material cools to form a clinker composed of the crystals of a number of calcium compounds which, when ground, form the cement. The addition of water hydrates these minerals and causes the cement to set, a complex process which involves a number of mineralogical reactions which take place at different rates.

4.3.2 Concrete

Concrete is effectively a synthetic rock which is composed of a mineral filler or aggregate bound together with a cement binder (Figure 4.10). The aggregate, in the form of sand and coarser materials, typically makes up between 60 and 80% of the concrete mix; the rest consists of the cement (6–18%), and the water needed to hydrate the material (14–22%). The success of concrete as a structural material is largely dependent on the mineralogy of the cement and on the petrology, size distribution and shape of the aggregates, as well as the composition of the water used in the production process. This means that it is extremely difficult to predict the final properties of the concrete.

The composition of the cement is most easily controlled as the mixture of raw materials, limestones and clays is well known in its production phase. The uniformity of the aggregate is less easy to control, and in some crushed rock aggregates, for example, there may be considerable variation in the quality and extent of weathering of igneous rocks. Aggregates which are unsound can result in problems in the finished concrete, leading to cracking, spalling and other detrimental processes. During the curing phase the concrete undergoes shrinkage, leading to microcracking in the cement paste between the aggregate or along the aggregate–cement boundary. If the aggregate is strong this will enhance the overall strength of the concrete. However, excessive shrinkage can occur where the aggregates used have a high clay or mica content, as these appear to absorb water which they subsequently lose during the drying process, enhancing cracking. Quartz-rich aggregates are the best as they are impervious,

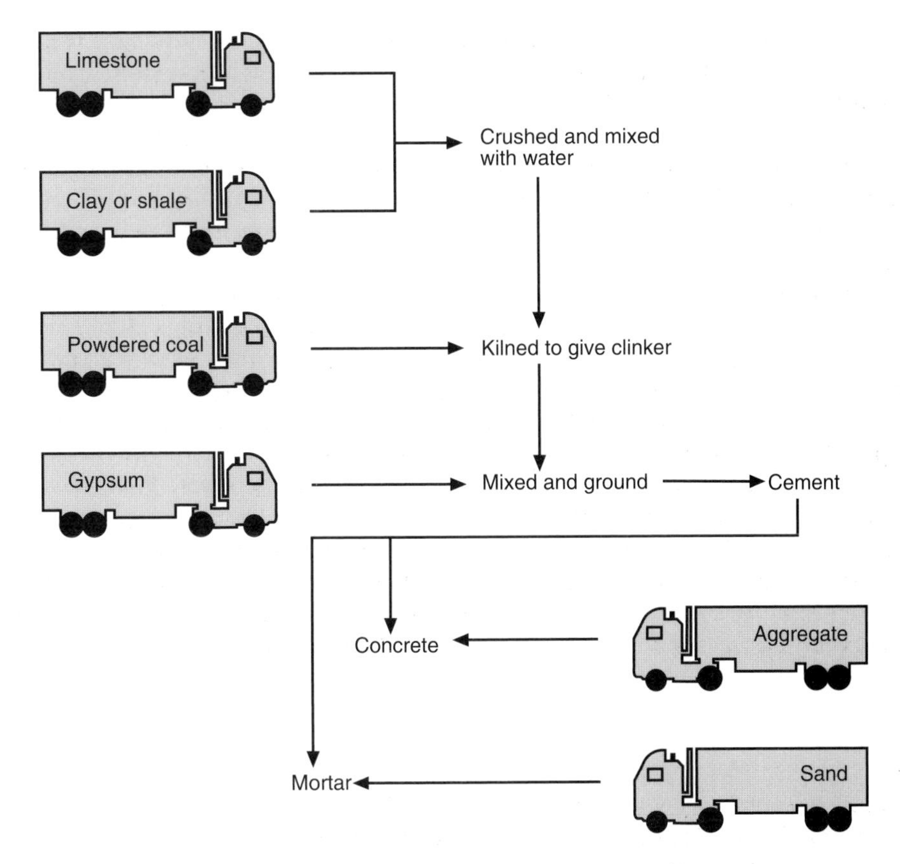

Figure 4.10 *Typical constituents of concrete and mortar*

resistant to weathering and provide the lowest shrinkage percentages. Despite this, strained quartz grains are known to have contributed to the rapid decay of some concrete structures due to reaction with alkalis in the cement. Aggregate shape and size are also difficult to standardise, but are among the most important factors in the success of the concrete. Natural aggregates are usually well rounded and provide a concrete mix which is easy to work when wet; angular fragments of crushed stone form aggregates which are less easy to work but pack well to produce a denser concrete mix. Both produce a concrete with a high degree of compressive strength. Crushed fissile and well cleaved rocks produce tabular aggregates which are unsatisfactory as they do not pack well in the concrete mix, thereby reducing its compressive strength.

4.3.3 Concrete Failures

Concrete failure is highly publicised as it ultimately leads to the destruction of costly buildings or large engineering structures. The majority of failures result from adverse aggregate–cement reactions, or through the decay of steel reinforcing rods. Reactions with the aggregate are caused by the action of alkali fluids within the concrete. These fluids are derived from reactions associated with the setting of the cement, most of which release sodium, potassium and hydroxyl ions into solution. These alkali solutions trigger reaction with various aggregate types, particularly where crushed rock aggregates have been used. These are damaging primarily because they promote physical expansions which can be very harmful.

The most important reactions are **alkali–silica**, **alkali–silicate** and **alkali–carbonate**. The reaction between alkali fluids and microcrystalline silica, such as opal or chert, present in the aggregate produces a silica gel which has the property of being highly expansive. The growth of this gel exerts a great deal of force which promotes internal failure of the concrete. The alkali–silicate reaction involves the expansion of platy silicate minerals on contact with the alkali solution, which again exerts internal pressure in the concrete. Finally, where dolomites contain a proportion of clay minerals, the alkali solution causes the process of dedolomitisation with subsequent microcracking. This allows further water penetration which in turn causes the contained clay minerals to swell. Mitigation of these reactions is only possible through the application of preventative measures, such as the exclusion of water from the concrete by the use of waterproof coating, and the mixing of materials intended to neutralise the alkalinity of the waters. The most appropriate measure is the use of non-reactive aggregates.

The use of steel reinforcing rods in concrete has allowed the development of extremely strong and complex engineered structures. However, reinforcing rods may be the cause of major concrete failures, particularly where percolating fluids cause the metals to rust and lead to the failure of large surface areas of the concrete structures.

4.4 STRUCTURAL CLAY

Bricks are primarily blocks of clay which have been extracted and fired after a degree of processing, and are used to create a building material (Figure 4.11). The process of brick and tile manufacture is extremely ancient, with examples from 1200 BC in Egypt, but it is the Romans who really perfected the art. Brickmaking became widespread, with mass production during the Industrial Revolution, which continues to the present day.

Brick clays are widespread and variable, and examples ranging in age from the Palaeozoic to the Quaternary are used. Brick clay is usually extracted from open-cast pits of varying size, from the small local pits often associated with small towns and villages in the last century, through to the extensive brick clay operations needed to service the construction industry. After extraction, the clay is screened for unwanted material and then formed into blocks, either by hand or by machines using set moulds, and from there the brick is created through firing in a kiln. The method of firing varies in detail, and it is this process,

Figure 4.11 *A typical Victorian brick-built building in East London. The bricks are Kentish Stocks from Quaternary loess (brickearths) in northern Kent*

together with the mineralogical content, which determines the nature and form of the brick, ensuring sufficient variation for architects and builders. There are five basic processes involved in the formation of a typical brick: (1) clay extraction; (2) clay processing; (3) brick forming; (4) brick drying; and (5) brick firing. In all of these stages the geology of the brick clay is of utmost importance, as discussed below.

Clay extraction is usually achieved through a variety of open-cut quarry methods, simply because of the widespread availability of brick clays of all ages at the surface in most brick-producing countries of the world, and because the low price of most bricks precludes extraction through costly mining methods. Clay strata are rarely uniform in composition and it is therefore important that the extraction process does not introduce irregularities in the clay through mixing, as this will directly influence unwanted variation in brick finish. This was less of a problem in the days when clay was extracted manually, but is now more problematical with mechanical methods of extraction. Processing of the clay is carried out after a period in which it has been stockpiled, usually exposed to the elements. This stage ensures the breakdown of unwanted components, such as pyrite, and the development of a uniform moisture content. It is important to remove or break down included material which can form a hazard during firing, such as the calcite-rich fossil belemnites of the English Oxford Clay which are prone to explode when heated. Processing itself is by grinding with steel rollers; this is intended to continue the process of homogenisation which was commenced during the stockpile phase, but complete mixing is rarely achieved. Additives at this stage may include: extra water, if necessary; **grog**, which is inert, usually ground rock fragments or sand intended to open the structure of the clay; and mineral colourants. Next, the brick is shaped, either by hand, which is rare today, or by machine. Hand-made bricks are usually formed by pressing a ball of processed clay into a sanded mould; machine-formed bricks can be dry-pressed or squeezed into steel moulds dependent on the water content of the clay, or extruded through a die as a rectangular column which is then cut to the required length. Drying of the shaped bricks enables them to be handled easily before firing. This was previously achieved through leaving them to dry in covered rows – clearly a time-consuming business – but today most formed bricks are dried in heated chambers. Finally, the moulded bricks are fired. The traditional method involves stacking the bricks to form a **clamp**, with air spaces between the bricks, and adding fine coal as fuel. The coal is lit at one end of the clamp and allowed to burn through the structure to fire the bricks. **Chambered kilns** are more complex: a series of brick-built chambers with interconnecting flues allow the fuel to be added from the top, and the bricks are fired in succession. Finally, **tunnel kilns** are formed from long refractory-brick tunnels through which carriages carrying the bricks pass, the centre of the tunnel being the firing area. In all the firing methods, variation in the appearance of the bricks may be achieved through the spacing of the brick stacks, and through the variation of the air supply.

The mineralogical constituents of brick clay control the success of the clay in producing bricks and enable the production of the wide variety of commercially available products. Brick clays, in common with most mudrocks, are usually a mixture of four clay minerals: **kaolinite**, **illite**, **smectite** and **chlorite**. These minerals ensure the plasticity of the clay in order to form brick, and to hold this form during the drying process; they also provide the fusibility during firing to create the brick itself. **Quartz** grains are a common component of brick clays, and this mineral assists in the separation of the brick from its mould and in providing an open texture to the clay which assists in the drying process. Quartz can undergo transformations to different mineral phases during firing which can cause some problems in the brick, such as microcracking, but in the main, quartz assists in making the brick strong and durable. A variety of iron minerals are present in most clays, and these provide the colour of the brick. Commonly, all of these convert to the single mineral **haematite** under the oxidising conditions of the firing process. This mineral is characteristically red-brown in colour and becomes successively darker at temperatures over 1000°C, and redder bricks can therefore be produced by increasing the temperature of the firing. Reducing conditions can be created by restricting the free flow of oxygen in the kiln through setting the bricks close together. This means that the iron combines with the silicates in the clay, which become liquid at high temperatures and form a blue skin on the surface of the brick on cooling. Some brick clays produce pale bricks through a complex process in which the iron is taken into the lattice of the different phases of kaolinite formed at temperatures in excess of 1000°C; this reduces the amount of haematite in the brick and prevents the characteristic reddening of the brick. Similar colours may be achieved by adding **calcite** to the clay, as on firing the iron minerals combine to form iron carbonates with a much lighter colour.

4.5 GYPSUM

Gypsum provides the main raw material from which domestic plaster is derived. Plaster owes its name to the term plaster of Paris, so-called because of the long history of mining of Tertiary gypsum deposits beneath the city of Paris. Gypsum is derived mostly from marine deposits, and important sources include the famous deposits of Miocene (Messinian) age which are associated with the development of a landlocked Mediterranean and its subsequent hypersalinity. In addition, large quantities of gypsum are derived as a by-product of the flue-gas desulphurisation process in coal-burning power stations.

Gypsum ($CaSO_4 \cdot 2H_2O$) is calcined or heated at between 150 and 165°C in order to produce the plaster. This simple process releases water from the compound to produce hemihydrate or bassanite ($CaSO_4 \cdot 0.5H_2O$). When mixed with water, bassanite rehydrates to form a network of gypsum crystals which constitute the strong but soft material we know as plaster. This may be mixed with sand, other aggregates or drying retardants in order to increase the versatility of the material.

4.6 GLASS SAND

Glass is manufactured through a process of melting quartz-based sands and some other vital mineral ingredients and cooling them in such a way that they are prevented from crystallising. Suitable sand for the manufacture of glass is less common than the sand deposits commonly used for aggregates. This is because the presence of unwanted impurities in the sand would lead to the manufacture of imperfect glass. For example, iron and other metals would produce unwanted colour in the glass, while minerals such as corundum or zircon would remain unmelted during manufacturing and would therefore create imperfections in the glass. The best and most extensive glass sands are found in the USA, Holland and Argentina, and include Quaternary sands which have undergone several cycles of reworking which helps to reduce impurities. However, processing is still necessary and includes crushing and grinding to promote even melting, and, where necessary, gravity separation and flotation to remove the mineral impurities.

Glass production relies on mixing of quartz glass sand with appropriate amounts of other minerals. Typical minerals include: sodium carbonate and boron, mostly derived from lacustrine evaporite deposits – boron is used as a flux in glass production; feldspar, derived from igneous rocks and also used as a flux; lithium, derived from the mineral spodumene, which is used to create ceramics that are capable of resisting temperature change without expansion or contraction; and strontium, derived from celestite, which is used in colour television tubes to help reduce damaging radiation.

4.7 SUMMARY OF KEY POINTS

- There are six main types of geomaterial: construction stone, aggregates, structural clay, cement/concrete, gypsum and glass sand.
- There are four main types of construction stone: dimension (architectural) stone, decorative stone, armourstone and rockfill.
- In choosing dimension stone the following properties need to be considered: structural strength, durability, appearance, ease of working and availability.
- Aggregates are the most important of all geomaterials and may be either derived from natural sand and gravel deposits or processed from crushed rock. When combined with cement to produce concrete, aggregates form the commonest building material.
- Clay, when fired, produces bricks which are among the commonest of all geomaterials because of their low unit cost.
- Gypsum, calcined to produce plaster, and high purity sand, used in glass manufacture, are important geomaterials worked across the world.

4.8 SUGGESTED READING

Overviews of geomaterials are provided in a number of standard reference works; particularly important are the readable accounts by Prentice (1990), Kesler (1994) and Manning (1995). The collection of papers edited by Hawkins (1991) also provides a lot of useful information, particularly in defining the scope of the term geomaterials. An article in *Geology Today* (Anon. 1990) provides a good introduction to building stones, while more detailed information can be obtained from Ashurst & Dimes (1977), Jefferson (1993) and Miglio (1996). The conservation and restoration of building stone is covered in the volume edited by Ashurst & Dimes (1990). Dibb *et al.* (1983) discuss armourstones as do Fookes & Poole (1981). The geology and application of aggregates are covered in detail by Lees & Kennedy (1975) and Smith & Collis (1993), while Hartley (1974) reviews aggregates from the perspective of their use as road surfaces.

Anon. 1990. Building stones. *Geology Today*, **6**, i–iv.

Ashurst, J. & Dimes, F.G. 1977. *Stone in building, its use and potential today*. The Architectural Press, London.

Ashurst, J. & Dimes, F.G (Eds) 1990. *Conservation of building and decorative stones*, two volumes. Butterworth-Heinemann, London.

Dibb, T.E., Hughes, D.W. & Poole, A.B. 1983. Controls of size and shape of natural armourstone. *Quarterly Journal of Engineering Geology*, **16**, 31–42.

Fookes, P.G. & Poole, A.B. 1981. Some preliminary considerations on the selection and durability of rock and concrete materials for breakwaters and coastal protection works. *Quarterly Journal of Engineering Geology*, **14**, 97–128.

Hartley, A. 1974. A review of the geological factors influencing the mechanical properties of road surface aggregates. *Quarterly Journal of Engineering Geology*, **7**, 69–100.

Hawkins, A.B. (Ed.) 1991. Papers presented at the conference 'Geological materials in construction'. *Quarterly Journal of Engineering Geology*, **24**, 1–168.

Jefferson, D.P. 1993. Building stone: the geological dimension. *Quarterly Journal of Engineering Geology*, **26**, 305–319.

Kesler, S.E. 1994. *Mineral resources, economics and the environment*. Macmillan, New York.

Lees, G. & Kennedy, C.K. 1975. Quality, shape and degradation of aggregates. *Quarterly Journal of Engineering Geology*, **8**, 193–209.

Manning, D.A.C. 1995. *Introduction to industrial minerals*. Chapman & Hall, London.

Miglio, B.F. 1996. The integrity of building materials. In: McCall, G.J.H., De Mulder, E.F.J. & Marker, B.R. (Eds) *Urban geoscience*. A.A. Balkema, Rotterdam, 215–234.

Prentice, J.E. 1990. *Geology of construction materials*. Chapman & Hall, London.

Smith, M.R. & Collis, L. (Eds) 1993. *Aggregates: sand, gravel and crushed rock aggregates for construction purposes*, second edition. Geological Society Engineering Geology Special Publication 9, London.

5
Water Resources

Water is probably the most fundamental resource in sustaining life, and yet most of the Earth's water is unsuitable for drinking or agricultural uses. This is because 98% of the planet's water is collected in the large ocean basins and is saline. Of the remaining 2%, only 0.4% is readily available for drinking, since 1.6% is locked into the Earth's ice sheets. Fresh water supplies are also unevenly distributed across the globe, with, for example, a country such as Iceland having around 558 times more water per head of population than Belgium, or 20 000 times more water per head than Egypt. It is not surprising, therefore, that the provision of fresh drinking water is a major preoccupation of many third world and developing countries, and that water supply has become a major factor in territorial disputes.

Delivery of water to the point of demand is as important as resource acquisition (Box 5.1). A further component of water resource management is the treatment of waste water, particularly since water is increasingly being recycled through the human use system several times. As early as 1965 it was estimated that in the USA at least 100 million people were using water that had been used at least once before. Water resource management, in the context of human use, therefore involves three components: (1) resource acquisition; (2) redistribution; and (3) water treatment and disposal. Issues of resource acquisition and redistribution are dealt with in this chapter while problems of water treatment are covered in Chapter 9.

5.1 SURFACE AND GROUNDWATER HYDROLOGY

A conceptual model of surface water hydrology is shown in Figure 5.1. On the surface the basic unit is the drainage basin, an area of land confined by **watersheds** over which surface water cannot pass (Figure 5.1A). Input into the drainage basin comes from two sources: from meteorology in the form of rain,

BOX 5.1: WATER SUPPLY IN BRITAIN: WHY IS THERE A WATER SHORTAGE?

For some areas within the British Isles the 1990s have been a period of water restrictions and threatened shortages, yet Britain is traditionally perceived as a very damp country. There are two components to this problem: (1) most of the rainfall falls in northern and western parts of Britain; and (2) the greatest demand is in the southeast where the majority live. In the north and west of Britain annual demand is less than half the effective drought rainfall, but in East Anglia demand reaches over 60% and in the Thames region demand may exceed supply. In recent years, however, it is not just the southeast which has experienced problems, as water restrictions have been introduced in both northwest England and, most notably, in Yorkshire. One of the principal causes has been a succession of relatively dry winters which have failed to replenish water stocks. To restore this deficit would require a sustained period of higher than average annual rainfall. Many of the reservoirs built early in the century were not designed to anticipate such prolonged periods of water deficit in conjunction with the high water demands of modern society. The wastage of water through leakage and inefficient transfers is also of increasing importance in reducing stock levels. There are two types of leakage loss in Britain. Firstly, company/distribution losses, which occur in Britain: (1) at service reservoirs; (2) on the 34 000 km of trunk mains; (3) on the 240 000 km of distribution mains; or (4) on the 20 million underground communication pipes which link individual buildings to the mains system. Secondly, customer losses which occur on the 21 million underground supply pipes within individual properties. At present water loss through leakage is well in excess of 12%, and some estimates place it as high as 30%. Normal domestic consumption is around 1 m^3 per person per week; in comparison a dripping tap will lose 0.1 to 1 m^3 per week, while a typical burst on a service pipe will lose between 200 and 250 m^3 per week and a burst on a 100 mm pipe will lose 500 to 3000 m^3 per week. There is on average in England and Wales one mains burst each year for every 10 km of mains.

Demand will outstrip supply in the southeast, southwest and east coast early in the next century unless: (1) new sources are brought on line; (2) water use is reduced, for example consumption could be reduced by over 10% if the amount of water flushed down the WC was reduced to that which is actually necessary to do the job; (3) leakage loss is reduced dramatically; or (4) investment is made to transfer water from wetter regions. This example illustrates the problem not only of acquiring water resources but of supplying them to the area of maximum demand.

Source: Environment Select Committee, 1996. *Water conservation and supply.* HMSO, London.

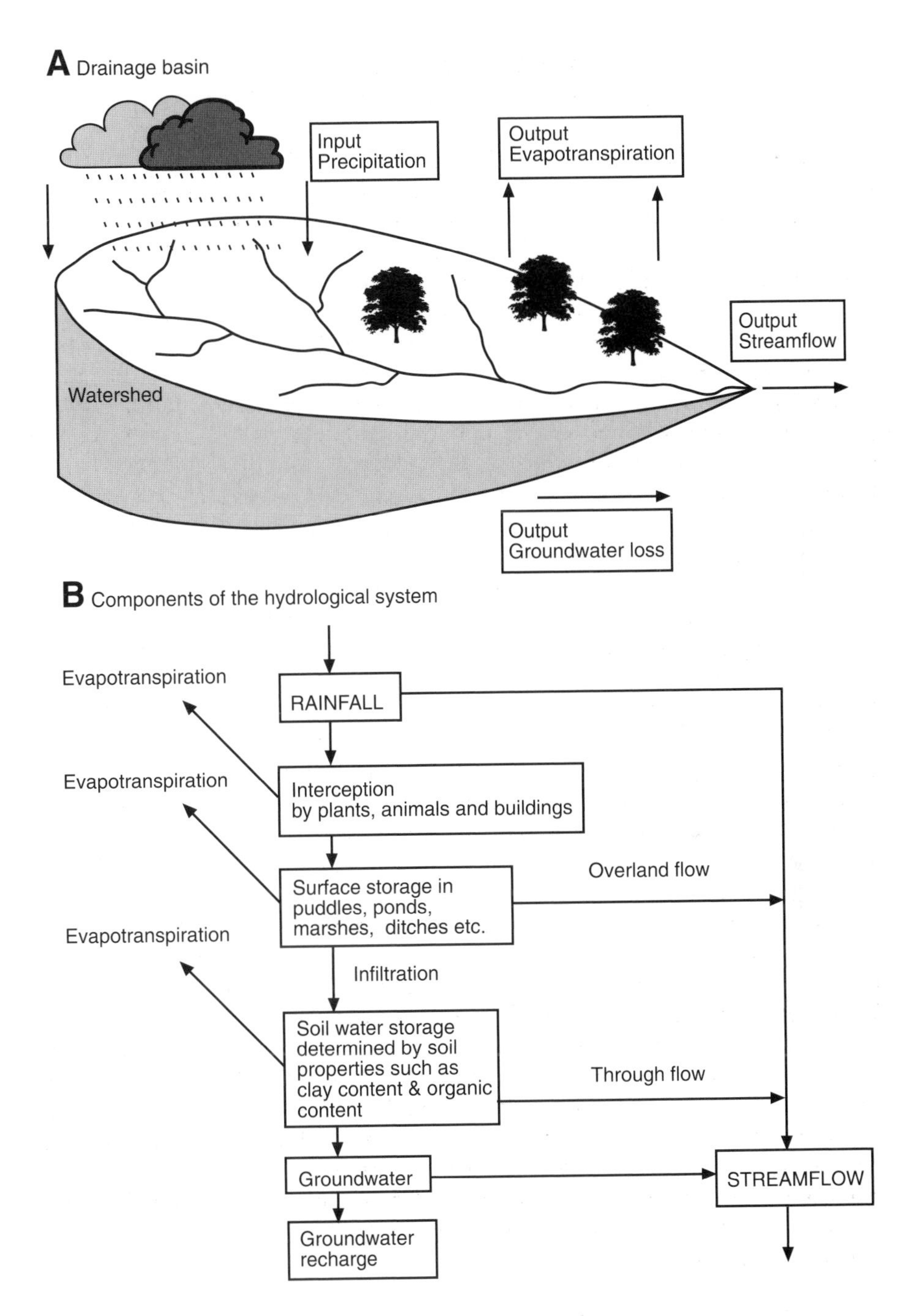

Figure 5.1 *Elements of the hydrological system.* ***A.*** *The drainage basin.* ***B.*** *The components of the hydrological system*

snow and hail, and from groundwater. Outputs from this drainage basin consist of stream or river flow and groundwater flow. The volume of precipitation and the rate at which it is transferred into a stream or river depend on such variables as: (1) interception and evapotranspiration by vegetation; (2) the magnitude of surface runoff; (3) infiltration and throughflow within surface soils and rocks; and (4) rate of water loss to groundwater. These variables are determined by the vegetation cover, soil characteristics, drainage system, geology, and antecedent weather conditions (Figure 5.1B).

Groundwater conditions are determined by the porosity and permeability of the rocks present within an area. Porosity is a measure of space and volume within a rock body, and in terms of groundwater can be considered as the amount of water a body of soil or rock can hold within itself. Two types of porosity can be identified: (1) **primary porosity**, which is a function of the pores or gaps between grains in sedimentary rocks and is therefore usually absent in crystalline rocks; and (2) **secondary porosity**, which is a function of fractures, joints or solution cavities within a rock and is therefore potentially present in all types of rock. In contrast, **permeability** expresses the ease with which water or fluid can pass through a rock and is a function of the connectivity of pores, joints, fractures and other water-bearing voids within a rock. It is important to emphasise that some rocks may have a high porosity but low permeability, such as clay.

Water entering the ground infiltrates through soil and rock under gravity, first through an unsaturated zone (**vadose zone**) until it reaches a depth at which all the voids are water-filled (**phreatic zone**). The upper level of saturation is known as the groundwater table; here the water flows under the influence of a hydraulic head determined by the shape of the groundwater table or **piezometric surface** (Figure 5.2A&B). Groundwater flow is always directed perpendicular to lines of equal hydraulic potential known as equipotential surfaces. The surface of the groundwater table, in uniformly permeable rock, follows in a subdued fashion the surface topography. In such uniform conditions, the pattern of equipotential surfaces and therefore groundwater flow lines are as shown in Figure 5.2B. In practice, however, the condition of uniform permeability is rarely met, and the pattern of groundwater flow may be much more complex and need not reflect surface water flow.

Groundwater is in constant motion, flowing from areas of high hydraulic potential to areas of low potential, and without recharge the surface of the groundwater table will therefore become more subdued over time (Figure 5.2C). The rate at which this occurs and the steepness of the water table surface will depend on the rate of groundwater flow. The rate of water flow within a permeable horizon is determined by **Darcy's law**, which at its simplest states that the rate of flow is determined by the piezometric or hydraulic gradient and

---

Figure 5.2 *(opposite) The mechanics of groundwater.* **A.** *Hydraulic head and groundwater flow; flow is always from areas of high hydraulic potential to areas of low potential.* ***B.*** *Groundwater flow in relation to equipotential surfaces.* **C.** *Fluctuations in the water table due to natural recharge around a chalk well in eastern England.* **D.** *Artesian aquifers [C Modified from:. Petts & Foster (1985)* Rivers and landscape. *Arnold, Fig. 2.5, p. 29]*

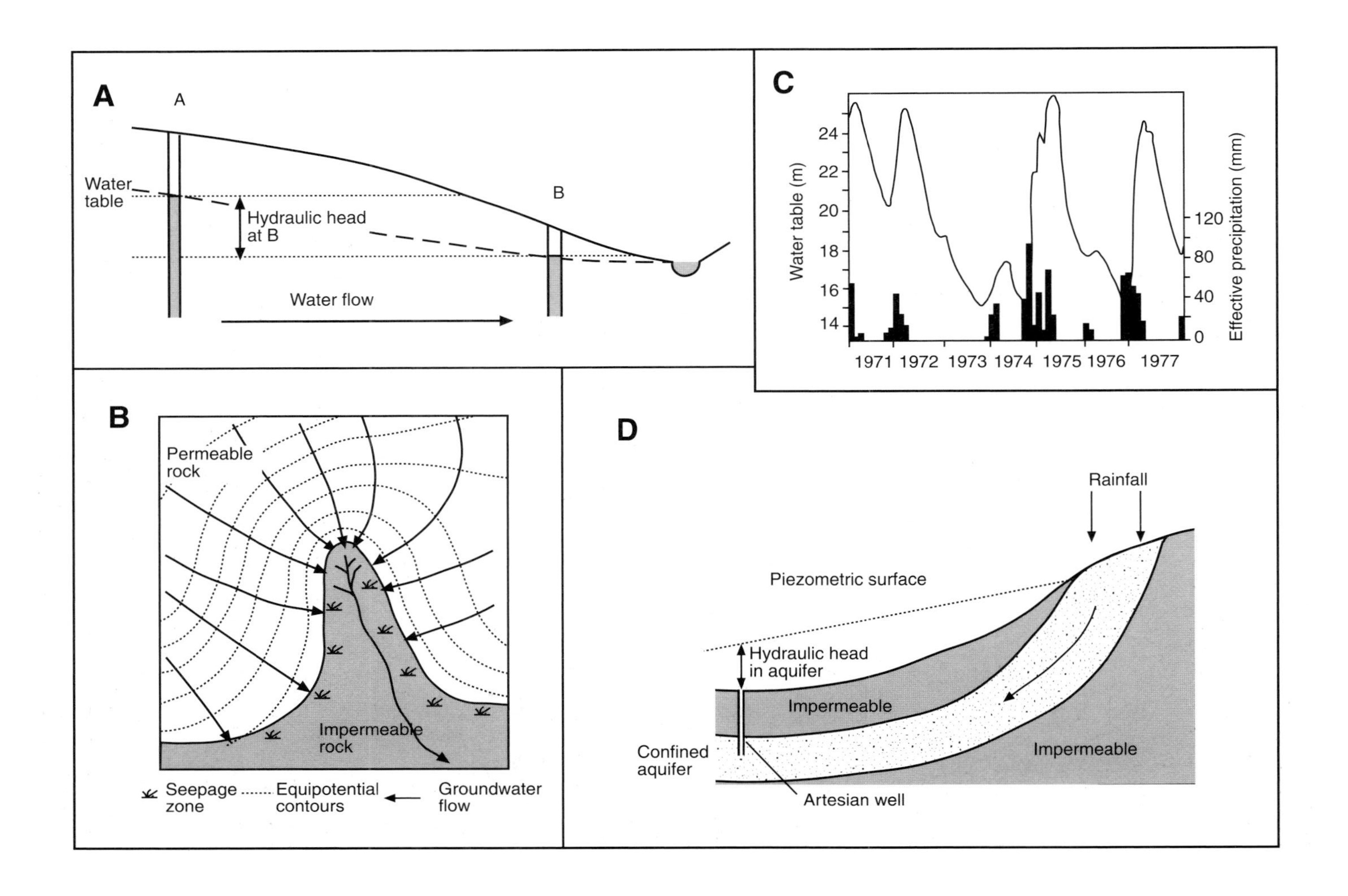
A
A
B
Water table
Hydraulic head at B
Water flow
B
Permeable rock
Impermeable rock
Seepage zone
Equipotential contours
Groundwater flow
C
Water table (m)
24
22
20
18
16
14
Effective precipitation (mm)
120
80
40
0
1971 1972 1973 1974 1975 1976 1977
D
Rainfall
Piezometric surface
Hydraulic head in aquifer
Impermeable
Impermeable
Confined aquifer
Artesian well

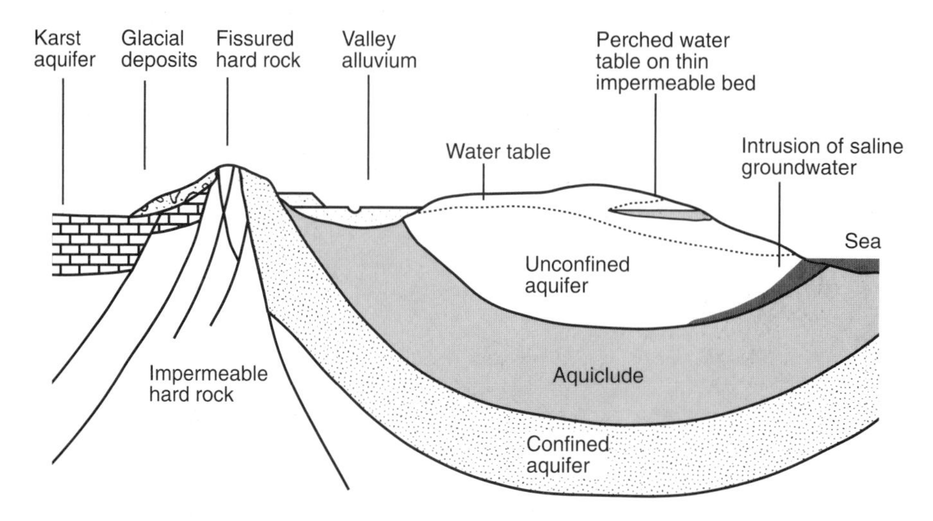

Figure 5.3 *Schematic diagram of some of the principal hydrological units or aquifers*

the intrinsic permeability of the sediment or rock involved. The greater the hydraulic gradient or the greater the permeability of the rock or sediment concerned, the faster the rate of water flow. This empirical relationship is a practical tool with which groundwater flow rates can be established and is therefore of considerable importance in a variety of environmental problems. It is important to note, however, that it is limited by a number of assumptions, in particular the uniform existence of laminar flow, which may not always hold, especially around pumped wells and in limestone terrain.

Permeable and water-bearing rocks are known as **aquifers**, while impermeable horizons are referred to as **aquicludes**. Groundwater movement is restricted to aquifers and groundwater basins (hydrological units), and is confined and defined by aquicludes. In certain situations the juxtaposition of aquifers and aquicludes may result in pressurised groundwater, and in certain situations groundwater may rise to the surface under its own pressure when released via wells known as **artesian wells** (Figure 5.2D).

As shown schematically in Figure 5.3, there are various types of hydrological unit or aquifer. Common aquifers include: (1) alluvial deposits, usually restricted to narrow belts along watercourses; (2) glacial deposits, such as moraines, eskers and kames (Box 5.2); (3) fissured hard rock, particularly fracture zones; (4) confined or artesian groundwater, in which an aquifer is sandwiched between two aquicludes; (5) karst (limestone) aquifers; (6) sedimentary basins or permeable and porous rock; and (7) thermal and mineral water rising along faults from either active or inactive tectonic or volcanic zones; such water is usually of considerable economic value owing to its perceived medicinal benefits. Defining and assessing different hydrological units is one of the principal tasks of the hydrogeologist and is essential in determining the groundwater resource present within a region.

BOX 5.2: THE UPPSALA ESKER AQUIFER: SUSTAINABLE GROUNDWATER MANAGEMENT

The city of Uppsala has a population of 160 000 and lies 65 km NNW of Stockholm in Sweden. The municipal water supply is derived from groundwater, both natural and, more recently, artificial. The aquifer exploited is located in an esker on which Uppsala is partly built. An esker is a sinuous

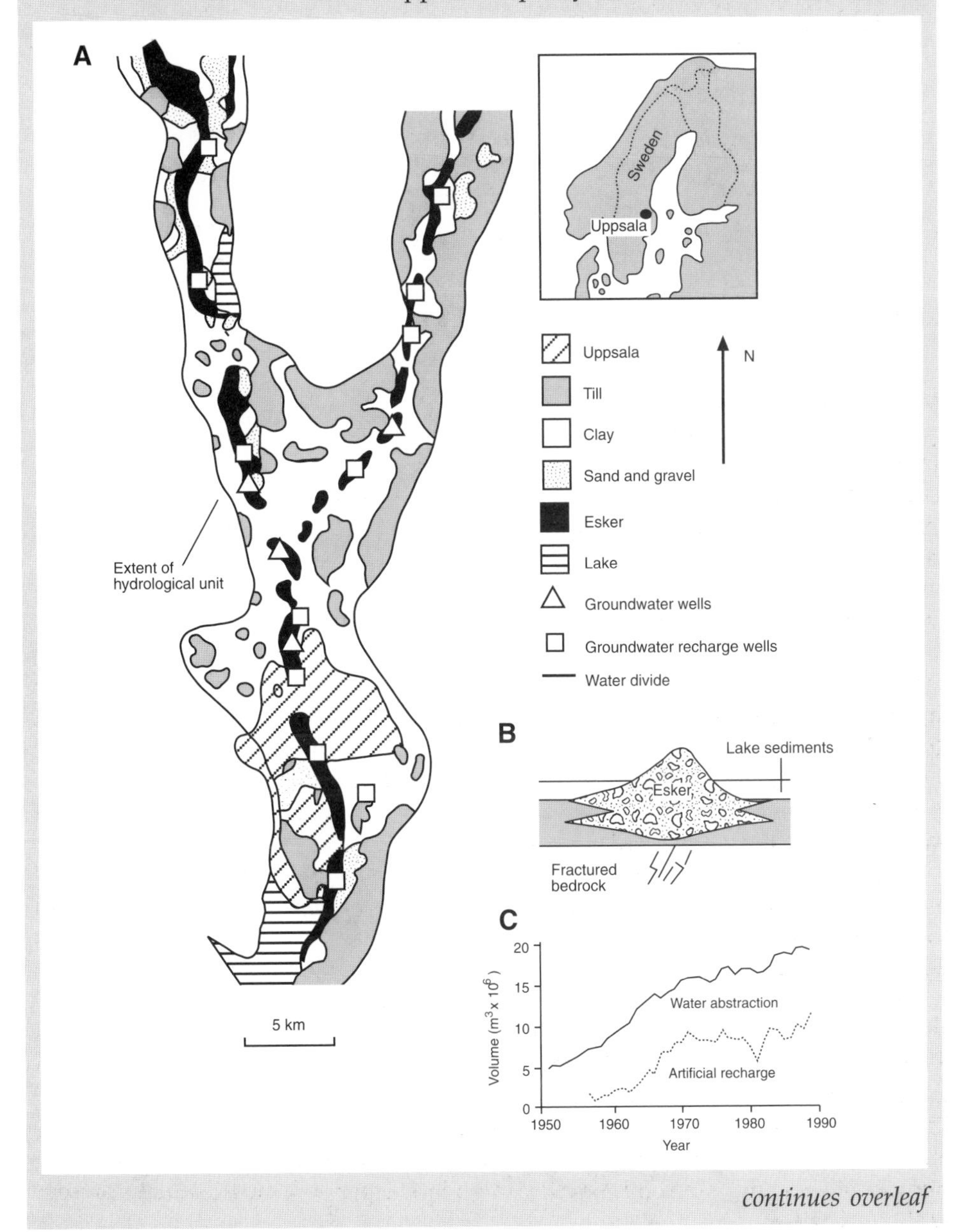

continues overleaf

BOX 5.2: *(continued)*

ridge of sand and gravel deposited beneath a glacier by meltwater. Geologically the area consists of Archean basement which has been glacially eroded and covered by glacial till. During deglaciation at the close of the last glacial cycle, meltwater feeding a glacial lake deposited a large esker. The sand and gravel of the esker is underlain by till and partially covered by glacial lake clays. In places the continuous ridge of sand gravel is over 150 m thick. High points on this ridge stand out above the glacial clays as shown in the map and section below. Uppsala was originally located on the esker owing to its defensible position and the presence of springs emerging from the flanks of the esker. The esker forms a discrete hydrological unit (Diagram A&B). The first pipeline system to tap these springs was constructed in the 1640s to supply the royal castle located on the esker crest and the modern supply system was initiated in 1872. Today the water demand is over 60 000 m^3 of water per day. Water quality is excellent since the sands and gravels naturally filter the water and purification is not required. In the 1950s calculation began to show that the aquifer was in danger of becoming severely overdrawn. To combat this problem the groundwater was recharged using water abstracted from the adjacent River Fyris and pumped into wells (Diagram C). The time lapse for the artificial recharged groundwater to reach the pumping wells was six to eight months, sufficient time for the aquifer to have filtered and purified the water. By recharging the aquifer the groundwater resource has become a sustainable resource. Recent developments have focused on protection of the well areas from commercial activities, such as the use of chemicals and sewage treatment work, which could pollute the aquifer. This provides an excellent example of the effective and suitable management of a groundwater resource and also illustrates how an aquifer can be used to store, transfer and distribute water resources.

Source: Kelk, B. 1994. Natural resources in the geological environment. In: Lumsden, G.I. (Ed.) *Geology and the environment in Western Europe*. Oxford University Press, Oxford, 34–138 [Diagrams modified from: Kelk (1994) In: Lumsden (Ed.) *Geology and the environment in Western Europe*, Oxford University Press, Figs 3.4 & 3.5, pp. 41 & 42]

5.2 WATER SUPPLY

Water is used not only for domestic consumption, but for agriculture, industry, power generation and waste disposal. In Britain about 12% is lost in leaks, 11% is used for industry, 36% for generating electricity, 1% in agriculture, while domestic supply accounts for 40% – 32% to flush toilets and 7% for baths and showers. These proportions may vary dramatically globally, particularly with respect to the climate and socio-economic character of a country. For example,

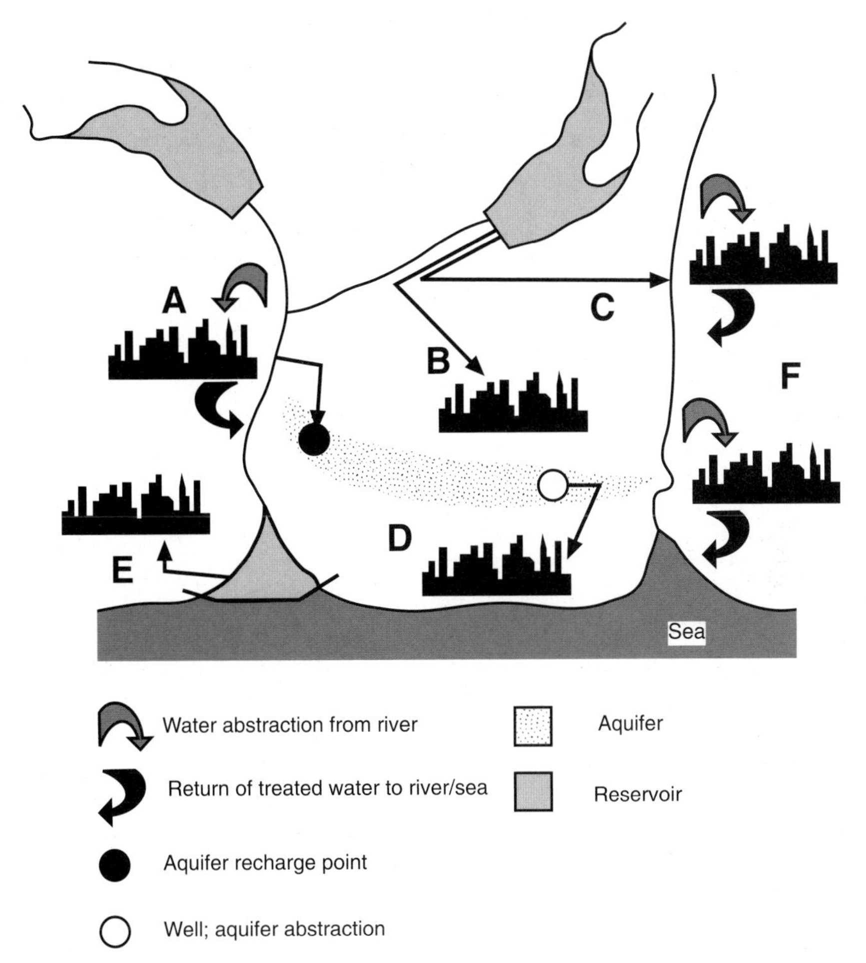

Figure 5.4 *Schematic diagram of the various methods of water transfer. Natural rivers may be used for water transfer (**A**); alternatively it may be pumped along pipes to urban areas (**B**) or along a combination of pipes and natural rivers (**C**). Water may also be used to recharge aquifers used for water abstraction (**D**); in this case the aquifer is used as a mechanism of water transfer. The mouth of estuaries can be dammed to create reservoirs (**E**), a strategy which has been used in Hong Kong (Figure 13.7). Water may be abstracted, used, cleaned and returned to a river several times along its length (**F**); multiple use of water is now commonplace*

over 90% of water is used for agriculture in many African countries and consequently failure of supply has major social and economic implications.

Figure 5.4 shows a schematic model of the available sources and methods of water transfer. Sources include: (1) artificial reservoirs; (2) river abstraction; (3) natural and artificial groundwater abstraction; and (4) desalination. More exotic sources, such as icebergs towed from polar regions, have also been considered seriously as a source of fresh water in arid regions (Box 5.3). Each of these sources is reviewed below.

BOX 5.3: WATER FROM ICEBERGS: SCIENCE FICTION OR REALITY?

The idea of towing large icebergs from polar regions to provide fresh water in arid areas is an attractive concept. As long ago as 1890 small icebergs were captured and either towed or sailed from Laguna San Rafael, Chile (45°S) to Valparaiso and even to Callao, in Peru (12°S) to provide a source of ice at a distance of over 3900 km. In Japan, glacier ice is sought after for medicinal purposes and as an expensive luxury for drinks.

However, until the early 1970s the idea of transporting large icebergs to provide fresh water did not receive serious appraisal. Since then the problem has periodically been the subject of intense research, although no attempt has yet been made to put the developing ideas and theories to the test. In considering this problem there are five main issues:

Problem one: identifying a supply of suitable icebergs. They need to be big enough to survive melting en route and should be stable (i.e. not likely to capsize). The most suitable bergs are the large tabular bergs derived from the ice shelves of Antarctica. In particular the Amery and Ross ice shelves produce almost any size of berg which could be desired.

Problem two: towing of the bergs would require a super-tug far bigger than any available today, although the technology exists to build one. The drag from a berg increases with towing velocity and therefore large bergs can only be moved slowly. Recent work has focused on reducing this drag by pre-shaping the berg. Although suitable super-tugs could be built, attaching the tow rope to the berg is a significant problem, since ice melts when placed under pressure. Wire hawser would simply melt through the berg under the applied strain.

Problem three: melting in transit will occur as the berg is moved into warm water. The degree of loss depends on the transit time which is a function of towing velocity and of ocean currents. The greater the transit loss the greater the size of berg required to start with in order to result in an economic delivery of ice. Certain routes take advantage of ocean currents and therefore transit times are reduced. For example, ice could be delivered to western Australia or to the Atacama desert relatively easily.

Problem four: processing of water on arrival. The berg will ground offshore and continue to melt and some form of processing plant will be required to melt the ice and transfer the fresh water onshore without it being unduly contaminated by sea water. These problems have yet to be examined in detail.

Problem five: the economic feasibility of the idea relative to the cost of other water sources. The problem here is that the cost of a super-tug and of the processing plant required are unknown and therefore the level of capital investment required is uncertain. However, provisional estimates suggest that water could be delivered more cheaply than if produced by desalination, or by the importation of liquid water.

Although the idea sounds like science fiction, the technology exists to exploit this water source, should it prove economically viable in the future. At present, many of the most populated arid regions are located in oil-rich countries, and therefore cheap energy exists for desalination. However, as the value of these energy reserves increases, attention may well turn to alternative sources of water.

Sources: Weeks, W.F. & Campbell, W.J. 1973. Icebergs as a fresh-water source: an appraisal. *Journal of Glaciology*, **12**, 207–233. Weeks, W.F. 1980. Iceberg water: an assessment. *Annals of Glaciology*, **1**, 5–10.

5.2.1 Artificial Reservoirs and Lakes

Reservoirs are constructed for a wide range of uses including water supply, power generation, irrigation and flood control, and vary from local structures to store treated water, to large reservoirs on major continental rivers. The aim is to buffer the natural variation either in supply from a river or catchment or in demand, by storing surplus water. Today, dams are being completed at a rate of two per day and it has been suggested that by the year 2000, 66% of the world's stream flow will be controlled by dams. Reservoirs may be constructed in a variety of ways including: (1) dam construction across a valley; (2) construction of banks or bunds completely enclosing an area of flat-lying terrain; (3) excavation of depressions and underground caverns; and (4) enclosing estuary mouths with dams. The most efficient and cost-effective method is the construction of dams across constricted valley sections, although bunded reservoirs and excavation may be required to facilitate water supply to low-lying regions.

In designing and planning reservoirs, the following factors need to be considered: (1) water availability in relation to short-term and long-term demand; (2) reservoir location and geometry; (3) hydrogeology of the reservoir floor and walls; (4) engineering geology of the dam site; (5) reservoir lifetime; (6) hazards to the integrity of the reservoir; (7) capital cost and operational costs; and (8) environmental impacts. Engineers constructing reservoirs for the purposes of power generation also need to consider the water head in their design (Box 5.4). We will consider each of these principal factors in turn.

1. **Water supply and demand**. It is vital to have a clear understanding of the water demand – both its seasonal variation and projected long-term variation – for which the reservoir is being constructed. Equally, it is important to understand the water supply and its variation with climate. Reservoirs are primarily filled by natural runoff and the volume of runoff is controlled by the catchment area and the effective rainfall it receives. Clearly, the input must exceed the output from the reservoir plus the losses due to evaporation from its surface and leakage. Evaporation loss may be considerable in certain climates; for example, River Dnieper in the Ukraine experiences a

BOX 5.4: DAMS CONSTRUCTED FOR POWER GENERATION

Dams and reservoirs are not just constructed for water storage and flood control, but are also important in generating power. Hydroelectricity is used widely in maritime mountainous regions such as Norway. The downhill flow of water is used to drive turbines which generate electricity. In Britain there are two main types of hydroelectric scheme: (1) pump storage schemes, in which excess electricity available at night is used to pump water to a high-level reservoir which is then released to generate electricity at short notice to meet a surge in demand, for example when a large proportion of the population put the kettle on simultaneously during the adverts in a popular TV programme; and (2) continuous power generation where water supply exceeds output through the turbines. Only in parts of Scotland is there

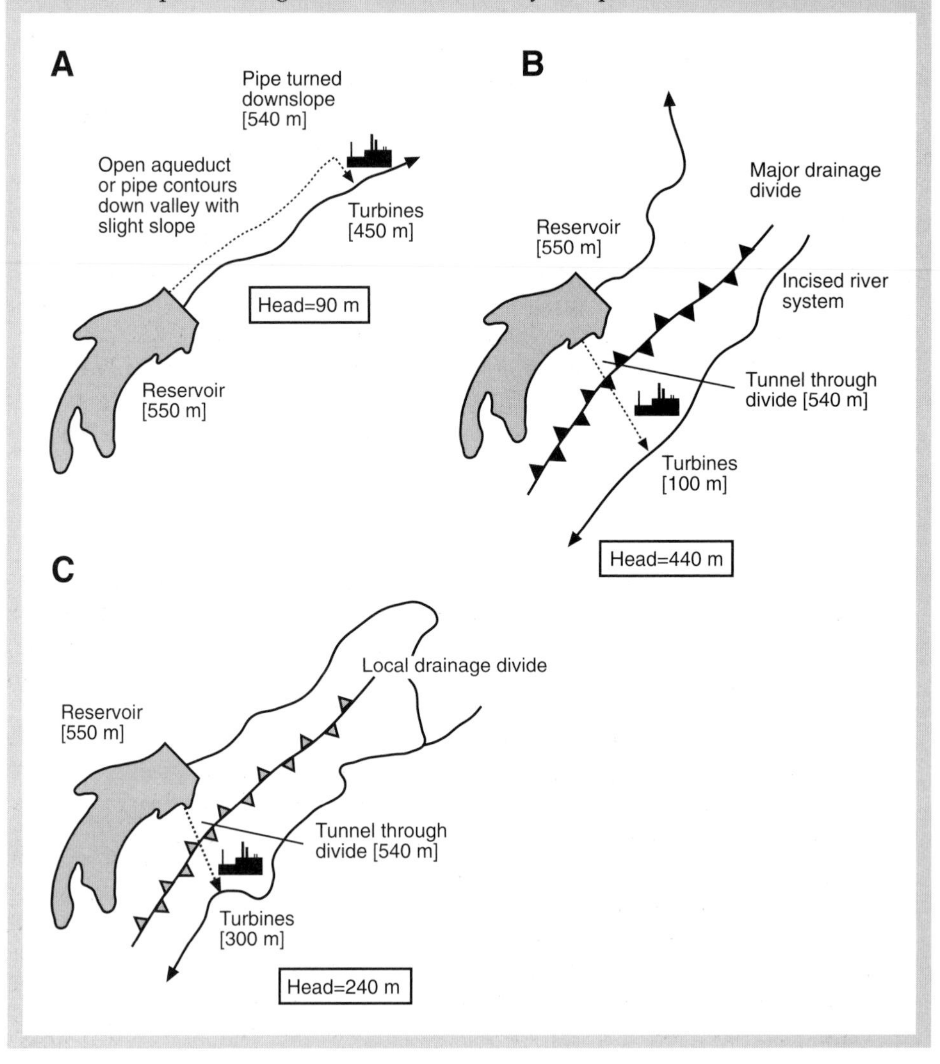

enough rainfall to make continuous hydroelectric power generation feasible in Britain. Pump storage schemes do not require particularly high rainfall and several important schemes are located in Wales.

In locating dams and reservoirs for hydroelectric schemes, many of the same variables and problems apply as for water supply reservoirs; however, in addition the problem of water head or fall distance must be considered. The output from a hydroelectric power station is a function of: (1) the head difference across the turbines; and (2) the volume of water available in unit time. The head is the difference in elevation of the water surface in the reservoir to that in the tail race below the turbines. This is simply controlled by the height of the dam relative to the power house. The volume of water available in a unit time is controlled by the diameter of the pipes supplying the water and ultimately the size of the reservoir and the rate at which it is recharged.

There are four different solutions to the problem of maximising the head of water above the turbines. The first solution is to locate the turbines at the base of the dam wall or adjacent to it. This limits the head of water to the height of the dam and is a design which is normally restricted to deeply incised valleys with very tall dams. Low head schemes can compensate for this if there are very large volumes of water available. The second solution involves transporting the water in pipes or open leats down-valley at a gradient which is less than that of the valley floor, as in this way water head is gradually increased (Diagram A). The third solution involves locating the power house in an adjacent, but much more deeply incised valley (Diagram B). The final solution involves transferring water down-valley more directly than the natural river, perhaps by piping water through spurs or intervening interfluves (Diagram C). These last two solutions involve increasing expense associated with tunnelling.

loss of 20% due to evaporation from reservoirs. Assessment of this must be made from the examination of long-term hydrological records in order to record all scales of variation in flow likely to be experienced during the lifetime of the reservoir. However, records are not always available, making such assessments difficult. In general, the greater our understanding of runoff generation and its variation, the more reliable the assessments of reservoir water budget are likely to be. Where shortfalls in water supply are identified, water may be diverted from other catchments. This may be done at a variety of scales including: the construction of hillside leats (lined ditches); the use of intersecting streams in adjacent catchments; the transportation of water from adjacent lakes or reservoirs; and tunnelling through drainage divides and the pumping of water from adjacent rivers and lakes. Pumping of water into reservoirs is expensive, but may be offset by the advantage of constructing reservoirs in ideal topographic, geological or land-use locations, thereby minimising capital costs and environmental damage. This is true of Grafham Water and Rutland Water in Britain, which receive most of their water supply by pumping; the sites were chosen due to the suitability of their location as opposed to the hydrological suitability of their catchments.

Comparison of the input hydrograph to a reservoir with demand is critical in determining the size of reservoir required; that is, the amount of storage capacity needed. This can be illustrated with reference to a series of hypothetical examples. Figure 5.5A shows the relationship between annual cycles in the input hydrograph and the rate of water demand, which is assumed to be constant throughout the year for simplicity. Storage is required if demand exceeds the rate of input. The volume of water removed on demand cannot exceed the excess water flow above the demand line at peak flows, otherwise the reservoir would never fill (Figure 5.5B) The downstream impact on flow levels is shown in Figure 5.5C. Reservoir size and consequently storage must be calculated for considerable time periods to ensure that variation in climate and therefore input from year to year is accounted for (Figure 5.5D). This relies, of course, upon predicting the frequency and magnitude of periods of low flow. Water shortages in many reservoirs in Britain during the mid-1990s resulted from a sustained period of low input which was not fully predicted. Rainfall during this period has not been sufficient to replenish the reservoirs, leading to a progressive depletion of the stored water.

2. **Reservoir location**. The geometry of a reservoir controls the volume of stored water, the area of land flooded and the pressure of water at the dam. This is a product of the topography of the site and can be assessed from topographic maps and surveys for a range of possible dam sites when choosing a location. In capital terms the dam is the most expensive item and the site chosen needs to minimise dam length and have as much relief as possible in order to maximise water storage. Consequently, constrictions in valley cross-profiles, especially where they are associated with deep valleys, are often an initial target at the reconnaissance stage. Water levels in a reservoir are usually controlled by the maximum height to which a dam can be constructed at a site, although in some areas low cols may restrict water levels unless additional secondary dams are constructed. Ideally, above a chosen dam site a valley should widen and the valley long profile should remain flat for as long as possible to maximise the water depth. In warm climates it may be desirable to restrict the surface area of a reservoir to minimise evaporation. A similar condition may apply if land values are high and the flooded area needs to be minimised. The storage required will be an important control in the choice of initial dam location, although subsequently the engineering geology of the proposed dam site will be critical in determining its practicality.
3. **Hydrogeology of reservoir floors and walls**. The reservoir floor and walls must not leak water. Impermeable rocks are ideal but not essential. Permeable rocks may be acceptable if the water table slopes down towards the reservoir and will still do so when the reservoir is full (Figure 5.6). Of greater importance is the permeability in the vicinity of the dam. Impermeable rocks, or rocks which can be rendered impermeable by grouting (injection of

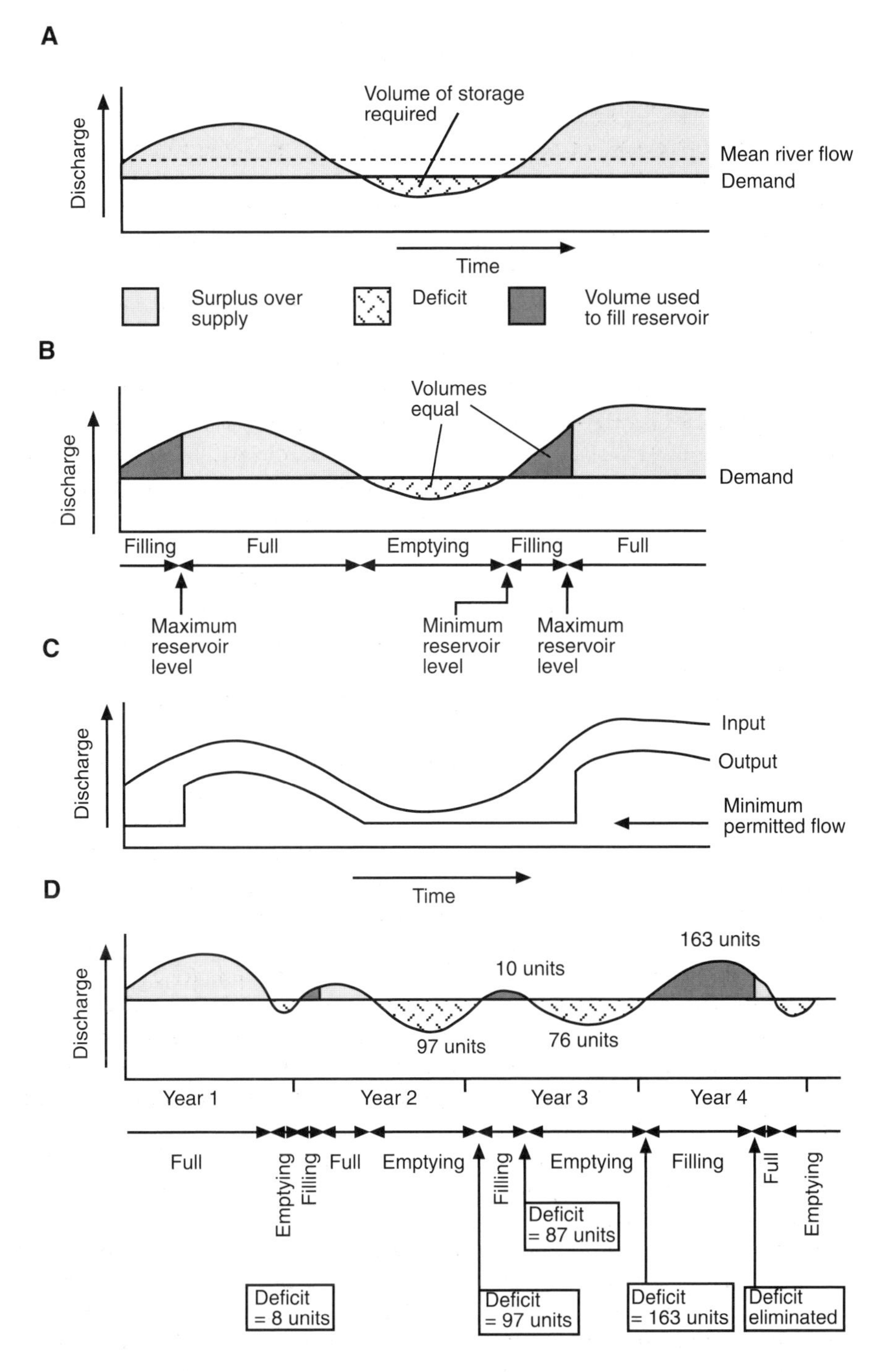

Figure 5.5 *Hypothetical hydrographs showing the relationship between discharge, storage and demand [Based on an original diagram by I.M. Platten]*

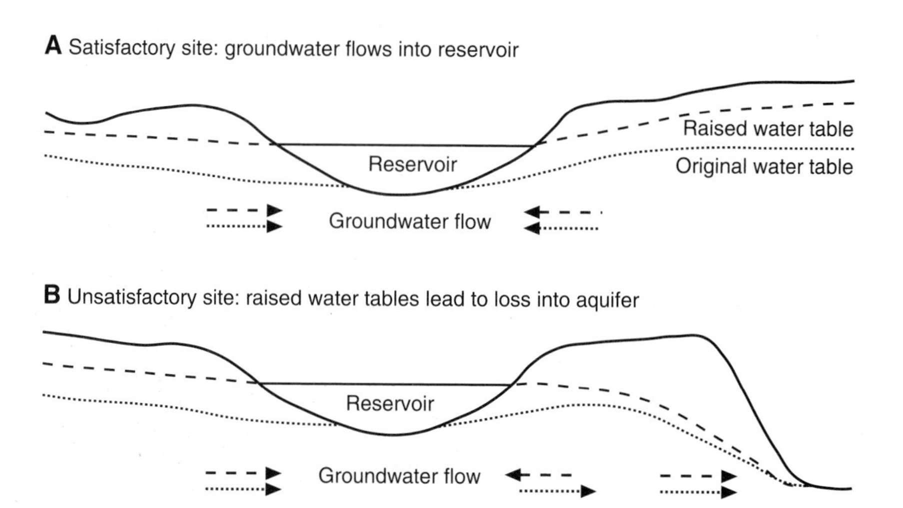

Figure 5.6 *Modification of the water table due to the construction of a reservoir.* ***A.*** *Good hydrological reservoir location.* ***B.*** *Poor hydrological reservoir location*

slurried clay or cement), are essential. The hydraulic gradient between the rock upstream, which is submerged in the reservoir, and the rock downstream will be large, as will the gradient around the ends of the dam. This will result in an extremely large water flow in even slightly permeable rocks, which can lead to erosion of pipe conduits in unconsolidated rocks, and eventual dam failure (Figure 5.7). Clays, mudstones and shales are generally impermeable and therefore acceptable, but gravels, sands and weakly cemented sandstones are not. Igneous rocks, metamorphic rocks and well cemented sandstones often contain numerous fractures and joints which can transmit water. If the fracture density is low (less than one per square metre) these may not pose a problem. Deeply weathered rocks should also be avoided. Problems may be overcome, where necessary, by excavation and grouting, although this adds substantially to the cost of the scheme.

4. **Engineering geology and dam design**. There are many different designs of dam, which include two principal types: (1) **gravity dams**, which rely on their weight to hold them in position and thereby impound the water, and which are usually made from reinforced earth, masonry or concrete (Figure 5.8); and (2) **arch dams**, which transmit the stress of the impounded water horizontally into the rock of the valley sides and are consequently made from thin concrete walls. Where foundations are weak (e.g. on clays and mudstones), gravity dams are usually made of earthwork in order to spread the weight; however, on firmer foundations (e.g. igneous or metamorphic rocks), concrete and masonry structures may be used. The foundations for gravity dams should show limited compaction when loaded with the weight of the dam and consequently deeply weathered sites or poorly consolidated Quaternary sediments should be avoided. Existing fractures or bedding

Figure 5.7 Photograph of a failed dam at Keppel Cove, Lake District, Britain. Dam failure was caused by piping of water beneath the dam which rests on Quaternary sediments. This pipe flow undermined part of the concrete dam causing it to fail

Figure 5.8 An example of a concrete and masonry gravity dam at Loch Sloy in the Scottish Highlands. This dam is founded on ancient crystalline terrain

should be minimal and if possible they should dip upstream. Downstream-dipping discontinuities may provide failure surfaces when loaded with the dam. The construction of arch dams relies on the rock mass strength of the valley sides (see Section 7.1.3). Where there is sufficient strength, thin arch dams, which are cheap and economical to build, may be adequate. However, since the stresses imposed in such situations are horizontal instead of vertical, an absence of fractures parallel to the valley sides is essential. In general, only rocks with a very high rock mass strength are suitable, and in all cases uniformity in the rock body is desirable.

5. **Reservoir lifetime**. Reservoir lifetime is controlled by the rate of sedimentation within the water body and therefore the length of time required to reduce the storage volume to unacceptable levels. Rivers entering a reservoir will deposit all or part of their sediment load as flow velocities are checked. The traction and saltation loads of the streams and rivers are deposited in a delta or fan on entry into the reservoir. Sand-sized suspension load settles quickly in the still reservoir water and residence times are sufficient in most cases to dramatically reduce the suspended silt and clay content. Control of sedimentation involves reducing erosion within the catchment by careful land-use management in which the proportion of bare soil is minimised by maintaining a continuous vegetation cover. Sediment traps located prior to water entry into the reservoir may work for small reservoirs if regularly cleared, and in a few exceptional cases dams have been designed to allow complete emptying and periodic sediment flushing. In general, sedimentation is a major problem, particularly in semi-arid regions with discontinuous vegetation cover.
6. **Hazards to dam integrity**. In the USA over 2000 settlements are currently threatened by catastrophic flooding from unsafe dams. Dam failure may result either from progressive dam weakening or from transient hazards such as landslides and earthquakes. Progressive dam weakening may occur through concrete deterioration, by weathering of foundation rocks, and by water seepage and associated piping through both the dam and its foundations. Transient hazards include earthquakes, mass movements, and overtopping by extreme discharges. Minor earthquakes may be associated with reservoir filling as the crust accommodates the increased load and these events may trigger small landslides. Large variations in water level will affect the porewater pressure of adjacent valley sides, and may also lead to slope failures. Provided the spill-ways and relief channels are adequate, water should not overtop the dam; if it does, erosion of the dam wall and exposure of the dam to stresses for which it was not designed may cause failure.
7. **Capital and operational costs**. The construction of large dams involves high capital costs. Not only is there the cost of the dam, but the purchase of the land to be flooded and the relocation of essential infrastructure may lead to considerable costs. The cost of typical dams may range from $1.5 million to in excess of $750 million. Operational costs are usually considerable and centre not only on dam maintenance but also on maintaining the water distribution network.

8. **Environmental impacts**. The construction of any reservoir has major economic and social impacts, including: (1) loss of agricultural land; (2) loss of natural habitats and indigenous valley floor flora and fauna; (3) loss of all infrastructure within the flooded area, such as roads, railways, pipelines, cables and buildings; (4) aesthetic impact, both adverse and desirable; (5) raised water tables in adjacent areas causing problems of drainage; and (6) downstream river impacts such as changes in flow regime, sediment transport, water quality and channel geometry. The downstream impacts of dam construction on rivers are considerable, spreading the impacts of reservoir construction over a large area, and they are therefore worth considering in some detail.

   Three scales of channel response to river impounding have been recognised. First-order impacts include changes in flow duration, flood frequency and sediment load. This leads to second-order impacts, such as changes in channel cross-sectional geometry and planform, while third-order impacts are associated with the interaction of changes in channel morphology with river ecology. Dam closure causes a reduction in: (1) the duration of flood flows; (2) the magnitude of flood peaks by between 20 and 75%; and (3) sediment discharge, since most sediment is trapped within the reservoir. Channel response to these changes may be very complex, involving both deposition and erosion of reaches, and depends on the existing channel characteristics and on whether both water and sediment discharge is reduced (Figure 5.9). In gravel-bed rivers, such as those commonly affected in Britain, the flows are no longer able to transport the sediment within the channel. In more fine grained alluvial channels the sediment-depleted flows below the dam may cause rapid channel floor erosion and scour as sediment is entrained by the flows. The nature of the response is very variable and depends not only on the degree of flow regulation, but also on the nature of the channel boundary.

   The possible adjustments to a decrease in both flow and sediment discharge include: (1) decrease in channel width; (2) increase or decrease in channel depth; (3) decrease in channel capacity; (4) change in channel planform; and (5) change in channel gradient. The commonest morphological response appears to be a decrease in channel capacity brought about primarily by a decrease in channel width. This results because discharge is the product of velocity times cross-sectional area; if discharge falls, velocity must therefore also fall, leading to deposition and a reduction in the channel cross-section. Channel reduction of the order of 50% is not uncommon. Width reduction is caused by the deposition of channel side bars or berms (Figure 5.10A). Such channel modifications commonly occur at river confluences where sediment is introduced into the sediment-starved main channel by unregulated tributaries (Figure 5.10B). This coarse tributary sediment cannot be flushed downstream by the reduced flow in the main channel and consequently builds up as a tributary bar, downstream of which channel side berms may be deposited (Figure 5.10B; Box 5.5). As Figure 5.11 illustrates, the effects of river regulation decrease downstream, as unregulated tributaries return the discharge regime to a more natural state.

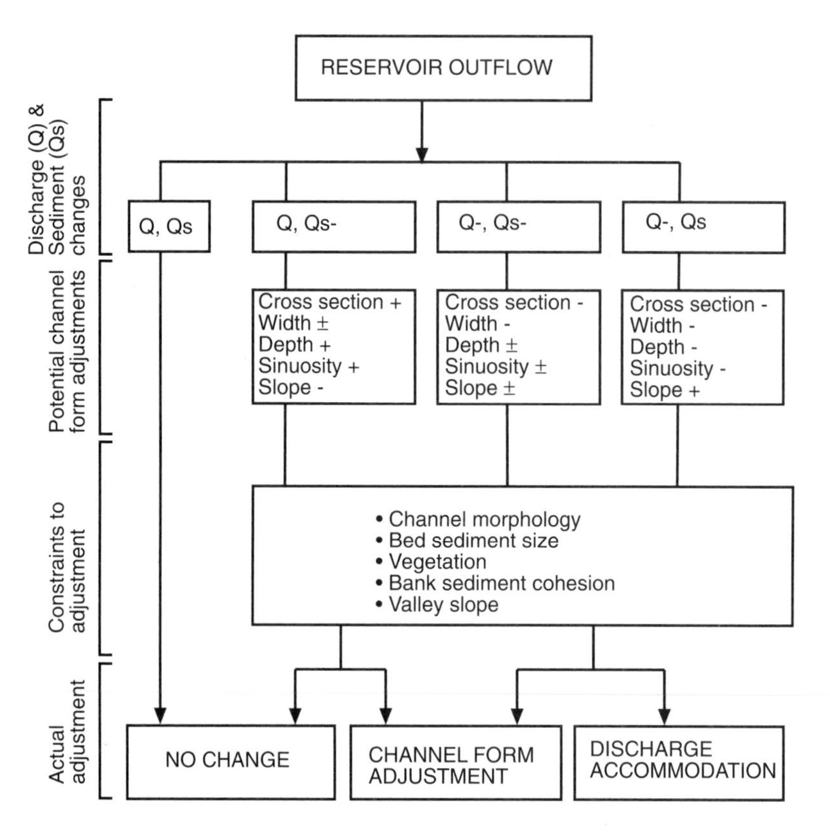

Figure 5.9 *Flow chart showing downstream fluvial response to the construction of a dam [Modified from: Petts (1979)* Progress in Physical Geography, **3**, *Fig. 1, p. 339]*

The impact of river impounding is not simply restricted to channel morphology but also has a major impact on water quality. Most large reservoirs develop thermal stratification seasonally. A highly stable summer stratification develops with a warm surface layer (**epilimnion**) separated from the cooler, denser water of the **hypolimnion** by a rapid temperature gradient known as the **thermocline**. During the autumn, the surface water cools and settles, and overturning of the water column occurs leading to isothermal conditions. The degree of thermal stratification which develops varies with geographical location and reservoir depth, but is common to the vast majority of reservoirs. Within shallow reservoirs, seasonal water mixing is sufficient to maintain aerobic conditions throughout the water column. However, in deeper reservoirs, little mixing occurs below the thermocline and sunlight does not penetrate far into the hypolimnion; deprived of reaeration, oxygen is consumed by the breakdown of organic matter leading to anaerobic conditions. Hydrogen sulphide gas is produced in these reducing conditions, along with carbon dioxide. In such cases, pH levels may fall and the solution of iron and magnesium from bottom sediments may

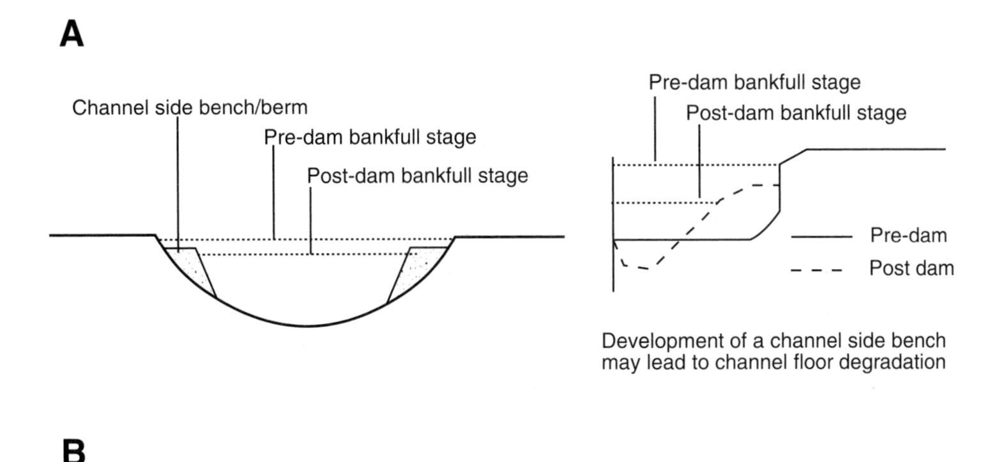

Figure 5.10 *Schematic model of channel response to flow regulation.* ***A.*** *Channel side berms or benches formed by bank-side deposition associated with a fall in discharge within a regulated flow.* ***B.*** *The formation of tributary mouth bars and channel side benches. Channel adjustment to flow regulation is often most marked below confluences between regulated and unregulated rivers. The unregulated tributary transports a lot of sediment which the main, regulated flow cannot move*

occur. Reservoirs act as nutrient traps, and phosphates and nitrates may also become concentrated in the hypolimnion water, adding to its density and helping to maintain the thermal stratification well into the autumn.

Dams may tap water from a reservoir at different levels, and if the reservoir is stratified, outflowing water may be drawn from several levels. The downstream impact on water quality depends on the type of water released: hypolimnion (release at depth) or epilimnion (surface release). Water release may affect the river downstream by changing its temperature, dissolved gases, solutes and turbidity content. Perhaps the best documented impact is

BOX 5.5: DOWNSTREAM IMPACT OF RESERVOIR: THE CUDGEGONG RIVER, AUSTRALIA

Benn & Erskine (1994) describe the downstream impact on the Cudgegong River in New South Wales caused by the completion of the Windermere Dam in February 1984. The dam was constructed for water supply and irrigation. The Cudgegong River has a gravel bed and flows in a confined bedrock valley. After closure, the duration of different mean daily flows was changed (Diagram A), the duration of high magnitude flows was reduced and the duration of low magnitude ones increased. For example, mean daily flows with duration less than 70% of the time were reduced by between 23 and 73%, while low magnitude flows, with a duration greater than 70%, were increased by between 58 and 96%. The dam also reduced downstream flood magnitudes for all recurrence intervals by between 57 and 72%. In addition, over 95% of all sediment within the river was trapped by the dam reducing sediment discharge. The river's response to these changes in flow and sediment regime was monitored at eight stations over 13 km downstream of

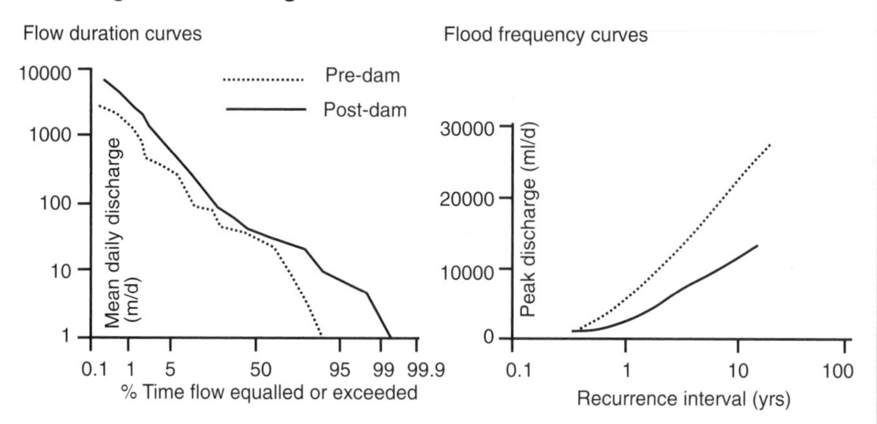

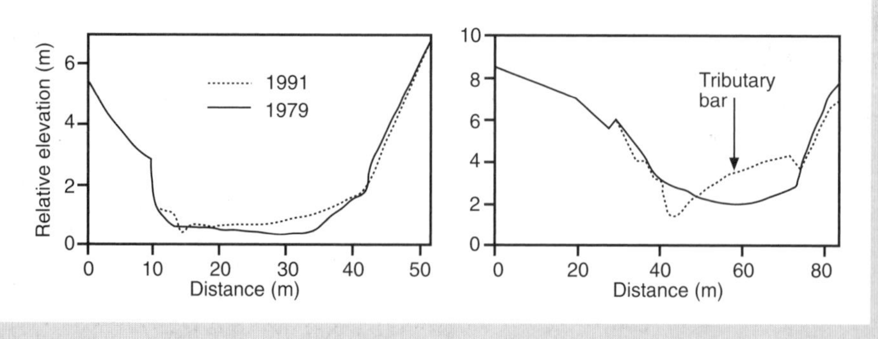

the dam (Diagram B). Cross-sectional area of the channel decreased by more than 10% at six of the eight monitored stations. Reduction in channel capacity was most marked below tributary confluences where well developed channel side berms were noted. Vegetation encroachment was also recorded along the channel, particularly in association with the development of channel side berms. Only one of the monitored sites showed no adjustment, which reflects the fact that it had a vegetated, armoured and therefore stable cross-section. Channel degradation (bed scour) occurred at several locations due to the entrainment of stream bed gravel by the sediment-free water. This work provides a very good example of the complex morphological changes which occur within a river channel in response to dam construction.

Source: Benn, P.C. & Erskine, W.D. 1994. Complex channel response to flow regulation: Cudgegong River below Windermere Dam, Australia. *Applied Geography*, **14**, 153–168. [Diagram modified from: Benn & Erskine (1994) *Applied Geography*, **14**, Figs 3, 4 & 5, pp. 156, 158 & 160]

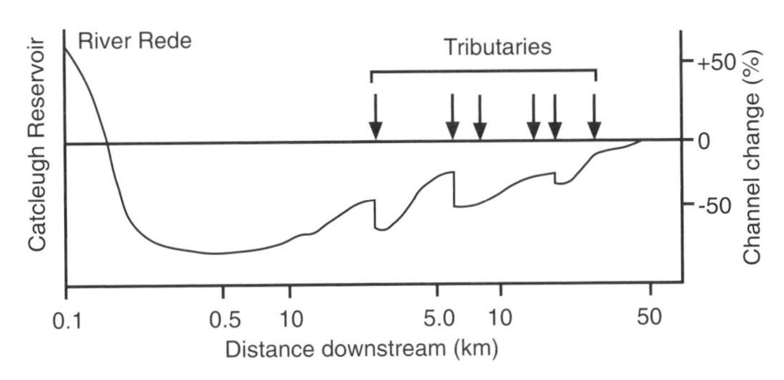

Figure 5.11 *Downstream decrease in fluvial impact associated with dam construction [Modified from: Petts (1979)* Progress in Physical Geography, *3, Fig. 4, p. 353]*

on downstream water temperature. Mean monthly summer temperatures have been reduced by 10°C on some rivers, while increased winter temperatures have prevented ice formation on some Canadian rivers and have damped temperature fluctuations. The release of water from an anoxic hypolimnion can cause low dissolved oxygen levels downstream. The release of hypolimnial water may also lead to precipitation of iron and manganese on the stream bed, a phenomenon reported on the River Elan below the Craig Goch Dam in Britain. The nature of the solute concentrations, relative to the input values above the reservoir, depends on whether the water release is from the hypolimnion, and therefore saturated with solutes, or from the epilimnion, which may be relatively solute-free (Table 5.1). Pulses of turbid water may also occur downstream of reservoirs. These are due either

Table 5.1 *Comparison of water quality both upstream and downstream of a dam for two rivers, one with a hypolimnion water release and one with release from surface epilimnion water. Note the decrease in impact downstream (all values are in milligrams per litre) [Modified from: Petts (1985)* Progress in Physical Geography, *Table 1, p. 500]*

	River Svratka, Czech Republic Hypolimnion or deep release			River Tees, UK Epilimnion or surface release		
	Upstream of dam	0.1 km downstream	10.5 km downstream	Upstream of dam	0.1 km downstream	22 km downstream
Sodium (Na)	6.4	11.4	8.7	3.6	2.6	4.2
Potassium (K)	3.7	6.2	4.3	1.0	0.52	0.96
Magnesium (Mg)	1.8	4.2	5.5	2.3	0.86	2.2
Calcium (Ca)	12.3	19.9	14.3	36.3	8.47	19.4
Sulphur dioxide (SO_2)	19.4	31.1	18.5	5.2	4.0	5.4
Nitrogen dioxide (NO_2)	5.0	7	8	0.13	0.2	0.35

to reservoir flushing to reduce sedimentation or to the export of large numbers of limnoplankton. Low quality water pulses may also be caused by flood events which stir reservoir bottom waters. Reduced downstream flow may also decrease dilution of waste products and effluent introduced into a river downstream, increasing toxicity levels particularly at low flows. All of these changes in water quality have an impact on the downstream fluvial ecology, although in most cases it does not lead to a reduction in species diversity but simply a change in composition. In summary, therefore, dam construction and associated river regulation has a significant downstream impact on the geomorphology, water quality and ecology of the river.

A wide range of variables needs to be considered in the construction of dams and reservoirs. They are, however, the most important components in the water supply system. Of less environmental impact is the modification of existing natural lakes. Raising lake levels by damming or controlling the exit gives additional storage with the minimum increase in flooded area, especially where the adjacent valley sides are steep. Clearly, where lake shores are undeveloped such schemes may have a significant cost advantage, both in environmental and capital terms, over the construction of new reservoirs.

5.2.2 River Abstraction

River abstraction involves the direct removal of water from rivers by either diversion or pumping. To be reliable as a source of water, the average stream discharge must exceed the rate of demand for water. If there is storage capacity within the supply system, such as holding reservoirs, then a degree of seasonal variation can be tolerated. If there is no storage capacity then the river must be able to supply demand even at low flows if it is to be the only source of water. Along a given river, water may be abstracted, used and returned to the river several times (Figure 5.4). The consequences of abstraction may be considerable on a river which is deprived of its natural discharge, although this depends on the amount and timing of the abstraction. Impacts of water abstraction include: (1) channel sedimentation and reduction in channel size due to decreased transport capacity; (2) encroachment of vegetation into the channel; (3) effects on the river ecology and river bank habitats; (4) an increased concentration of solutes and pollutants due to a decrease in dilution; (5) adverse aesthetic impacts; and (6) downstream impacts on fisheries, navigation and other river-based activities. Many of these impacts are similar to those caused by the construction of reservoirs (see Section 5.2.1). The level of impact depends on the amount of water removed and the minimum level of water flow set for the river. Calculating the amount of water which can be removed from a river without causing adverse impacts depends on a detailed knowledge of the discharge regime of the river and its variation over a number of years. In most cases a minimum discharge level, sufficient to minimise downstream impacts, is set for a river, and the volume of water abstracted is carefully controlled to ensure that this level is always maintained. However, problems arise from the fact that in most cases the length of hydrological record is insufficient to document all variation present, and prolonged periods of low discharge may not be anticipated.

5.2.3 Groundwater Abstraction

In Britain, approximately 18% of water supply is derived from groundwater abstraction. In southeast England, this figure rises to as much as 50%. The majority of this water, 75%, is derived from Cretaceous chalks and Permo-Triassic sandstones. Groundwater may be obtained either directly from springs or by pumping from saturated aquifers. In the case of springs the rate of supply is limited by the rate of spring discharge. Where extraction occurs directly from below the water table, either by pumping or by pressurised outflow in the case of artesian wells, the water withdrawal should be in equilibrium with the rate of groundwater recharge if the shape and form of the water table is to be preserved. If removal exceeds discharge over a prolonged period, the water table will fall. Removal of groundwater by pumping causes a draw-down in the water table in the vicinity of the well (Figure 5.12A); this has the potential to cause settlement and in coastal areas may lead to contamination of the aquifer by saline groundwater (Figure 5.12B). Large-scale removal of groundwater in excess of recharge may cause regional subsidence. For example, historically the over-exploitation of the chalk aquifers of the London Basin has reduced the height of the water table by over 125 m, causing widespread subsidence, although the water table is now rising. The link between groundwater abstraction and subsidence has also been demonstrated clearly elsewhere (Figure 5.13). Artificial **groundwater recharge** is practised in some parts of the world (Box 5.2) both as a mechanism of water transfer (water is added at one location and withdrawn at another) and to combat problems of subsidence. Aquifer depletion may also reduce stream flow during periods of low flow. In the USA, groundwater contributes about 30% of water to all stream flow, and aquifer lowering is a significant cause of low stream flow in many areas.

In assessing the feasibility of groundwater supply, a combination of geological and hydrological information is required. The first step is to determine the regional geology and both the porosity and permeability of the principal lithological units. Using this information it is possible to identify hydrological units, that is, those units of aquifer bounded by aquicludes. Typical aquifers are shown in Figure 5.3 and may vary in size from individual glacial landforms or fracture zones, to huge sedimentary basins such as the chalk aquifers of western Europe. The potential groundwater supply from a given hydrological unit is determined by means of a groundwater inventory. This simply involves evaluating the components of the hydrological balance in which total precipitation is equal to the amount lost by evaporation and evapotranspiration, the amount lost due to stream flow and the amount which infiltrates into groundwater. By measuring each of these components it is possible to determine the groundwater increment over a sufficiently long time period to remove small-scale variability.

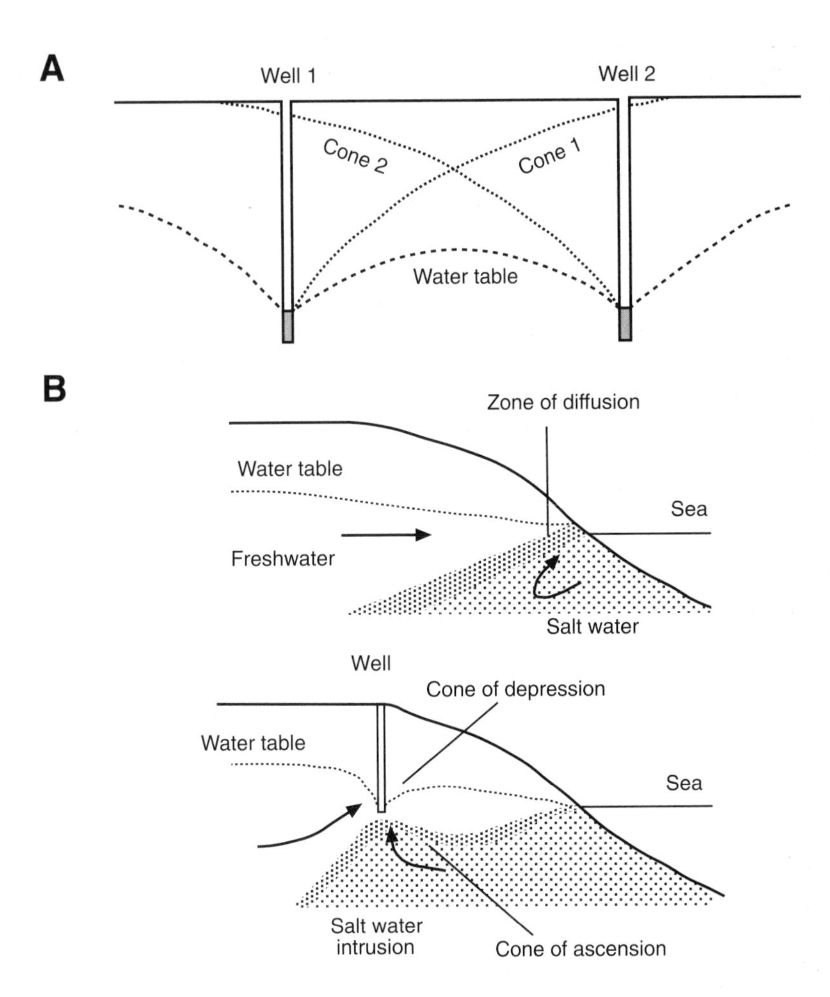

Figure 5.12 *Downdraw of water table in the vicinity of a well (**A**) and possible groundwater contamination by saline groundwater as a consequence of this downdraw in coastal regions (**B**)*

The amount of natural recharge determines the safe groundwater yield for a given hydrological unit, that is, the amount which can be removed without depleting the finite groundwater reserve. In practice, determining the groundwater increment and therefore the safe yield is difficult, not least because in most countries different statutory agencies are responsible for surface hydrology and subsurface or hydrogeology. For example, in the UK, surface hydrology is documented and managed by the Institute of Hydrology, while hydrogeology is managed by the British Geological Survey. As a consequence, aquifers in some parts of the world have been overexploited.

Despite concern about overexploitation and depletion of reserves, groundwater is a resource which is significantly underutilised in many countries, although it is probable that with increased demand from an expanding population it may

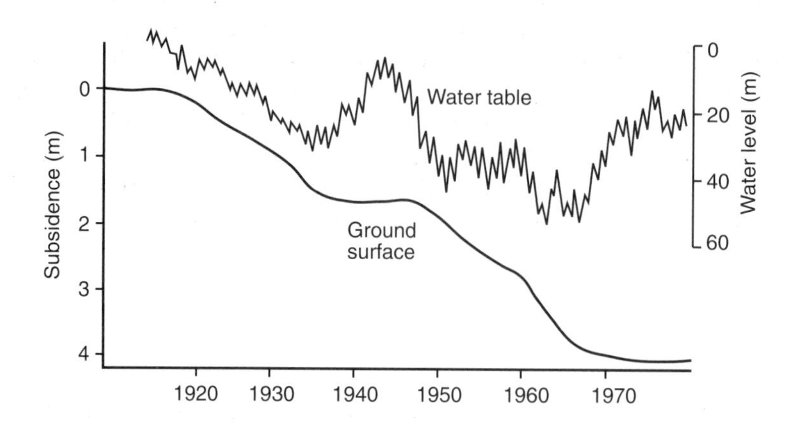

Figure 5.13 *The link between ground subsidence and water abstraction in the Santa Clara Valley, California [Modified from Waltham (1994)* Foundations of engineering geology. *Blackie, p. 57]*

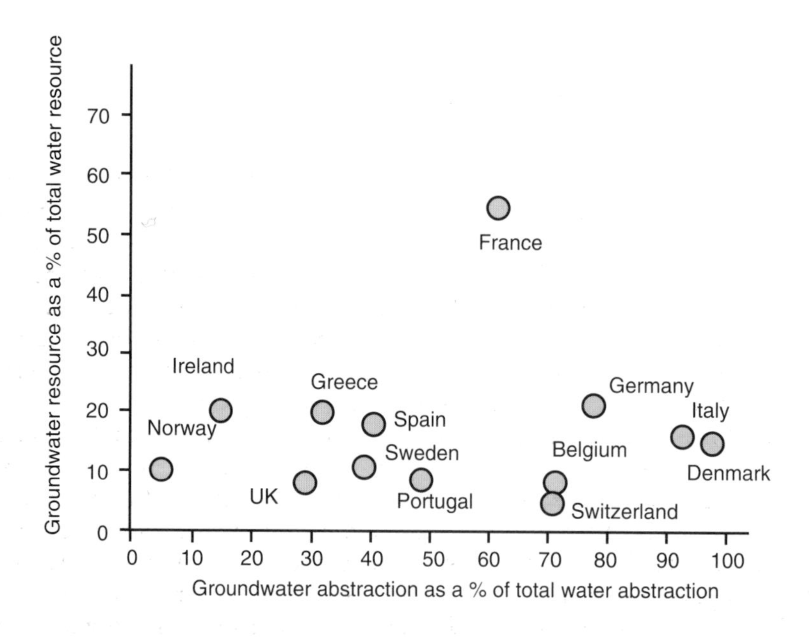

Figure 5.14 *Degree to which European countries exploit their groundwater reserves, illustrating the potential reserves available if exploited in a sustainable fashion [Modified from: Kelk (1992) In: Lumsden (Ed.)* Environmental geology in Western Europe. *OUP, Fig. 3.11, p. 53]*

grow in importance. Consequently effective future management of this resource is essential. Figure 5.14 shows the relative proportion of groundwater as a percentage of the total water resource in a given country, plotted against the percentage of water actually obtained from groundwater abstraction. Certain countries, such as Italy and Denmark, clearly exploit their groundwater resource to the full, while others, such as France, have a very large but underexploited resource. Britain also underutilises its total groundwater resource, although specific aquifers, such as those under London, are overexploited.

5.2.3 Desalination

In areas with a surplus of cheap fuel, **desalination** of seawater to produce fresh water is a practical option. Seawater is evaporated and condensed on a large scale to produce fresh water. This is clearly a fuel-intensive process, but is economic in the oil-rich regions of the Middle East where oil is more easily obtained than water. Each cubic metre of seawater contains about 3.5% salt (40 kg), which must be reduced to 0.05% before it becomes acceptable for use. Large desalination plants are capable of producing 20 000 to 30 000 m^3 of water per day at a cost in the USA which is 10 times that paid for water derived from any other source.

5.3 WATER TRANSFER

In developed countries, water reserves are frequently divorced from population centres and water transfer is of particular importance. Methods of water transfer include (Figure 5.4): (1) pipelines, canals and other artificial waterways; (2) groundwater recharge and abstraction; and (3) transfer within natural river systems. The aim is to deliver a constant supply of water sufficient to meet the demand in a specific location and to minimise leakage or other loss. The more complex the methods of water transfer and the greater the travel distances involved, the more costly and wasteful the scheme may become.

Water transferred into a drainage basin from elsewhere will ultimately need to be discharged via its river system unless pumped directly out to sea. Water imported and used by a town is usually pumped after treatment into a river for downstream transport. Therefore river discharges are often increased by water imports (Figure 5.15). This commonly leads to channel adjustment, increasing channel capacity, as well as having implications for the flood regime of the river. Natural rivers are also frequently used to transfer water from one location to the next, minimising the need for pipelines and pumping stations. All of these processes cause changes in the natural river system such as: (1) channel geometry; (2) planform; (3) riverside or riparian vegetation; (4) river navigation; and (5) the fluvial ecology.

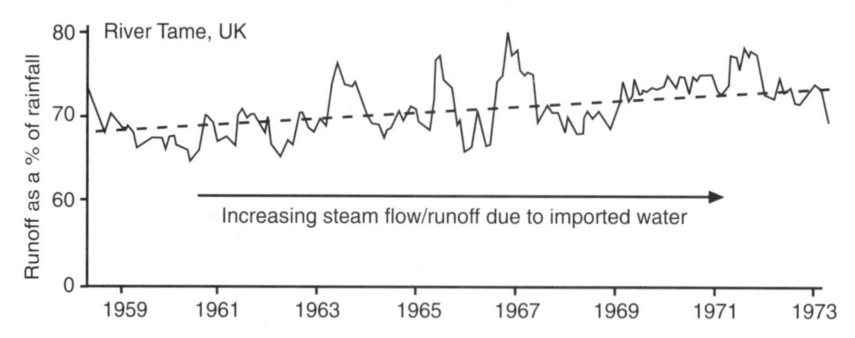

Figure 5.15 *Increased stream flow due to water imported into the catchment of the River Tame in Britain [Modified from: Petts & Foster (1985)* Rivers and landscape. *Arnold, Fig. 2.4, p. 25]*

5.3 SUMMARY OF KEY POINTS

- Water is essential to life. It varies widely in availability from one location to the next, both globally and regionally. Principal problems are the storage and transfer of water from areas where it is abundant to meet the demand in those areas where it is not.
- The rate at which water is delivered from a drainage basin either to stream flow or to groundwater depends on the topography and land use within the catchment, both of which can be changed by human action.
- Principal water sources are artificial reservoirs, river abstraction, groundwater withdrawal and desalination. Water storage within the supply system is essential to buffer variability in supply from variability in demand. Reservoirs are essential in this respect.
- Water can be transferred from one location to the next by the use of pipes, canals and tunnels, natural rivers, and recharge of groundwater.
- Water resources are sustainable and should ideally be managed accordingly. The environmental consequences of excessive groundwater abstraction are of particular note.

5.4 SUGGESTED READING

Clarke (1991) provides an overview of global water issues, while Pearce (1982) and Kirby (1984) provide a UK perspective. Basic information on surface hydrology and hydrogeology can be obtained from Brassington (1988) and Price (1985). Further information on water supply is given in Downing & Wilkinson (1991). The downstream impact of reservoir construction is covered in several papers and books including Petts (1979, 1980, 1984, 1985), Kondolf (1993) and Benn & Erskine (1994). European examples of different types of aquifers are provided by Kelk (1994), who also covers some of the management issues associated with groundwater.

Benn, P.C. & Erskine, W.D. 1994. Complex channel response to flow regulation: Cudgegong River below Windermere Dam, Australia. *Applied Geography*, **14**, 153–168.
Brassington, R. 1988. *Field hydrogeology*. Open University Press, Milton Keynes.
Clarke, R. 1991. *Water: the international crisis*. Earthscan, London.
Downing, R.A. & Wilkinson, W.B. 1991. *Applied groundwater hydrology*. Oxford University Press, Oxford.
Kelk, B. 1994. Natural resources in the geological environment. In: Lumsden, G.I. (Ed.) *Geology and the environment in Western Europe*. Oxford University Press, Oxford, 34–138.
Kirby, C. 1984. *Water in Great Britain*. Penguin Books, Harmondsworth.
Kondolf, G.M. 1993. Channel adjustment to reservoir construction and gravel extraction along Stony Creek, California. *Environmental Geology*, **21**, 256–269.
Pearce, F. 1982. *Watershed: the water crisis in Britain*. Junction, London.
Petts, G.E. 1979. Complex response of river channel morphology subsequent to reservoir construction. *Progress in Physical Geography*, **3**, 329–362.
Petts, G.E. 1980. Morphological changes of river channels consequent upon headwater impoundment. *Journal of the Institute of Water Engineers and Scientists*, **34**, 374–382.
Petts, G.E. 1984. Sedimentation within a regulated river. *Earth Surface Processes and Landforms*, **9**, 125–134.
Petts, G.E. 1985. Water quality characteristics of regulated rivers. *Progress in Physical Geography*, **10**, 492–516.
Price, M. 1985. *Introducing groundwater*. Allen & Unwin, London.

6
Aesthetic and Scientific Geological Resources

In this chapter we define the aesthetic, cultural and scientific importance of geology, and examine the need for its conservation. Geology has long been the subject of study and a source of inspiration, even before its definition as a physical science. For example, geological materials, such as precious gems, attractive stones and interesting fossils, have long been prized for their great beauty. Similarly, landscapes, sculpted by the forces of nature from rock, have been the inspiration of countless poets, painters and writers. Finally, scientifically, it is often only field sites which provide a reliable test of many geological theories. These aspects together demonstrate the need to protect a diminishing resource threatened by human development and construction (Figure 6.1). In this chapter we present the rationale for the conservation of the geological resource, together with the practical aspects of its protection.

6.1 THE RATIONALE FOR CONSERVATION

The conservation of any aspect of heritage, whether it be human history, as represented by historic sites or buildings, or elements of the natural world, such as the biological and geological resources of the planet, relies upon four basic convictions (Figures 6.2 & 6.3): (1) that such aspects be conserved for their own sake; (2) that they provide a basis for economic exploitation; (3) that they form a basis for research, training and education; and (4) that they have aesthetic or cultural value.

The first of these tenets is most commonly understood with reference to wildlife and ecological aspects of the natural world. It has its basis in many of the major cultural and spiritual traditions of the world, particularly in the major religions of both the modern and ancient world which have traditionally

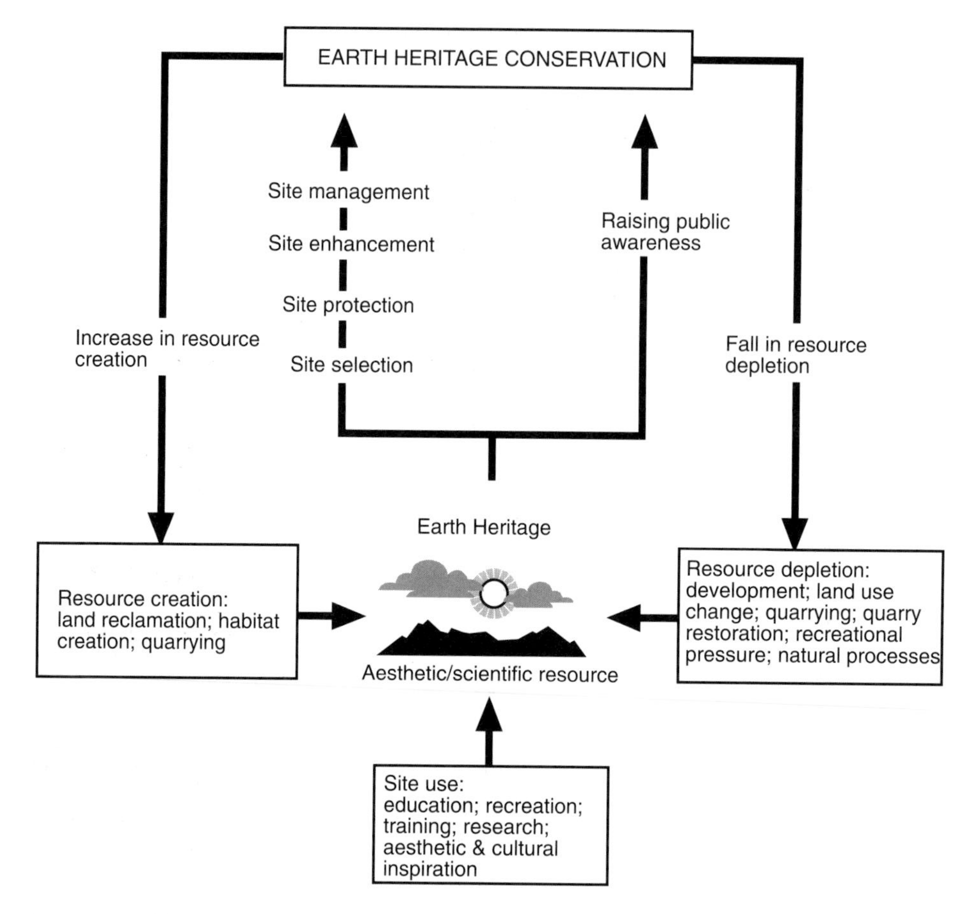

Figure 6.1 *The conservation resource, a balance between resource creation and depletion*

espoused a principle of empathy with, and respect for, our fellow creatures. However, in practical terms the conservation of wildlife for its own sake was really only achieved in the late nineteenth century following a change in emphasis from exploitation to conservation, which manifested itself in Britain with the development of the Royal Society for the Protection of Birds, and other similar organisations. This fundamental principle of wildlife conservation is, however, more difficult to reconcile with inanimate or non-sentient features of a landscape, such as landforms and exposures of rock. Unlike wildlife, these features convey an appearance and impression of solidity and permanence which is at odds with their actual vulnerability. In general this aspect is the most difficult to convey to the public in general, and developers in particular, and therefore the conservation of such features has traditionally lagged behind that of wildlife. However, despite this, the development of the National Parks concept in many countries in the late nineteenth and twentieth centuries is in part associated with this idea.

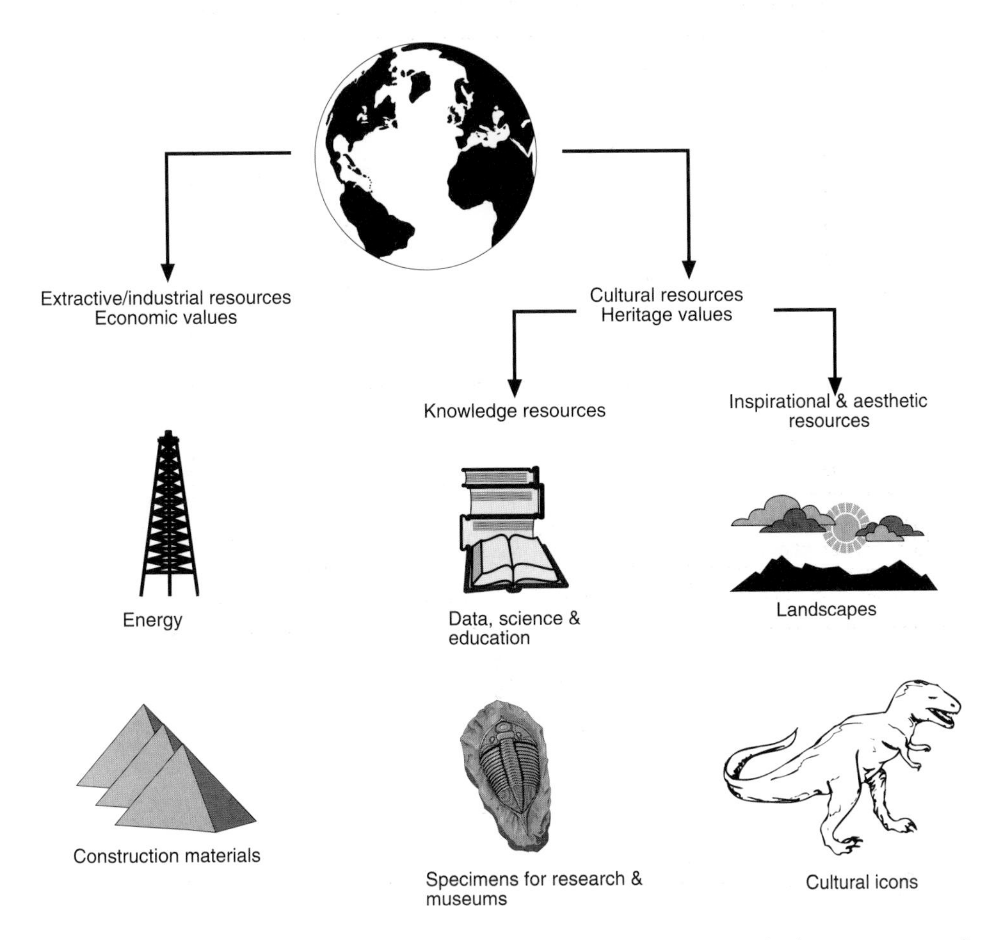

Figure 6.2 *The rationale for Earth heritage conservation, based on both economic and heritage values [Modifed from: Wilson* et al. *(1994)* Earth heritage conservation. *Geological Society, Fig. 5.1, p. 157]*

The other guiding principles of conservation are broadly linked and all illustrate the way that aspects of the natural or geological resource may be used in some way by humans. Of these, economic exploitation is the most common, and a clear understanding of the nature of a physical, biological or even historical resource is important in order to exploit it. Scientific research, and associated with this the need for the training and education of the next generation of scientists and technologists, is also of great importance to the furtherance of human progress. Finally, the Earth's physical resources provide inspirational and aesthetic assets in the form of beautiful landscapes and attractive specimens, both of which contribute to the spiritual and emotional well-being of people. It was on this principle that parks and green open spaces in towns and cities, and national

Figure 6.3 Earth heritage as an aesthetic and educational resource. **A.** The mountain landscape of Snowdonia [Photograph: A. F. Bennett]. **B.** The educational resource

parks and other large areas in the country were protected from at least the nineteenth century onwards in many industrialised countries, such areas providing the 'lungs of the city' or an escape from the harsh industrial landscape.

6.2 THE NATURE OF THE RESOURCE

Earth science encompasses the formation, constituents, structure and history of the Earth. It draws upon the other sciences in the explanation of the physical components and history of the Earth. It is divisible into two broad areas which in effect are two end members of a single continuum: geomorphology, concerned with understanding the processes of the Earth which have shaped its surface; and geology, concerned with the historical development, constituents and structure of the Earth. Together these provide the framework for understanding the Earth's physical resources.

The Earth's physical resources are only directly accessible at the surface of the Earth, although quarrying and mining activities can determine the nature of the resource at levels within the crust. The aesthetic and scientific Earth science resource is similarly confined to the surface and upper levels of the crust and is composed of two basic components: landforms, comprising physical features of the landscape which are primarily the subject of geomorphology; and rock outcrops, comprising rock units which are present, but not necessarily exposed, at the surface, and which are primarily the subject matter of geology. In most cases there is a direct relationship between these two individual components; for example, lithology often has an important influence in the formation of certain landforms, and the exposure of rocks at the surface is in many cases associated with geomorphological activity. In addition, the built environment is also fundamentally an expression of the geological and geomorphological resource. All three are discussed below.

6.2.1 Landforms and Landscape

Landforms are the physical expressions of internal (endogenic) and external (exogenic) processes which have operated to shape the surface of the Earth. Landforms are varied in form and reflect the diversity of the processes which have formed them, but a convenient way of classifying them is by considering them as either static or active. **Static** landforms are those in which the activity that produced them no longer operates, although other processes may be acting upon them (Figure 6.4). Typical examples include the landforms of the last glacial interval in Britain and North America. **Active** landforms are those in which processes are still operating to form and transform the landscape. Typical examples include coastal cliffs and spits, and rivers. Together these landforms make up the **landscape** which is effectively a collection of both active and static landforms. Landforms and landscapes provide an important resource for aesthetic, cultural and scientific reasons.

Figure 6.4 *Landform and landscapes as an Earth heritage resource: limestone pavement at Hutton Roof in Cumbria, England [Photograph: A.F. Bennett]*

Aesthetically, landscape can be defined on the basis of a series of criteria which seek to determine relative worth in terms of **landscape perception**. Landscape perception is a way of determining landscape value in terms of beauty, taste, regional association and other subjective values. It is usually assessed through the application of a series of criteria applied at a number of

Table 6.1 *Landscape perception criteria [Modified from: Tait* et al. *(1988)* Practical conservation. Site assessment and management planning. *Hodder and Stoughton, Table 2.2, p. 28]*

Criterion	Descriptions
Scale	Intimate, small, large, vast
Enclosure	Tight, enclosed, open, exposed
Variety	Uniform, simple, varied, complex, surprising, patchy
Harmony	Well balanced, harmonious, discordant, chaotic
Movement	Dead, calm, lively, busy, frantic
Texture	Smooth, rough, coarse grained
Naturalness	Wild, unmanaged, remote, undisturbed
Tidiness	Untidy, neat, overmanaged
Colour	Monochrome, subtle, muted, colourful, garish
Smell	Pleasant, unpleasant, obnoxious
Sound	Intrusive, noisy, quiet
Rarity	Ordinary, unusual, rare, unique, familiar
Security	Comfortable, safe, intimate, unsettling, threatening
Stimulus	Boring, monotonous, bland, interesting, surprising, invigorating
Beauty	Ugly, uninspiring, pretty, attractive, majestic, picturesque

carefully selected viewpoints. Each criterion is defined by a series of descriptors which allows for relative values to be given and comparisons to be made (Table 6.1). Effectively, landscape perception is rated on the following criteria: naturalness, variety, scale, familiarity, regional characteristics and associations.

Naturalness is predominant and reflects a desire to observe a truly natural landscape unfettered by human activity. In most cases, particularly in the developed countries, very few landscapes have escaped the influence of human activity, and therefore this is more often than not an impression rather than a real aspect of the landscape. In other words, people value a landscape which appears to be natural, and which does not have visually intrusive buildings out of harmony with the setting. Similarly, areas which experience a great deal of noise from road, rail or air traffic, or unpleasant smells from industry or agriculture are also generally not considered 'natural', thereby affecting the value judgement placed upon them. **Variety** determines the interest levels in the landscape. For example, this can reflect the amount of upland and lowland areas, the presence of water in lakes and rivers, the amount of cultivated and uncultivated land, and the presence of exposed rock or forested areas. **Scale** refers both to the size and openness of the area under study, and to the nature of the landforms; for example, the difference between rivers and streams, and mountains and hills. **Familiarity** reflects a human fondness for the familiar and a desire for the unexpected. Finally, **regional character** is a function of setting, usually determined by the local geological and geomorphological conditions. This is also often translated into the regional character of the construction of unobtrusive buildings which are 'in harmony' with the landscape.

Culturally, landscape is important for at least three reasons: (1) as an 'icon' which reflects national character; (2) because of an association with a specific historical event; and (3) as an inspiration for writers, artists and musicians. Landscapes are most commonly utilised as national icons by governments, tourism agencies, airlines and others who wish to demonstrate something of their national character. Typical examples include: the rolling chalk downland and 'white cliffs' of England; the flat glacial outwash plain of the Pampas of Argentina; the Butte and Mesa country and the Grand Canyon of the western United States; Victoria Falls in Africa; and Ayers Rock in Australia. Historically, landscape may be associated with specific events or periods of history. For example, the outcome of many famous battles was dependent on the landscape: Waterloo (1815) was fought on a clay plain in Belgium, while the Somme (1916) took place in the dissected chalk upland of northern France. Both of these are commemorated on the ground and are intimately associated with the local landscape. Finally, landscape is an extremely important source of inspiration for artists and writers. The poems of Wordsworth and Coleridge were inspired by the English Lake District; the paintings of Constable and Turner are famous for depicting the landscape; and the writings of Thomas Hardy are known for their intimate relationship with the gentle landscape of Dorset ('Wessex') in England. These three famous examples are just a few from the large body of men and women who have been inspired to depict landscape in words, music or on canvas.

Scientifically, landforms and landscape are the subject matter of geomorphology, which seeks to explain the processes which are acting or have acted in order to create the Earth's surface. As defined above, landforms fall into two categories: active and static. Active process landforms are studied in order to understand and observe the actual operation of the processes which are forming them. This entails observation of the process in the 'natural laboratory' of the landscape, and relies upon accurate measurement and observation of landforms which have been largely unaffected or changed by human activity. Apart from advancing knowledge of how natural processes operate, such studies have an extremely important practical application, as a full understanding of active processes will help, for example, in the design of flood defences for rivers and coastal defences to protect against marine erosion, in the mitigation of mass movements, and in the production of volcanic hazard maps (see Chapters 11 & 12). Static landforms are studied because they demonstrate how processes have operated in the past to form the landscape we know today. Typical of such landforms are the depositional and erosional landforms associated with the last ice age in Britain. Understanding them involves a detailed analysis of their pattern, form and composition and ideally involves a comparison with modern active process glacial sites in places such as Iceland, Greenland or Spitsbergen. Apart from simply advancing knowledge, there is an important practical dimension to this type of study. Glacial deposits provide extensive reserves of aggregates for the construction industry; understanding the extent of the material allows for appropriate mineral planning activities.

6.2.2 Rock Outcrops

Rock outcrops, by strict definition, are those rock units which come to the surface of the Earth and may be marked on a geological map. However, it is

important to distinguish between outcropping rock units which are **exposed** as visible rock breaking the surface of the landscape in a variety of settings, and those which are **obscured**, the outcrop being covered by surface deposits, soils, or human developments such as towns and cities. Exposure of rock is dependent on a wide variety of activities, both natural and human-induced (Figure 6.5). Typically, exposed rocks are found in one of two basic settings: (1) as a result of human activity, such as in quarries, and road and rail cuttings; or (2) as a result of natural processes, in upland, mountainous areas, or in coastal or river cliffs.

Exposed rock outcrops have the greatest practical use, as they provide the basis for detailed geological study at the surface. In most cases, natural exposures have provided the means for interpreting the obscured geology and for extrapolating the underground relationships. They have also provided the basis for exploration which has ultimately led to the creation of artificial exposures in quarries, pits and cuttings, which are the subject of mineral exploitation or the result of engineering projects. Obscured geology provides potential for future exposure, in the context of future exploitation in quarries or through the activity of natural processes, such as continued coastal erosion.

Aesthetically and culturally, the main value of rock outcrops lies in their contribution to the formation of specific landforms, such as the granite tors of Dartmoor in southern England, or the sandstone units of the Butte and Mesa country of Montana, USA. In general, natural exposures enhance the aesthetic appeal of a landscape, and will form a component part of any landscape assessment. However, exposures created by human activity are mostly considered an eyesore, and are, to many people, unwelcome intrusions in the landscape. This explains the desire for screening during the working life of quarries, for example, together with the procedure for restoration back to the original landscape which is often a condition of the granting of planning permission in many countries. The main importance of rock outcrops is in their scientific and educational importance. Exposed rock allows for the testing of geological hypotheses in their correct setting, which is important for a proper assessment of geological structure, age, history and setting. Exposed rock, whether in natural exposures or in quarries and cuttings, also provides the best opportunity of training and education, as it is in such settings that a true appreciation of geological structure and relationships can be examined, providing a context for laboratory work.

6.2.3 The Built Environment and Urban Landscape

The built environment of our towns and cities is a reflection of the Earth science resource. It has two basic components: (1) remnants of the primary geomorphology of the region prior to construction; and (2) the buildings, roads and other constructions which are composed primarily of materials derived from the geological resource, in the form of building stones and other construction materials (see Chapter 4).

In many cases, the original siting of a town or city was controlled by the nature of the geomorphology. For example, Roman, Norman or medieval defensive

Figure 6.5 Geological exposure sites: quarries **(A)**, cuttings **(B)**, mines **(C)** and special excavations **(D)**

B

D

positions on high ground or within river meanders are typical in Britain. The growth of urban centres which engulfed the countryside following the Industrial Revolution has left a legacy of remnant geomorphology in the centre or suburbs of many towns and cities. This is often preserved in a seminatural state in city parks or in the grounds of large private houses which have since been incorporated into the fabric of the city. These remnants were retained primarily as the 'lungs' of the great industrial cities in the nineteenth century and after, and have an aesthetic and cultural significance which is in many cases equivalent to that of the naturally occurring landscape, despite the fact that often very little in the way of unaltered landscape is left. The scientific importance of such areas is muted, but there is no doubt that there is some educational benefit of having such open spaces located within the inner city, close to schools and other educational centres.

Building stones similarly have an important aesthetic and cultural association. Often the early, pre-Industrial Revolution and transport-age building stones were locally derived and are in harmony with the local landscape and geology. The built environment therefore mirrors the local geology, and the vernacular architecture of villages, towns and cities, using local stone, has a strong cultural connection and aesthetic appeal. Educationally, the built environment offers a great challenge, in displaying a wide range of geological materials which are subjected to stringent weathering effects of the often hostile urban atmosphere (Figure 6.6).

6.3 RESOURCE MANAGEMENT

Effective management of the geological and geomorphological aesthetic/scientific resource is necessary in order to protect vulnerable sites and enhance the value. Effective management is primarily a function of five processes: **assessment, awareness, enhancement, protection** and **management.** These are discussed below.

6.3.1 Assessment of Value

Assessment of value is primarily subjective and is a function of the comparison of selected areas of the Earth with others. It assumes that some areas are more important than others on the basis of aesthetic, cultural or scientific grounds, and in most countries this has led to the selection of a series of areas of varying size, usually referred to as **sites** (Table 6.2). **Geotope** is an alternative name coined for those spatial areas which are characterised by the possession of distinctive and outstanding geological and/or geomorphological features. The recognition of such spatial units is central to the concept of conservation.

The process of selecting conservation sites, whatever their size and nature (aesthetic or scientific), is a critical component in the conservation process. For example, the inclusion of a weak site within a series or network of sites may undermine the integrity of them all. In defending a site or area it is necessary to

Figure 6.6 *The urban geological resource: a granite drinking fountain, Holborn Hill, London*

state the criteria which make it worthy of conservation, relative to adjacent or other areas, particularly where conservation may involve expense. If selection of each site and of the site network as a whole cannot be justified rigorously then it will not be conserved when threatened by development.

The problem with site selection is one of subjectivity: what is of value to one person may not be to another. This is particularly true when dealing with

Table 6.2 *Types of national geological, biological and aesthetic conservation areas. The emphasis here is on Britain, but many of these are also used in the rest of Europe and the world*

Designated area	Description
National Parks	Large areas designated by Parliament in recognition of their scenic importance and use for open-air recreation; 11 parks in England and Wales
National Scenic Areas	Areas of outstanding scenic interest in Scotland that form part of the national heritage; no statutory protection
Areas of Outstanding Natural Beauty	A wide range of nationally significant and traditional landscapes; careful control on land-use planning and management
Environmentally Sensitive Areas	This scheme aims to promote farming methods which preserve and enhance wildlife habitats, characteristic landscapes and historic features. Farmers are offered payments for following environmentally beneficial farming practices. An European Union scheme
Forest Parks	Large areas of attractive forest and open country managed by the Forestry Commission. Established to give visitors access to forests
Heritage Coasts	Selected areas of undeveloped coastline that are given a special designation to protect the coast's vulnerable beauty and enhance enjoyment in accord with conservation
World Heritage Site	Natural and cultural treasures of exceptional interest and universal value that are safeguarded for the future (e.g. St. Kilda and the Giant's Causeway)
Biosphere Reserves	Protected areas of land and coastal environment representing significant examples of biomes throughout the world, particularly valuable as benchmarks for measurement of long-term changes. A biome is a major or regional ecological community of plants and animals extending over large natural areas
Special Protection Areas	Established to conserve bird species that are endangered, have restricted or local populations, or which require special habitat protection measures. An European Union scheme
Ramsar Sites	Wetland sites that were designated in accordance with a convention signed at Ramsar, Iran, in 1971 to protect important wildfowl habitats
National Nature Reserves (NNR)	Some of the most important biological and geological sites preserved for research and study
Marine Nature Reserves (MNR)	Important Natural Areas covered by tidal waters or parts of the sea; protected to conserve marine wildlife or geological features, or provide opportunities for their study
Sites of Special Scientific Interest (SSSI)	Designated solely on rigorously defined scientific criteria; statutory protection; usually small in scale

landscape aesthetics and has led to attempts to develop objective scoring systems (Box 6.1). These do not remove the subjectivity completely, but at least allow it to be audited and the conclusions reached to be judged. Where selection criteria are less clearly expressed, the subjectivity, and therefore the validity, of a site is often difficult to evaluate externally.

The principle of site selection and designation is illustrated by the case of the conservation process in the United Kingdom, which has a system of statutory protected sites which date back to the years immediately following the second world war (Box 6.2). The National Parks and Access to the Countryside Act of 1949 enabled the government to protect areas of the British countryside which were both scenically and scientifically important. The aim was not to 'moth-ball' these areas or to prevent development, but simply to exert positive control on the planning process to enhance their aesthetic and recreational potential. The National Parks were designated on the basis of scenic beauty and regional character, without direct consideration of scientific attributes. The parks were concentrated in England and Wales, reflecting the pressure from expanding industrial centres such as the Midland, northwestern and northeastern conurbations; to date, there are no National Parks in Scotland. The National Parks were established on the basis of an overall assessment of scenic beauty and regional character, which included the buildings and small villages. Visually intrusive towns and other built-up areas were excluded.

The 1949 Act also provided for the establishment, by Royal Charter, of the Nature Conservancy which was empowered to define areas which were worthy of protection by virtue of their scientific value. The Nature Conservancy was expected to select and designate both ecological and geological sites of importance as Sites of Special Scientific Interest (SSSI) and these were assessed on the basis of strictly defined scientific criteria, but with no recourse to natural beauty or cultural significance. The designation of an SSSI did not afford a site direct protection, but simply ensured that a full consultation process between a developer and the Nature Conservancy took place. The aim was to obtain a degree of planning control and thereby achieve compromise between the needs of the developer and those of the conservationist. Until the early 1970s, sites were selected on an ad hoc basis by 'experts'. In 1977 a national survey of all potential Earth science sites was undertaken by the Nature Conservancy Council, the successor body to the Nature Conservancy. This Geological Conservation Review aimed to place site selection on a national basis and thereby strengthen the integrity of the SSSI network and its conservation (Box 6.3). In 1981 the SSSI system was legally strengthened by the Wildlife and Countryside Act which allowed the Nature Conservancy Council to list a series of Potentially Damaging Operations (PDOs) as part of the SSSI designation. Typically, the PDO list included all those activities which, if carried out in an uncontrolled manner, could lead to a diminution in the value of the site. For geological features, particularly harmful activities include such things as the dumping of waste materials, major construction works, deep ploughing, the modification of site drainage, the large-scale storage of materials, and the planting of forests. Under the 1981 Act, owner/occupiers wishing to carry out any of these activities are

BOX 6.1: AESTHETIC SITE EVALUATION: LEOPOLD'S METHOD

A proposal to build a dam in Hell's Canyon on the Snake River, Idaho, was strongly opposed by conservationists. Leopold (1969a,b) attempted to quantify the aesthetic value of this site relative to 11 other possible dam sites in Idaho and to other famous beauty spots in the USA in order to provide an objective assessment of the aesthetic value of Hell's Canyon. Leopold's approach was based on the assumption that there is some benefit to society from the existence of 'unchanged' landscapes, that a unique landscape has more value to society than a common one, and that the unique qualities enhancing landscape value are those having some aesthetic, scenic or human interest. Three sets of factors were selected to embody this definition: (1) physical factors, such as river depth, width, slope, bank erosion, valley height/width

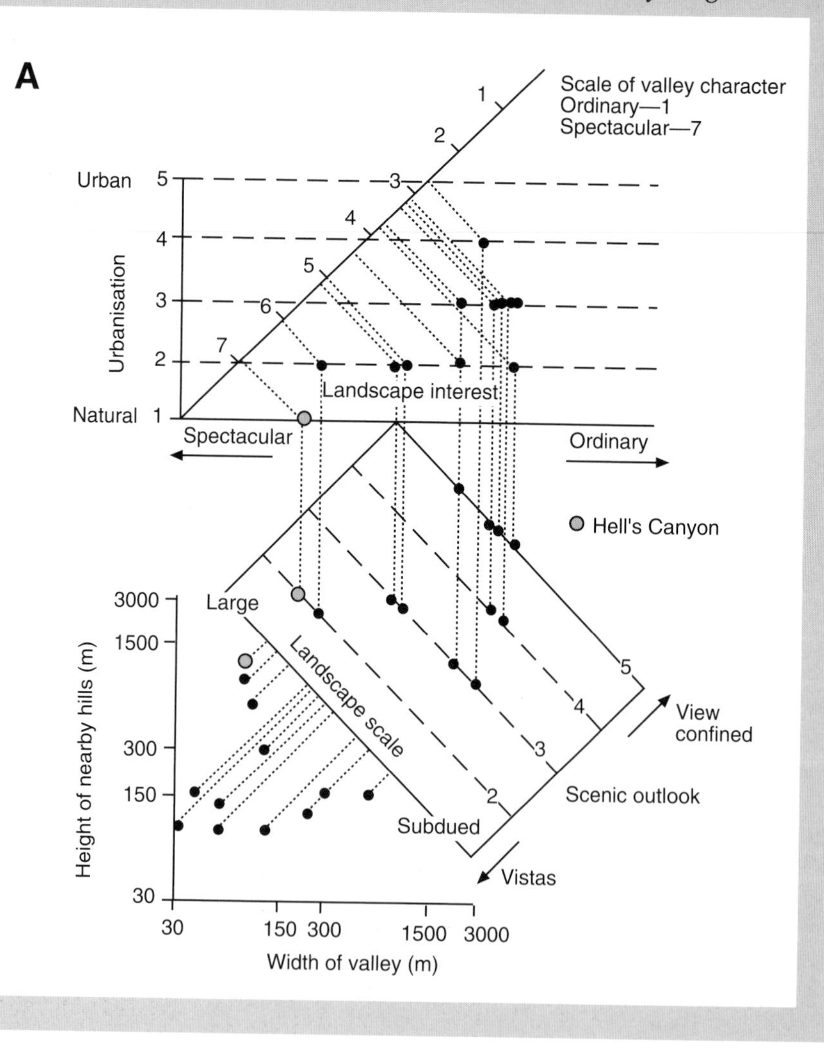

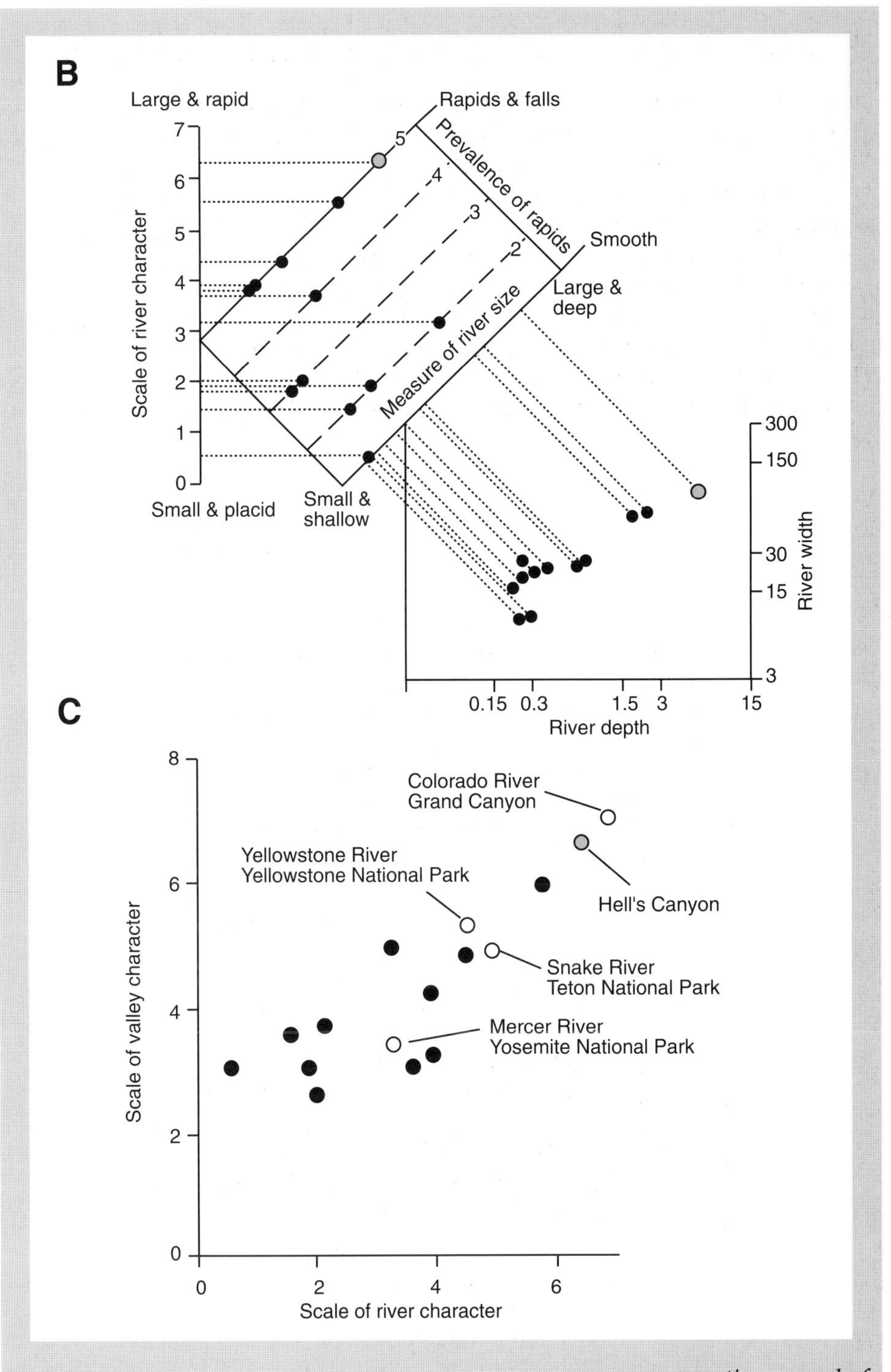

continues overleaf

BOX 6.1: *(continued)*

and valley width; (2) biological factors, such as water colour, turbidity, the presence of algae, water plants and river fauna; and (3) human interest factors, such as accessibility, land use, degree of urbanisation, utilities and the presence of rubbish. In total, 46 factors within the three groups were recognised. The landscape value of each factor was ranked 1 to 5, in some cases using specific measurements (e.g. stream width) while in others using qualitative criteria. Site 'uniqueness' was defined as the reciprocal of the number of sites sharing a particular evaluation number for a factor. For example, if all 12 Idaho sites shared the same number, the 'uniqueness ratio' for each site would be 1/12 or 0.08. In contrast, if only one site of the 12 had a particular number, if would be 'relatively unique' and would have a uniqueness ratio of 1. The uniqueness ratios for each site can be summed and the total can be compared with the totals for other sites. The equal weighting given to each site does not distinguish between sites which are unique for positive (attractive) reasons and sites which are unique for negative reasons. Despite this Leopold (1969a,b) was able to establish the aesthetic value of Hell's Canyon.

Leopold (1969a,b) exploited the data further by defining two variables affecting a person's perception of a river valley: valley character and river character. Leopold suggested that valley character was a function of the availability of distant views (landscape interest) and the degree of urbanisation. By combining these two factors as shown in Diagram A, he was able to define a single value for landscape interest. In a similar way he generated a score for river character (Diagram B), based on the assumption that grander rivers can be attributed to size, apparent speed of flow, and the extent of broken water. Diagram C shows the two scores obtained for Hell's Canyon relative to the 11 other sites and four other famous sites in the USA.

This approach to landscape evaluation is attractive because it is largely based on simple observations and provides a semiquantitative result. More importantly, the score system allows the evaluation to be subject to both internal and external audit. It should be noted, however, that the system is not free from subjectivity.

Sources: Leopold, L.B. 1969a. Quantitative comparison of some aesthetic factors among rivers. *US Geological Survey Circular*, 620; Leopold, L.B. 1969b. Landscape aesthetics. *Natural History* (October), 35–46. [Diagrams modified from: Cooke & Doornkamp (1990) *Geomorphology and environmental management*. OUP, Figs 2.36 & 2.37, pp. 61 & 62]

first required to consult the appropriate conservation agency. The period of consultation allows both parties to air their views, usually leading to a compromise solution and the development of a suitable management plan with, in some cases, compensation payable to the owner/occupier for loss of revenue. In recent years this type of conservation legislation for elite sites in the United

BOX 6.2: THE HISTORY OF NATURE CONSERVATION IN BRITAIN

Modern nature conservation in Britain started in the mid-nineteenth century with the Seabird Protection Act of 1869, and with a general growth in concern for the protection of natural history. However, a national conservation framework was only established in 1949, with the National Parks and Access to the Countryside Act. The Act created the National Parks, conferred powers on local authorities to create local nature reserves, and required the identification and conservation of Sites of Special Scientific Interest (SSSI) 'by reason of flora, fauna or geological or physiographical features'.

To oversee the selection, notification and conservation of SSSI, the Nature Conservancy was set up. During the 1950s, 1960s and 1970s, Earth science staff of the Nature Conservancy (1949–73) and its successor, the Nature Conservancy Council (1973–1991), contributed to the development of the SSSI and National Nature Reserve (NNR) networks.

The protection afforded to SSSI was improved with the passage through Parliament of the Wildlife and Countryside Act 1981, which enhanced arrangements for the effective management of conserved land.

The Environmental Protection Act 1990 and the Natural Heritage (Scotland) Act 1991 saw the most significant changes since 1949 in the Government's nature conservation strategy. The splitting of the Nature Conservancy Council into three country-based organisations – English Nature, Scottish Natural Heritage and the Countryside Council for Wales – reflected the desire to achieve conservation aims much closer to the country area. These arrangements, however, did not allow concerted action by the conservation agencies across Britain where common concerns and issues arose. In such circumstances, the agencies operate through the Joint Nature Conservation Committee (JNCC).

Source: Wilson, R.C.L., Doyle, P., Easterbrook, G., Reid, E. & Skipsey, E. 1994. *Earth heritage conservation*. Geological Society, London.

Kingdom has been supplemented by a wider range of conservation schemes at the local or regional level through the co-operation of owner/occupier and conservationist. There has also been a move away from the identification of conservation networks based on specific sites, to a mechanism by which a more holistic view of Earth heritage conservation can be obtained, which integrates wildlife, geological and aesthetic interests. In Britain this has led to the definition of **Natural Areas** as a means of creating regional focus in conservation aims. As currently defined, Natural Areas are tracts of land unified by landform type, lithologies and soils, which display characteristic vegetation types and wildlife species, and support broadly similar land uses and settlement patterns. In all, 76 Natural Areas have been identified within Britain. There are two broad objectives for this scheme: (1) to foster a holistic approach to conservation by emphasising

BOX 6.3: GEOLOGICAL SITE SELECTION IN BRITAIN: THE GEOLOGICAL CONSERVATION REVIEW

In 1977 the Geological Conservation Review (GCR) process was started to provide a rigorous process of site selection to give strength to the geological SSSI network. The aim is to identify a series of localities of national or international importance which reflect the range and diversity of Britain's Earth heritage. Site selection is undertaken with reference to three basic principles.

1. National or international importance. The term 'national' in this context refers to Great Britain, and means that any site chosen for the GCR has been assessed, wherever possible, against comparable features across the whole of Britain. In this way, only the most important Earth science sites within Britain are selected, and a national network of sites is created.
2. Minimum site selection. This is reflected in the restriction of GCR site numbers to that which is sufficient to conserve the interest, and in the restriction of the size of individual sites to the minimum area necessary to include the feature of importance. The emphasis on keeping the number and size of GCR sites to a minimum reflects two factors. First, to help ensure that GCR site status is confined to sites of bona fide national importance. Second, to ensure that only land of the highest scientific value is included within an SSSI. These features are important in practical conservation. For example, the scientific case for conserving a given site is much stronger if it is the only SSSI for a given feature of its kind in the country, as opposed to one of five sites.
3. Current understanding. This reflects the fact that geological knowledge is constantly changing and evolving and that site selection is based on knowledge at that time.

A total of 2992 GCR sites has been selected in a series of 'blocks', each representing a particular geological time period or type of interest, across Great Britain by a team of specialist academics and advisors. These have been incorporated into 2200 SSSI. This approach illustrates the importance of rigorous site selection within a conservation strategy, giving strength to the whole site network.

Source: Ellis, N., Bowen, D.Q., Campbell, S., Knill, J.L., McKirdy, A.P., Prosser, C.D., Vincent, M.A. & Wilson, R.C.L. 1996. *An introduction to the geological conservation review*. Joint Nature Conservancy Council, Peterborough.

the linkages between geology, wildlife and land use; and (2) to set objectives to enhance and conserve the key natural characteristics of each area, by involving both the conservation agencies, and the people who live and work within each area.

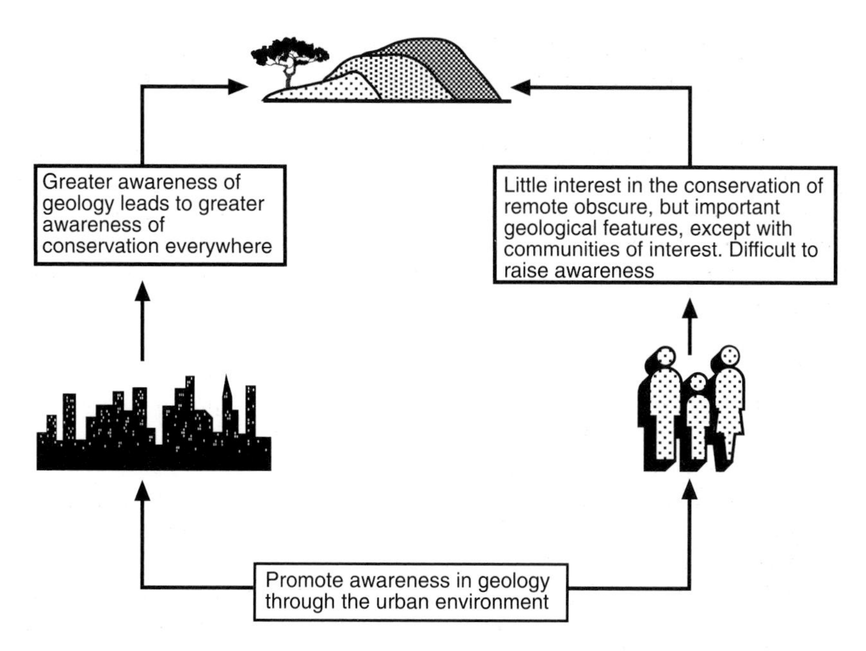

Figure 6.7 *The role of urban geology in promoting wider Earth heritage conservation. Interesting people in urban geology, where the majority of people live, increases involvement in the subject and enhances awareness, leading to the greater conservation of all geological features [Modified from: Doyle & Bennett (1997)* Geology Today *13, Fig. 2, p. 30]*

6.3.2 Promotion of Awareness

Awareness is in most cases the most important aspect of successful conservation. The successful conservation of wildlife and the promotion of the 'green' movement over the last 20 years is a function of the increase in awareness of the need for the protection of fragile habitats. Unfortunately, the conservation of geological features suffers from a lack of awareness of the ultimate fragility of the resource, and this has hampered its effective conservation. In most developed countries, and especially in Britain where 80% of the population live in urban areas, awareness of the decline in the quality of the Earth science resource in countryside areas is low, and consequently there is no national mood for its effective conservation. The answer lies in the effective demonstration of the worth of geology to the public, which in most cases can be achieved through raising the profile of the geological materials and residual geomorphology of the built environment (Figure 6.7).

Awareness can also be enhanced through an appropriate use of signage and of interpretative boards, guided tours, leaflets and other literature. In such cases, the links between the human experience and the site in question must be emphasised, drawing upon aspects of resource exploitation, scenic explanation, historical and cultural themes and the demonstration of the 'geological story' which may be fascinating in its own right.

6.3.3 Site Enhancement

Effective management and conservation of a site also needs the development of site enhancement techniques. These can vary from the promotion of a greater awareness of a site through better access, to physical actions such as, for example, the clearance of unwanted vegetation or dumped rubbish. Site enhancement is often necessary at the small scale, and particularly in exposure sites, where the overriding conservation principle is that of the maintenance of the exposure. Here, particularly where the material is unconsolidated and prone to movement, the periodic cleansing and clearance of the most important faces is an ongoing necessity.

6.3.4 Site Protection and Management Principles

The effective protection of spatial areas which encompass geological or geomorphological features of importance (geotopes) is dependent on the definition of an appropriate management strategy. In some cases this is a parallel to the difference in strategy between **preservation**, protection in an unchanged state or 'moth balling', and **conservation**, protection with a recognition that the resource is bound by nature to evolve and change with time. Preservation is most commonly associated with the protection of historic monuments, where management aims are intended to keep the monument almost in a state of suspended animation, unchanging and as like the past as is humanly possible. In contrast, conservation allows the natural or continued evolution of a landscape or habitat, but may involve some active management, for example, to prevent encroachment by unwanted flora or fauna.

Areas selected on the basis of their scenic beauty and cultural associations are mostly managed using both preservation and conservation principles. Active conservation management is used to maintain the character of the environment and to prevent unwanted intrusions, while specific components within it may be subject to stringent preservation orders. If the balance between conservation and preservation is wrong, this can have the effect of trivialising the landscape so that it takes on an unreal air which is detached from the surrounding, unprotected and 'working' landscape.

Areas selected on the basis of their scientific importance are also protected with reference to preservation and conservation principles, but their application depends on the nature of the scientific interest in the site. Here, the distinction is drawn between those sites which consist of a finite resource, **integrity sites**, and those which represent a much greater resource which is otherwise obscured elsewhere. These are referred to as **exposure sites** as they commonly include the exposures of rock outcrops which are not common above surface. This concept has been widely used in Britain and other parts of Europe as the guiding principle of Earth science site protection. Integrity sites include those with a rare fossil or mineral assemblage, such as the Devonian Rhynie Chert in Aberdeenshire, Scotland, which contains exceptionally rare and well preserved examples of the Devonian land-based flora and fauna, and those which are unreproducible and unique, including most landform sites. Exposure sites include most geological sites selected for conservation on the basis that they represent the typical expression of a particular geological unit. By their nature they include most stratotypes, that is, those sites which contain the most typical or best exposed examples of stratigraphical units. In effect, the management principle to be applied at integrity sites is the maintenance of the integrity of the site, clearly akin to the concept of preservation. Here, the principle is the maintenance of the integrity of the site. At exposure sites, the management aim is to keep the exposure open; if threatened, the site could be duplicated through the selection of an alternative exposure, or through the creation of a new exposure in the same material which is otherwise plentiful underground. Here, the most important principle is the maintenance of the exposure.

Threats to the Earth science resource are considerable. For example, the SSSI network of 2200 Earth science sites in Britain generates over 400 applications for development or substantial change each year. Of these proposals, over 93 would, if carried out, cause deterioration to the site, and about 20 would cause serious damage or destruction. Site threat is therefore significant. Threats may be defined on the basis of their potential impact in impairing the quality of specific areas of the geological landscape. Their mitigation is dependent on the scale of the problem, the area to be conserved and the rationale behind its conservation, and whether the feature is of aesthetic/cultural or scientific importance. The threats associated with aesthetic/cultural sites differ from those associated with scientific interests and are listed below.

1. **Threats to aesthetic/cultural sites**. In general, such areas fall into two basic categories. The first is associated with the large-scale land-use areas such as National Parks or similar areas of the landscape which are grouped together on the basis of shared characteristics, such as Snowdonia National Park in Wales, Yellowstone National Park in the USA and the Parque de los Glaciars in Patagonia. The second is the small-scale, limited geographical area, usually associated with a specific feature or set of features which constitute an important or unusual feature of significance. In both cases the main threats are: (A) large-scale development such as industrial, housing, transport links and other large-scale construction projects; (B) large-scale mineral extraction, particularly from newly opened quarries; (C) waste management, which can significantly alter the landscape; (D) recreational pressure from the volume of visitors and their activities, such as mountain biking, walking, causing footpath erosion, and driving four-wheel-drive vehicles; and (E) incremental loss through small-scale operations, such as new building works, and change in agricultural practices.
2. **Threats to scientific sites**. The management of most sites or geotopes selected on scientific criteria is dependent on whether they are deemed to be integrity or exposure sites. The main threats to both types of site are associated with large-scale development, and are dependent on the setting of the site: whether, for example, it is situated on a coastal cliff or within a disused quarry (Figure 6.8). The main threats are: (A) major developments such as the

construction of industrial plants or roads, which may obscure exposures and destroy a site's integrity; (B) coastal defence schemes, which obscure exposures and damage the integrity of coastal sites; (C) waste disposal, which is a damaging activity ultimately leading to loss of exposure in former quarry sites, which are commonly targeted as landfill voids; and (D) quarrying, which although beneficial in terms of the creation of exposure, is particularly damaging to integrity sites such as cave systems, static landforms and exceptionally rare mineral and fossil sites. Fossil and mineral collecting, whether for scientific, commercial or recreational reasons, is rarely a major threat to any but the most sensitive sites, although many countries forbid collecting in sensitive areas under the banner of the preservation of national heritage (Box 6.4).

There are two approaches to conservation: (1) conflict, and (2) compromise. **Conflict** deals in direct opposition to any form of development and is a high-risk strategy. In most cases, this is employed by the local 'green lobby' with varying success, and is largely dependent on winning the support of both local and national opinion. Unfortunately, direct opposition is usually an 'all-or-nothing' approach in which the stakes are consequently very high. Win, and the site is saved; lose, and the site itself may be totally lost to development. **Compromise** is often viewed as being inconsistent with conservation, but is the strategy most commonly used by statutory conservation agencies. As a mechanism for effective conservation it recognises both the need for development and the need for conservation of the natural environment. In most cases, compromise solutions will inevitably lead to some damage to the site, and its limitation is therefore of some importance; the extent of the damage will be a function of how far the conservation organisations will go before a position of outright opposition is reached. In practice, effective practical conservation is more about compromise than conflict, since in the majority of cases a policy of total conservation cannot be justified at the expense of economic development.

Effective conservation is assisted by awareness, and by effective legislation which facilitates the search for compromise. An enhanced awareness of the value of the Earth's geological or geomorphological heritage provides the best possible counter to the majority of the threats discussed above, and is demonstrated most effectively in its counterpart, wildlife conservation, which can muster a powerful 'green' lobby in defence of threatened biological sites. This lobby is also efficient in the protection of large-scale landscape areas, such as in the fight for the protection of Antarctica and other major wilderness areas, as well as in the protection of the National Parks of many countries across the world. However, it is currently less effective in the protection of rock outcrops or landforms where the rationale for their conservation is purely scientific. Here the need for effective legislation is stronger. The best legislation is that which facilities the search for a compromise within the planning process before a development takes place. Typical compromise solutions include the following.

1. **The conservation void**. Geological interests within quarries are threatened by their infill and restoration. A compromise solution is to leave a conservation

Figure 6.8 *Typical threats to exposure sites.* ***A****. Coastal defence obscuring coastal cliff exposures.* ***B****. Fly-tipping within disused quarries*

void or corridor within the quarry, adjacent to part or all of one face, which is not infilled and thus provides permanent access to the rock exposure. The majority of the quarry is infilled leaving access to only a small part of the original face (Box 6.5).

BOX 6.4: THE US PALAEONTOLOGICAL NATIONAL MONUMENTS

The United States National Parks Service currently protects 360 areas as National Parks. Many of these are geotopes, designated because of their unique geomorphological or geological features, and some, such as Yellowstone, are well known across the world. Several of the parks in the Rocky Mountains of Utah, Wyoming and Colorado are protected as National Monuments on the basis of their palaeontological importance.

The most famous is the **Dinosaur National Monument**, the headquarters of which is located near the town of Dinosaur, in Colorado. The park was designated in 1915 and extended to include areas of landscape interest in later years. The initial site was a quarry close to the Green River in Utah. This was first worked for dinosaur bones by Earl Douglass of the Carnegie Museum in Pittsburgh in 1908, yielding many famous Jurassic genera, including *Brontosaurus* (*Apatosaurus*), *Stegosaurus* and *Allosaurus*. Since the 1950s the quarry has been worked to expose the dinosaur bones, and in the summer months the public can view, in a specially constructed museum, conservators working to expose dinosaur bones in the quarry face.

Florissant Fossil Beds National Monument is close to the town of Florissant in Colorado. It is protected simply on the basis of its poorly exposed but world famous Tertiary fossil beds which were first discovered in 1873. These contain exceptionally preserved fossil plants and insects of great diversity and are an extremely important illustration of a land-based fossil ecosystem. The area was designated a National Monument in 1969 following the threat of housing development. Visitors to the site come mostly to see the visitor centre which diplays many of the exceptional fossils.

Fossil Butte National Monument in Wyoming is a National Monument by virtue of its fish and other aquatic creatures. These fossils are found in a butte which exposes around 100 m of the Eocene Green River Formation. The site was first worked for fossils in 1897 and became a National Monument in recognition of its extraordinary diversity of fossils in 1972.

In all these parks, fossil collecting is prohibited by the Antiquities Act of 1906 (for vertebrate fossils) and the Organic Act of 1916 (all fossils) which prevent removal of 'objects of antiquity' from federal lands. Collecting in the parks can only be carried out by permit, usually as part of a specific research project by bona fide scientists. However, the collection and disposal of fossils for profit or otherwise from non-federal lands outside the parks is not prohibited by law. At Florissant, it is the integrity of a finite resource that is at stake; at Fossil Butte and Dinosaur, it is the exceptional nature of the exposure which is threatened. In general, except at very special sites, fossil collecting does not pose a threat to geological conservation.

Sources: McGrew, P.O. & Casilliano, M. 1990. *The geological history of Fossil Butte National Monument and Fossil Basin*. National Park Service Occasional

Paper No 3; Meyer, H.B. & Weber, L. 1995. Florissant Fossil Beds National Monument. Preservation of an ancient ecosystem. *Rocks & Minerals*, **70**, 234–239; West, L. & Chure, D. 1994. *Dinosaur. The Dinosaur National Monument Quarry*. Dinosaur Nature Association, Utah; National Parks guides to: *Dinosaur National Monument; Florissant Fossil Beds National Monument* and *Fossil Butte National Monument*. National Parks Service, US Department of the Interior.

BOX 6.5: THE POTENTIAL CONFLICT BETWEEN QUARRY RESTORATION AND GEOLOGICAL CONSERVATION: CONSERVATION VOIDS

Quarrying has led to some of the most important geological discoveries, by providing geological exposures where none existed before. In Britain a large number of quarries, both active and disused, have been notified as Sites of Special Scientific Interest, and thereby given legal protection, due to the geological interest they contain. In recent years this has brought the conservation lobby into conflict with site owners wishing to infill and thereby restore their quarry. This is particularly true of sites scheduled for infill with domestic landfill, which is often extremely profitable. Infilling a site in this way will destroy and prevent access to the geological features. In order to resolve this conflict, the idea of the conservation void has been developed (Wright 1989). As shown in the diagrams below, a small part of the quarry is left unfilled to provide geologists with an exposure for study. These conservation voids need to be carefully designed to avoid problems of flooding and cliff failure. They have been used widely during the last 15 years within Britain to preserve some of the most important geological interests in the country.

The effectiveness of this conservation strategy has been reviewed by Bennett *et al.* (1997), who monitored a series of sites where this approach has been used. This work suggests that the technique is only applicable to hard rock sites. Loose Quaternary and clay-rich sites suffer from slumping and drainage problems and in many cases the conservation void does not work. In contrast, at hard rock sites conservation voids are more successful, although they require maintenance to control vegetation and fly-tipping. The conservation void is widely used in Britain and is a practical compromise between the educational and research needs of the geological community and the requirement for landfill sites. It illustrates the need for compromise in environmental management.

continues overleaf

BOX 6.5: *(continued)*

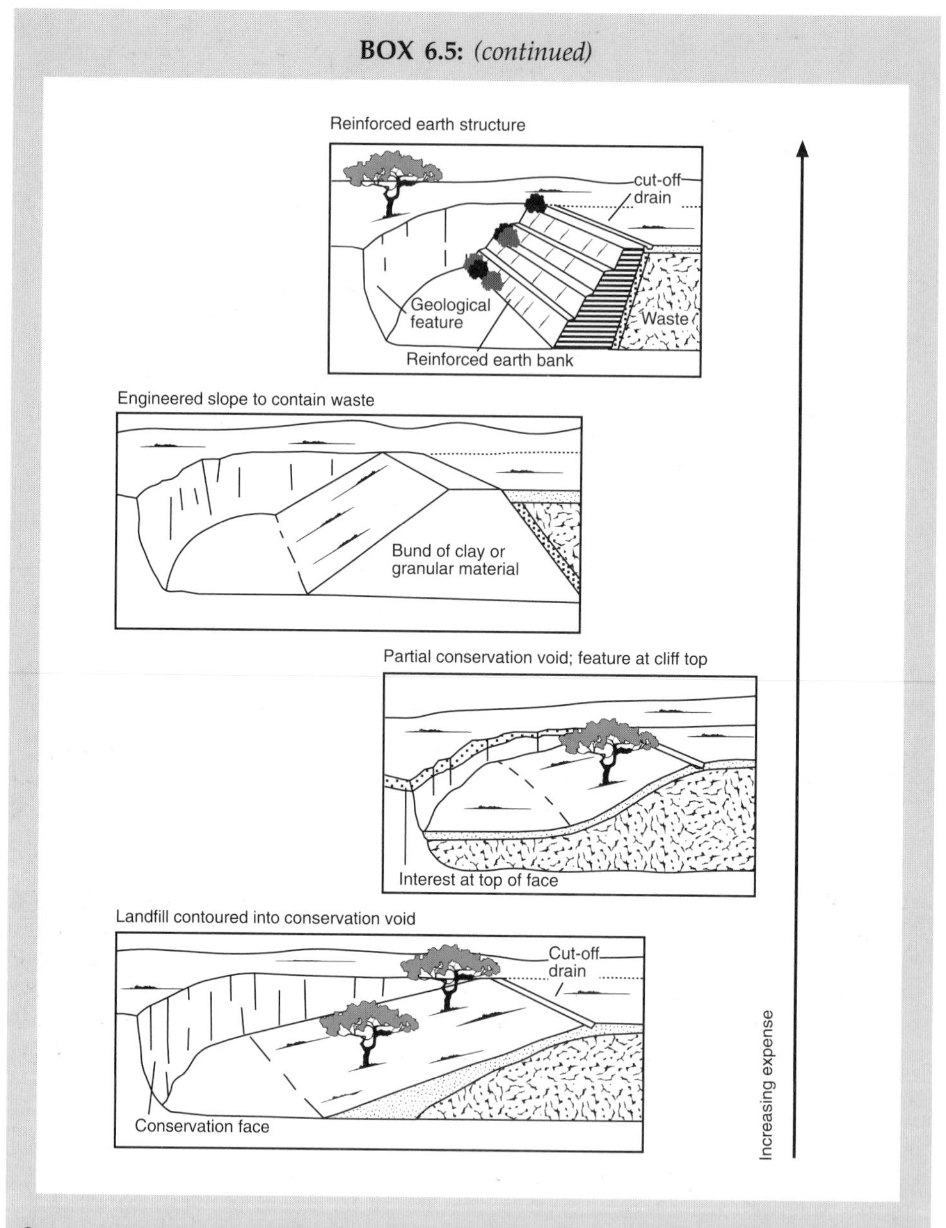

Sources: Wright, R. 1989. Of rocks and rubbish: conserving geological faces in refuse sites. *Earth Heritage Conservation*, **27**, 9–11; Bennett, M.R., Doyle, P., Glasser, N.F. & Larwood, J.G. 1997. An assessment of the 'conservation void' as a management technique for geological conservation in disused quarries. *Journal of Environmental Management*, **49**. [Diagrams modified from: Bennett *et al.* (1997) *Journal of Environmental Management*, Fig. 1]

"It serves you right. Go and dig up someone else's fossils"

Figure 6.9 *Cartoon illustrating the conflict between conservation and coastal defence. Geologists eroding cliffs to maintain exposure, while local residents are often concerned about loss of property and land*

2. **Offshore or soft-engineered coastal protection**. Traditional hard engineering solutions to coastal erosion, such as sea walls, obscure geological exposures. Conflict develops between the need to preserve exposure and the need to prevent coastal erosion and loss of cliff-top infrastructure (Figure 6.9). A compromise solution is to use offshore breakwaters or crenulated bays to reduce coastal retreat rates, while maintaining some erosion to allow the creation of fresh exposure (Box 6.6).

6.3 SUMMARY OF KEY POINTS

- Earth heritage conservation is based on four convictions: (1) that rocks and landforms should be conserved for their own sake; (2) that they provide a basis for economic exploitation; (3) that they form a basis for research, training and education; and (4) that they have aesthetic or cultural value.
- The scientific and aesthetic Earth science resource is varied, but typically comprises surface landforms and processes, rock exposures created by human activity or natural processes, and the unusual or rare rock, mineral or fossil deposits.

BOX 6.6: COASTAL CONFLICT: THE EXAMPLE OF WALTON-ON-THE-NAZE, BRITAIN

Walton-on-the-Naze is a small residential resort on the Essex coast north of London. The Naze is a peninsula of land which extends to the north of the town for 3 km and separates the salt marshes of Hamford Water, made famous by Arthur Ransom in his children's book *Secret Water*, from the North Sea. The Naze Tower, built as a navigation aid in 1720, occurs just to the north of the village and is a listed (i.e. a protected) building. The peninsula is over 20 m above sea level and the seaward face is eroding at between 0.5 and 1.5 m a^{-1} by rotational slumping driven by coastal erosion at the base of the cliff. The cliffs contain London Clay at the base, over which shallow marine sands of the Red Crag occur before the sequence is capped by a layer of loess. The sea cliffs were notified in 1961 as a Site of Special Scientific Interest (SSSI) for the Pleistocene fossil-rich Red Crag deposits, and the bird fossils within the London Clay. Cliff erosion in front of the town has been halted by battering-back of the cliffs and drainage, along with the construction of a sea wall to reduce basal erosion. Groynes are also employed to maintain the beach. The cliffs of the Naze, however, remain unprotected.

The local residents have been campaigning for a number of years for a similar defence scheme to that in front of the main town to prevent erosion of the Naze and preserve what is for them a valuable open space. More importantly, for many the eroding cliffs look unsightly and hinder beach access. Apart from the tower, no houses are under imminent threat from erosion. Conservation of the tower as a historical building is the responsibility of a government body known as English Heritage. It is also in their interest to see cliff erosion stopped. In contrast, English Nature, the government body responsible for SSSI in England, wishes to preserve the cliffs in their current eroding state. Not only are the rotational failures of educational interest, but the rock units and their fossils are of international scientific importance. Erosion of the cliffs maintains the geological exposure, so that geologists can gain access. As a consequence, two different conservation interests are in conflict on the Naze.

Over a number of years a succession of coastal defence proposals have been put forward by the local authority responsible for this stretch of coast, causing conflict with the conservation agency English Nature. This led to a public inquiry in 1978, which saw the proposals turned down on this occasion due to opposition from the conservation agencies and because the cost of the work was not justified by the cost of the land which would be lost by erosion. This conflict between conservation and the wishes of the local residents has continued and led in the late 1980s to the appointment of a consultant to find a compromise solution which would reduce the rate of land loss, but preserve some erosion and therefore maintain the geological exposure. The agreed compromise was based on the idea of a crenulated bay. A series of coastal hard points was to be established using large armourstone blocks. With time, erosion either side would generate embayments. Periodically, the hard points would have to be extended in order to avoid them being outflanked as erosion proceeded. Over time a series of artificial headlands and embayments would result. The rate of cliff recession would be reduced gradually as the embayments deepened due to the shoaling effect on storm waves. A stable coastline would result, in which erosion was completely reduced after 100

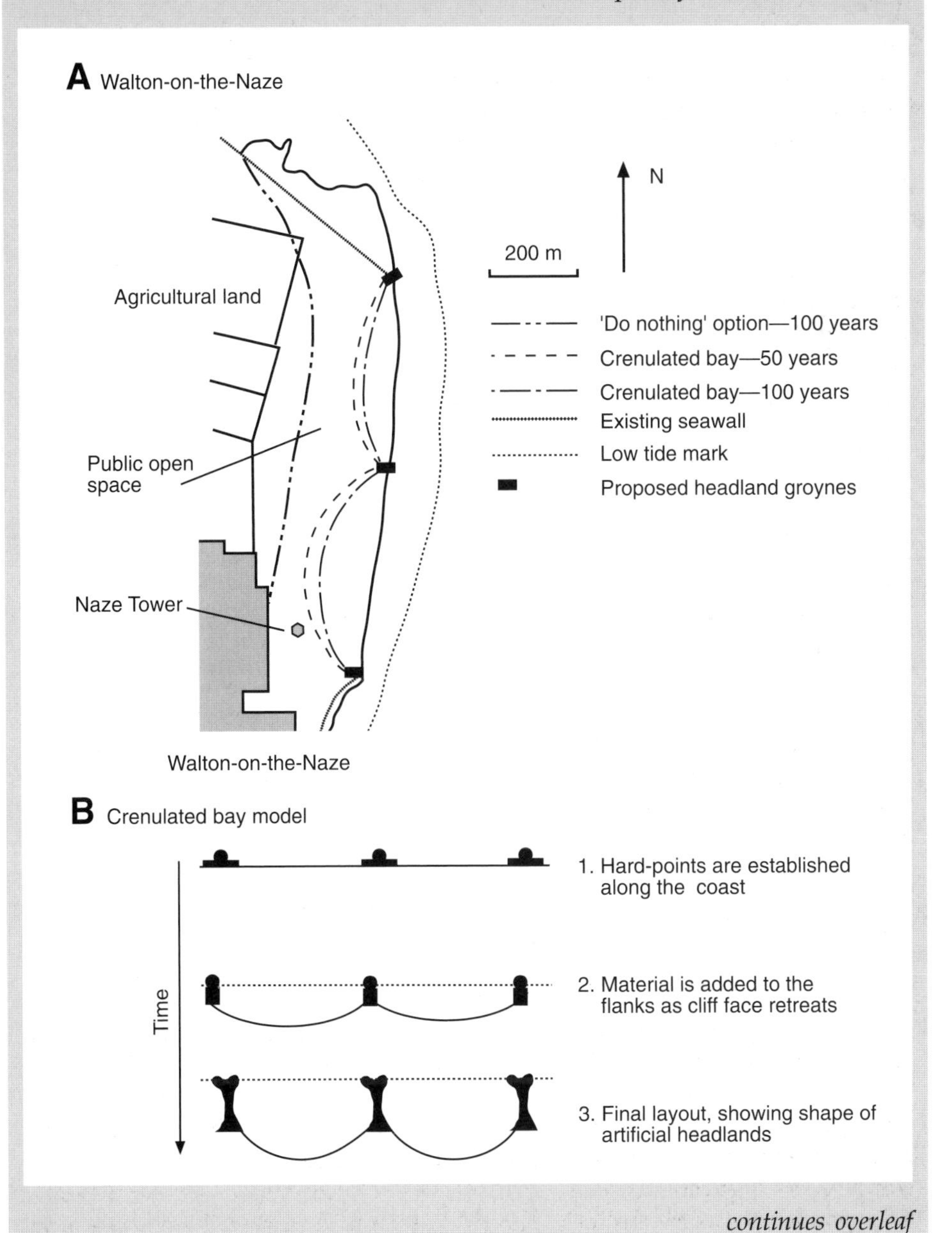

continues overleaf

BOX 6.6: *(continued)*

years. This compromise reduces land loss, maintains geological exposure for the next 100 years and thereafter it could be excavated from the degraded cliffs. Although this solution was agreed between the various interested groups, the local authority cannot currently raise the grant aid from central government to fund the work, since the cost benefit analysis does not justify the work and at present the cliffs remain untouched. This example illustrates the type of conflict which can occur on a coast between different interest groups and also demonstrates the type of compromise that is required to resolve such conflicts.

- Effective Earth heritage conservation has four features: (1) site selection and assessment; (2) raising public interest and awareness; (3) site enhancement and promotion; and (4) site protection and management.
- Effective conservation involves: (1) raising public awareness and commitment to conservation; and (2) seeking a compromise between developers and conservationists which recognises the need for development while at the same time recognising the scientific value of a site.

6.6 SUGGESTED READING

General information about Earth heritage conservation can be obtained from the edited volumes by Stevens *et al.* (1992) and O'Halloran *et al.* (1994). Evans (1997) provides a good overview on the history of nature conservation in Britain. Wilson *et al.* (1994) is the only book which covers the whole of the subject in an accessible manner. Stevens *et al.* (1994) introduces the fundamentals of the integrity and exposure site concept. Information about techniques useful in the management of geological sites can be obtained from the works by McKirdy (1990), Nature Conservancy Council (1990) and Badman (1994), while fossil collecting issues are discussed by Norman (1992). Strüm (1994) defines the term geotope and provides a useful perspective on Earth heritage conservation. The papers collected in Bennett *et al.* (1996), and that of Doyle & Bennett (1997), discuss the importance of urban geology in fostering awareness of earth heritage conservation. Finally, Tait *et al.* (1988) provide information about practical conservation, with an emphasis on the role of land-use planning.

Badman, T. 1994. Interpreting earth science sites for the public. In: O'Halloran, D., Green, C., Harley, M., Stanley, M. & Knill, J. (Eds) *Geological and landscape conservation*. Geological Society, London, 429–432.

Bennett, M.R., Doyle, P., Larwood, J. & Prosser, C.D. (Eds) 1996. *Geology on your doorstep. The role of urban geology in earth heritage conservation*. Geological Society, London.

Doyle, P. & Bennett, M.R. 1997. Earth-heritage conservation in the new Millennium: the importance of urban geology. *Geology Today*, **13**, 29–35.

Evans, D. 1997. *A history of nature conservation in Britain*. Second edition. Routledge, London.

McKirdy, A.P. 1990. *A handbook of earth science conservation techniques*. Appendices to *Earth science conservation in Great Britain. A strategy*. Nature Conservancy Council, Peterborough.

Nature Conservancy Council, 1990. *Earth science conservation in Great Britain. A strategy*. Nature Conservancy Council, Peterborough.

Norman, D.B. 1992. Fossil collecting and site conservation in Britain: are they reconcilable? *Palaeontology*, **35**, 247–256.

O'Halloran, D., Green, C., Harley, M., Stanley, M. & Knill, J. (Eds) 1994. *Geological and landscape conservation*. Geological Society, London.

Stevens, C., Gordon, J.E., Green, C.P. & Macklin, M.G. (Eds) 1992. *Conserving our landscape*. Proceedings of the conference *Conserving our landscape: evolving landforms and Ice Age heritage*, Crewe.

Stevens, C., Erikstad, L. & Daly, D. 1994. Fundamentals in earth science conservation. *Memoire de la Société Géologique de France*, **165**, 209–212.

Stürm, B. 1994. The geotope concept: geological nature conservation by town and country planning. In: O'Halloran, D., Green, C., Harley, M., Stanley, M. & Knill, J. (Eds) *Geological and landscape conservation*. Geological Society, London, 27–31.

Tait, J., Lane, A. & Carr, S. 1988. *Practical conservation. Site assessment and management planning*. Hodder and Stoughton, London.

Wilson, R.C.L., Doyle, P., Easterbrook, G., Reid, E. & Skipsey, E. 1994. *Earth heritage conservation*. Geological Society, London.

7
Engineering Geology

Engineering geology is the application of geology in the design, construction and performance of civil engineering works. Its focus is the behaviour of different geological materials with regard to problems encountered in construction (Figure 7.1). Specifically, engineering geology involves: (1) evaluating the strength and stability of soil and rock; (2) identifying areas susceptible to failure; (3) recognising potentially difficult ground conditions; and (4) designing remedial measures to combat these problems. The professional engineering geologist will therefore be involved with many of the aspects covered in this book, particularly in the design of protection methods associated with natural hazards, in the examination of stability of mineral workings, or in the determination of the strength and durability of different types of building stone. In some respects, engineering geology can be considered as the technical side of human exploitation of the geological environment, and as such is therefore a particularly important branch of environmental geology, since questions of safety are often at stake. Engineering geology is a vast and technical subject, and much of its component detail is beyond the scope of this chapter. However, the aim here is to: (1) provide an overview of the tools used in the investigation of potential construction sites (site investigation); (2) illustrate the types of engineering problems associated with different types of terrain; and (3) discuss two major engineering tasks in which geology has an important influence – rock excavations and tunnelling.

7.1 SITE INVESTIGATION

The single most important task of the engineering geologist is to characterise and predict ground conditions prior to detailed design and construction. This process is referred to as **site investigation**. This task involves: (1) documenting the solid and drift geology; (2) documenting the strengths and behavioural characteristics of the rocks and engineering soils present; and (3) recognising

Figure 7.1 The importance of engineering geology in the built environment. A housing development on Mesozoic mudstones and limestones near Bilbao, northern Spain, which are being actively eroded by the sea

potentially hazardous ground. This information is vital in: (1) the selection of construction sites or transport routes; (2) establishing the design specifications; (3) the design and choice of materials; (4) establishing the work programme; and (5) ultimately in determining the cost. If problems go undetected, accidents, delay, expensive redesign during construction, and worse, after completion, may occur. Site investigation is therefore one of the most fundamental parts of engineering geology. In Britain over 30% of all major civil engineering projects are delayed by poor ground conditions and over 50% of projects exceed the original cost estimates due to poor site investigation. Ideally, site investigation should continue until it is safe to carry out the work, but this rarely happens. The economy of brief site investigation is usually false since it is cheaper to pay for further site investigation than to contend with the financial burden associated with the delay and redesign of projects. In this discussion we recognise three scales of activity: (1) regional site investigation, involving the comparison of alternative sites or the choice of route corridors; (2) site documentation; and (3) material testing. Each is considered below.

7.1.1 Regional- or Reconnaissance-scale Site Investigation

This scale of exercise is conducted essentially to identify the relative merits, in engineering terms, of alternative sites, and is especially important in the choice of route corridors for roads and railways. It is not usually relevant to projects in densely populated areas such as Britain, where the location is often predetermined by other factors. However, it is of considerable importance in less populated countries, particularly in areas where there is the possibility of hostile terrain.

Regional-scale investigation is essentially an exercise in terrain evaluation (see Section 1.5.3), and involves the use of published maps and remote sensing. The type of assessment at this scale will depend largely on the nature of the project and the type of terrain. At one level, it may simply involve the recognition of potential problems such as abandoned mineral workings, areas of unstable ground or sensitive lithologies. At a more sophisticated level it may involve detailed geomorphological mapping to identify terrain types and their associated engineering problems. This last type of investigation is well illustrated in the evaluation of route corridors for roads in mountain ranges such as the Himalayas. For example, in locating the route corridor for the Dhahran to Dhankuta road in eastern Nepal, a simple terrain model was used to identify potential route corridors (Figure 7.2). This model was based on a combination of experience and published literature and identifies five landsystems, of which systems 2 and 4 contain numerous mass movement hazards, while system 5, located on the valley floor, is prone to fluvial hazards. Landsystem 3 consists of ancient valley floors and degraded debris slopes and is the most stable terrain element within the landscape. An ideal route corridor should cover most ground within landsystem 3 and minimise the run across slope, and therefore across the path of the numerous active landslides, in landsystem 4. Similarly, landsystem 2 is best avoided where possible because of slope hazards. Rapid terrain analysis allows the terrain to be classified and the route corridor selected to maximise the relative stability offered in landsystem 3. In general, this example illustrates the importance of reconnaissance mapping in engineering geology, which is made increasingly possible by the growth in sophistication and availability of remote sensing.

7.1.2 Site Documentation

Once a site has been chosen or a route corridor identified, the next stage is detailed **site documentation**. This involves two complementary phases: (1) the desk study, and (2) the field survey. The **desk study** involves assembling all published and unpublished maps, reports and papers relevant to the site, along with all the archival data which can be obtained. The type of information available will vary from one country to the next and on the type of job being undertaken. However, it usually includes the following.

1. **Published and unpublished material**. Typical material includes maps, memoirs, reports and papers detailing the geology, geomorphology and soil cover.

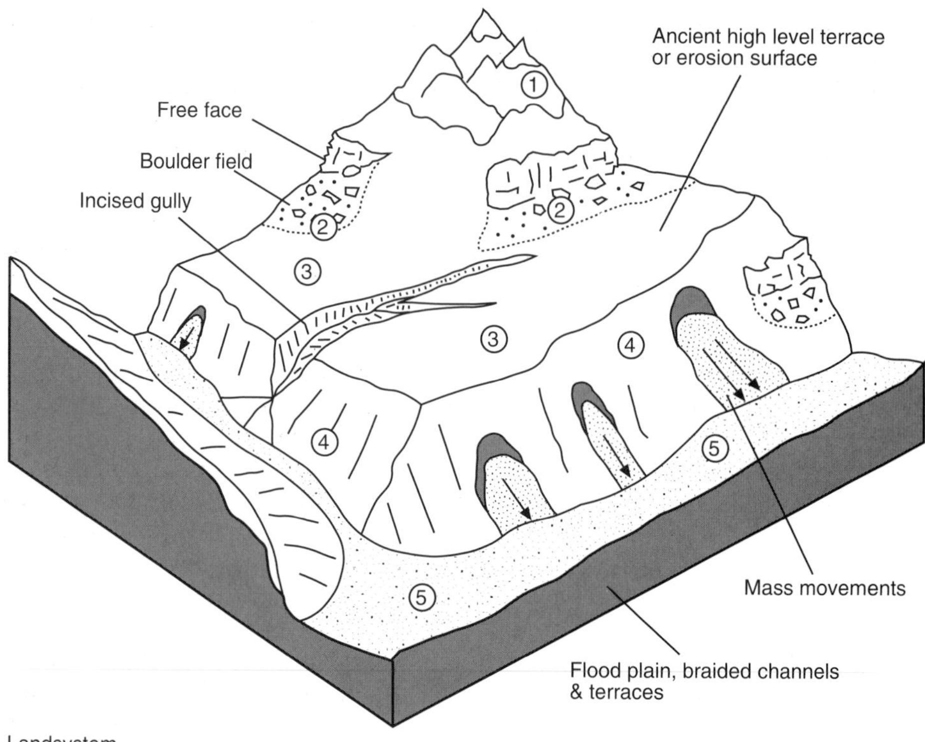

Figure 7.2 *Block diagram showing the principal landsystems within high mountain regions such as the Himalayas [Modified from: Fookes* et al. *(1985)* Engineering Geology, ***21***, *Fig. 3, p. 18]*

2. **Historical maps**. These provide information about rates of landscape change, mining history and land-use change.
3. **Topographic maps**. Such maps give basic slope and drainage information.
4. **Air photographs** and **satellite images**. These are often held in commercial or national archives, and often give greater detail than topographic maps.
5. **Miscellaneous maps**. For example, cave plans exist for most cave passages actively explored in Britain. Detailed site plans, including the original planning submissions, should be available for all redevelopment sites.
6. **Borehole logs**. In Britain it is a legal requirement that all borehole logs greater than 30 m deep are lodged with the National Geosciences Data

Centre, and this data set also includes information about many shallow excavation records and some original cores.

7. **National surveys or databases**. For example, in Britain in recent years the Department of the Environment has commissioned surveys/databases for landslides, mining instability, natural cavities, and erosion and flooding.
8. **Mining records**. In mining areas, the layout of active and abandoned mines may be recorded along with a plan of all shafts and adits. However, these records may be of variable quality. For example, in Britain coal mining records are good for all post-1947 activity, but earlier workings may be unrecorded.

In this way a large amount of information can be assembled about a site, and the potential geological problems identified. This is particularly true for sites which are being redeveloped and therefore not only have a history of past site investigation but may also contain hazardous ground conditions, such as made-ground or contaminated land, associated with previous land uses. From this synthesis, areas requiring detailed **field survey** can be identified and the data requirements for such work established. Fieldwork may involve the following tasks.

1. **Geological mapping**. This may not be necessary if the geology of a site is well understood. Where the geology is structurally complex, detailed mapping and recording of the structures present may be required to identify, for example, planes of structural weakness susceptible to failure following loading.
2. **Geomorphological mapping**. In certain types of development, such as road design, detailed mapping may provide information about the surface sediment cover, and identify relevant geomorphological hazards such as mass movements.
3. **Geophysical surveys**. These may be used to interpolate geological information between boreholes or as an aid to locating suitable areas for boreholes and trial pits. They may also be important in the search for hidden, but suspected, hazards such as sinkholes and mine shafts. Techniques include ground-penetrating radar, electrical resistivity surveys, seismic surveys, gravity surveys and magnetic surveys. The choice of technique depends upon the information requirement or the nature of the suspected problem (Table 7.1).
4. **Trial pits and trenches**. These provide excellent information about the top 2–5 m within engineering soils (i.e. any unlithified rock). They allow detailed three-dimensional facies variations to be examined and provide good access for sampling soils for laboratory testing.
5. **Boreholes**. These are used to probe the subsurface and recover samples for laboratory investigation. In Britain, three main types of borehole drilling techniques are commonly used: light percussion drilling, rotary coring, and rock probing. **Light percussion drilling** utilises a mobile A-frame with a power winch which is used to raise and drop a steel shell in order to recover

Table 7.1 *Geophysical methods of site investigation. Costs, based on 1992 prices, show the length or area covered by a survey costing £500, the approximate cost of one 20 m borehole [Based on information in: Waltham (1994)* Foundations of engineering geology. *Blackie]*

Method	Principle	Application	Cost
Electrical methods, including resistivity and conductivity survey	Works on the principle that different sediment, soil or rock types have different and characteristic electrical properties	Used widely in mineral exploration, but not used in site investigation since results are often difficult to calibrate against specific material and structures and therefore difficult to interpret	n/a
Ground-penetrating radar	Trolley-mounted transmitter and receive record microwave electromagnetic radar signals reflected from contrasts in ground properties	Restricted in use due to high capital cost and need for trained operator. Principal limitation is the limited depth of penetration, 10 to 20 m in dry soil but only 1 to 3 m in wet clay. Used to identify shallow sinkholes and surface sediment structures	700 m of line profile
Seismic surveys	Shock waves produced by hammer blows or explosives are reflected or refracted on geological boundaries/ contrasts. Reflection seismics provide little information on shallow structures and refraction surveys are more commonly used in site investigation	Widely used due to low capital cost of equipment, two-person operation and easy interpretation, although artefacts and other distortions may complicate interpretation	Six soundings to 20 m deep
Magnetic surveys	A magnetometer is used to record the distortion of the Earth's magnetic field	Simple to use; one-person operation; dipole, positive next to negative, magnetic anomalies pick out vertical linear features such as buried shafts or sinkholes; metal fences, drains or power lines may limit use	0.5 ha on a 3 m grid
Gravity surveys	Record minute variations in the Earth's gravitational force	High-cost, delicate instrument; one-person operation; negative anomalies may record underground voids or low density rock/soil infilling sinkholes or buried valleys; not very sensitive, voids need to be large to stand out from background noise; shape and nature of void difficult to interpret from the shape of the anomaly	0.2 ha on 4 m grid
Electromagnetic surveys	Electromagnetic field is generated in the ground and its intensity is measured; record electromagnetic conductivity; high conductivity of clay, basalt and water contrasts with low conductivity of sand and limestone	Low-cost equipment; one-person operation; useful in mapping shallow lateral variation in clay-filled fissues, filled sinkholes, rockhead steps, channel fills and fracture zones	0.5 ha on 3 m grid

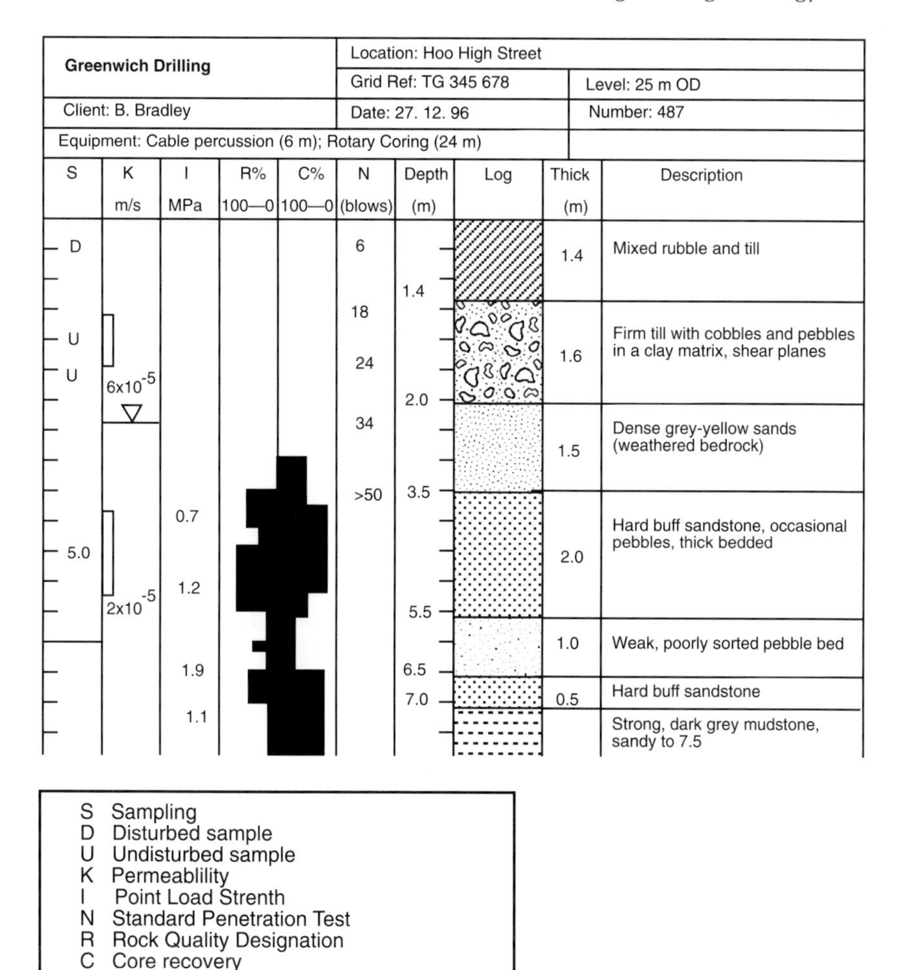

Figure 7.3 *An example of an engineering borehole log [Modified from: Waltham (1994)* Foundations of engineering geology. *Blackie, p. 43]*

a short length of core from engineering soils. **Rotary coring** is carried out using a truck-mounted rig which drives a rotary drill-bit with a tungsten carbide or diamond-studded tip into rock and soil in order to recover core sections in a hollow drill bit. Typical cores may be 1 to 3 m long at depths of over 100 m if required. **Rock probing** involves the use of a large rotary percussion rig with a hammer action and a tricone roller or drag-bit to examine rocks at depths up to and beyond 100 m. There is no core recovery, and penetration rates are used to indicate rock strength and provide information about the presence of voids. Chippings flushed out from the drill-bit may also be used to provide information. Its principal advantage is that it is cheaper than rotary coring. All boreholes are carefully logged (Figure 7.3) and the following information is typically recorded: lithological descriptions, thicknesses, penetration test data, point load strengths, water table, presence of iron staining, field permeability test data, fractures, joint density or rock quality designation, percentage of core recovery and sample points. Point load strength, penetration tests and rock quality designation are material strength tests which are described in the following section. Core recovery provides information about weak zones, since the more broken the rock or soil, the less the core recovery.

The spacing of boreholes depends upon the specific construction project. Typically, for buildings they are between 10 and 30 m apart, while they are more usually spaced 30 to 300 m apart along road lines. However, in areas with problems such as landslides or abandoned mine shafts, boreholes may be placed as little as 1 to 5 m apart. Generally, boreholes should penetrate to a depth beneath the building foundation of at least one and a half times the foundation width. In addition, if sound bedrock or **rockhead**, is not met, then at least one deep control hole is required to establish the state and depth of the rockhead. In areas where rock cavities are predicted or expected, closely spaced rock probes are usually required.

Consequently, through a combination of deskwork and targeted fieldwork, the principal engineering problem on a site can be identified, located and in many cases quantified.

7.1.3 Material Testing

Testing of rock and soils taken from boreholes and trial pits can be divided into three types of activity: (1) hard rock testing; (2) rock mass strength determinations; and (3) soil testing.

Hard rock testing involving laboratory analysis of the intact strength of the lithologies present is rare, since most have sufficient bearing capacity and the strength envelope is well known for most common rock types. Intact rock strength ignores the importance of fracture density and weathering in determining the strength of the rock mass as a whole. Consequently, in most cases rock types are simply identified and strength values are read from published tables. Where direct testing is necessary the most common method is the **unconfined compressive test**, in which a cube of rock is loaded between two metal plates. In the field, a point load test or Schmidt hammer test may be used to determine strength. In the **point load test**, cylinders of rock are loaded across their diameter between two 60° steel points with a tip radius of 5 mm until they fail. Portable equipment is available for testing cores. Point load strength is approximately one-twentieth of unconfined compressive strength. The **Schmidt hammer** is a hand-held, spring-loaded hammer which measures the rebound of its tip from the rock surface. Values usually correlate well with the unconfined compressive strength; the stronger the rock, the greater the rebound.

In practice, the effective **rock mass strength** is determined by a combination of the fracture pattern and the weathering state. It is therefore important to take a holistic view of the entire rock mass in determining strength characteristics and not base assessments simply on point-specific testing of intact rock samples. The geological components which determine rock mass strength include fracture density, fracture orientation with respect to the slope, presence of fracture infills, groundwater characteristics and weathering state. In cores, the fracture density is given by the **rock quality designation** (RQD), in which the length of each core piece that comes out is measured and used to assess the rock quality; the smaller the fragments, the poorer the rock quality. An RQD value of over 70 generally indicates a sound rock. These variables are combined in rock mass strength classifications which recognise the cumulative effect of each geological component of rock strength. Components are given different weightings which depend upon their relative importance in determining rock mass strength. There are two commonly used classification systems: (1) the **geomechanics system** (Table 7.2A), and (2) the **Norwegian system** (Table 7.2B). A rock mass or borehole core is scored against each element within the classification to obtain an indication of its rock mass strength.

Soil testing is a routine part of site investigation. Engineering soils are defined as any unlithified sediment and may include both soil as studied by pedologists, and unlithified rocks as studied by geologists. The aim of testing is to determine not only the strength of a soil, but also its sensitivity to failure. Engineering soil properties are determined by grain size, mineralogy and water content. Soils can be broadly divided into: (1) fine grained ($< 63\ \mu m$) cohesive soils; and (2) coarse grained non-cohesive soils. For fine grained soils, consistency is an important concept; this is the moisture content at which a soil behaves in a plastic or liquid fashion. This is measured with regard to a series of limits known as **Atterberg limits**. As water content increases in a soil it will shift from a solid, to a semiplastic, to a plastic and finally to a liquid state (Figure 7.4A). A series of limits and indices help define this process. The **plastic limit** (PL) is the minimum moisture content at which a soil can be rolled into a cylinder 3 mm in diameter. This approximately corresponds to a shear strength of 100 kPa (Figure 7.4A). The **liquid limit** (LI) is the minimum moisture content at which a soil will flow under its own weight. The difference between these two limits is known as the **plasticity index** (PI) which gives an indication of the amount of soil moisture required to change a soil from a semiplastic to a liquid state. The mobility of a soil at a particular moisture content (W) is given by the **liquidity index** (LI) expressed by the equation $LI = (W - PL)/PI$, which is a measure of both the consistency of a soil and its strength; the higher the liquidity index, the more unstable a soil is. The Atterberg limits can be used to classify fine grained soils as shown in Figure 7.4B, and to derive consistent descriptions using the consistency index, as shown in Table 7.3. The plastic properties of different clay soils vary depending on the amount of clay and the type of clay minerals present. In particular, the presence of montmorillonite is associated with very active or unstable clays.

The shear strength of soils is an important property and is frequently determined for engineering soils. The shear strength of an engineering soil is expressed by the **Coulomb failure envelope**, in which:

Table 7.2 *Engineering rock mass strength classifications.* ***A.*** *The geomechanics system of rock mass rating.* ***B.*** *The Norwegian Q system.* ***C.*** *Rock mass classes, general guidelines [Modified from: Waltham (1994)* Foundations of engineering geology. *Blackie, p. 50]*

A. Geomechanic system of rock mass rating

Parameter	Assessment of values and rating				
Intact rock uncombined compressive strength (MPa)	>250	100–250	50–100	25–50	1–25
Rating	15	12	7	4	1
RQD (%)	>90	75–90	50–75	25–50	<25
Rating	20	17	13	8	3
Mean fracture spacing	>2 m	0.6–2 m	200–600 mm	60–200 mm	<60 mm
Rating	20	15	10	8	5
Fracture conditions	rough & tight	open <1 mm	weathered	gouge <5 mm	gouge >5 mm
Rating	30	25	20	10	0
Groundwater state	dry	damp	wet	dripping	flowing
Rating	15	10	7	4	0
Fracture orientation	Very favourable	Favourable	Fair	Unfavourable	Very unfavourable
Rating	0	–2	–7	–15	–25

Rock mass rating is the sum of the six ratings

Table 7.2 *(continued)*

B. Norwegian (Q) system

Ratings are determined visually (see tables) for six parameters and the Q value is then calculated from:

$Q = (RQD/J_n) \times (J_r/J_a) \times (J_w/SRF)$

Q values range from <0.01 to >100

RQD, Rock quality designation

Values from borehole data	10–100	
If RQD <10, use 10 to calculate Q	10	
J_r, Joint roughness		
Discontinuous joints	4	
Rough or irregular, undulating	3	
Smooth & undulating	2	
Rough & planar	1.5	
Smooth & planar	1.0	
Slickensided & planar	0.5	
No rock wall contact across gouge	1.0	
J_w, Joint water factor		
Dry or minor flow	<100 kPa water pressure	1.0
Medium inflow	100–250 kPa	0.66
Large flow in sound rock	250–1000 kPa	0.5
Large flow washing out joint infills	250–1000 kPa	0.33
Very high flows	>1000 kPa	0.2–0.05
J_n, Joint set number		
Massive, no or few joints	0.5–1.0	
One joint set; if random also, add 1	2	
Two joint sets; if random also, add 2	4	
Three joint sets; if random also, add 3	9	
Four or more joint sets	15	
Crushed rock, earth-like	20	
J_a, Joint alteration number		
Rock wall in contact:		
Clean, tight joints	0.75–1.0	
Slightly altered joint walls	2	
Silty or sandy clay coatings	3	
Soft clay coatings	4	
Gouge or filling <5 mm thick:		
Sandy particles or fault breccia	4	
Stiff clay gouge	6	
Soft or swelling clay gouge	8–12	
Thick or continuous clay zones	10–20	

continues overleaf

Table 7.2 *(continued)*

SRF, Stress reduction factor	
Fractured rock prone to loosening:	
Multiple weakness zones with clay, loose rock	10
Multiple weakness zones, no clay, loose rock	7.5
Single weakness zones, cover depth >50 m	5
Single weakness zones, cover depth <50 m	2.5
Loose open joints	5
Sound rock:	
Low stress, near surface	2.5
Medium stress	1.0
High stress, tight structure	0.5–2.0
Mild–heavy rock bursts	5–20
Mild squeezing or swelling rock	5–10
Heavy squeezing or swelling rock	5–20

C. Guideline properties and rock mass classes

Class	I	II	III	IV	V
Description	Very good rock	Good rock	Fair rock	Poor rock	Very poor rock
Rock mass rating	80–100	60–80	40–60	20–40	<20
Q value	>40	10–40	4–10	1–4	<1
Safe cut slope	>70°	65°	55°	45°	<40°
Tunnel support	None	Spot bolts	Pattern bolts	Bolts + shotcrete	Steel ribs
Stand-up time for span	20 years for 15 m	1 year for 10 m	1 week for 5 m	12 h for 2 m	30 min for 1 m

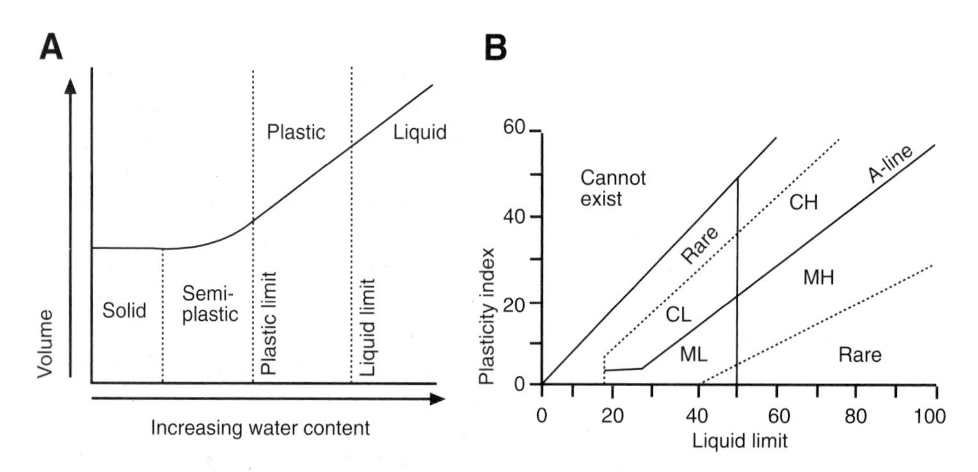

Figure 7.4 *Atterberg limits used to classify and describe the sensitivity of fine grained soils.* **A.** *Variation in volume and state with increasing moisture content.* **B.** *Classification of fine grained soils (CL=clay; ML=silt; MH=clayey silt; CH=plastic clay)*

Table 7.3 *Soil classification.* **A.** *Based on the consistency index for fine soils.* **B.** *Based on standard penetration test N-value for coarse soils (sands and gravels) [Modified from: Dearman (1991)* Engineering geological mapping. *Butterworth-Heineman, Tables 4.20, 4.22 & 4.24, pp. 42 & 43]*

A. Consistency index for fine soils

$$\text{Soil consistency index} = \frac{\text{Liquid limit} - \text{Moisture content}}{\text{Liquid limit} - \text{Plastic limit}}$$

Term	Consistency index	Field description
Very soft	<0.05	Exudes between fingers when squeezed in hand
Soft	0.05–0.25	Can be moulded by light finger pressure
Firm	0.25–0.75	Can be moulded by strong finger pressure
Stiff	0.75–1.00	Cannot be moulded by finger pressure; can be indented by thumb nail
Very stiff or hard	>1.00	Can be indented by thumb nail

B. Standard penetration test (SPT) for coarse soils

Term	SPT N-value (blows per 300 mm penetration)
Very loose	1–4
Loose	4–10
Medium dense	10–30
Dense	30–50
Very dense	>50

$$\text{Shear strength } (\tau) = c + (\sigma - \text{PWP}) \tan \Phi$$

where c is cohesion derived from the interparticle bonds, which is high in clays but zero in coarse sands; σ is the normal stress or load; PWP represents pore-water pressure; and Φ is the angle of internal friction due to the interparticle roughness. These properties are illustrated in Figure 7.5A. It is clear from this equation that changes in normal stress, or in porewater pressure, due to drainage changes have a critical role in determining the strength of a soil. Weathering may reduce both cohesion and the angle of internal friction. The shear strength of clays changes with remoulding of the sediment (Figure 7.5B). As a clay is reworked, its peak strength declines to a residual value due to the realignment of clay particles. In very sensitive clays this realignment may reduce shear strength dramatically. The shear strength of soils is determined in the laboratory using a **triaxial test**, in which a cube of soil is loaded axially, top and bottom, while the pressure around the cube is maintained using a pressure bath to prevent lateral failure. The experiment is repeated for different loads, and the applied loads and associated pressures within the containing bath are used to calculate the shear strength (Figure 7.5C).

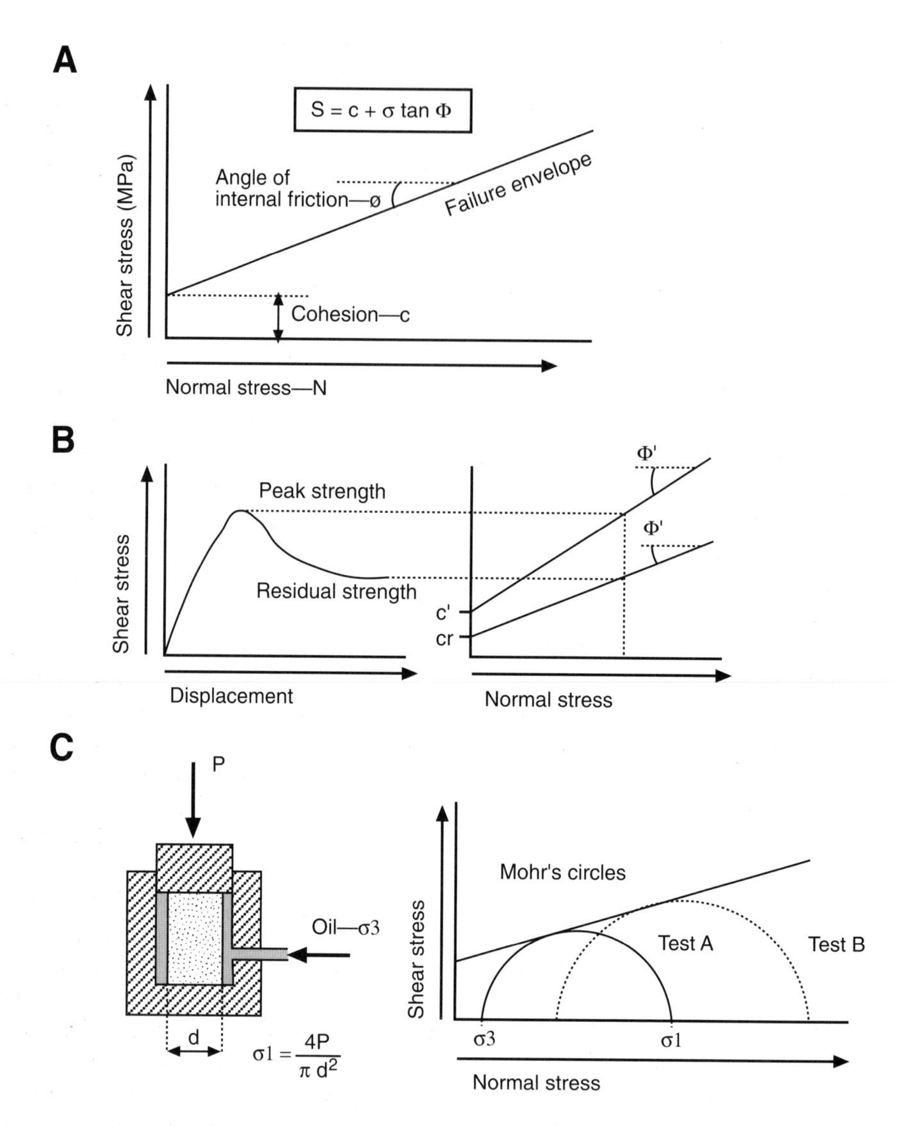

Figure 7.5 *Shear strength of rock and engineering soil.* ***A.*** *Components of shear strength.* ***B.*** *The effect of remoulding of engineering soils on shear strength.* ***C.*** *The triaxial shear test*

In the field, **cone penetration tests** are used to estimate the bearing capacity of engineering soils. Most cone penetrometers consist of a 60° steel cone, 36 mm in diameter, inside a steel sheath or cylinder. Both the cone and sheath are driven into the soil and the resistance to penetration of both the cone and sheath are recorded, providing an indication of material strength. Alternatively, a **standard penetration test** is used, particularly as part of borehole investigations. In this test, a 51 mm split tube is driven into the sediment for 150 mm, then a 64 kg hammer is dropped on to it over a distance of 760 mm and the number of blows required to drive the tube a further 300 mm is recorded. This provides a rough indication of the strength of a soil (Table 7.3). Strength determinations are important in designing cut slopes and for assessing the sensitivity of soils to movement, consolidation and failure. Other tests, such as estimates of permeability, may then be made within boreholes and trial pits; the range of tests will depend primarily on the nature of the proposed project.

7.1.4 Site Reports and Engineering Geology Maps

Information obtained from site investigations is collated and synthesised either in the site report or in some form of engineering geology map or plan. The type of output produced depends largely on the aims of the site investigation and its scale. For example, in some areas, regional engineering geology maps have been produced which identify those areas with hazardous or problematic ground conditions. Similarly, maps may be produced for transport corridors, identifying problem areas where more detailed investigation or design work may be required. However, if the site is small the information is more commonly summarised either on an engineering plan or in a site report.

An **engineering geology map** is any geological or topographic map or plan which has been annotated with information about ground conditions relevant to construction. There are many different types of maps used at a wide range of scales. However, there are two broad categories of map which form end members of a continuum: (1) the landsystem map; and (2) the detailed site plan. The engineering **landsystem map** is thematic in nature and its purpose is to identify and classify land parcels with similar engineering properties. These properties have usually been investigated in detail at typical sites within each land parcel, providing generalised reference information for the whole area (Figure 1.7). This approach is commonly used at a regional scale. In contrast, **detailed site maps** or **plans** consist of maps containing geomorphological, geological and topographic data with varying amounts of annotation which contain point-specific data (Figure 7.6). This type of map is essential in the design of foundations or cut slopes in difficult terrain. Detailed site maps may take a variety of forms, from simple annotation of site plans to comprehensive surveys.

Although conventions exist about the production of engineering geology maps and plans, their essential spirit is the annotation of a map with information about lithology, structure, geomorphology and groundwater which are relevant to a particular construction problem, and the degree of generalisation depends upon the scale and purpose of the map.

7.2 ENGINEERING PROBLEMS IN COMMONLY OCCURRING TERRAIN

In this section, the principal engineering problems associated with different types of geological terrain are discussed. The main types of terrain from an engineering

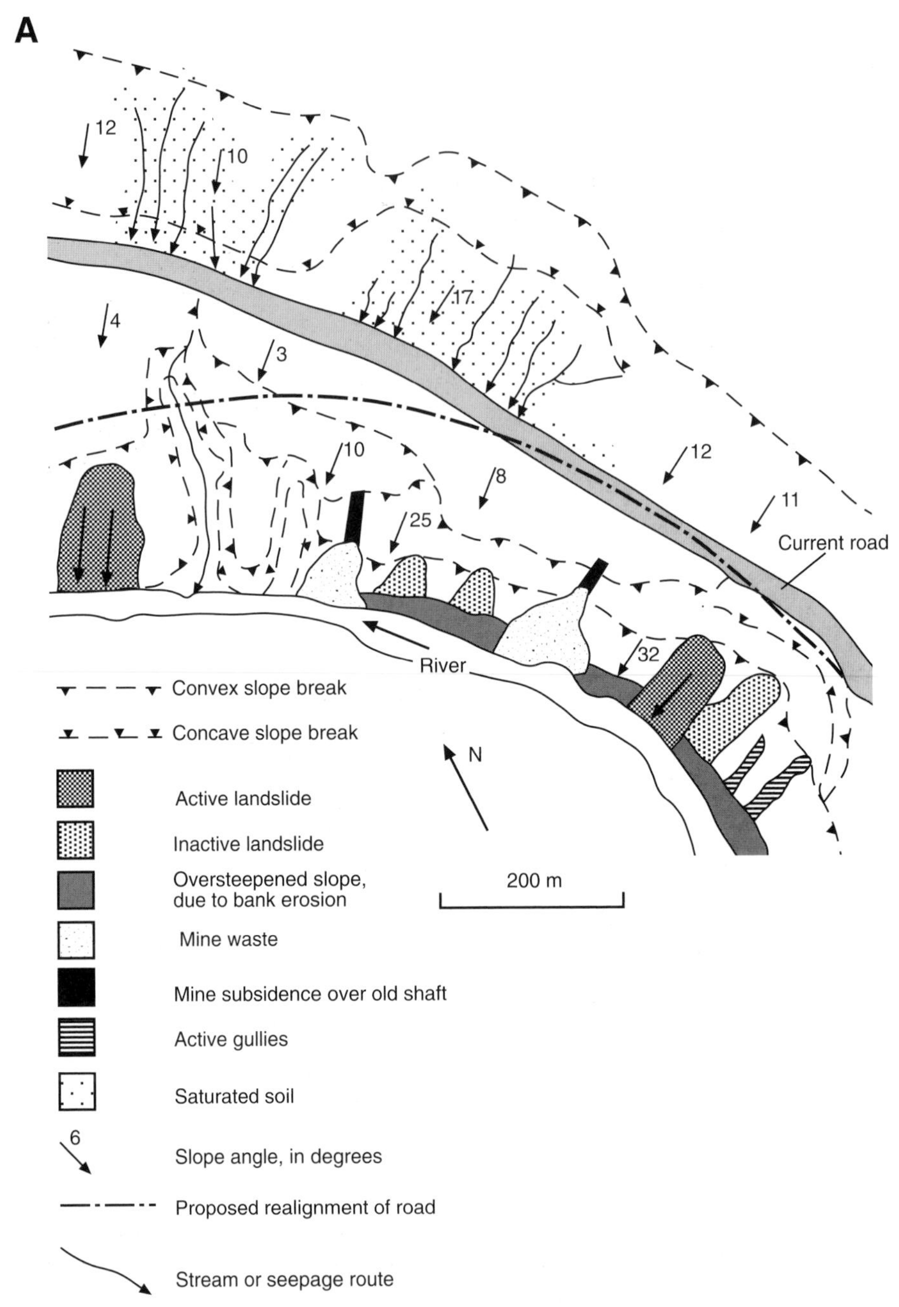

Figure 7.6 *Examples of engineering geology maps, plans and sketches.* **A.** *Typical map produced for the realignment of a road, showing breaks of slope, geomorphological processes and other potential problems.* **B.** *Detailed map of a proposed dam site showing variation in rock mass strength of the foundation rocks.* **C.** *Sketch, based on a photograph, showing the stability works neccessary on a section of cliff. [Based on information in: Dearman (1991)* Engineering geological mapping. *Butterworth-Heinemann]*

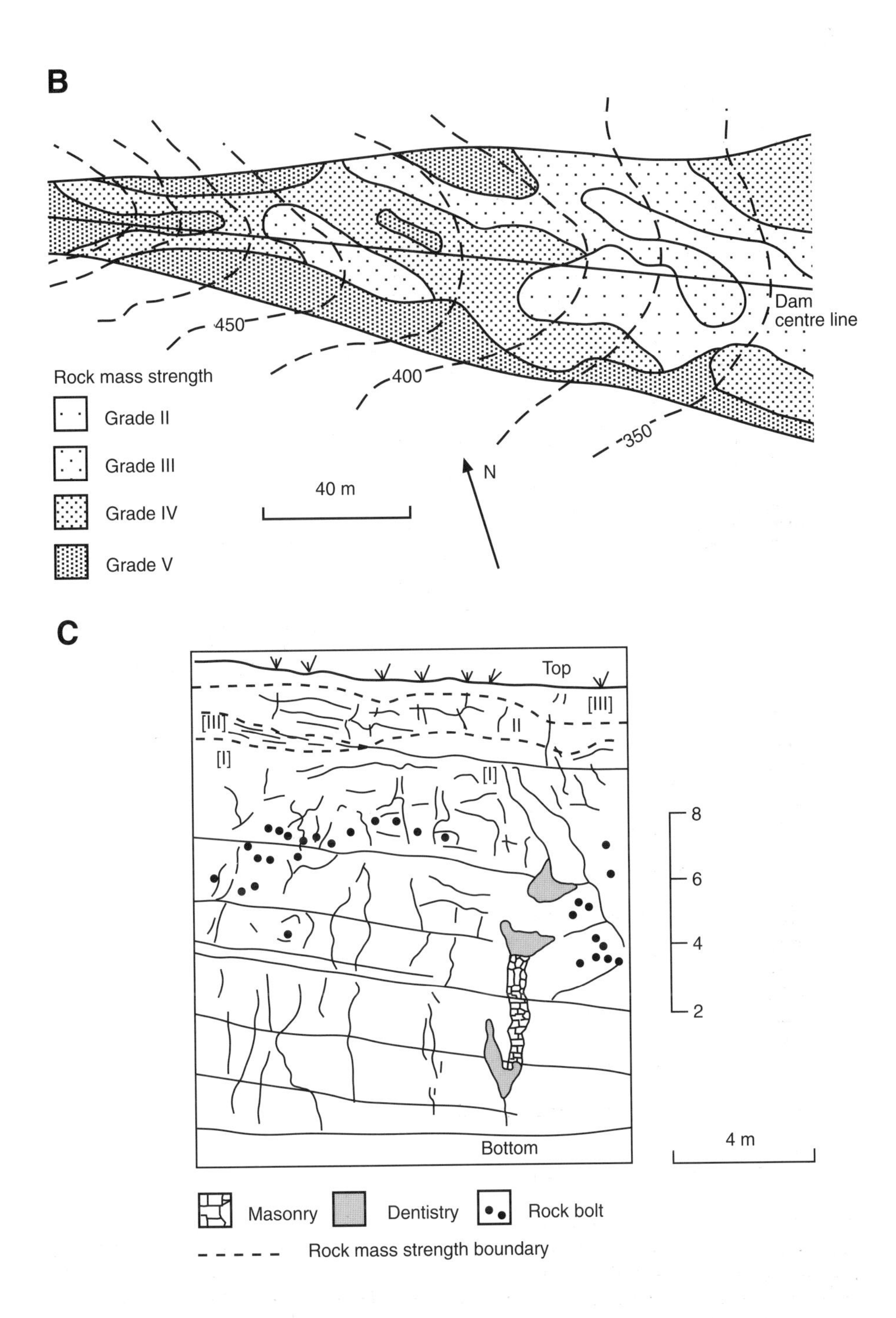
B
450
400
350
Dam
centre line
Rock mass strength
Grade II
Grade III
Grade IV
Grade V
40 m
N
C
Top
[III]
[III]
II
[I]
[I]
8
6
4
2
Bottom
4 m
Masonry
Dentistry
Rock bolt
Rock mass strength boundary

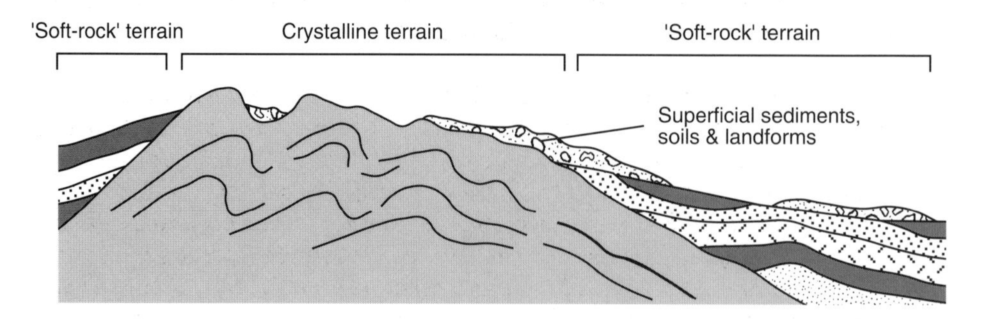

Figure 7.7 *A conceptual cross-section through the three main types of engineering geology terrain types discussed in the text: crystalline terrain, 'soft rock' terrain and superficial landforms and sediments*

geology perspective are: (1) crystalline ('hard rock') terrains; (2) sedimentary or 'soft rock' terrains; and (3) surficial sediments and landform terrains. The distribution of these three broad terrain types is shown conceptually in Figure 7.7, and each is associated with a different range of engineering problems or challenges.

7.2.1 Crystalline Terrains

Most areas containing crystalline rocks can be considered relatively stable with acceptable load-bearing capacities. This is particularly so where the rock has a uniform, unfoliated texture, is not folded or faulted, and is not weathered. This ideal is not easily reached, as many crystalline terrains, particularly those of great antiquity, often have very complex tectonic histories and may have been repeatedly folded, faulted and fractured. Two basic types of crystalline terrain may be recognised: (1) metamorphic basement, usually composed of foliated schists and gneisses; and (2) unroofed igneous intrusions.

Metamorphic terrains are common in orogenic belts of Precambrian, Palaeozoic and Mesozoic age, and comprise regional metamorphic rocks of high grade, such as schists and gneisses, which are commonly associated with metasedimentary cover sequences of phyllites, psammites and marbles. This type of assemblage is common in the Precambrian–Palaeozoic orogenic belts at the heart of most of the major continents, although it also occurs in the Mesozoic–Cenozoic orogenic belts of the Alpine–Himalayan mountains, the west coast of the Americas, and much of Asia.

Unroofed intrusive rocks are also often associated with orogenic belts, but may be of more variable setting and size, dependent on the type of intrusion. Batholiths and other plutons provide the most extensive crystalline terrain of this type, although smaller intrusives can also provide problems for engineering within otherwise sedimentary successions. In all cases, the cover sequences have been stripped by erosion to reveal the intrusive rocks beneath. Two factors are of great importance when engineering within crystalline terrains.

1. **Structural discontinuities and foliations**. Discontinuities may often determine the strength of the rock mass and provide the engineering challenge in crystalline rock terrains. A clear understanding of structural geology therefore has an important role to play during site investigation, in order to identify potential failure surfaces and block size. In metamorphic terrains, an understanding of the foliation direction is important, as these surfaces are prone to failure when loaded. For example, in New York, loading of the Manhattan Schists, weakened by plant root growth along foliations, gave rise to catastrophic failure in which a block with a total volume of 430 m^3 failed along the inclined foliation plane.

   Undeformed igneous rocks are less likely to have significant foliation, and are generally the most suitable materials for the founding of large or complex engineering projects. However, joint spacing can be a determinant in the overall strength of the rock mass, as closely spaced joints can help promote failure and deep weathering. This was a particular problem in the construction of the command centre of NORAD – the North American Air Defence Command – which was constructed in the 1960s in a granite body beneath Cheyenne Mountain in Colorado. This centre was intended to be resistant to nuclear attack, and the granite permitted the excavation of the extremely large chambers necessary to contain the underground centre. However, after excavation, close joint spacing was observed in places, casting doubt on the safety of the complex and necessitating the construction of reinforced concrete domes in some of the chambers.
2. **Depth to rockhead**. In some cases, sound rock may be at considerable depth beneath a weathered regolith; this laterally variable cover sequence is formed through a variety of processes, not least of which is the chemical leaching of feldspar-rich granites and gneisses. In this process, feldspars break down to form an often structurally weak clay-mineral complex, passing at depth into isolated 'corestones' of weathered crystalline rock lying above an irregular rockhead surface (Figure 7.8). Deep chemical weathering of crystalline terrain is most common in tropical areas with high rainfall, and has caused a significant problem in the construction of high-rise buildings and underground rapid-transit systems in Hong Kong (Figure 7.9). Here, deep leaching of an exposed granite intrusion means that large buildings require extensive piling, and excavations for the underground system have to be deep in order to reach structurally sound crystalline rock basement.

   The weathering profile of crystalline terrains may also be irregular causing lateral continuity problems, with rapid shifts from sound crystalline material to structurally weak sediments. This is compounded in ancient terrains by an uneven erosional unconformity surface overlain by superficial sediments. For example, the schists capable of supporting high-rise blocks on Manhattan Island in New York are buried in places beneath a considerable depth of laterally discontinuous superficial sediments, requiring the construction of

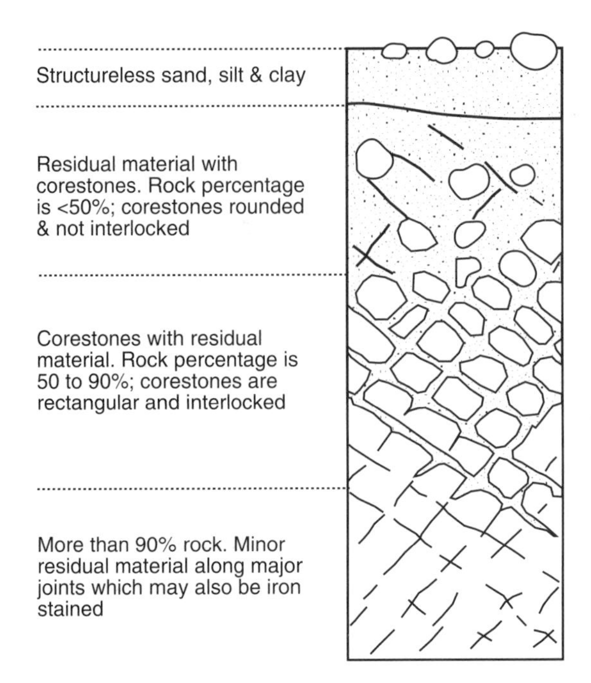

Figure 7.8 *Zones of weathering of crystalline rocks [Modified from: McFeat-Smith (1989)* Bulletin of the Association of Engineering Geologists, **26**, *Fig. 12, p. 44]*

piling keyed to the schist rockhead. This was the case with the World Trade Centre, once the world's tallest building, in which the schist rockhead was approximately 7 m beneath water-lain silts and fill materials. Variability in the rockhead surface topography, and therefore in the relative thickness of superficial sediments, can also cause problems in tunnel construction, as a variety of engineering techniques will be necessary to handle the transitions between crystalline rock and different types of engineering soils (Figure 7.10).

7.2.2 'Soft Rock' Terrains

This terrain component is primarily composed of sedimentary rocks such as clays, sandstones, chalks and other limestones, shales and other similar rock types. As Figure 7.11 illustrates, the majority of these have sufficient bearing capacity for most construction projects, with perhaps the exception of clays and some shales. However, a wide range of engineering problems can be encountered in such terrains, from mass failures to compaction and subsidence problems. We illustrate two typical soft rock terrains, and discuss their associated problems: (1) foundation problems on carbonate rocks; and (2) problems associated with clay compaction. On carbonate terrains the engineer must focus on problems of solution, subsidence and cave collapse, while the compaction and flow of clays may cause significant problems in the design of foundations.

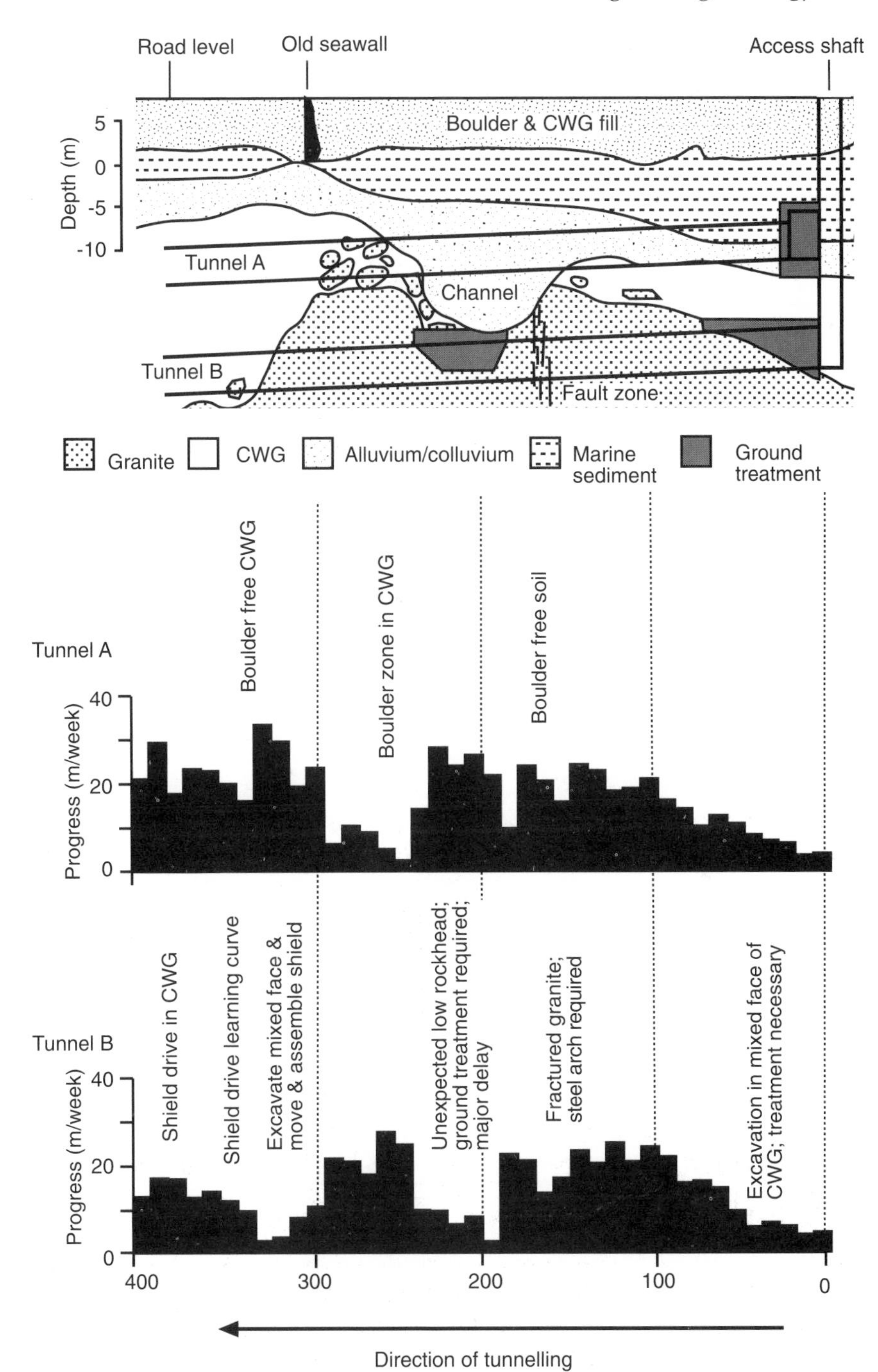

Figure 7.9 *Problems of tunnelling in weathered cystalline terrain in Hong Kong; tunnels cut through rapidly changing geological materials (CWG=weathered granite consisting of silts and weathered boulders) [Modified from: McFeat-Smith (1989)* Bulletin of the Association of Engineering Geologists, ***26****, Fig. 35, p. 90]*

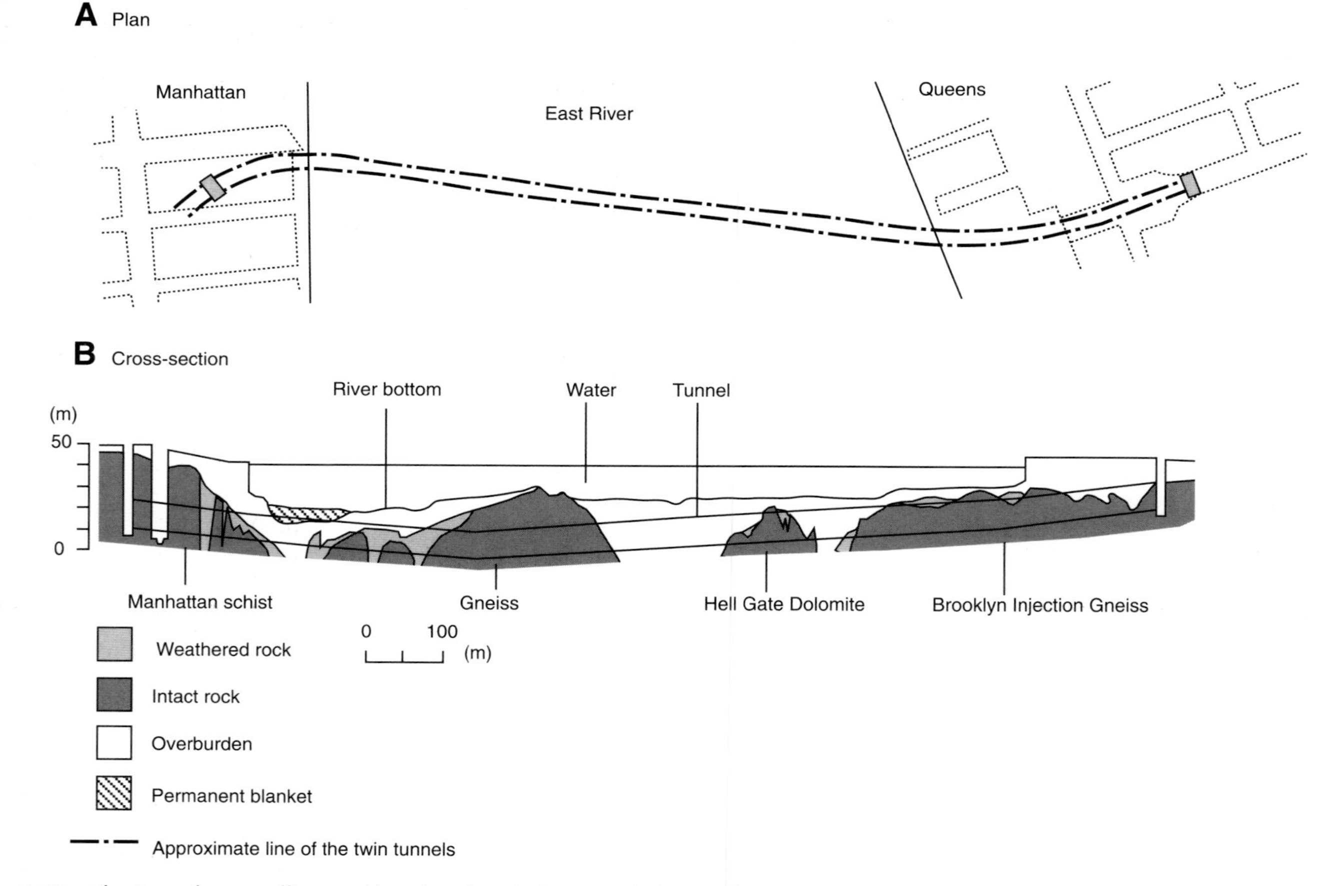

Figure 7.10 *The irregular cystalline rockhead surface in New York City and its impact on tunnelling [Modified from: Fluhr (1957)* Engineering Geology Case Studies, **1**, *Fig. 1, pp. 6–7]*

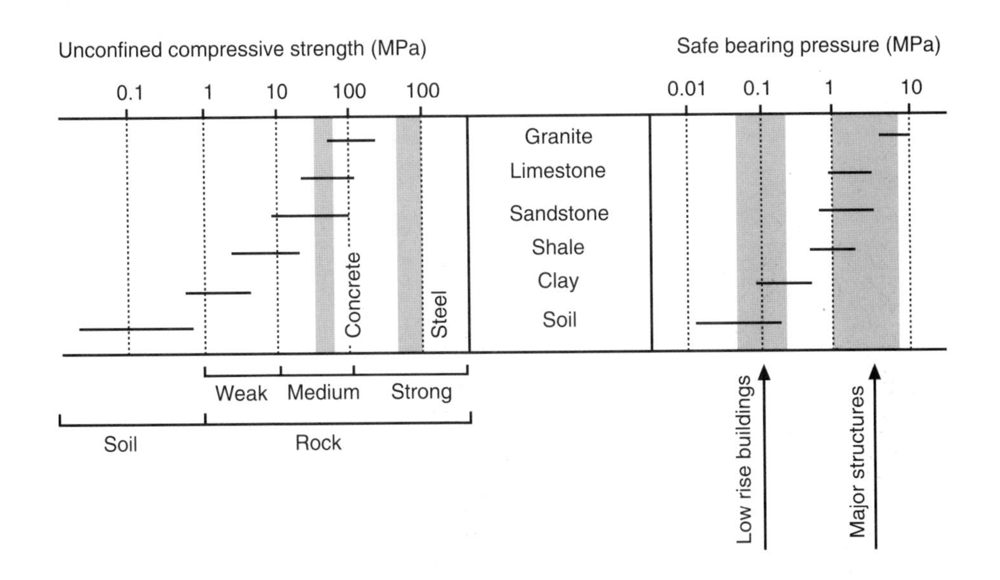

Figure 7.11 *Bearing capacity of common rock types [Modified from: Waltham (1994)* Foundations of engineering geology. *Blackie, p. 3]*

Carbonate terrain

Carbonate rocks dissolve in contact with weakly acidic rain and soil water. This acidity is derived from solution of carbon dioxide in the atmosphere and in the soil. In limestone areas, solution of joint networks leads to caves, caverns and underground passages, while in chalks solution is often less focused and cave systems are less common. Figure 7.12 provides a conceptual model of the types of phenomena associated with carbonate terrains. In engineering terms, the following problems may be encountered.

1. **Pinnacled or irregular rockhead**. Solution of the rockhead, and in particular the exploitation of major joints, may create an irregular and uneven rockhead surface (Figure 7.12A&B). In Britain this may be exacerbated by the presence of thick layers of frost-shattered chalk. Careful site investigation involving densely spaced boreholes is essential within such regions (see Section 7.2.3). Figure 7.13A illustrates the types of foundation design which may be used to overcome these problems.
2. **Sinkholes and subsidence**. Sinkholes are closed surface depressions, which may or may not be visible on the ground surface. They may be conical, cylindrical or irregular, isolated or clustered, and are typically 1 to 50 m deep and between 1 and 200 m wide (Figure 7.12A). Larger examples may be referred to as **dolines**. They may form in a variety of ways (Figure 7.12C&D): (1) by ground solution (**solution sinkholes**); (2) by the collapse of subterranean caverns (**collapse sinkholes**); or (3) by the loss of soil through

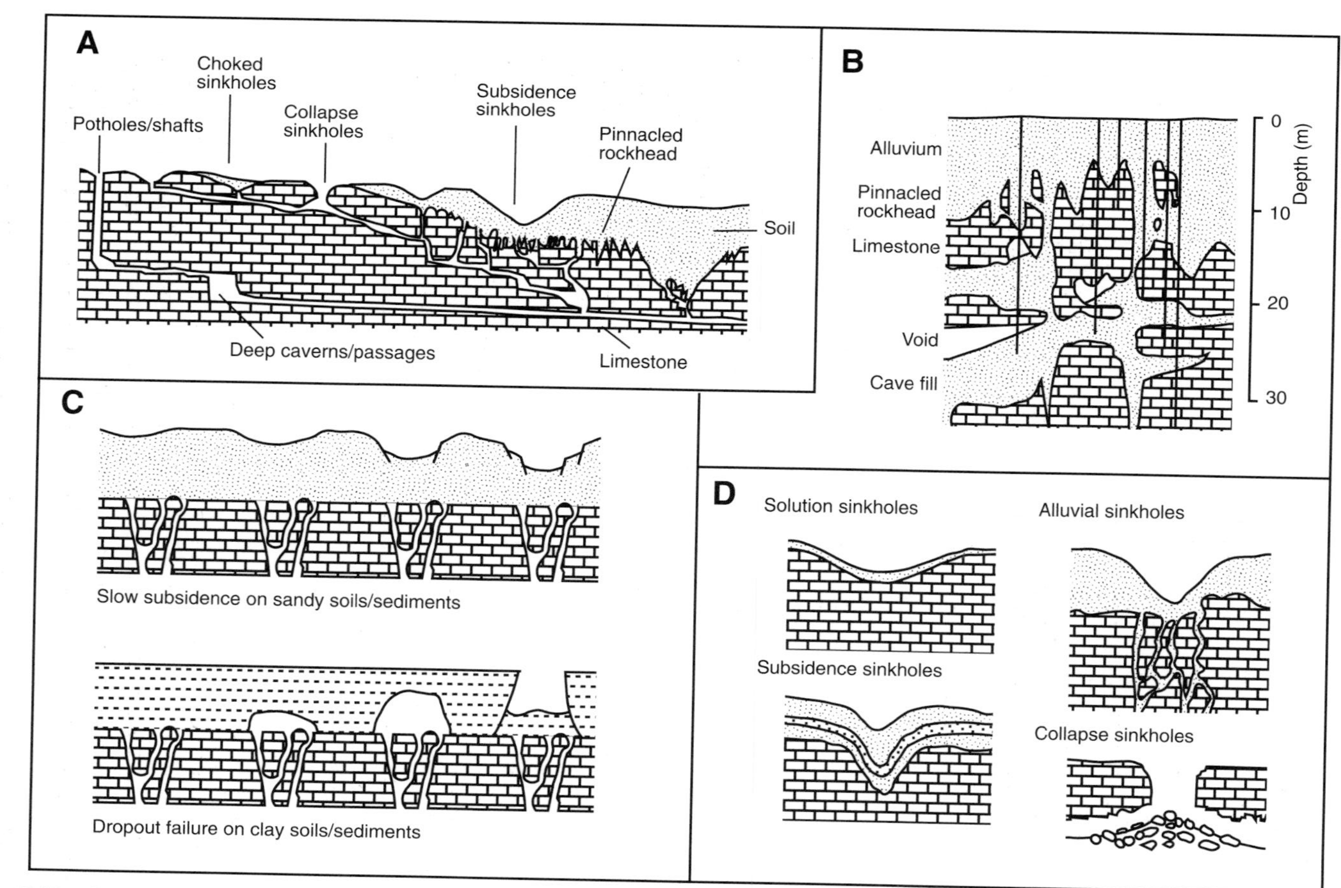

Figure 7.12 *Engineering geology problems associated with carbonate terrain.* **A.** *Common hazards in carbonate terrain.* **B.** *Pinnacled rockhead at a site in Kuala Lumpur, Malaysia.* **C.** *Types of ground collapse depending on the nature of the clay soil cover.* **D.** *Four mechanisms of sinkhole formation [A & B Modified from: Waltham (1994)* Foundations of engineering geology. *Blackie, p. 58]*

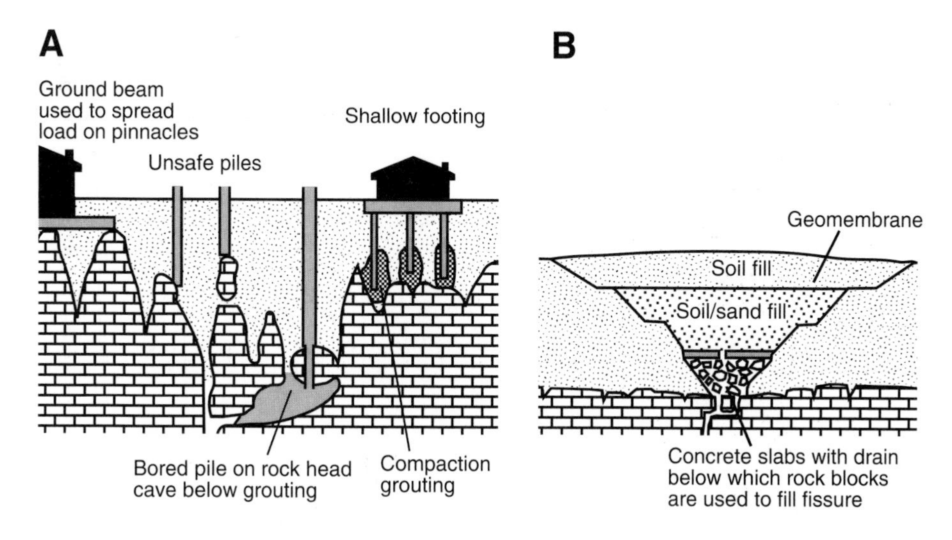

Figure 7.13 *Engineering solutions to carbonate rockhead hazards.* ***A.*** *Foundation design.* ***B.*** *A method of infilling a sinkhole [Modified from: Waltham (1994)* Foundations of engineering geology. *Blackie, p. 59]*

downward water drainage to the subterranean drainage system (**subsidence sinkholes**). Subsidence sinkholes, which account for 99% of all ground collapses on limestone, may form either by slow subsidence of sandy soils or by collapse of more clay-rich soils (Figure 7.12C). Buried sinkholes pose a significant foundation problem, and predicting their occurrence is difficult. They can, however, be recognised on geophysical surveys (Box 7.1). Sinkholes are natural phenomena but their formation may be accelerated by human action. Lowering of the water table, especially beneath the rockhead, through abstraction may increase the drainage gradient inducing accelerated downward movement of water and therefore solution and soil movement. Uncontrolled drainage, especially associated with soakaway drains, may accelerate the process of solution. As shown in Figure 7.13B, sinkholes may in some cases be safely infilled.

Clays

Clays and clay-rich rocks occur widely in lowland terrains. Clays have a high porosity and are formed from clay minerals whose sheet-like crystal lattice structures make them weak and deformable as well as compactable. Consolidation of clay may lead to subsidence and settlement of structures. Subsidence is greatest on thick clays with a high smectite content, low silt content and of a young age without a history of overconsolidation. The following problems can be identified.

1. **Shrinkage**. Water loss may cause clays to consolidate or shrink. This may occur locally due to moisture extraction by tree roots or by artificial pumping.

BOX 7.1: THE SEARCH FOR BURIED SINKHOLES: USE OF GEOPHYSICAL TOOLS

Sinkholes can pose an important engineering hazard, especially when buried or concealed beneath sediment. Where buried sinkholes are suspected, detailed site investigation is required. This could involve probing for the rockhead using a network of closely spaced boreholes. However, a more cost-effective solution is to use some form of geophysical survey.

There are a variety of types of geophysical survey ranging from seismic surveys, to various electrical methods in which the resistance or conductivity of the ground is measured on the premise that different materials have different conductivities. However, most successful in the recognition of sink-holes are magnetic surveys (McDowell, 1975). Here the distortion of the Earth's magnetic field by the ground materials and their structure is recorded by a simple hand-held instrument (Proton Precession Magnetometer) placed either at a network of regular points or along a traverse over the study area. The diagram below shows a magnetic survey across a sinkhole at Upper Enham in Hampshire, England. The magnetic anomaly clearly identifies the sinkhole. This technique is limited by the presence of buried powerlines or other metallic materials which cause distortions; it is also limited to near-surface features. This example illustrates the importance of geophysical techniques in site investigation.

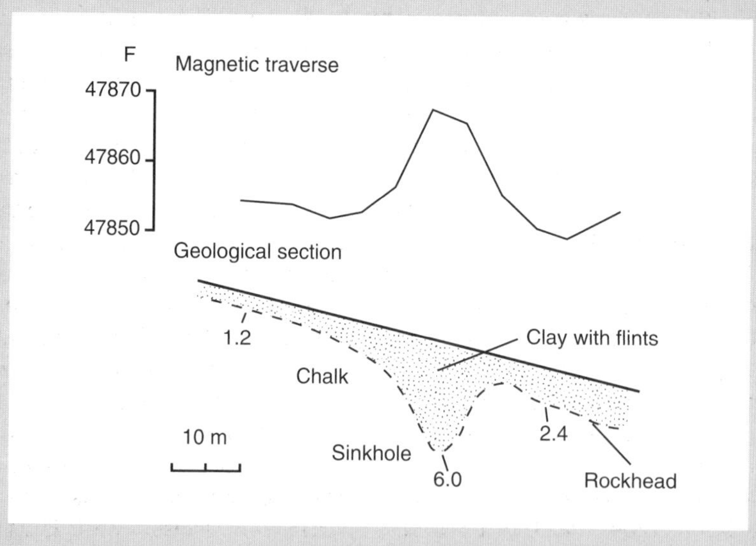

Source: McDowell, P.W. 1975. Detection of clay filled sink-holes in the chalk by geophysical methods. *Quarterly Journal of Engineering Geology*, **8**, 303–310. [Diagram modified from: McDowell (1975) *Quarterly Journal of Engineering Geology*, **8**, Fig. 2, p. 308]

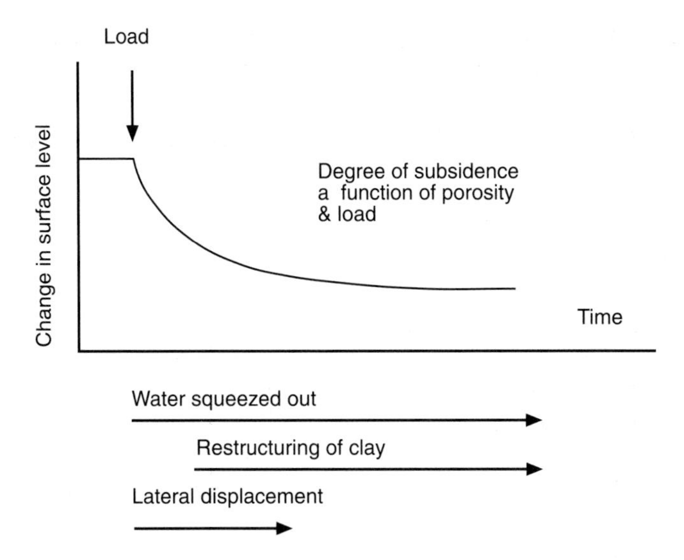

Figure 7.14 *Changes in the physical properties of clays during consolidation [Modified from: Waltham (1994)* Foundations of engineering geology. *Blackie, p. 56]*

For example, during the droughts of 1976 and 1995, building damage was widely reported on structures located on London Clay in southern England.

2. **Settlement**. All clays are prone to some settlement. The applied load causes water to be squeezed out, and consolidation of the clay occurs. The degree of consolidation depends on the consolidation history, the water content and the load applied (Figure 7.14). Consolidation tests can be undertaken in the laboratory in order to estimate the magnitude of these problems. The remedy is to avoid the loading of clays by placing foundations beneath the clay layer or, where this is not possible, the construction of underground infrastructure sensitive to settlement (e.g. pipes) is postponed until settlement has occurred. Settlement is a particular problem in Mexico City which is constructed on a drained lake bed composed of highly compressible clays with very high moisture contents. The construction of the Palace of Fine Arts on a huge concrete raft, in order to minimise the foundation problems of settlement, imposed a load of 110 kPa and caused settlement of over 3 m. In contrast, the Latino-Americana Tower was constructed on deep foundations which penetrated to a more stable sublayer of sand. The foundations were also constructed to be as 'buoyant' as possible with numerous basements and in-built voids, and the resultant subsidence was less than 0.25 m.

   Settlement causes most problems in construction where it occurs in a differential fashion. This may be due to differential loading or to lateral variations in the silt content of the mudrock. Once differential settlement occurs, it is reinforced through positive feedback: slight differential settlement → differential loading → increased differential settlement → increased differential loading. The problem is illustrated by two famous examples. The first is the construction of a grain elevator at Transona in Canada, which tilted by 27° in a single morning in 1912. This was probably caused by differential compaction of the clay subsurface due to variation in depth to the rockhead. The second example is the Leaning Tower of Pisa in Italy. The cathedral bell tower, which is 58 m tall with a weight of 14 000 tonnes, imposes a load of 500 kPa and has tilted 4 m from the vertical. Differential settlement was probably initially caused by slight facies variations within the clays and was subsequently accelerated by the differential loading caused by the tilting. Settlement may occur not only on a local scale but also regionally.

3. **Clay expansion**. Certain mudrocks that are rich in clay minerals such as montmorillonite may expand dramatically when hydrated. Montmorillonite clays form primarily from the weathering of volcanic rocks in warm climates and are a particular problem in parts of the USA, where damage caused by clay expansion exceeds the combined costs of earthquakes and floods. Remedies include liming of the clay to form more stable, calcium-rich varieties of the clay mineral species or, more commonly, maintaining the existing water state through careful drainage control.

7.2.3 Surficial Sediment and Landform Terrains

Within much of the northern hemisphere there are many areas covered by superficial sediments which were the product of the Cenozoic Ice Age, during which a succession of large ice sheets grew and decayed within mid-latitude regions. Beyond the limits of these ice sheets, intense periglacial action occurred. This activity left a legacy of surficial landforms and sediments which have important implications for engineering geology, not least due to the rapid changes in facies and ground characteristics associated with such terrain. Owing to this complexity it has become commonplace to identify a number of glacial and periglacial landsystems in order to summarise the complexity. Within northern Europe or North America it is possible to recognise four types of glacial/periglacial landsystem: (1) subglacial landsystems; (2) supraglacial landsystems; (3) glaciated valley landsystems; and (4) periglacial landsystems. Each of these landsystems is associated with a different set of engineering problems which are outlined below. In addition, they may all be covered by postglacial organic and inorganic soils sediments, which may also be of engineering significance.

Subglacial landsystems

The subglacial landsystems are dominated by sediments which are deposited either subglacially from debris transported at the base of a glacier (lodgement tills) or by the reworking of sediment overridden by the glacier (deformation tills) (Table 7.4; Figure 1.7). Traditionally, lodgement tills have been assumed to be generally homogeneous and therefore to form ideal ground conditions. However, practice has shown that they are far from homogeneous and may

Table 7.4 *Types of glacial till and their properties [Modified from: Bennett & Glasser (1996)* Glacial Geology. *John Wiley & Sons, Table 8.1, p. 174]*

Type of till	Characteristics
Lodgement till	Bimodal or multimodal size distribution – boulders to clay particles; subrounded to subangular clasts, striated and faceted; boulder pavements or concentrations may occur; deformation structures including fractured clasts common; consolidated, fissures and shear planes/fractures; clast fabric is usually strong
Subglacial meltout till	Bimodal or multimodal size distribution – boulders to clay particles; subrounded to subangular clasts, striated and faceted; deformation structures and shear planes/fissures usually absent; clast fabric is usually strong; may contain crude stratification
Supraglacial meltout till	Coarse unimodal size distribution; angular to very angular clasts, may contain some striated, faceted and more rounded examples; crude stratification may reflect scree-like accumulation of slopes or glacier structure; fluvial reworking common; low bulk density, rarely consolidated; very variable in composition
Supraglacial flow tills	Very variable properties reflect the fluidity of the sediment flow from which they are deposited; low bulk density, rarely consolidated and very variable in composition

have many associated engineering problems. Four commonly recognised engineering problems are associated with lodgement tills.

1. **Deformed rockhead**. Subglacial erosion may have fractured bedrock surfaces which have subsequently become buried beneath lodgement till. Large rafts of bedrock may be incorporated into the till. For example, chalk bedrock rafts at Trimmingham, on the Norfolk coast of England, are 300 m long and over 20 m thick. Defining the depth of the rockhead beneath tills during site investigation using boreholes may therefore be difficult. Rockhead deformation may also affect strength and stability. Undetected rockhead deformation caused the redesign of the Stwlan Dam at Ffestiniog in North Wales. Here, dam construction along the lip of a glacial corrie was delayed due to rockhead deformation. Ice flowing from the corrie had detached bedrock rafts from the lip, moving them downslope. These rafts were only discovered during construction and this caused a project delay while more detailed site investigation and project design were conducted.
2. **Soft zones**. Thin layers of sand and other water-lain sediments may occur within lodgement tills, deposited by subglacial meltwater streams (Figure 1.7). These layers may significantly affect the strength of the till and are sufficiently localised to be missed during borehole surveys. The problem of soft zones in lodgement till is illustrated by the construction of a 14-storey twin-tower block on 107th Street in Edmonton, Canada. On the assumption that the till beneath the site was uniform, just three boreholes were sunk initially. However, very low strength values were recorded using the standard penetration test (Figure 7.15), corresponding to a soft sandy zone

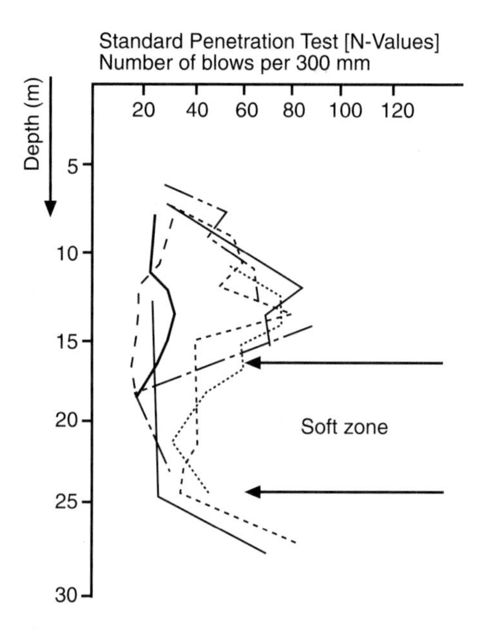

Figure 7.15 *Standard penetration test data from five boreholes beneath a proposed 14-storey twin-tower block in Edmonton, Canada [Modified from: Thomson* et al. *(1982)* Canadian Geotechnical Journal, ***19***, *Fig. 4, p. 179]*

within the till. Further boreholes were sunk and a change in foundation design was required to prevent differential settlement of the two towers.

3. **Joints and fissures**. Lodgement tills often have well developed shear fabrics in which fissures and joints formed by the overriding ice are a common component. Joints and fissures may also develop as subglacial tills which are unloaded during deglaciation. The presence of these fissures may affect the directional strength of tills. This was illustrated in roads near Hurleford in Ayrshire, Scotland, which cut into lodgement till. Both banks of the cutting were initially dug and contoured to the same slope angle, believed to be stable given the strength of the till. However, one bank continually failed via a series of shallow slips, while the other one remained stable. The key difference proved to be the orientation of the fissures in the till with respect to the slope. Where the fissures were inclined out of the slope it was unstable, but where they inclined into the slope it was stable. Consequently the banks had to be recontoured to different angles, despite having similar shear strengths. Fissures have also proved of importance in examining the permeability of lodgement tills used in the design of waste-disposal sites (Box 9.3).
4. **Weathering**. Tills weather quickly through a range of different processes including oxidation, hydration, leaching of soluble minerals (carbonates), and through the mechanical disintegration of larger clasts. The effects of weathering are summarised in Figure 7.16, and include: (1) increase in clay and silt content due to mechanical disintegration, resulting in increased

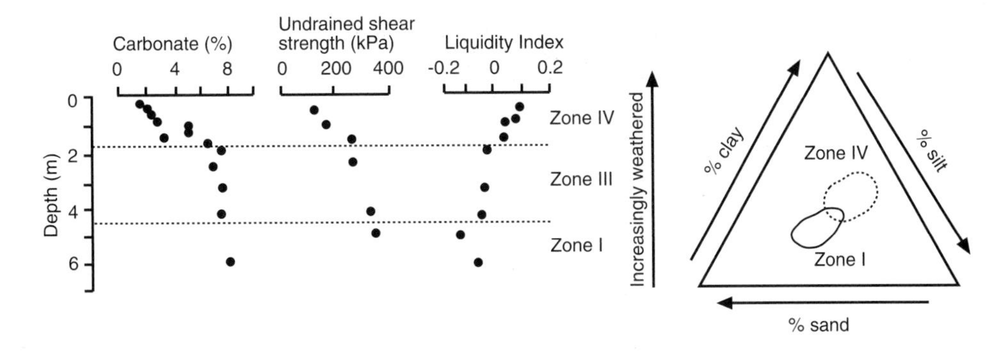

Figure 7.16 *The effect of weathering on the geotechnical properties of lodgement till [Modified from: Eyles & Sladen (1982)* Quarterly Journal of Engineering Geology, **14**, *Figs 3 & 5, pp. 134 & 137]*

plastic and liquid limits and an increased plasticity index; (2) increased clay content and increase in clay activity due to the formation of secondary minerals, again increasing plasticity; and (3) increased moisture contents – moisture increases faster than the plastic limits thereby increasing the liquidity index.

Other considerations within the subglacial landsystem are the surface occurrence of sand and gravel accumulations associated with outwash sedimentation as the glacier retreats. These may be highly faulted where they were originally deposited onto buried ice and are therefore very variable in composition.

Supraglacial landsystems

At the margins of some cenozoic ice sheets, flow compression caused the basal layers of glacier ice to rise upwards, in some cases along thrust faults. As a consequence, the ice margin may have become covered by large amounts of surface or supraglacial debris. This prevented the ice from melting and a hummocky topography of ice-cored hills resulted. This surface sediment quickly became saturated and unstable and flowed into low areas to form flow tills. In such areas, a hummocky relief of mounds and hollows resulted on deglaciation. Sedimentologically this type of terrain is extremely diverse, containing several different types of till including lodgement tills, meltout tills and flow tills (Table 7.4), and may include considerable quantities of glacio-fluvial sands and gravels. These sediments are highly variable and may be extensively faulted and folded due to subsidence caused by the meltout of buried ice. The following engineering problems can be identified within such terrain.

1. **Stability of cut slopes**. Owing to the very variable nature of the sediments within such terrains, the prediction of stable slope angles is extremely difficult. Local weak spots frequently occur due to the washing out of fines.

2. **Drainage routes**. Sediment variability may cause problems in predicting drainage routes and pathways within sediment bodies, causing significant problems in the design of drains.
3. **Low and variable bearing capacity**. Sedimentary variability causes variation in bearing capacity. Meltout tills and flow tills are generally much more porous and less well consolidated than lodgement tills, and consequently are weaker and have lower bearing capacities.
4. **Liquefaction**. Some sandy meltout tills are prone to liquefaction caused by the vibrations which result from the movement of heavy site machinery.

Glaciated valley landsystems

In confining valleys, complex landform assemblages with variable sedimentary composition occur. All types of till may be found in a small area in association with glacio-fluvial sand and gravels (Table 7.4). In addition, large thicknesses of valley-side debris, including talus and slumped till from unstable valley-side deposits/landforms, are present. These infill the bedrock topography carved by glacier ice. Engineering problems include the following.

1. **Sedimentary variability**. This causes problems in the design of cut slopes, drainage works, and in the prediction of bearing capacity.
2. **Oversteepened slopes**. The steep valley sides of many glaciated valleys may be particularly prone to failure (see Section 11.4; Figure 11.27). Removal of the lateral support provided by glaciers may be significant, and glacial unloading both by erosion and by the removal of ice may cause valley bulges and sheet joints to develop. Sheet joints form parallel to the glacial valley slope and are therefore particularly prone to failure.
3. **Valley floor problems**. Glaciated valleys may have a complex rockhead topography which is buried and therefore obscured by deposition. Incised channels and basins are common. Unlike fluvial valleys, glaciated valleys do not always have a rockhead gradient which is continuously downvalley. The valley floor may consist of a series of basins and rock bars. Prediction of the rockhead geometry is therefore very difficult without detailed site investigation.

Periglacial landsystems

The principal engineering problems associated with former periglacial regions can be divided into those due to near-surface disturbance and those due to more deep-seated disturbance. In addition, the deposition of wind-blown dust or loess, and cover sands may cause engineering difficulties.

1. **Near-surface disturbance**. Typical disturbances include involutions, frost shattering, solifluction and ice wedges, all of which may be a cause of concern for the engineer designing shallow foundations. Involution may lead to surface sediment diversity due to surface sediment mixing. Frost shattering

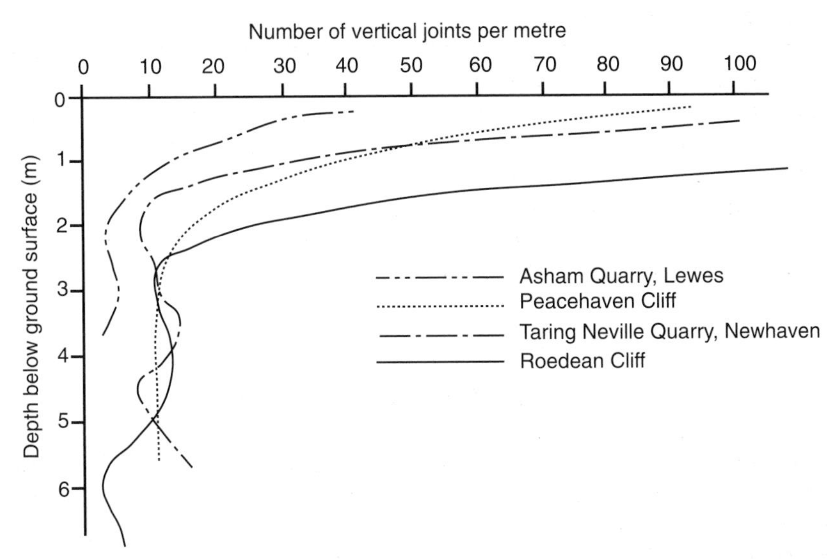

Figure 7.17 *Near-surface disturbance in chalk due to frost action opening joints during former periods of periglaciation [Modified from: Williams (1987) In:* Periglacial processes and landforms in Britain and Ireland. *Cambridge University Press, Fig. 10.1, p. 128]*

may have disrupted and weakened rock surfaces, encouraging failure. This is particularly a problem with the chalk rocks in southern England. Figure 7.17 shows the rapid increase in joint density towards the surface in chalk rocks at four sites in southern England, caused by freeze–thaw on the chalk surface during the last glacial period. Shattered chalk may form a putty which is thixotropic when saturated, turning it into a slurry. This is a particular problem for the movement of site machinery in winter months and in the driving of piles. Reactivation of slip surfaces beneath solifluction lobes has been associated with some famous engineering problems. For example, during the construction of the Sevenoaks by-pass in southern England, the engineering geologists failed to recognise the potential problem posed by crossing a zone of former solifluction lobes. During road construction, loading of the palaeo-slip faces beneath the lobes caused their reactivation, delaying road construction significantly. Fossil ice wedges infilled with sand may also pose ground stability problems (Box 7.2).

2. **Deep-seated disturbances**. These include problems of cambering, valley bulges and the occurrence of fossil pingos. Cambering is the downslope movement of blocks of competent rock over weaker, more ductile rocks as a result of periglacial disturbance. It is a common phenomenon along the Cretaceous and Jurassic escarpments of southeast England where limestone, chalk and sandstones overlie clays. As movement occurs, tension cracks known as **gulls** open and are usually infilled with sediment. If the infilling is incomplete, linear runnels parallel to contour lines may be visible on the surface. Although movement is no longer taking place, the presence of gulls

BOX 7.2: NEAR-SURFACE DISTURBANCES: FOSSIL ICE WEDGES

In active periglacial environments, frozen ground contracts as it cools and cracks form in which water collects and freezes to form ice wedges. These ice wedges grow in size over time. Ice wedges formed in lowland Britain during the Cenozoic Ice Age. As climate warmed at the close of the last glacial cycle the frost wedges melted and sediment replaced the ice to form an ice wedge cast, as shown in the photograph below. In plan view ice wedges form polygonal nets and fossil examples can be picked out in crop patterns when viewed from the air.

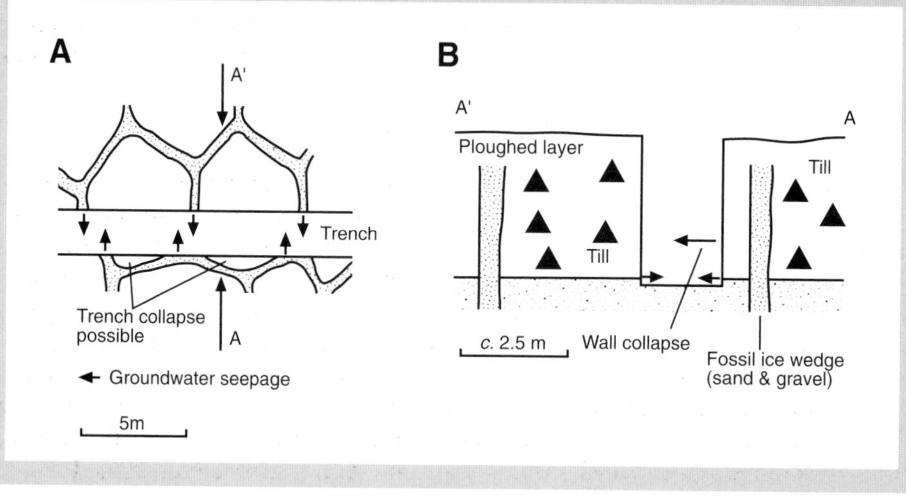

Morgan (1971) described the geotechnical problems encountered in digging a gas pipeline through an area of ice wedge casts near Wolverhampton in the English Midlands. The casts were infilled with sand, causing variations in the strength of the glacial till in which the trench was being dug. The trench walls collapsed locally where they intersected a wedge, and collapsed on a much larger scale where the trench ran parallel to a wedge, as shown in the diagram opposite. This example illustrates the type of problems which might be encountered in an area formerly subject to periglaciation, and emphasises the need for careful site evaluation.

Source: Morgan, A.V. 1971. Engineering problems caused by fossil permafrost features in the English Midland. *Quarterly Journal of Engineering Geology*, **4**, 111–114. [Diagrams modified from: Morgan (1971) *Quarterly Journal of Engineering Geology*, **4**, Fig. 2, p. 113]

has important implications for the stability of foundations, particularly where their presence is undetected by site investigation (Box 7.3). The discovery of fossil pingos buried by fluvial sand and gravel beneath London illustrates the difficulty in predicting the shape of the rockhead, in this case the surface of the London Clay (Box 7.4).

3. **Problems associated with loess or cover sands**. Loess and cover sands occur extensively within some regions, particularly in parts of China where they are often over several hundred metres thick. The engineering problems associated with loess are numerous and stem from its low density due to loose particle packing, and consequently high porosity and permeability. The development of columnar jointing may also cause problems. Typically, the deposits are metastable and collapse on wetting or during loading. Problems include settlement and fluid mass failure when saturated either artificially or during extreme rainfall events. Pre-consolidation under temporary earthworks may be used to reduce the problems of settlement, while careful drainage design is required to reduce the problems of failure.

Postglacial sediments

Organic accumulations such as peat may pose considerable engineering problems, owing to their high moisture content and compressibility. Where possible, peat is best removed and backfilled. Where this is not possible, an alternative solution is pre-loading and consolidation. This involves spreading a temporary solid surcharge over the area, the total weight of which is greater than that of the construction; in this way the peat settles and consolidates, and construction takes place following removal of the temporary soil cover. Both of these solutions only work where the peat occurs as a relatively continuous cover. However, such deposits are often interbedded in pockets in river alluvium and similar sediments, in which case it may be difficult to either consolidate or

BOX 7.3: PERIGLACIAL CAMBERING

The engineering problems associated with valley cambering are well illustrated by Hawkins and Privett (1981). They describe the problems encountered in the construction of a housing estate at Radstock in Avon. The site was located on a 4° to 7° slope composed of Blue Lias Limestone overlying Rhaetic and Keuper Marl. Cambering of the limestone downslope over the marl had opened 1 m wide gulls and 5 m wide zones of fracturing and

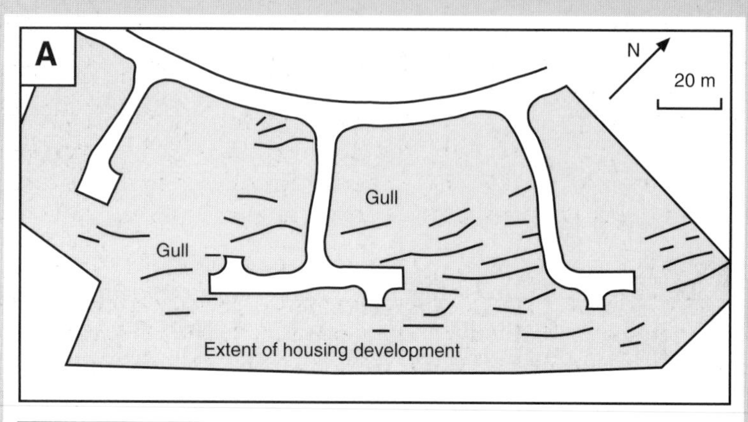

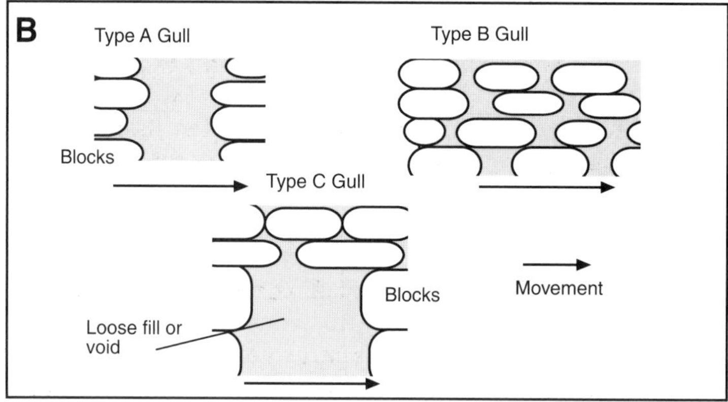

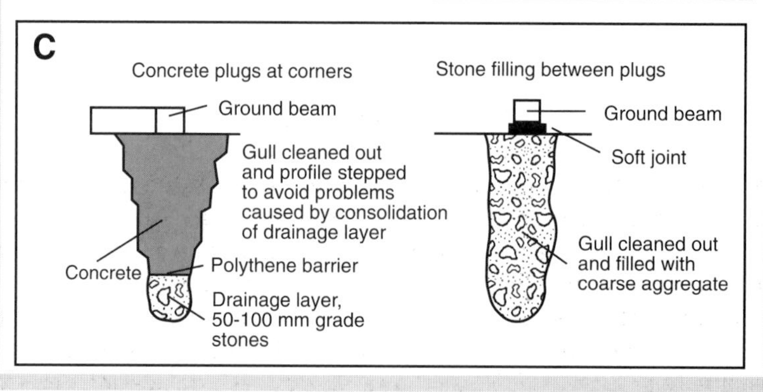

disturbance. These gulls and zones of disturbance were not found during the initial site investigation and only became apparent during the excavation of foundation trenches (diagrams A&B). Although the site is stable, the presence of gulls beneath foundations poses a significant problem. As a consequence the site had to be redesigned twice to relocate house foundations away from gulls. This reduced the number of houses on the site from 92 to 78. In some cases it was, however, impossible to avoid construction on gulls. In these situations gulls were excavated out and filled with either concrete of gravel. Garages were placed over gulls on thick concrete rafts. Where gulls could not be avoided by house foundations these were placed on ground beams above excavated and filled gulls (diagram C).

This example clearly illustrates the problems of engineering in cambered terrain, but also illustrates the importance of good site investigation to identify problems in the first instance. Engineering geologists should have been able to predict the possibility of cambering at this site.

Source: Hawkins, A.B. & Privett, K.D. 1981. A building site on cambered ground at Radstock, Avon. *Quarterly Journal of Engineering Geology*, **14**, 151–167. [Diagrams modified from: Hawkins & Privett (1981) *Quarterly Journal of Engineering Geology*, **14**, Figs 4, 10 & 19, pp. 154, 159 & 165]

remove it. This was a particular problem in the construction of the A55 road near Conway in North Wales (Figure 7.18). The solution involved constructing the road on a 0.8 m thick concrete raft supported by 12 000 piles emplaced on a 2 m by 2 m grid. Clearly such problems and solutions add considerably to the expense of a project.

7.3 ROCK EXCAVATION AND TUNNELLING

Geology has a particularly important role in the excavation of rock surfaces and in the construction of tunnels or similar underground structures, each of which is discussed below.

7.3.1 Rock Excavation

There are three commonly used excavation methods: (1) **direct excavation**, using a mechanical digger of some description; (2) **ripping**, using either a tractor-mounted ripper or a pecker which has a hydraulic pick mounted on a mechanical excavator (Figure 7.19); and (3) **blasting**. The appropriate method depends on the rock mass strength (Figure 7.20A), a product of fracture/joint density, shear strength and the degree of weathering, all of which may vary across a single face (Figure 7.20B). Where blasting is not possible, excavation is

BOX 7.4: FOSSIL PINGOS BENEATH LONDON

Pingos are giant ice blisters which form in active permafrost regions, either along spring lines or on areas of saturated ground. Fossil pingos consist of shallow craters with ramparts. A series of enclosed craters were found during

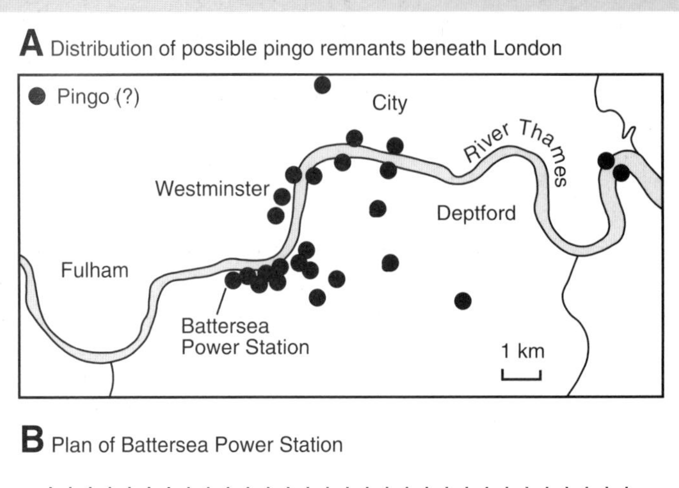

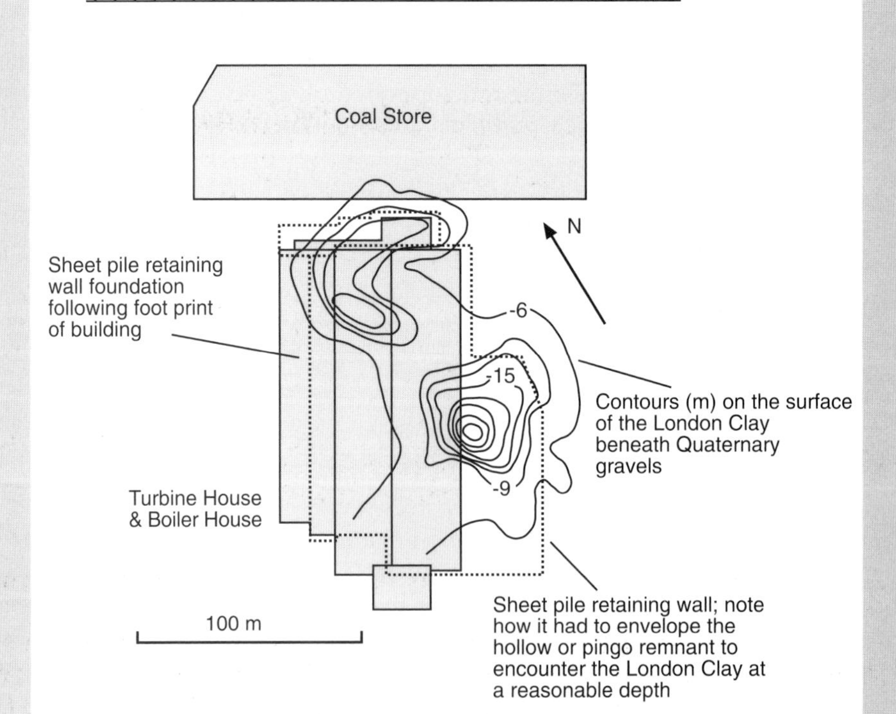

investigations beneath London (diagram A). Central London is underlain by London Clay which is covered by sand and gravel deposited by the River Thames during the Cenozoic Ice Age. At first these craters at the surface of the London Clay which are buried by sand and gravel were interpreted as fluvial scour features. However, their occurrence along a possible former seepage line has led to their reinterpretation as fossil pingos (Hutchinson, 1980).

In engineering terms they pose a significant problem if undetected. For example, during the construction of Battersea Power Station one of these pingos was encountered, causing a costly diversion of a retaining wall (diagram B). The unpredictability of the rockhead is the problem.

Sources: Higginbottom, I.E. & Fookes, P.G. 1971. Engineering aspects of periglacial features in Britain. *Quarterly Journal of Engineering Geology*, **3**, 85–117; Hutchinson, J.N. 1980. Possible late Quaternary pingo remnants in Central London. *Nature*, **284**, 253–255. [Diagrams modified from: Higginbottom & Fookes (1971) *Quarterly Journal of Engineering Geology*, **3**, Fig. 7, p. 97; Ballantyne & Harris (1994) *The periglaciation of Great Britain*, CUP, Fig. 5.22, p. 80]

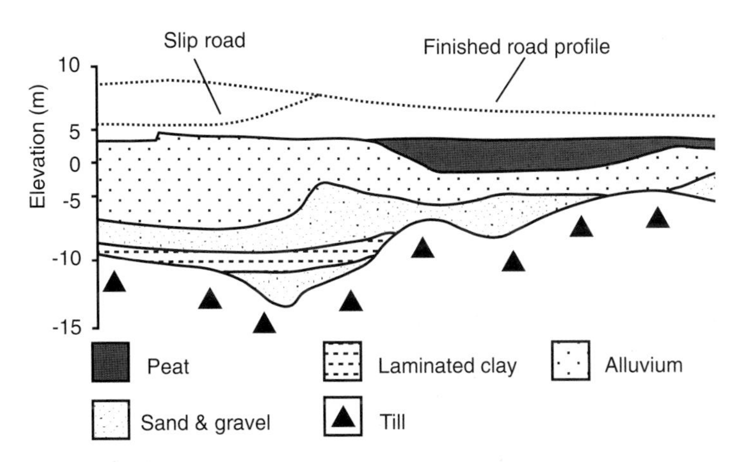

Figure 7.18 *The occurrence of a peat layer within Holocene valley sediment caused problems during the construction of the new A55 trunk road near Conway in North Wales [Modified from: Arber (1984)* Quarterly Journal of Engineering Geology, **17**, *Fig. 2, p. 336]*

usually undertaken by a pecker or hydraulic breaker, but this method can be very slow.

A variety of blasting techniques exist. In most cases some form of open-face blasting is used. This involves drilling a series of holes between 50 and 100 mm in diameter parallel to the open face and dipping at 70° to 80° towards the face. Each line of holes is set back at between 2 and 4 m from the face (Figure 7.20C).

Figure 7.19 A hydraulic pick mounted on a mechanical excavator, known as a pecker. In this case it is excavating Ordovician sedimentary rocks in Wales

This distance is known as the **burden** and is ideally 30 to 40 times the drill hole diameter. Holes are spaced at about 50 times the drill hole diameter, generally between 3 and 5 m apart. Two or more lines of holes may be fired at once, and a delay between each line of 5 milliseconds per metre of burden is usually used. This ensures that each line is effective. The choice of explosive will depend on the type of rock and the conditions in the shot hole. For example, if the shot holes are water-filled then water-resistant explosives are required. All these variables can be altered to produce different effects, such as the degree of rock fragmentation, and to minimise such adverse effects as blast vibration. Blast faces should ideally be parallel to major joint sets since blast energy tends to travel along such surfaces. Blasting is only the first phase of the exercise and secondary blasting of large blocks is sometimes required to reduce blocks into manageable sizes, although this may also be done mechanically. In the excavation of cuttings, the opening up of an initial **free face** from which blasting can proceed is often achieved by drilling small galleries and blasting from these.

Figure 7.20 (opposite) Methods of rock excavation. **A.** The link between rock mass strength and the excavation mechanism: blasting, ripping or digging. **B.** Variation in weathering, fracture density and strength across a face causing variation in excavation methods. **C.** Blasting methods [Modified from: Waltham (1994) Foundations of engineering geology. Blackie, pp. 74 & 75]

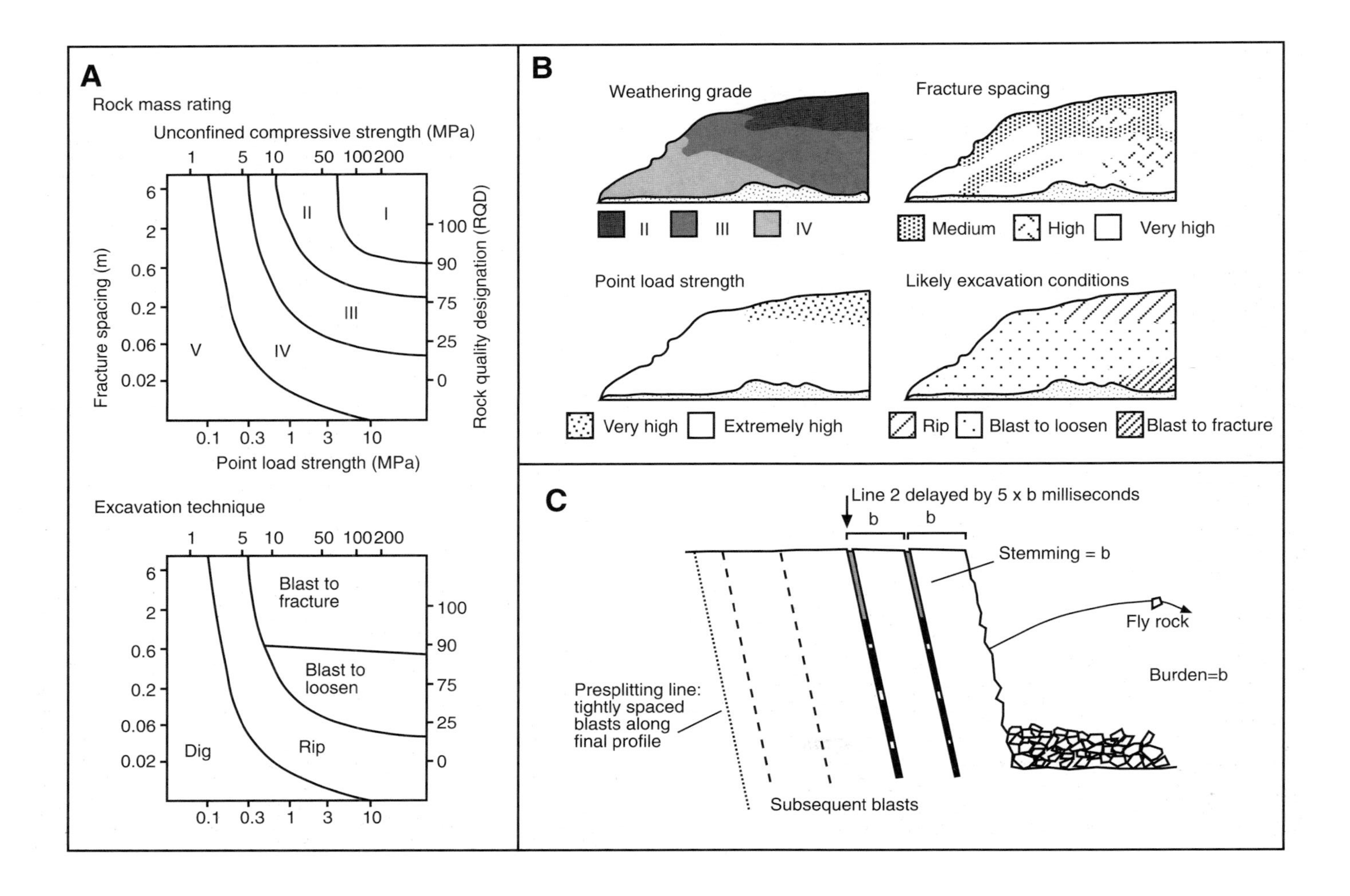
A
Rock mass rating
Unconfined compressive strength (MPa)
1 5 10 50 100 200
Fracture spacing (m)
6 2 0.6 0.2 0.06 0.02
Rock quality designation (RQD)
100 90 75 25 0
I II III IV V
0.1 0.3 1 3 10
Point load strength (MPa)
Excavation technique
1 5 10 50 100 200
6 2 0.6 0.2 0.06 0.02
100 90 75 25 0
Blast to fracture
Blast to loosen
Rip
Dig
0.1 0.3 1 3 10
B
Weathering grade
II III IV
Fracture spacing
Medium High Very high
Point load strength
Very high Extremely high
Likely excavation conditions
Rip Blast to loosen Blast to fracture
C
Line 2 delayed by 5 x b milliseconds
b b
Stemming = b
Fly rock
Burden=b
Presplitting line: tightly spaced blasts along final profile
Subsequent blasts

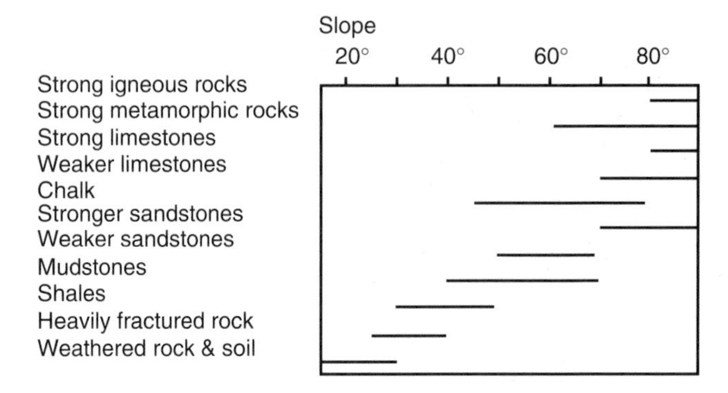

Figure 7.21 *Stable cutting slopes for selected rocks. The terms strong and weak refer to rock mass strength characteristics [Modified from: Waltham (1994)* Foundations of engineering geology. *Blackie, p. 74]*

Shaping the final form of blast face requires **controlled blasting**. The aim is to produce a stable face with as few fractures as possible and with correct design specifications such as slope and height. This may be achieved through **pre-splitting**, in which drill holes with a spacing of only 10 to 20 times the hole diameter are placed along the line and charged with low density explosives and fired with a high unexploded burden. This produces a fracture parallel to the desired line, up to which normal blasting may proceed. In very fragile locations **line drilling** may be used, in which a line of holes spaced at only twice the hole diameter is used to limit the extent of blasting. These holes are filled with charge and fired. This is expensive and consequently rarely used. Most slopes are constructed with steps or breaks so as to minimise slope height and improve safety.

Wherever possible cut slopes should be excavated back to sound rock. Stable slope angles for different lithologies are shown in Figure 7.21. In practice, these are frequently reduced due to problems associated with outward-dipping fractures, weathering or where slopes may receive a high load. On clays and other soft sediments, as well as on highly weathered rock, slope design is based on information about strength of the engineering soil, such as its cohesion and weight, the geometry of the slip faces present and on the presence of tensional cracks or other similar features. A knowledge of the strength and origin of these engineering soils is essential to the design of stable slopes. Where it is not possible to reduce a slope to a stable angle, remediation and stabilisation works may be required. These may include (Figures 7.22 & 11.25): (1) the construction of retaining walls (concrete, masonry, gabions); (2) the use of engineered soil slopes, which are only successful/economic for relatively small slopes and may require careful anchoring; (3) careful drainage design to remove excess pore-water; (4) the use of reinforced earth and gabion walls, along with a variety of geomembranes to reduce erosion, and vegetation may also play an important role in slope stabilisation; (5) dental masonry, in which weak zones are removed

Figure 7.22 Rock stabilisation work on a new road cutting along the A5 near Ty-nant in North Wales. **A.** The rock face is benched to reduce height, to collect fallen blocks, and to reduce surface runoff. **B.** Catch fence to restrain falling blocks. **C.** Wire mesh to catch falling blocks, plus rock bolts/anchors used to increase the integrity of factured rock. **D.** Close-up of rock bolts/anchors used to increase the integrity of factured rock [Photographs: W. Whale]

Figure 7.22 *(continued)*

and infilled by masonry, brickwork, gabions or concrete – small overhangs may be dealt with in this way; (6) sprayed concrete (**shotcrete**) used over wire mesh to stabilise loose friable ground; (7) rock bolts which may be used to 'bolt in' loose blocks and improve stability; (8) wire mesh used to contain loose blocks; (9) the construction of ditches and fences to catch loose blocks; (10) scaling down either by hand or by blasting to remove loose blocks. Where slopes intersect active or inactive mass movements, more complex stabilisation works may be required (see Section 11.3.2).

7.3.2 Tunnelling

Tunnelling, or the excavation of large underground caverns, is another important engineering task in which geology plays a critical role. There are two stages in the construction of a tunnel: the first involves excavation and the second involves stabilisation. Excavation methods include: (1) mechanical excavation; (2) drill and blast; (3) use of a roadheader, which is a machine mounted with a rotary milling head; and (4) use of a tunnel-boring machine. The choice of method is largely determined by the length of tunnel to be excavated and the nature of the ground conditions. For example, the use of a tunnel-boring machine is rarely economical where the tunnel is less than 1 km long. The second aspect is the choice of internal support used within the tunnels. A variety of methods exist among which there are perhaps two broad types: (1) **passive support**, in which steel ribs or precast concrete segments are placed to support the tunnel walls; (2) **active support**, in which the rock is strengthened, by using rock bolts and shotcrete, to create a stable self-supporting arch within the tunnel. A particularly popular variant of active support is the New Austrian Tunnelling Method in which the rock is allowed to deform enough to redistribute the stress and reach a new stable state around the tunnel. Shotcrete and a few rock bolts are placed in the tunnel initially, but are not designed to take the whole of the weight; adjustment then occurs within this flexible lining and a stable state is reached before any further lining work is undertaken. The choice of support system is largely determined by the rock mass strength of the ground, as illustrated in Figure 7.23.

Geological problems encountered in tunnelling include: (1) **faults**, which affect ground conditions and hydrology; (2) **groundwater**, in which high flow rates may pose a particular problem; (3) **squeezing ground**, where plastic deformation of clays and shales occurs in response to the overburden closing the opened void; (4) **rockbursts**, in which sudden bulging of the tunnel walls occurs at depth in rocks with high unconfined compressive strength; (5) **swelling ground**, which involves wall closure due to swelling of clays through hydration; (6) **rockhead hazards**, for example, breaking through the rockhead into sands and gravels in valley floors or on the sea floor may cause a major inrush of water and sediment; and (7) **stress release** within the tunnel roof which may lead to roof failures. In areas of difficult ground it is necessary to probe carefully in front of the advancing tunnel face. In some cases grouting –

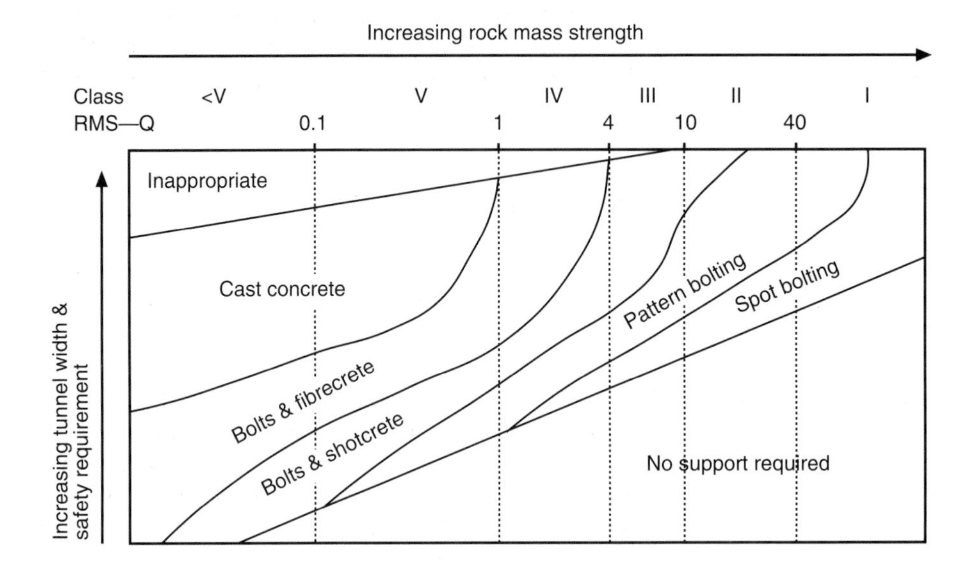

Figure 7.23 *The link between rock mass strength and the method of tunnel support [Modified from: Waltham (1994)* Foundations of engineering geology. *Blackie, p. 77]*

the infilling of potentially difficult or problematic fractures prior to tunnelling – may be required to minimise problems.

7.4 SUMMARY OF KEY POINTS

- Engineering geology is an important part of environmental geology and involves the application of geological knowledge to the design, construction and performance of civil engineering works. Its focus is the behaviour of geological materials during and after construction.
- Site investigation involves the pre-development identification of difficult ground conditions and other building hazards. It may take many different forms, from regional land-use planning surveys, to site selection and site-specific analysis.
- Within site investigation, four main stages, not always present, can be identified: (1) regional reconnaissance to identify the most problem-free location or route corridor; (2) site documentation; (3) material testing; and (4) data presentation either as engineering geology maps or in detailed site reports.
- In engineering terms one can recognise three types of geological terrain, each with its own engineering hazards: (1) ancient hard rock terrain; (2) soft rock terrain; and (3) superficial sediments and landforms. Of particular note in a country such as Britain are the problems associated with the glacial and periglacial landforms and sediments which cover much of the surface.

- Of the range of engineering geology tasks, rock excavation and tunnelling involve more geological input than any other. Choice of excavation techniques is controlled by rock mass strength, as is the method of tunnel stabilisation used.

7.5 SUGGESTED READING

The engineering geology literature is vast, and the following is a selection of some key texts, review papers and papers on the specific examples referred to in the text. Excellent overviews of the subject of engineering geology are provided by Beavis (1985), Johnson & DeGraff (1988), Bell (1993), Goodman (1993), Lumsden (1994) and Waltham (1994). Rock mass strength classification, in particular the Norwegian Q system and its application to tunnelling, is discussed by Barton *et al.* (1974). The alternative rock mass rating scheme is discussed by Bieniawski (1989). Weltman & Head (1983) provide information on the process of site investigation. Dearman & Fookes (1974) and Dearman (1991) provide authoritative accounts of the use of engineering maps, and a guide to their preparation is given by the International Association of Engineering Geologists (1976). Clark *et al.* (1979) provide a good example of a site investigation, in this case for a dry dock in western Scotland, while Brunsden *et al.* (1975), Jones *et al.* (1983) and Fookes *et al.* (1985) discuss the role of geomorphology and reconnaissance-scale site investigation in Himalayan road design and route selection.

Numerous case histories are contained within the *Quarterly Journal of Engineering Geology*, *Engineering Geology* and the *Bulletin of the Association of Engineering Geologists*. The engineering problems associated with weathering of hard rock terrains are covered by Fookes *et al.* (1971) and Baskerville (1982). Engineering aspects of sinkholes and carbonate terrains are well covered by Edmonds (1983), Culshaw & Waltham (1987) and Cooke & Doornkamp (1990). Burnett & Fookes (1974) discuss the regional engineering problems associated with the London Clay. Gray (1992) provides an excellent and accessible account of the problems associated with engineering on glacial and periglacial terrain. Eyles & Dearman (1981) present the main glacial landsystems and discuss their engineering significance. Anderson (1970) discusses the problems of rockhead deformation associated with the construction of the pump storage scheme at Ffestiniog in Wales. Eyles & Sladen (1981) discuss the role of weathering in changing the engineering properties of tills, and Thomson *et al.* (1982) provide a case study on the problems of soft zones within lodgement tills. The problems caused by fissures in glacial till are illustrated by McGown *et al.* (1974). Case histories of problems associated with engineering on periglacial terrain can be obtained from Morgan (1971), Hawkins & Privett (1981), Jones & Derbyshire (1983) and Penn *et al.* (1983).

Anderson, J.G.C. 1970. Geological factors in the design and construction of the Ffestiniog Pumped Storage Scheme, Merioneth, Wales. *Quarterly Journal of Engineering Geology*, **2**, 183–194.

Barton, N., Lieu, R. & Lunde, J. 1974. Engineering classification of rock masses for tunnel design. *Rock Mechanics*, **5**, 189–236.

Baskerville, C.A. 1982. The foundation geology of New York City. In: Legget, R.F. (Ed.) *Geology under cities*. Geological Society of America Reviews in Engineering Geology, Volume 5, 95–118.

Beavis, F.C. 1985. *Engineering geology*. Blackwell Scientific Publications, Oxford.

Bell, F.G. 1993. *Engineering geology*. Blackwell Scientific Publications, Oxford.

Bieniawski, Z.T. 1989. *Engineering rock mass classifications*. Wiley-Interscience, New York.

Brunsden, D., Doornkamp, J.C., Fookes, P.G., Jones, D.K.C. & Kelly, J.M.H. 1975. Large scale geomorphological mapping and highway engineering design. *Quarterly Journal of Engineering Geology*, **8**, 227–253.

Burnett, A.D. & Fookes, P.G. 1974. A regional engineering geological study of the London Clay in the London and Hampshire basins. *Quarterly Journal of Engineering Geology*, **7**, 257–295.

Clark, A.R., Hawkins, A.B. & Gush, W.J. 1979. The Portavadie dry dock, west Scotland: a case history of the geotechnical aspects of its construction. *Quarterly Journal of Engineering Geology*, **12**, 301–317.

Cooke, R.U. & Doornkamp, J.C. 1990. *Geomorphology in environmental management*, second edition. Clarendon Press, Oxford.

Culshaw, M.G. & Waltham, A.C. 1987. Natural and artificial cavities as ground engineering hazards. *Quarterly Journal of Engineering Geology*, **20**, 139–150.

Dearman, W.R. 1991. *Engineering geological mapping*. Butterworth-Heinemann, Oxford.

Dearman, W.R. & Fookes, P.G. 1974. Engineering geological mapping for civil engineering practice in the United Kingdom. *Quarterly Journal of Engineering Geology*, **7**, 223–256.

Edmonds, C.N. 1983. Towards prediction of subsidence risk upon the Chalk outcrop. *Quarterly Journal of Engineering Geology*, **16**, 261–266.

Eyles, N. & Dearman, W.R. 1981. A glacial terrain map of Britain for engineering purposes. *Bulletin of the International Association of Engineering Geology*, **24**, 173–184.

Eyles, N. & Sladen, J.A. 1981. Stratigraphy and geotechnical properties of weathered lodgement tills in Northumberland, England. *Quarterly Journal of Engineering Geology*, **14**, 129–141.

Fookes, P.G., Dearman, W.R. & Franklin, J.A. 1971. Some engineering aspects of rock weathering with field examples from Dartmoor and elsewhere. *Quarterly Journal of Engineering Geology*, **4**, 139–185.

Fookes, P.G., Sweeney, M., Manby, C.N.D. & Martin, R.P. 1985. Geological and geotechnical engineering aspects of low-cost roads in mountainous terrain. *Engineering Geology*, **21**, 1–152.

Goodman, R.E. 1993. *Engineering geology; rock in engineering construction*. John Wiley & Sons, New York.

Gray, J.M. 1992. Applications of glacial, periglacial and Quaternary research: an introduction. *Quaternary Proceedings*, **2**, 1–16.

Hawkins, A.B. & Privett, K.D. 1981. A building site on cambered ground at Radstock, Avon. *Quarterly Journal of Engineering Geology*, **14**, 151–167.

International Association of Engineering Geologists, 1976. *Engineering geological maps: a guide to their preparation*. UNESCO Press, Paris.

Johnson, R.B. & DeGraff, J.V. 1988. *Principles of engineering geology*. John Wiley & Sons, Chichester.

Jones, P.F. & Derbyshire, E. 1983. Late Pleistocene periglacial degradation of Lowland Britain: implications for civil engineering. *Quarterly Journal of Engineering Geology*, **16**, 197–210.

Jones, D.K.C., Brunsden, D. & Goudie, A.S. 1983. A preliminary geomorphological assessment of part of the Karakoram Highway. *Quarterly Journal of Engineering Geology*, **16**, 331–355.

Lumsden, G.I. (Ed.) 1994. *Geology and the environment in Western Europe*. Oxford University Press, Oxford.

McGown, A., Sali-Salivar, A. & Radwan, A.M. 1974. Fissure patterns and slope failure in till at Hurleford, Ayrshire. *Quarterly Journal of Engineering Geology*, **7**, 1–26.

Morgan, A.V. 1971. Engineering problems caused by fossil permafrost features in the English Midlands. *Quarterly Journal of Engineering Geology*, **4**, 111–114.

Penn, S., Royce, C.J. & Evans, C.J. 1983. The periglacial modification of the Lincoln Scarp. *Quarterly Journal of Engineering Geology*, **16**, 309–318.

Thomson, S., Martin, R.L. & Eisenstein, Z. 1982. Soft zones in the glacial till in downtown Edmonton. *Canadian Geotechnical Journal*, **19**, 175–180.

Waltham, A.C. 1994. *Foundations of engineering geology.* Blackie Academic & Professional, Oxford.

Weltman, A.J. & Head, J.M. 1983. *Site investigation manual.* Construction Industry Research and Information Association, London.

8
Engineering Geology in Extreme Environments

In this chapter we examine how the application of geological knowledge is essential in the construction of human infrastructure within extreme geological environments. Many environments may be considered as extreme (Table 8.1); however, here we identify two as illustrations of the problem: polar non-glacial deserts, and subtropical deserts. Polar deserts experience problems of seasonal drainage, frozen ground and intense frost action, while subtropical deserts experience problems of flash flooding, saline groundwater, rapid weathering and drifting sand. In both cases, careful choice of location, and modifications in foundation and structural design of buildings are required. These two environments give an illustration of how geological information is essential in planning and designing for development in difficult geological terrain.

8.1 DEVELOPMENT IN ACTIVE PERIGLACIAL ENVIRONMENTS

Periglacial environments occur extensively in the polar regions of both the Eurasian and North American continents. They are defined here as those areas which experience cold climates and intense frost action, irrespective of their proximity to glaciers. The presence of exploitable oil and gas reserves on the margins of the Arctic Basin have dramatically increased the level of development and construction within these areas during the last 30 years. Construction of housing and associated infrastructure, as well as oil and gas pipelines, has required careful engineering and design to combat this hostile geological environment. In North America particularly, the oil crisis of the early 1970s did much to focus attention on arctic oil reserves as a politically stable fuel supply, and stimulated a number of large development projects in Alaska and northern Canada.

Table 8.1 *Extreme climatic, geomorphological and geological locations for urban settlements [Based on information in Marker (1996) In: McCall, G.J.H., De Mulder, E.F.J. & Marker, B.R. (Eds)* Urban geoscience. *A.A. Balkema]*

Environment	Principal problems
A. Climatic	
Periglacial	Permanently frozen ground requiring special construction and foundations for buildings and infrastructure
Arid	Water supply problems; may be subject to soil erosion; may have chemically reactive groundwaters which can damage construction materials and roads
Tropical	Humid environment may cause deterioration in construction materials; bedrock may be deeply and unevenly weathered causing problems for foundation works
B. Geomorphological	
Mountainous	Liable to slope instability, rockfall and avalanches; can be subject to flash flooding
Flood plains	Liable to periodic flooding; variable ground conditions for foundation works
Coastal plains	Liable to periodic flooding, increasing in probability with rising sea levels; thick sedimentary units may be of variable quality for foundation works; may be subject to tsunamis
Coastal settings with weak-rock cliffs	Liable to rapid coastal erosion with undercutting and landsliding; other areas may be subject to deposition of sediments causing problems for ports, etc.
Islands	Liable to storm and flood events and rising sea levels.
C. Geological	
Active plate margins	Characteristic of the Pacific rim; associated with active endogenic hazards from earthquakes and volcanoes; currently at least 0.5 billion people live in such zones
Poor ground conditions	Caused by weak or variable engineering soils and bedrock conditions; require expensive foundation works

There are four geological problems associated with development in areas of active periglaciation: (1) permafrost or perennially frozen ground; (2) problems of frost heave; (3) the seasonal and variable nature of river discharge; and (4) the problems of lake and sea ice. Of these, permafrost is the most significant.

Permafrost is perennially frozen ground, that is, ground frozen continuously for two or more years. Figure 8.1 illustrates the principal components of permafrost. On the surface there is a layer of ground known as the **active layer** which is seasonally unfrozen and subject to repeated freeze–thaw cycles. Characteristically the active layer is water-logged, since drainage is impeded by the frozen ground below, and is therefore often unstable and subject to flow. Beneath the active layer the ground is permanently frozen and may contain substantial quantities of ground ice, although in free-draining areas it may be ice-free. The upper limit of the permafrost is known as the **permafrost table**. Permafrost may

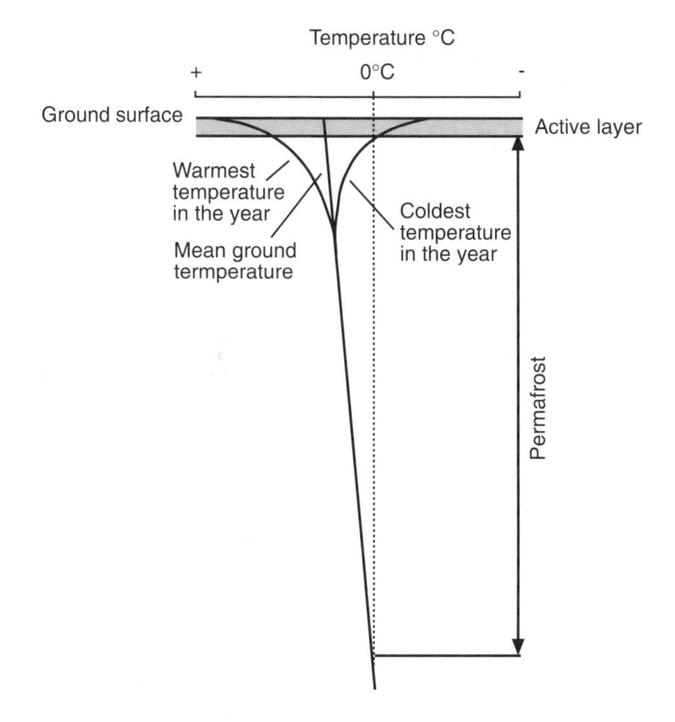

Figure 8.1 *Ground temperature profiles showing the extent of permafrost and of the active layer*

be laterally continuous, usually where the mean annual temperature at a depth of 10 to 15 m is less than −5°C, or may be thinner and discontinuous, with typical mean annual temperatures at 10 to 15 m of between −5°C and −1.5°C (Figure 8.2). Areas of unfrozen ground within an area of permafrost are referred to as **talik**. Over 80% of Alaska, 50% of Canada, 47% of Russia and 22% of China are underlain by permafrost (Figure 8.2). Its widespread occurrence makes permafrost of considerable importance in civil engineering within these countries and it may cause increasing problems from permafrost thawing with global warming (Box 8.1).

The thermal structure of permafrost is often highly sensitive to change in vegetation or to the soil surface. The removal of vegetation (Figure 8.3) or the construction of buildings (Figure 8.4) has a dramatic effect on permafrost. The permafrost table is lowered and the thickness of the unstable active layer is increased. This causes subsidence of the ground surface. The degree of ground subsidence varies with the original ground ice content and leads to broad depressions with irregular or hummocky relief known as **thermokarst**. Flow and creep within the active layer may also occur.

The second problem associated with periglacial environments is that of **frost heave**. As thawed ground within the active layer refreezes, it expands causing the ground and objects embedded within it, such as foundations, to rise. The processes of frost heave are not well understood, but the impact on human

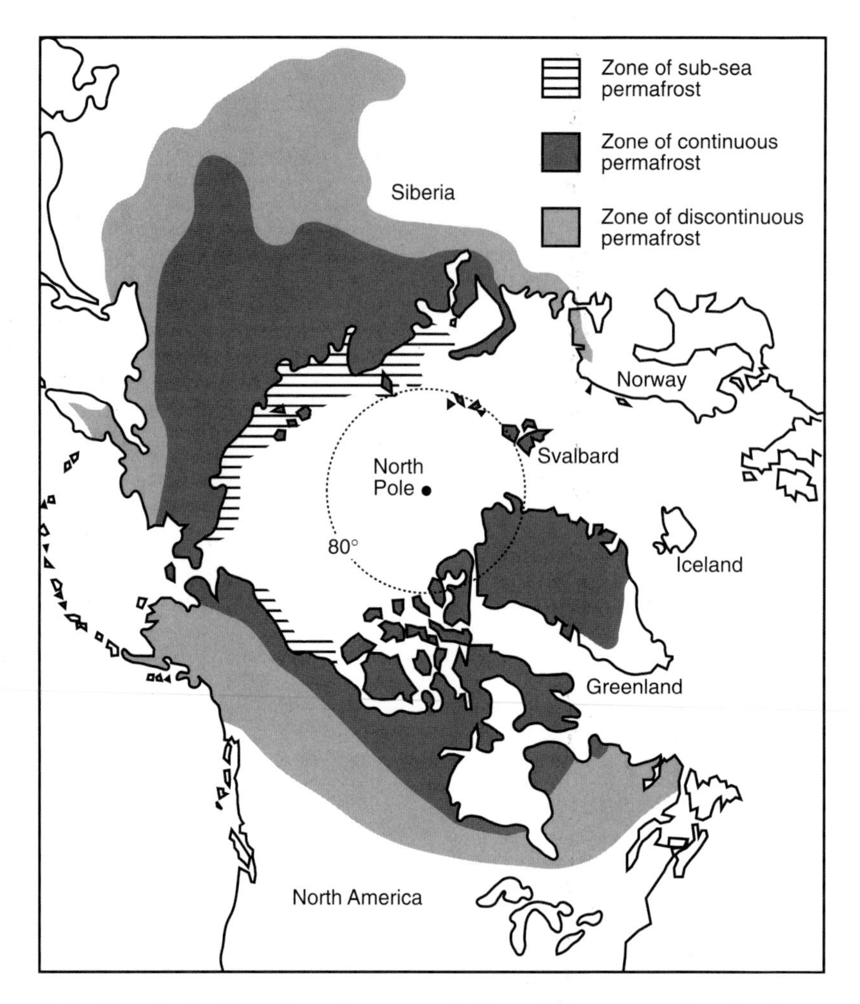

Figure 8.2 *The distribution of continuous and discontinuous permafrost in the northern Hemisphere [Modified from: Perrins* et al. *(1969)* US Geological Survey Professional Paper, *No. 678, Fig. 1, p. 2]*

infrastructure is considerable. Piles or blocks sunk into the ground will rise upwards as the ground freezes; this process will occur to different degrees within a small area as the ground freezes and expands to varying amounts depending on the quantity of soil moisture present. Susceptibility to frost heave varies with grain size (Figure 8.5).

The seasonal nature of river discharge may also cause significant problems. Rivers, streams and springs become frozen during the winter months to slabs of ice known as **aufeis, naled** or **icings**. During the spring and summer this ice melts and the river becomes swollen. At this time channels and culverts may still be blocked by icings, causing flooding. Typical discharges for arctic rivers are shown in Figure 8.6 and illustrate the extremely seasonal nature of the flow. As

BOX 8.1: GLOBAL WARMING AND PERMAFROST: A CATASTROPHIC SCENARIO

Global warming, through the accelerated emission of greenhouse gases, has major implications for permafrost regions of the Arctic. Most studies agree that global warming will be greatest in polar and subpolar regions and will also be associated with an increase in precipitation. Estimates vary as to the magnitude of these changes, but there is consensus that they will occur. This process may be accelerated by the widespread degradation of arctic permafrost. Frozen methane occurs within this permafrost, particularly in subsea permafrost of the arctic continental shelf. This methane will be released into the atmosphere as the permafrost decays, accelerating global warming. The decay of permafrost will also be accelerated by the increased precipitation predicted by most greenhouse scenarios. As standing water develops in thaw basins this will cause thermal erosion of the permafrost, accelerating its decay.

Permafrost degradation leads to the development of thermokarst, a topography of broad hollows and uneven ground, caused by ground subsidence as the frozen ground thaws. Due to the removal of ice, the ground subsides into an irregular network of depressions which will destroy all surface vegetation as well as any buildings or transport links on the surface. The engineering implications in permafrost regions of widespread permafrost degradation are considerable, particularly where investment due to the oil industry has been high. Degradation of the permafrost beneath the arctic coastal plain could cause the transgression of the Arctic Ocean over large areas. Decrease in sea ice on this ocean and increasing water temperature may accelerate this process by causing thermal erosion of the permafrost within coastal cliffs, leading to rapid coastal land loss.

It is important to remember, however, that large areas of permafrost today in the discontinuous permafrost zone are out of phase with contemporary climate and that there may not be a simple linkage between increased air temperature and precipitation and permafrost decay. There will also be a significant time lag between climate change and environmental response. Despite these reservations, global warming is likely to have the greatest environmental and landscape impact within permafrost regions.

Source: Demek, J. 1994. Global warming and permafrost in Eurasia: a catastrophic scenario. *Geomorphology*, **10**, 317–329.

a consequence, large flood events may have a short but dramatic effect on the river form and upon structures, such as bridges, which impinge upon them.

For settlements located on or close to the sea or lakes, additional problems may occur due to the development of ice. Flexure and compression of this sea or lake ice may cause it to ride onshore causing damage to adjacent structures. This is a particular problem for offshore oil and gas drilling installations in the Arctic

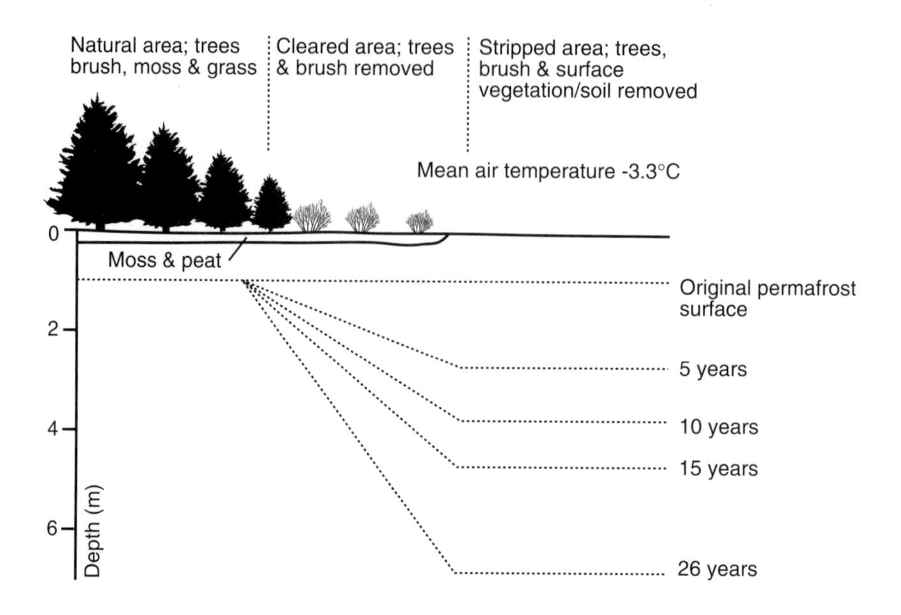

Figure 8.3 Lowering of the permafrost table by the removal of vegetation and soil [Modified from: Cooke & Doornkamp (1990) Geomorphology in environmental management. Oxford University Press, Fig. 8.15, p. 220]

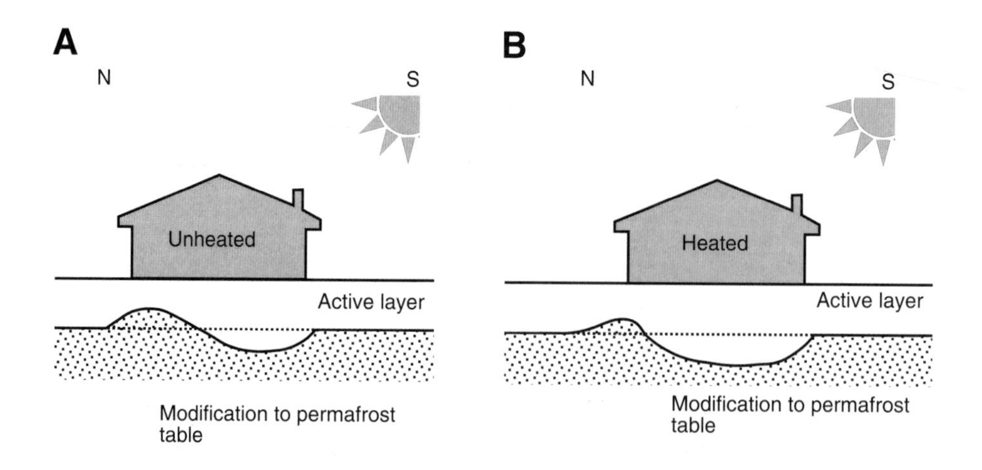

Figure 8.4 Modification of the permafrost table below heated and unheated buildings in the northern Hemisphere [Modified from: Cooke & Doornkamp (1990) Geomorphology in environmental management. Oxford University Press, Fig. 8.16, p. 221]

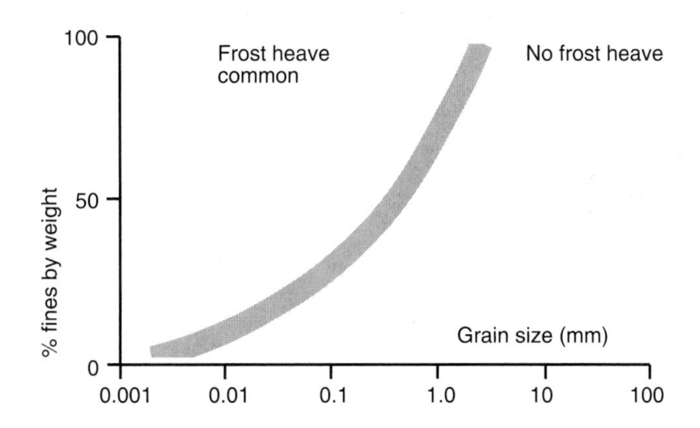

Figure 8.5 *Susceptibility of different grain size fractions to frost heave [Modified from: Cooke & Doornkamp (1990)* Geomorphology in environmental management. *Oxford University Press, Fig. 8.13B, p. 218]*

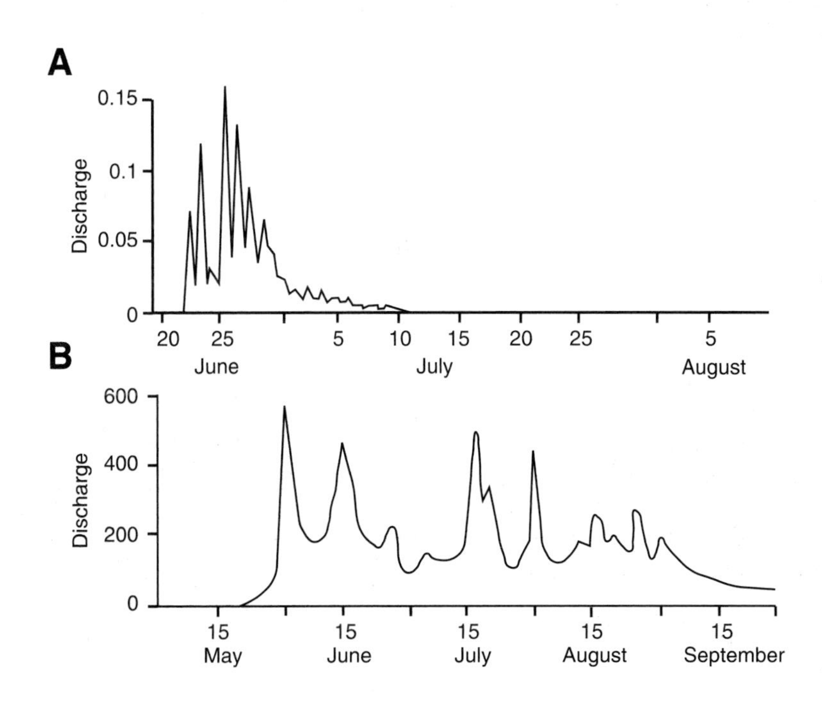

Figure 8.6 *Typical fluvial discharge variation in an arctic periglacial river [Modified from: Ballantyne & Harris (1994)* Periglaciation of Great Britain. *Cambridge University Press, Fig. 8.1, p. 142]*

BOX 8.2: THE FORMATION OF THERMOKARST BY HUMAN ACTION

The thermal balance within the ground of permafrost regions is delicate. Any action which disturbs the vegetation cover or ground surface may induce ground thaw and the development of thermokarst. French (1975) illustrates this with regard to the construction of the Sachs Harbour Airstrip, on Banks Island in the Northwest Territories of Canada. The area is underlain by continuous permafrost in which ice contents vary from 20% to 50%. The airstrip was constructed between 1959 and 1962 and was planned to be 1300 m long and 50 m wide. Owing to the presence of a dip in elevation in the centre of the proposed runway site, sand and gravel were stripped from shallow borrow pits to level the runway. Approximately 5 ha of material was removed to a depth of 1 to 2 m. Since 1962 the borrow pits on either side of the runway have undergone progressive subsidence and thermokarst modification. Today, a hummocky-type terrain composed of mounds and depressions with a relief of 1 to 1.5 m is present. The mounds are of variable shape and are typically 1 to 5 m in diameter, while depressions are more linear with widths of the order of 1 to 3 m. Standing water exists within many of the depressions and gullies have developed on parts of the site. Thermokarst modification was still occurring 10 to 12 years after the initial disturbance.

This irregular topography imposes severe limitations on access to the airstrip and will restrict any future expansion. This example clearly illustrates the consequence of ignoring the problems of permafrost in construction within the high Arctic.

Source: French, H.M. 1975. Man-induced thermokarst, Sachs Harbour, Banks Island, Northwest Territories. *Canadian Journal of Earth Science*, **12**, 132–144.

Sea which are often located on artificial islands made by dumping transported fill within coffer dams.

Ignorance of the difficulties of construction on permafrost and of related hazards can cause significant problems with the structural integrity of buildings and transport links (Boxes 8.1 & 8.2). We can identify four main construction problems within these regions: (1) the construction of stable foundations for buildings; (2) the provision of water and sanitation to these buildings; (3) the provision of communication links; and (4) the construction of oil and gas pipelines.

8.1.1 Stable Foundations

Foundation design depends primarily on the recognition of **thaw-susceptible permafrost**, that is permafrost with a significant quantity of ice within it. Dry or ice-free permafrost is not thaw-susceptible and little subsidence or soil movement will occur during thaw. Dry permafrost is associated with coarse, granular,

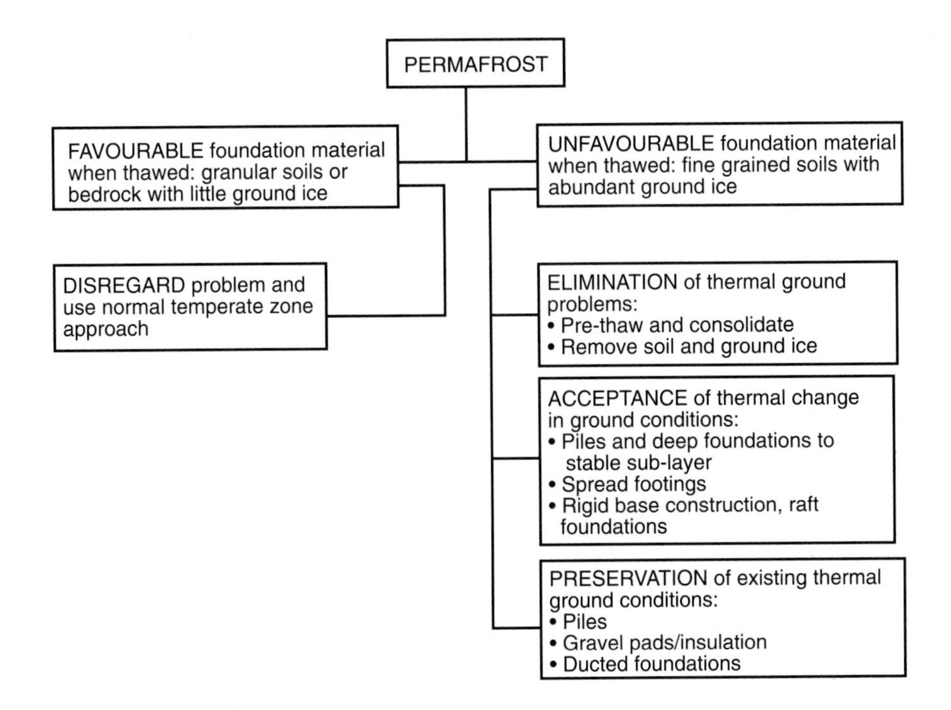

Figure 8.7 *Design options when building on permafrost*

free-draining soils, while wet permafrost is associated with fine grained soils. Fine grained soils are also more susceptible to frost heave (Figure 8.5). In practice, however, it is often very difficult to estimate even remotely the quantity of ground ice which is present, and consequently development should normally proceed on the assumption that ground ice is present. If permafrost is present then there are four main approaches (Figure 8.7).

1. **Disregard the permafrost**. If the permafrost is dry or if bedrock is at or close to the surface then it can be safely ignored, since any thaw caused by construction will not lead to serious subsidence.
2. **Eliminate the permafrost**. If the permafrost is discontinuous or very thin, pre-thaw and ground consolidation may be used prior to construction in order to eliminate the permafrost. This is usually done naturally by removing the vegetation and surface soil cover. Alternatively thin layers of permafrost may be removed mechanically and replaced by free-draining fill. Such methods are time-consuming and usually involve considerable expense.
3. **Preserve the permafrost**. The approach here is to maintain the ground in a frozen state beneath a building. This may be done either by ventilation or by insulation. With ventilation the principle is to raise the building above the ground surface to allow air circulation in order to maintain ground temperatures and dissipate any heat generated by the building. Buildings

Figure 8.8 *Typical foundations within an arctic city, in this case Longyearbyen, Svalbard.* **A.** *Foundations consisting of a thick gravel raft, interbedded with sheets of foam insulation, and concrete piles.* **B.** *Wooden piles placed into a thick layer of gravel. Both solutions aim to preserve the integrity of the permafrost beneath the building*

are usually placed on piles (Figures 8.8 & 8.9). This is the commonest approach for all heated buildings. The material used for the piles depends on the structural requirements of the building. Wooden piles are favoured for all light buildings since wood has a relatively low rate of heat conductivity. Piles are usually inserted either from a temporary gravel pad or other surface covering to prevent damage and thermal subsidence to the area around the pile. Typically piles are placed in pre-drilled holes. The minimum depth to which piles are placed should be twice the thickness of the active permafrost expected to develop during the lifetime of the structure. In certain circumstances piles may be refrigerated, particularly where they are made of steel for structural reasons. This refrigeration may be of

two types: (1) passive thermal piles, in which no external power is used and a coolant simply circulates within a sealed system; and (2) artificially refrigerated piles, in which an external power source is used to drive a coolant through the pile as in a refrigerator – this is only used where cheap power sources are available and very stable ground conditions are required.

The alternative approach to permafrost preservation is to place the building on a pad or raft of gravel and thereby insulate the ground beneath. Correctly installed, the pad will cause the permafrost table to rise into the pad (Figure 8.9A). The amount of material required within a pad may be reduced if an insulating layer is included; this may consist of peat, wood chips, sulphur foams or polystyrene foam. This is particularly common for roads, airfields, railways and large unheated structures. Large ventilation ducts, pipes or culverts may be introduced in some gravel pads to assist in the process. Equally, concrete ducts may be incorporated into foundation design (Figure 8.9 C&D). These ducts or culverts may have doors which are

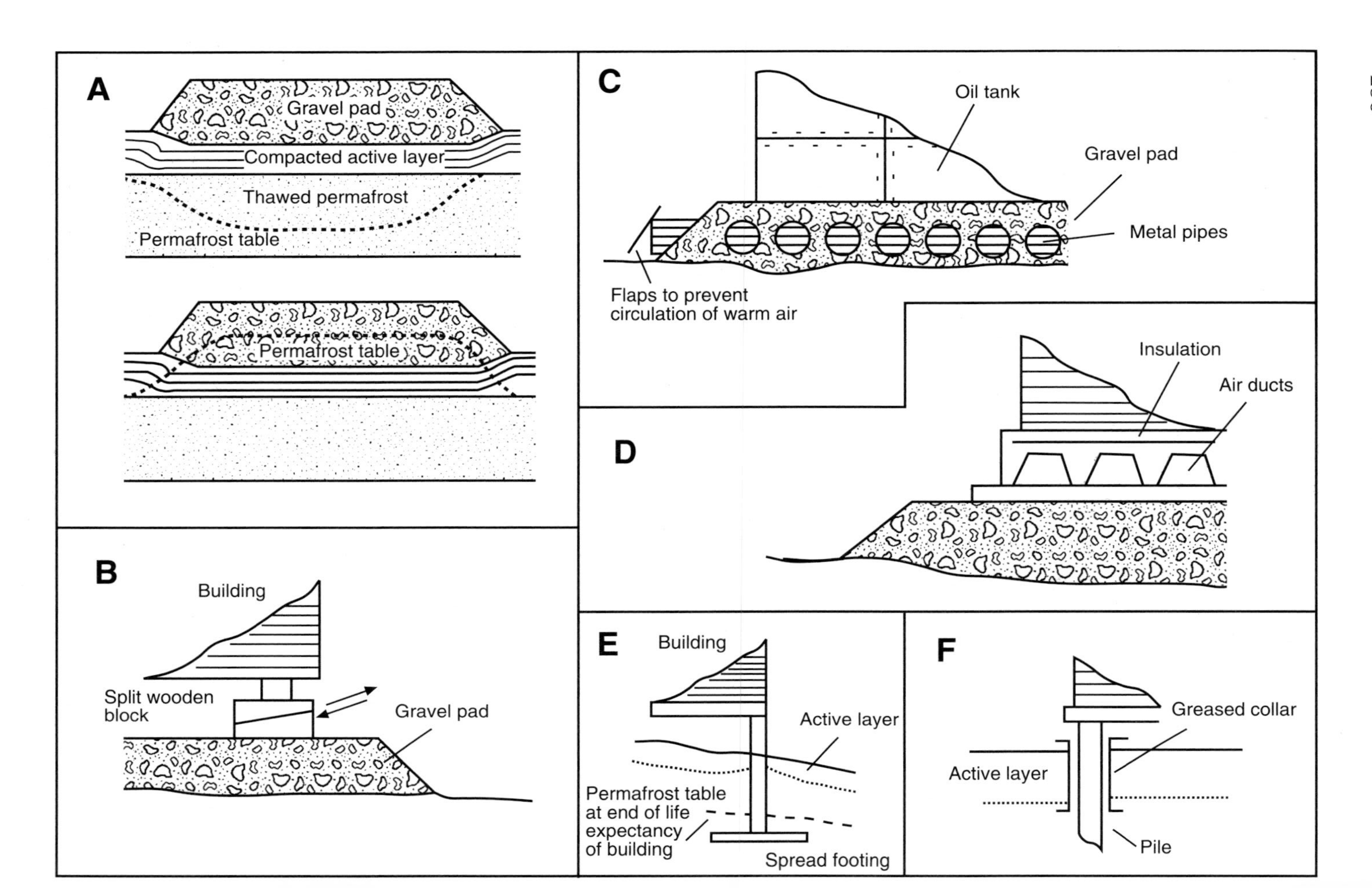
A
Gravel pad
Compacted active layer
Thawed permafrost
Permafrost table
Permafrost table
B
Building
Split wooden block
Gravel pad
C
Oil tank
Gravel pad
Metal pipes
Flaps to prevent circulation of warm air
D
Insulation
Air ducts
E
Building
Active layer
Permafrost table at end of life expectancy of building
Spread footing
F
Greased collar
Active layer
Pile

opened when the air temperature is typically below –3°C and closed during warmer temperatures. Ducts need to be carefully placed with respect to local wind directions in order to maximise the air flow through them. This method has been adopted extensively in the construction of large oil tanks.

4. **Design for subsidence.** In this case the foundations of buildings are placed on a stable substratum at depth or placed on large rigid foundation rafts or spread footings. The idea here is to maintain building stability while thawing occurs around it. Spread footings consist of reinforced concrete, wood or steel footings shaped like an inverted 'T', placed so that the base is anchored firmly in the permafrost (Figure 8.9E). The base of the footing should be placed at a depth equal to at least twice the eventual thickness of the active layer that will occur due to thawing in the lifetime of the building.

In addition to these measures, effective insulation within the buildings is essential. Some differential settlement or tilting is inevitable and most foundation designs include devices such as that illustrated in Figure 8.9B to maintain a horizontal surface within the building.

Accommodating frost heave is an important consideration, particularly with the placement of piles. Frost heave within the active layer may cause piles to rise and fall seasonally and will generate extreme tensile stresses within them, as they are literally stretched as the ground in which they are frozen expands and contracts. In particularly severe cases a collar may be placed around the pile so that the active layer may expand and contract around the pile (Figure 8.9F).

Where buildings are located close to a lake or seashore, the build-up of ice and its onshore movement may prove hazardous. Figure 8.10 illustrates two possible solutions to this problem.

8.1.2 Provision of Sanitation

The provision of basic utilities to houses and other buildings within areas of intense periglaciation is a particular problem. In most European countries this basic infrastructure is placed below ground, but this is usually not possible in arctic regions due to pipe freezing. There are three problems: (1) water supply; (2) fluid transfers; and (3) waste disposal.

Figure 8.9 *(opposite) Foundation design in permafrost regions.* ***A.*** *The thickness of the gravel pad is crucial in determining the modification of the permafrost table; if it is too thin, melting occurs. Ideally the permafrost table should rise into the gravel insulation.* ***B.*** *Detailed design of building foundations on a gravel pad; the split wooden blocks can be moved to relevel the building as the gravel pad settles, or as minor thaw adjustment occurs.* ***C.*** *Metal air ducts may be introduced into a gravel insulation pad to conduct heat away; these air ducts must be shut in warm air to prevent its introduction into the gravel pad.* ***D.*** *Concrete raft foundations may also be designed with air ducts within them.* ***E.*** *Spread-footing foundations are used when preservation of the permafrost surface is not possible.* ***F.*** *Greased collars are sometimes placed around pile foundations to minimise the effects of ground heave*

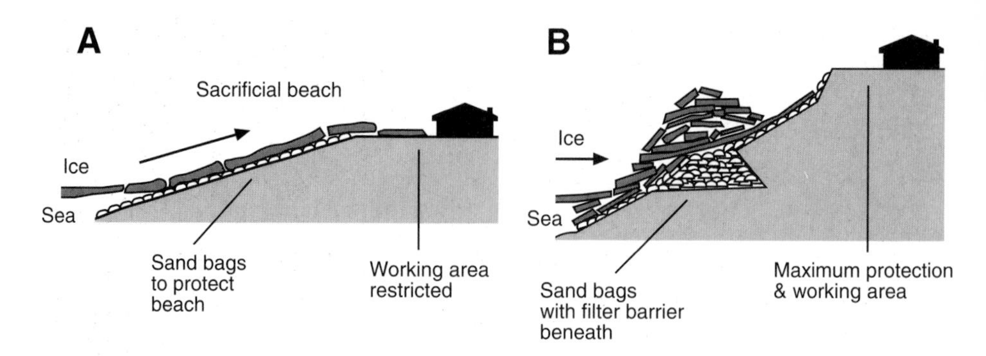

Figure 8.10 *Protection of buildings from seasonal sea or lake ice.* **A.** *Sacrificial beach, lined with sand bags to prevent erosion.* **B.** *Beach designed to compress and pile the ice up over a short distance [Modified from: Harris (1986)* The permafrost environment. *Croom Helm, Fig. 7.8, p. 150]*

Water supply is a severe limitation to urban development within periglacial regions. Water can be derived either from surface ponds and lakes or from groundwater. In areas of discontinuous permafrost, some groundwater recharge occurs, and access to it is possible. There are three main types of groundwater: (1) supra-permafrost groundwater in talik which exists above the permafrost table but below the depth of seasonal freezing, and may occur particularly in areas of relict permafrost which are out of phase with contemporary climate; (2) intra-permafrost groundwater contained within thaw zones confined by permafrost; and (3) sub-permafrost groundwater in thaw zones and conventional aquifers below the permafrost. Supra-permafrost and intra-permafrost groundwater are usually of limited volume and may be heavily contaminated by mineral components. In discontinuous permafrost, recharge of subsurface aquifers may occur and can provide valuable water reserves. However, in areas of continuous permafrost where it may be well over 100 m thick, sub-permafrost groundwater cannot be easily located or retrieved without it freezing. In these locations water supply is restricted to surface pools and ponds, which if sufficiently deep should not freeze completely. The availability of suitable ponds and lakes provides a major restriction on the location of settlements and may restrict their growth.

The transfer of fluids involves the delivery of fresh drinking water and the removal of waste products. There are three principal systems to deal with this: (1) insulated burial; (2) delivery and collection; and (3) utilidor systems. Insulated burial is possible in areas with discontinuous permafrost or with only a minor frost problem. Pipes are coated in thick layers of polyurethane foam and then buried between a sandwich of foam slabs. In most cases this is sufficient to prevent both freezing and permafrost. thawing.

An alternative is the delivery and collection of all fluids. Tankers are used to deliver water to each dwelling which is then stored in an insulated tank. Waste sewage is either stored in a heated and insulated tank which is periodically pumped clean or deposited in polythene bags which are left to freeze solid

Figure 8.11 *A utilidor system in Longyearbyen, Svalbard*

before collection. These bags are often referred to as **honey bags**. This type of system is widely used.

A more sophisticated and increasingly used system is to place all utilities above ground in a conduit elevated on piles known as a **utilidor** (Figure 8.11). In this way pipes are kept above the ground surface, which remains relatively undisturbed, and can be insulated and heated if required to maintain fluid flow. This has important implications for the construction of roads and pavements, which have to be elevated or bridged over the utilidor. This is particularly important since sewage pipes need to be organised so that gravity maintains flow. Fresh water systems can be placed under high pressure and therefore made to flow uphill; however, sewage cannot. Careful design to ensure a constant gradient is therefore essential.

Waste disposal depends on the type of sanitary system employed. Three types of waste can be identified, each of which is dealt with in a slightly different way: (1) undiluted or solid waste ('honey bags'), products of bucket toilets where piped systems are absent; (2) moderately diluted wastes, the products of using heating holding tanks; and (3) conventional strength waste water produced from flush toilets and transported via utilidors.

Undiluted waste is usually disposed of via a landfill site some distance from the settlement. Moderately dilute wastes are commonly disposed of in some form of holding lagoon where bacterial digestion or separation is employed, or alternatively the waste is simply flushed out to sea. Conventional strength waste water is either disposed of at sea or digested in a holding lagoon (see Section

9.7.1). Disinfecting the effluent is an essential part of the process owing to the high micro-organism content of the waste and the slow rates of decay/digestion. It is not surprising therefore that certain types of dysentery are 74% more common in arctic Canada than elsewhere.

8.1.3 Transport Links

The provision of communication links such as air strips, roads and railways is of particular importance within such a hostile environment. Any form of communication link may upset the thermal balance of thaw-sensitive permafrost. In certain cases the passage of a single vehicle may cause destruction of the vegetation and surface compaction, resulting in permafrost thaw and linear subsidence.

Critical to minimising the problems is effective route selection and therefore initial terrain analysis. In general it is advantageous to locate a road or railway along a drainage divide, thereby maximising drainage and minimising the number or river crossings, culverts and embankments required. When this is not possible the availability of suitable frost-stable construction material is critical. For example, a 1.6 m wide gravel pad, such as that typically used for roads in northern Alaska, involves over 23 755 m^3/km of gravel, while simply maintaining an unmetalled road involves over 65 m^3/km of gravel every year.

The main problems faced are: (1) maintaining the thermal equilibrium; (2) river crossings; (3) drainage and icing; and (4) stabilisation of adjacent embankments or slopes. Roads and railways are normally constructed on gravel pads to maintain thermal ground equilibrium. However, there is rarely sufficient material to prevent all thaw subsidence and a compromise is required between the cost of maintenance caused by subsidence and the cost of importing large amounts of free-draining fill. Regular regrading and maintenance of roads, which are largely unmetalled, is therefore essential. In the case of air fields, where metalled surfaces are desirable, runways are allowed to settle and thaw for up to five years before they are surfaced, to allow all subsidence to occur. A common problem is the slippage, particularly via rotational failures, of the sides of the pad and of other embankments. The construction of toe berms has become commonplace in order to counteract this, but in practice it is very difficult to eliminate and in most cases is tolerated and incorporated into the design.

River crossings pose a significant problem for roads and railways. Below the river the ground is usually unfrozen, while on either side permafrost may occur. In the transition between two areas of thaw is a zone in which seasonal freezing occurs. It is in this zone that bridge piers are most at risk of frost heave, as illustrated in Figure 8.12. A variety of alternative engineering solutions exist, but in most cases the need for bridge piers is minimised and bridge footings are placed well back from the banks on ice-rich permafrost.

Drainage may be a significant problem. Wherever possible, roads should be built parallel to the direction of groundwater movement to aid drainage. Cross-drainage is best dealt with via metal culverts which are able to accommodate

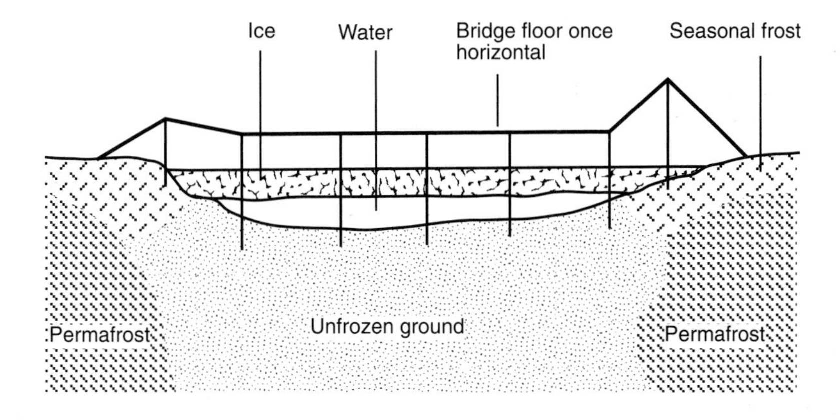

Figure 8.12 *Frost-heaved piling of bridge spanning an arctic river near Clearwater Lake, in Alaska [Modified from: Perrins* et al. *(1969)* Geological Survey Professional Paper, *No. 678, Fig. 31, p. 31]*

significant quantities of subsidence and settlement. Ponding of water near roads or in ditches may cause thermal erosion of the permafrost by accelerating the melting of ground ice. This water may also lead to stability problems. One of the causes of this water ponding is often the development of thick icings each winter within culverts and drainage channels. These are slow to melt, causing ponding of spring drainage. The growth of icing across the road surface is also a significant traffic hazard. The identification of significant areas of icing at the planning stage is essential, but avoidance is not always possible and does not deal with the problem of icing within culverts. Attempts to minimise the problem of icings revolve around monitoring their location and size so that oversized culverts and relief channels can be placed in key locations to carry excess water flow. Sealed pipes may also be placed in some culverts and only opened when required to remove drainage while the culverts remain iced up. Cut-off drains may be used to control seepage and icing growth away from the road. Similarly, if the volume of icing is known and is fairly constant then it is possible to place a dyke upslope, parallel to the road, often with an adjacent excavation that is of sufficient size to contain the entire icing. Alternative and more labour-intensive methods include the artificial thawing of iced culverts each year with steam jets. In parts of central Alaska, fire pots made of 200 litre drums kept burning continuously are used to keep culverts ice-free. These so-called 'moose-warmers' require regular maintenance and consume between 40 and 100 litres of fuel per day.

One of the major problems in permafrost regions is dealing with sections cut into ice-rich permafrost. In the case of bedrock, the only problems are those of perched water tables and associated seepage and icing problems. However, problems of stability, slumping and sediment flow occur within ice-rich cohesive sediments. The best approach is to cut slopes vertically and then allow them to thaw, slump and thereby attain a stable natural slope angle before road

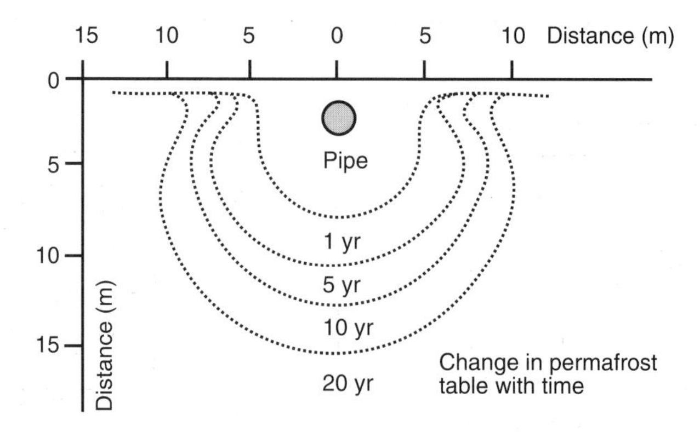

Figure 8.13 *Thaw zones around an oil pipeline approximately 2.5 m below the surface and with a temperature of 80°C [Modified from: Cooke & Doornkamp (1990)* Geomorphology in environmental management. *Oxford University Press, Fig. 8.17, p. 221]*

construction starts. Where this is not possible owing to restrictions on space, cut slopes are encased within a thick gravel layer in order to insulate them and prevent thaw, saturation and therefore sediment instability.

8.1.4 Oil and Gas Pipelines

When oil leaves the ground it is warm and, depending on its viscosity, must be kept warm in order to remain mobile. If buried, oil pipes will have a dramatic effect on the permafrost (Figure 8.13). By contrast, natural gas can be moved at almost any convenient temperature provided that a sufficient pressure gradient is retained; however, its explosive nature in confined spaces increases the need to ensure that leakages do not occur.

The largest oil pipeline project in North America is the Trans-Alaskan Pipeline from the oilfields of Prudhoe Bay on the arctic coast to Valdez on the Gulf of Alaska, from where the oil is shipped south by tanker (Figure 8.14). The pipeline is necessary since Prudhoe Bay is ice-bound for most of the year. The 1300 km pipeline resolved many of the design issues associated with the construction of pipelines in arctic regions and set the standards, at least in North America, for pipeline design in permafrost regions. As a consequence we shall use its design to demonstrate the general principles associated with such structures in the Arctic.

Figure 8.14 *(opposite) The Trans-Alaskan Pipeline, Valdez to Prudhoe Bay.* ***A.*** *Choice of route and types of permafrost encountered. Note the active faults (earthquake zone) in the vicinity of the Alaskan mountain range. The length of pipeline above ground and below is also shown.* ***B.*** *Design of elevated sections.* ***C.*** *Design for buried sections [Based on data in: Williams (1986)* Pipelines and permafrost. *Carlton University Press]*

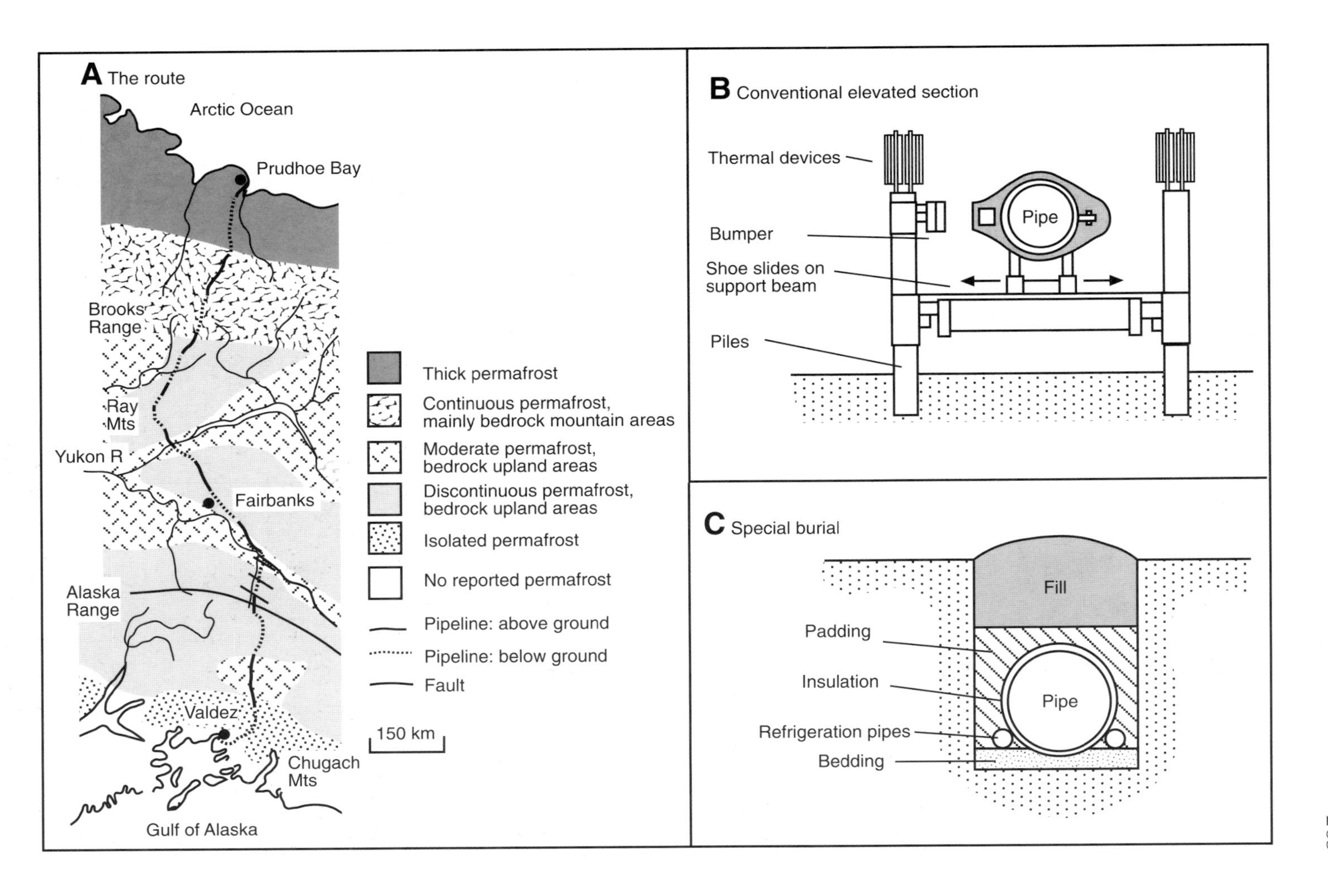

A The route
Arctic Ocean
Prudhoe Bay
Brooks Range
Ray Mts
Yukon R
Fairbanks
Alaska Range
Valdez
Chugach Mts
Gulf of Alaska
Thick permafrost
Continuous permafrost, mainly bedrock mountain areas
Moderate permafrost, bedrock upland areas
Discontinuous permafrost, bedrock upland areas
Isolated permafrost
No reported permafrost
Pipeline: above ground
Pipeline: below ground
Fault
150 km
B Conventional elevated section
Thermal devices
Bumper
Shoe slides on support beam
Piles
Pipe
C Special burial
Fill
Padding
Insulation
Pipe
Refrigeration pipes
Bedding

The Trans-Alaskan Pipeline was constructed between 1969 and 1977 at a cost of $7 billion and has a diameter of 1.22 m with a pipe wall thickness of 13 mm. The oil leaves the ground at Prudhoe Bay at a temperature of 80°C and, owing to its viscous nature, must be moved at a temperature of between 60 and 65°C. About a million barrels of oil a day were passing through the pipeline by 1978 (200 million litres), although this fell in the 1980s to only 600 000 barrels. Not only were there the problems of construction on permafrost, but also a significant threat from earthquakes.

The initial task was detailed terrain analysis in order to identity first a practical route, and second any significant permafrost bodies. Associated with this was a drilling programme involving 6000 boreholes to determine the ice content and stability of the permafrost. A further 2000 holes were drilled during construction to supplement this information. Approximately 640 km of the pipeline were buried conventionally through areas where the permafrost was either absent or relatively thin and therefore considered to be non-hazardous. Conventional burial involved laying the pipe at depths of between 450 mm and 4 m along with two zinc-ribbon anodes, to prevent electrolytic corrosion of the steel pipe. The rest of the pipeline was raised above the surface, with the exception of an 11 km section which was subject to special burial in areas where burial was necessary for aesthetic or practical reasons. Special burial involved either laying the pipe with insulation or, in some cases, burying it with refrigeration pipes attached to compressors in order to circulate coolant within them. Raised sections were placed on piles constructed of steel because of the weight of the pipeline and the stress generated within the active layer and by the thermal contraction and expansion of the pipe itself. As a consequence of heat conduction within the piles, the majority (80%) contained cooling devices. These devices consist of approximately 50 mm diameter sealed tubes which extend below the surface and contain anhydrous ammonia coolant. In the winter months this evaporates from the lower end of the tube and condenses at the top, where each pipe is fitted with metallic heat exchanger fins (Figure 8.14). The evaporation occurs because during the winter months, the ground is warmer than the air outside. This evaporation cools the lower end of the pipe and the surrounding ground; by cooling the permafrost in winter its temperature is sufficiently lowered to prevent it from thawing during the summer. Each pile typically extends 8 m into the ground.

Thermal contraction and expansion of the pipeline were accommodated by laying the pipeline in a trapezoidal zigzag fashion and anchoring it only at intervals of between 250 m and 600 m. In addition, the pipeline was allowed to move laterally on its pile foundation by up to 4 m. In the earthquake zones this lateral movement was increased to as much as 6 m. Over 142 stop valves were also fitted to limit oil escape to between 13 000 and 50 000 barrels, if any rupture occurred.

River crossings were also a potential problem. Wherever possible the pipeline was attached to existing bridges. In other cases it was buried 5–6 m beneath the river bed and encased in thick concrete to counter the problems of bed scour. The seasonal nature of the flow regime (Figure 8.6) can cause bed scour of up to 2 to 3 m in spring. Elsewhere, purpose-built pile bridges were used.

Since construction, there have been two significant leakages from the pipe; both were due to permafrost subsidence and over 50 km of the pipeline had subsided, especially in areas where it had been buried. Aerial infrared photographs were used to monitor the operation of heat exchange fins on the piles, and by 1980 it was discovered that many were not operating as efficiently as predicted owing to internal corrosion. In order to minimise the destruction of the permafrost during construction, temporary gravel pads were placed and thick foam plastic board used. Despite these problems, the Trans-Alaskan Pipeline has set the standard and illustrates the difficulties of construction within regions of active permafrost.

8.2 URBAN DEVELOPMENT IN SUBTROPICAL DESERTS

Rapid urban expansion and development, funded by oil revenues, during the 1970s in the Middle East emphasised many of the problems of construction within subtropical deserts. Unlike the periglacial environment these problems focus around three particular hazards to construction: (1) flash floods; (2) groundwater salinity and salt weathering; and (3) the drift of sand and dust. Each of these problems is considered in turn.

8.2.1 Flash Floods

Flood events occur infrequently within drylands and are therefore rarely perceived as a significant problem to construction. It is perhaps this fact more than any other that makes them a problem, coupled with the availability of easily erodible soil due to the lack of a continuous vegetation cover. The fluvial system within desert regions is often ephemeral, transitory and poorly defined. Not only does the flood water cause a particular problem but it is often associated with large sediment loads which are dumped rapidly by waning flows. In urban expansion and in the construction of roads and other communication links, there are four main approaches to deal with the flood problem: (1) prediction/warning; (2) avoidance or relocation; (3) minimising disturbance; and (4) construction works.

Flood prediction and warning are difficult owing to the infrequency of flood events and consequently their poor data record in arid regions. More significantly, the onset of such events is usually extremely rapid making warning of little practical value. Avoiding flood hazards is of greater value where control over future development and land-use planning can be exercised. Terrain analysis based on geomorphological mapping of the fluvial components present may provide a basis for hazard mapping. The lowland fluvial system in these regions is frequently dominated by a series of alluvial fans and deep wadi channels. It is possible to conceptualise the potential flood hazard on such alluvial fans on the basis of the proximity of the surface to those channels which have been active recently (Figure 8.15). For example, young fan surfaces, active

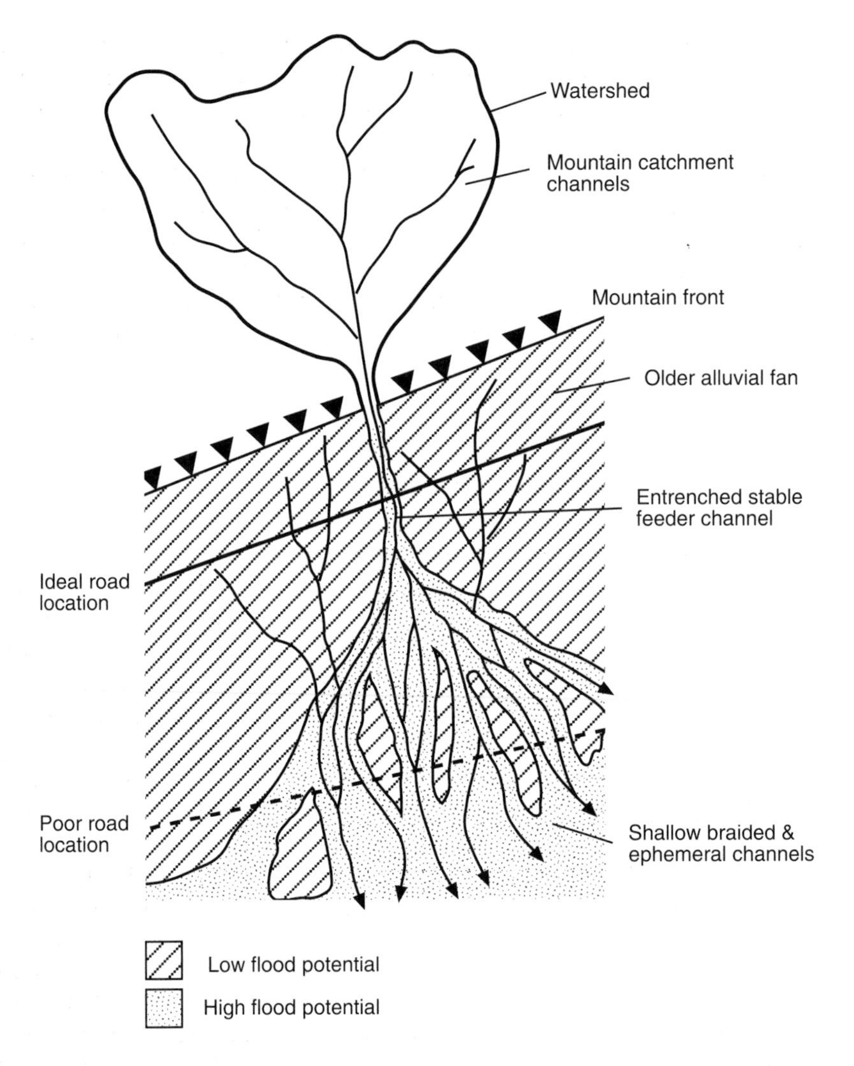

Figure 8.15 *Components of an alluvial fan and their hazard potential [Modified from: Cooke* et al. *(1982)* Urban geomorphology in drylands. *Oxford University Press, Fig. VI.5, p. 198]*

wadi channels and areas at a similar elevation to the active channel floor are all at greater risk than older fan surfaces which are removed in elevation from the active channel system. By mapping the geomorphology of the fluvial system it is possible to classify each surface on the basis of its flood hazard potential. Although subjective, such hazard maps can be used effectively to guide future urban expansion or development to avoid problem areas, and they have been used to assist urban development in the town of Suez in the Middle East.

Similarly, in the location of roads an effective understanding of the geomorphological systems present may be of particular value. For example, if a road is

to cross a series of alluvial fans then it is best located towards the apex of those fans where channels are usually entrenched and relatively stable (Figure 8.15), as opposed to the base of a fan where there are numerous poorly defined and transitory channels to be crossed.

A policy of minimising land disturbance may also be effective in reducing the sediment volumes transported within flood flows. Natural desert surfaces often attain a degree of erosional stability either through the formation of surface crust or through removal of easily erodible sediment. Disruption of these equilibrium surfaces may increase the surface erodibility and therefore the potential for sediment entrainment by flood flows. Ultimately the flood problem may be addressed by the construction of flood berms and barriers with which to deflect flood waters.

8.2.2 Groundwater Salinity and Salt Weathering

The breakdown of rock and concrete by salt weathering is an important geomorphological process in many environments, but is particularly destructive in semiarid and arid regions. Groundwater is often particularly saline owing to concentration of salts by evaporation and evapotranspiration coupled with low rainfall. In coastal regions, seawater intrusion may also significantly alter groundwater chemistry, increasing its salinity. In deserts, saline groundwater is particularly concentrated in two common components of the desert landscape: **sabkhas**, which are broad flats that arise from intertidal sedimentation; and **playas**, which are the floors of enclosed drainage basins. In both areas the water table is close to the surface and usually extremely saline. Groundwater rises above the water table into a zone known as the **capillary fringe**. High day-time air and surface temperatures, and low and sporadic rainfall coupled with high rates of evaporation result in marked capillary rise of water which causes the upward leaching of salts from the water table. This concentrates salts within the upper part of the soil profile and on the ground surface as **salt efflorescence**. The most commonly occurring salts include calcium carbonate, calcium sulphate and sodium chloride. The geometry of the water table and the limit of capillary rise, the capillary fringe, are important in determining the extent of the problem of salt weathering (Figure 8.16).

The precipitation of salts from groundwater causes accelerated weathering in four ways: (1) the growth of salt crystals through time due to evaporation and/or cooling of saline solutions causes the build-up of crystals in fissures, cracks and pores of rocks and other building materials; (2) the thermal expansion of salt crystals, causing internal stresses within the rock, occurs due to the strong diurnal temperature variation of hot arid regions; (3) the hydration of certain salts may cause expansion and therefore rock distress; and (4) chemical processes associated with salt attack may cause volume changes especially within such materials as concrete – both chloride and sulphate salts react with cement causing serious deterioration. Not only is the problem caused by salts deposited from rising groundwater, but it is exacerbated by the introduction of salts into building

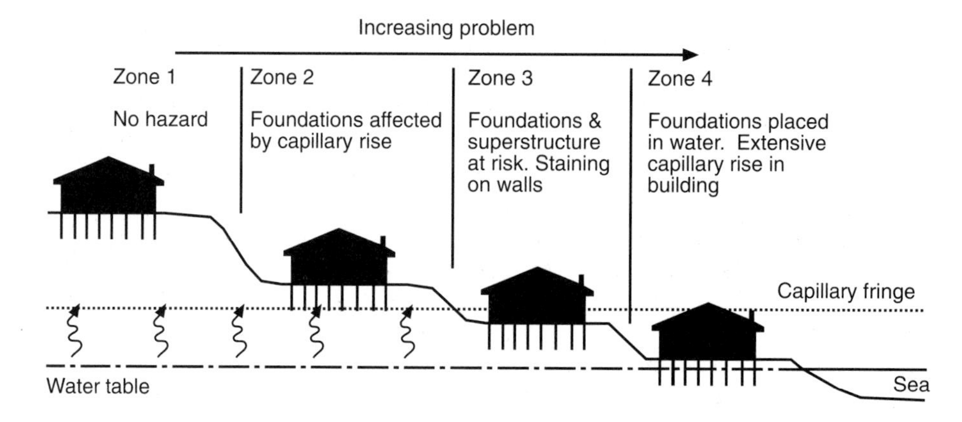

Figure 8.16 *Hazard zones associated with salt weathering*

material such as aggregates and in the water used to mix cement and similar products. The processes of salt weathering have been widely simulated in the laboratory, which has assisted in quantifying the relative importance of different salts and in evaluating the susceptibility of different rock types (Box 8.3).

In dealing with these problems, the best solution is to predict the occurrence of intense salt weathering and either plan to avoid such sites or alternatively ensure that foundations and buildings are adequately protected (Figure 8.16). Foundation protection includes: the use of sulphate resisting cements; careful screening and washing of all aggregates to remove salts; the use of very dense and therefore low porosity concrete; and the tanking (sealing) of foundations to prevent groundwater contamination.

Prediction of the salt weathering threat involves consideration of three variables: (1) the variation in groundwater geochemistry; (2) the geometry of the water table; and (3) determining the extent of the capillary fringe. Groundwater salinity can be gauged by examining its electrical conductivity, which increases with salt content. Similarly, boreholes can be used to record the geometry of the water table and its seasonal variation. Finally, the limit of the capillary fringe can in most cases be recognised by the extent of damp ground or wall surfaces, and the extent of salt efflorescence and 'puffy' ground caused by salt growths. These three variables can be combined to produce a hazard map as illustrated in Figure 8.17. As in most hazard maps, the choice of hazard classes is arbitrary, although it provides a means of combining groundwater salinity and water table depth data in a format which can be interpreted by non-experts.

8.2.3 Movement of Sand and Dust

Drifting sand and dust is a significant problem for some desert settlements. This is the product of three main factors: (1) rapid urban expansion in a potentially hazardous location, often historically determined by the availability of water;

BOX 8.3: LABORATORY SIMULATION OF SALT WEATHERING

Salt weathering involves the precipitation of salts in the pores and fissures of rocks. As the volume of salt increases the rock experiences distress and ultimately failure. This process has been simulated in the laboratory using environmental cabinets in which blocks of rock, of known weight, are repeatedly immersed in salt solution and then allowed to dry (e.g. Goudie *et al.* 1970; Cooke, 1979; Goudie, 1993). After each cycle of wetting and drying the weight of the rock, and therefore any mass lost by weathering, is determined. In this way it is possible to experiment with different salt solutions in order to find the most effective (Diagram A) and to examine how different types of rock respond to salt weathering (Diagram B). Experimental results suggest that sulphate salts are the most effective agents of decay and that rock porosity is an important consideration in the susceptibility of rocks to salt weathering. Such information is important in the choice of building materials in areas where salt weathering is known to be a particular problem. This work illustrates how pure academic laboratory-based research can have important implications in applied studies.

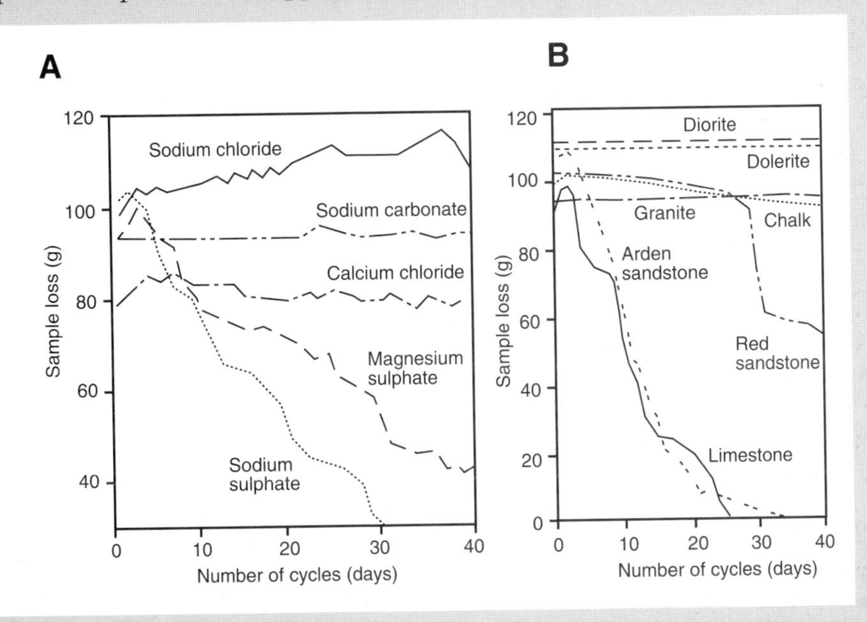

Sources: Goudie, A.S., Cooke, R.U. & Evans, R. 1970. Experimental investigation of rock weathering salts. *Area*, **4**, 42–48; Cooke, R.U. 1979. Laboratory simulation of salt weathering processes. *Earth Surface Processes and Landforms*, **4**, 347–359; Goudie, A.S. 1993. Salt weathering simulation using a single-immersion technique. *Earth Surface Processes and Landforms*, **18**, 369–376. [Diagram modified from: Brunsden (1979) In: Embleton & Thornes (Eds) *Process in geomorphology*. Arnold, Figs 4.16 & 4.17, pp. 124 & 125]

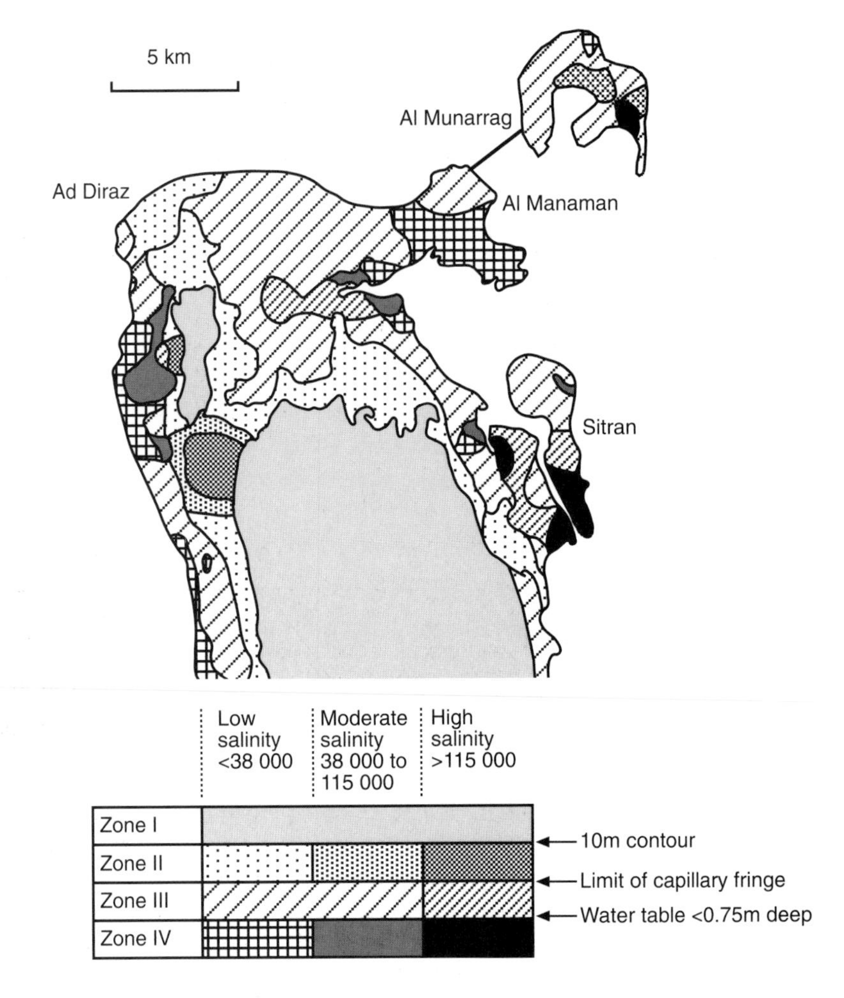

Figure 8.17 *Salt weathering hazard map for northern Bahrain constructed by combining groundwater salinity with depth to groundwater table data [Modified from: Cooke* et al. *(1982)* Urban geomorphology in drylands. *Oxford University Press, Fig. V.15, p. 180]*

(2) the growing aspiration of inhabitants for improvements in environmental quality; and (3) increasingly widespread disturbance of the ground surface by human activity. Ground disturbance leads to the removal of natural vegetation and the destruction of natural soil crusts and stabilised pavement surfaces. There are a number of different aspects to this problem: (1) sand blast problems, accelerated abrasion of buildings and human discomfort due to blowing sand; (2) deflation (erosion) around the base of buildings, telegraph poles and other structures; and (3) the deposition of sand and dust in the lee of buildings and on communication links.

Control methods for dealing with problems of sand and dust movement can be summarised as follows (Figure 8.18).

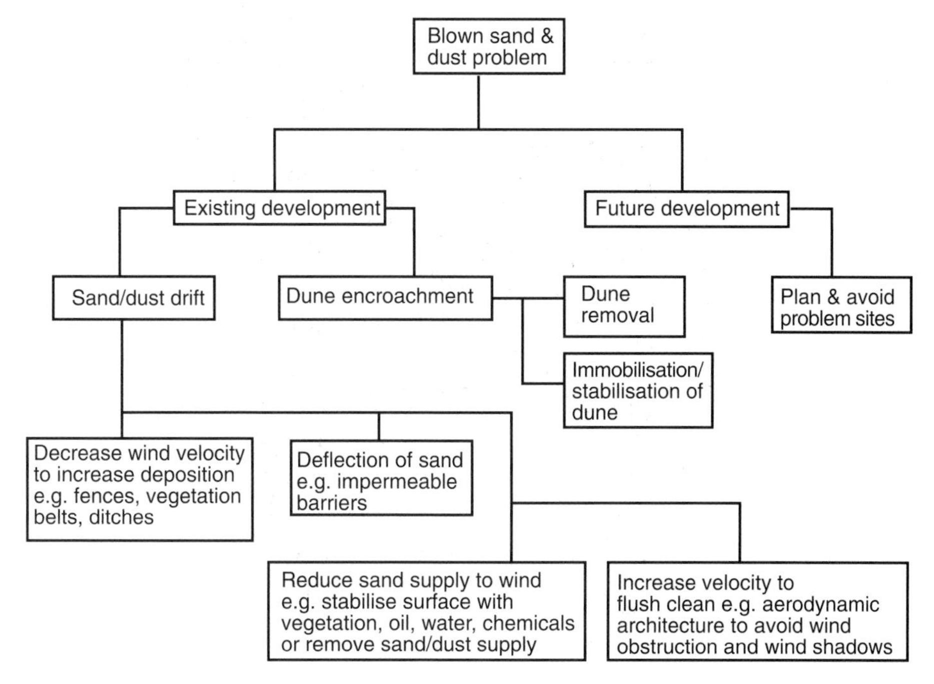

Figure 8.18 *Management solutions to the problem of drifting sand and dust*

1. **Avoidance**. Site surveys which identify areas of potentially difficult terrain can be used in the location of settlements, urban expansion and infrastructure. As with all environmental hazards, avoiding the problem is the most effective solution. Assessing the potential problem usually involves a combination of: (1) analysis of meteorological information – velocity, direction and frequency – to identify those wind directions with the greatest potential for sand transport (i.e. those strong enough and occurring with sufficient frequency to be a problem); (2) geomorphological mapping to identify potential sand and dust sources, as well as mobile sand dunes. This last point is particularly important since if strong winds blow over sand-free bedrock there is unlikely to be a sand drift problem, while if they blow over loose, unvegetated sand then there will be. Potential problems occur when there is a coincidence in sand/dust sources and the frequent occurrence of wind directions capable of sand/dust transport. Identifying these wind directions is critical in assessing the hazard distribution (Box 8.4). In practice, most settlements or communication links are historically located, usually nucleated around water sources, and this reduces the viability of avoidance methods. However, the direction of urban expansion can be based on effective site surveys so that potential problems are avoided.
2. **Removal**. This involves the removal of sand dunes and potential source areas of sand/dust supply, an extreme and rarely practical solution. Moreover, dune removal rarely works since wind flow over flattened areas tends

BOX 8.4: THE PREDICTION OF SAND AND DUST HAZARDS

Jones *et al.* (1986) examined the problem of blowing sand and dust at the town of Alfa in the Middle East. Their aim was to estimate the magnitude of the problem and its distribution around the town. They used three types of data: (1) LANDSAT images; (2) meteorological data; and (3) geomorphological mapping. LANDSAT images were used to identify the principal sand dunes and trends in sand movement.

Meteorological data available from the airport at Alfa were used to calculate sand and dust drift potentials. Only potential sand/dust movements can be calculated from wind data since actual transport will also depend on the availability of sand to be transported. Potential sand movement (q) is a function of $V^2(V-V_t)T$, in which V is the wind velocity measured at 10 m, T is the percentage of the total time during which a wind of particular velocity (V) blows from any one direction, and V_t is the sand-movement threshold taken as 12 knots for sand. Potential dust movement can be calculated in a similar way by simply using a lower threshold velocity. In this way sand and dust roses of potential transport can be constructed and used to delineate zones of potential sand/dust transport around the town.

This, however, only provides information about potential transport and in order to estimate the actual transport problem this information must be combined with information about the availability of sand/dust. If potential sand-transporting winds blow over debris-free rock surfaces little transport will occur, but if they blow over loose sand, transport may take place. This information can be provided by geomorphological mapping. At Alfa, Jones *et al.* (1986) not only mapped the geomorphology of the surrounding area but also classified each geomorphological surface in terms of its potential as a supply for sand and dust. By combining all this information it is possible to predict which areas around the town are likely to experience the greatest problem from sand and dust drifting. Such information can be used to plan future expansion of the town to ensure that development does not occur in areas of high sand/dust drifting. This example again illustrates the importance of geological site evaluation prior to development.

Source: Jones, D.K.C., Cooke, R.U. & Warren, A. 1986. Geomorphological investigation, for engineering purposes, of blowing sand and dust hazard. *Quarterly Journal of Engineering Geology*, **19**, 251–270.

to quickly re-establish bedforms. Continual maintenance involving the removal of sand from around buildings and roads may be an option in some cases.

3. **Vegetation stabilisation**. Vegetation may be used to bind loose sand and dust upwind of a settlement in order to reduce the drift problem. This can be achieved through either natural or artificial recovery. Natural recovery

involves the protection of an area, often by enclosure, where relics of the original vegetation exist. Effectiveness depends on the extent to which the natural vegetation has deteriorated. Artificial recovery involves planting in order to establish a new ecosystem and thereby stabilise the sand. Planting is usually associated with mulching and other soil conservation strategies.

4. **Surface stabilisation**. This can be achieved in several different ways. Water can be used to stabilise surface sand, but because of high evaporation rates this must be undertaken frequently and is therefore expensive, although it is an effective short-term expedient. Gravel, stone or crushed rock may be spread over mobile sand. This reduces movement, but spreading is often difficult owing to the unstable nature of the mobile sand. In many cases heavy machinery cannot be used, and an abundant source of appropriate aggregate material is also required. Oil may be used to stabilise the surface, and although ugly it is highly effective. Any oil used must not restrict plant growth or long-term stabilisation and should not cause a pollution problem. Chemical sprays offer a good alternative, especially where they are combined with seeding. These sprays tend, however, to promote the development of surface crusts which if locally removed may be quickly undermined, thereby causing widespread failure. Maintenance involving the respraying of damaged areas is therefore essential.
5. **Fences**. Fences exert a drag force on the wind field causing a net loss of momentum, and therefore lead to deposition. The design of fences needs to consider the balance between fence porosity and the turbulence that it causes. If the porosity of the windbreak is decreased then the drag it exerts increases, causing greater deposition. However, the more impermeable a fence, the greater the wind turbulence in its lee which can cause scour and remobilisation of the deposited sand. Windbreaks may take a variety of forms. First, vegetation windbreaks and shelter belts may be used where irrigation water is available or groundwater is close to the surface. Careful orientation with respect to dominant sand/dust-bearing wind directions and consideration of height, width and porosity are important. Plants are particularly good at trapping dust. Unfortunately such windbreaks compete with crops and other forms of agriculture for water. Secondly, diversion fences made from wooden panels, metal plates, stone walls or earthworks may be used to deflect sand/dust-bearing winds. The fences are normally set up in either a single diagonal or V-shaped pattern at an acute angle to the dominant wind direction. The more acute the angle of the wall to the wind, the longer the likely life-span of the structure, but the area protected is smaller. Thirdly, impounding fences can be used to trap sand in artificial dunes or drifts. The aim here is to reduce the wind transport capacity, causing deposition. The volume of sand trapped by a fence is proportional to its height. Fences are placed at right angles to the prevailing wind direction and need periodical raising, as shown in Figure 8.19A, to maintain sand retention. In practice, a series of fences may be required to prevent all sand reaching a structure (Figure 8.19B). The fences themselves may consist of wooden panelling, brushwood or stakes.

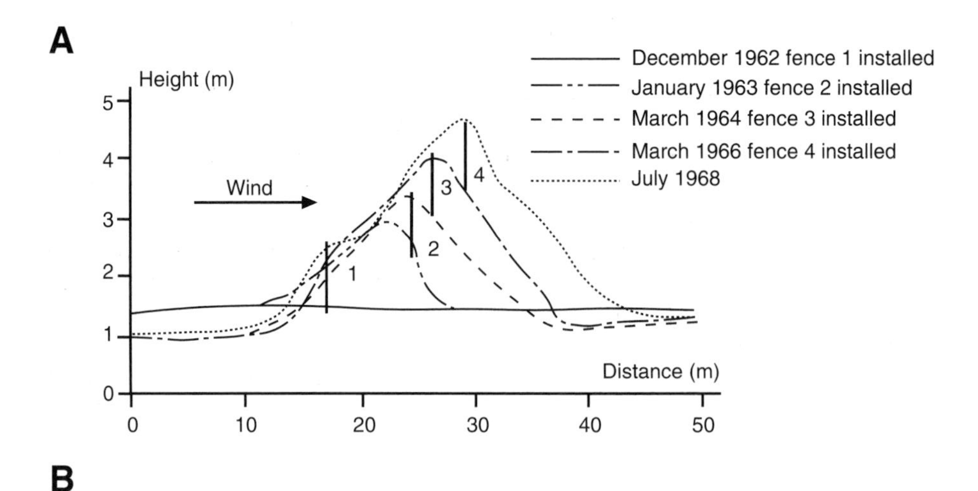

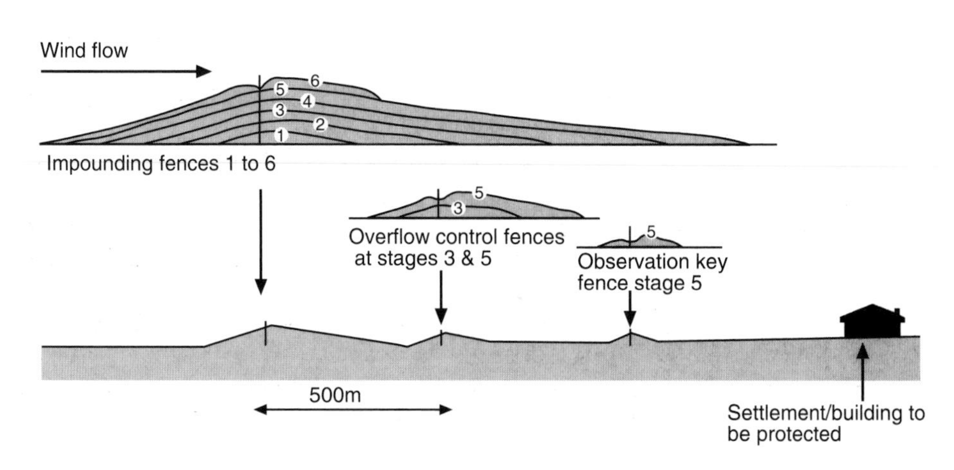

Figure 8.19 *The use of fences to impound blown sand and reduce drift hazards.* **A.** *A sequence of fences necessary to protect a small settlement from blown sand.* **B.** *A series of fences may be required to provide adequate protection. [A Modified from: Watson (1985)* Quarterly Journal of Engineering Geology, **18**, *Fig. 3, p. 240; B Modified from: Cooke* et al. *(1982)* Urban geomorphology in drylands. *Oxford University Press, Fig. VIIB, p. 284]*

6. **Architectural control**. The layout of buildings can be used to reduce sand drift problems. The separation between buildings should be such as to ensure that any wind blowing between them will be of sufficient velocity to achieve clearance, but without increasing human discomfort from sand blast. By reducing building height, tunnel and canyon effects in which the wind is accelerated between buildings, can be reduced, leading to a reduction in abrasion damage and human discomfort. Careful location of doors, windows and ventilation ducts with respect to the dominant sand/dust drift direction is also important.

8.3 SUMMARY OF KEY POINTS

- When engineering in extreme environments, careful terrain analysis is essential to identify problems prior to construction.
- In periglacial environments, problems are encountered with the thaw of permafrost or frozen ground, with frost heave and with the seasonal nature of the river regimes.
- Design strategies for permafrost include: (1) disregarding it; (2) eliminating it; (3) preserving it; and (4) designing for thaw subsidence. In most cases attempts are made to preserve the permafrost by constructing buildings on piles or rafts of gravel.
- Sanitation in periglacial regions is a problem since buried pipes may freeze. This can be overcome by: (1) insulated burial in marginal regions; (2) transportation of all water and waste; and (3) the use of utilidors.
- Care is required in the location and design of roads and railways. Drainage is a particular problem, especially where icings tend to form.
- Oil and gas pipelines pose a significant engineering challenge in periglacial regions.
- Subtropical deserts are characterised by a combination of three problems: (1) flash flooding; (2) salt weathering; and (3) drifting sand and dust.
- Hazard mapping and careful site selection are essential to reduce the impact of flash flooding.
- Salt weathering has a destructive effect on building material and rock. Foundation protection is essential where they intersect the water table or the capillary fringe.
- A variety of techniques can be used to reduce the problem of drifting sand and dust, including: (1) sand/dust removal; (2) vegetation; (3) surface stabilisation; (4) the use of fences and other structures to trap sand; and (5) careful architectural design to minimise the problem.

8.4 SUGGESTED READING

An overview of construction and development in periglacial regions is provided by Gray (1992). The problems of poor design in permafrost regions are illustrated in the work of Perrins *et al.* (1969), Kerfoot (1973) and French (1996). Kreig & Reger (1976) demonstrate the methods and importance of pre-construction terrain evaluation in such regions. A detailed synthesis of all aspects is provided by Harris (1986), while details on foundation design are given in Thomson (1980), Crawford & Johnston (1971) and Johnston (1981). Worsley (1992) provides case studies of how three different settlements in arctic Canada have dealt with the problems of the periglacial zone. The construction of pipelines in arctic regions is covered in depth by Williams (1986). Perhaps the best overview is supplied by French (1994).

Development in the hot drylands is covered in detail by Cooke *et al.* (1982), a book based primarily on consultancy projects undertaken by the authors. Cooke

& Doornkamp (1990) also provide good coverage of these issues. Detailed information on the control of drifting sand and dust can be found in the work of Watson (1985) and Jones *et al.* (1986).

Cooke, R.U. & Doornkamp, J.C. 1990. *Geomorphology in environmental management*, second edition. Clarendon Press, Oxford.

Cooke, R.U., Brunsden, D., Doornkamp, J.C. & Jones, D.K.C. 1982. *Urban geomorphology in drylands*. Oxford University Press, Oxford.

Crawford, C.B. & Johnston, G.H. 1971. Construction on permafrost. *Canadian Geotechnical Journal*, **8**, 236–251.

French, H.M. 1994. Living on ice: problems of urban development in Canada's North. *Geoscience Canada*, **21**, 163–175.

French, H.M. 1996. *The periglacial environment*. Second edition. Longman, Harlow.

Gray, J.M. 1992. Applications of glacial, periglacial and Quaternary research: an introduction. *Quaternary Proceedings*, **2**, 1–16.

Harris, S.A. 1986. *The permafrost environment*. Croom Helm, London.

Johnstone, G.H. 1981. *Permafrost engineering design and construction*. John Wiley & Sons, Toronto.

Jones, D.K.C., Cooke, R.U. & Warren, A. 1986. Geomorphological investigation, for engineering purposes, of blowing sand and dust hazard. *Quarterly Journal of Engineering Geology*, **19**, 251–270.

Kerfoot, D.E. 1973. Thermokarst features produced by man-made disturbances to the tundra terrain. In: Fahey, B.D. & Thompson, R.D. (Eds) *Research in polar and alpine geomorphology*. Third Guelph Symposium on Geomorphology. Geoabstracts, Norwich, 60–72.

Kreig, R.A. & Reger, R.D. 1976. Preconstruction terrain evaluation for the Trans-Alaskan Pipeline Project. In: Coates, D.R. (Ed.) *Geomorphology and engineering*. Dowden, Hutchinson & Ross, Stoudsberg, 55–76.

Perrins, O.J., Kachadoorian, R. & Greene, G.W. 1969. *Permafrost and related engineering problems in Alaska*. US Geological Survey Professional Paper, No. 678.

Thomson, S. 1980. A brief review of foundation construction in the Western Canadian Arctic. *Quarterly Journal of Engineering Geology*, **13**, 67–76.

Watson, A. 1985. The control of wind blown sand and moving dunes: a review of the methods of sand control in deserts, with observations from Saudi Arabia. *Quarterly Journal of Engineering Geology*, **18**, 237–252.

Williams, P.J. 1986. *Pipelines and permafrost*. Carlton University Press, Canada.

Worsley, P. 1992. Problems of permafrost engineering as exemplified by three communities in the North Western Canadian Arctic. *Quaternary Proceedings*, **2**, 67–78.

9
Waste and Pollution Management

The everyday activities of human life produce an immense amount of waste products, the majority of which are released into the environment. These wastes, produced as gases, liquids or solids, may not be directly harmful or damaging to the environment, but require active management in order to prevent overloading of natural systems. It is the management of wastes from our urban centres and industrial heartlands that presents the most significant problem for the future. As these products of our built environment continue to multiply, so greater efforts are needed to ensure that they are contained and do not cause significant pollution of the environment. This chapter defines and explores the meaning of waste, contamination and pollution, and examines the management of these by-products of human activity. We concentrate on: (1) the definition of waste, pollution and contamination; (2) the social and political aspects of waste management; (3) the disposal of solid wastes from households, industry and commerce in open dumps; (4) the disposal of solid wastes in landfill; (5) the containment of hazardous, particularly radioactive, wastes; (6) the treatment and release of effluents; and (7) the management of waste gases and particulate matter through their release into the atmosphere. In all of these issues, the geological environment has a central role.

9.1 WASTE MANAGEMENT AND THE GEOLOGICAL ENVIRONMENT

Waste is disposed of in three main environments: geological, atmospheric and marine (Figure 9.1). In this book the geological environment is defined as encompassing the geosphere, effectively the Earth's system from its surface through to its core. This definition encompasses the land surface, geological units and soils as well as groundwaters, all of which can be used in the disposal of liquid and solid waste. Waste in the geological environment forms the focus

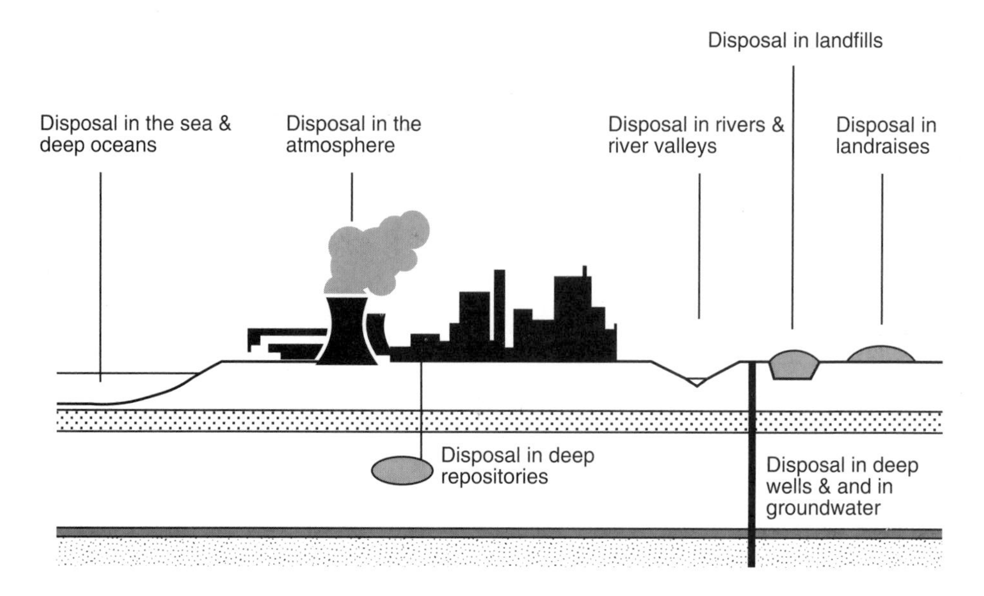

Figure 9.1 *Conceptual model of waste disposal options*

for this chapter, but it is important to remember that waste disposal within the atmosphere and marine systems is of equal importance and may also be associated with environmental hazards.

Waste gases and smoke derived from the activities of industry, the burning of hydrocarbons by power stations, and from other sources are dealt with by release into the atmosphere on the dilute-and-disperse principle, particularly in the developing world. These gases are not directly released into the geological domain, although it is important to realise that the resulting atmospheric pollution does have an impact upon it. For example, atmospheric pollution has had a dramatic effect on the durability of building materials and monuments in urban areas (Box 4.2), and has increased the acidity of rainwater causing a consequent increase in the acidity of soils and the pollution of groundwaters. Perhaps the most pressing environmental problem is that of human-induced climate change, leading in turn to sea level fluctuations which will have a direct effect on coastal environments. Waste products continue to be dispersed into the marine environment. Typical waste products which have been, or continue to be, released into the sea for dispersal are sewage sludge and waste waters, liquid wastes from industrial activity, dredge spoil, general solid waste, and low level radioactive wastes. Many other potential pollutants are released accidentally via oil spills and shipping accidents. Pollution caused by waste contamination of the sea has contributed to the decline in both shell- and finned-fisheries in coastal waters, and in damage to tourism through the pollution of beaches and the creation of hazards to human health.

The geosphere is utilised as a waste repository through the introduction of waste products into surface features of natural or human origin such as valleys, slopes and quarries, or through the creation of specially designed underground repositories in deep, specially drilled wells or underground mines (Figure 9.1). Landfill is the most important mechanism for the disposal of a wide range of domestic and industrial solids, although open dumps were once common. Disposal of such waste materials in landfills has been carried out according to both the dilute-and-disperse and concentrate-and-contain principles (see Section 9.2.3): the former for the slow dispersal of by-products from the decay of domestic wastes in the form of gases (into the atmosphere) and liquids (into the groundwater); the latter where such activities are undesirable. A detailed knowledge of the solid geology and hydrogeology of landfill sites is essential to prevent pollution. Geological repositories continue to be the best option for the containment of intermediate to high level radioactive waste (Figure 9.1). Once again, detailed knowledge of geological properties and groundwater conditions is important to prevent the breaching of the barriers intended to prevent pollution by radioactive waste products. Finally, the past activities of industry and commerce have left a legacy of derelict land contaminated with waste products, which are potentially hazardous to human health and safety.

9.2 WASTE AND POLLUTION

Human activity, particularly in the built environment, is the main producer of waste: mountains of domestic rubbish build up in our towns and cities; sewage farms cope with the treatment of increasing amounts of effluents; great stockpiles of spent materials are constructed in our industrial centres; and gases and particulate-rich smokes still pour out from power stations and urban conurbations. The cumulative effect of these wastes is damaging, leading to contamination of land, water and atmosphere, and ultimately to their pollution through overload of the natural recycling abilities of the environment. Pollution and contamination are therefore the products of poor waste management (Figure 9.2). The definitions of waste, pollution and contamination are discussed below in more detail.

9.2.1 Waste

As legally defined, in Britain at least, the term **waste** refers to any substance which constitutes a scrap material, effluent or unwanted surplus material arising from the application of a process, or any substance or article which requires disposal through being broken, worn-out, contaminated or otherwise spoiled. This definition is similar to many others across Europe and other countries of the world and encompasses most of the by-products of industrial, commercial and domestic activity. These by-products include representatives of all three physical states: (1) gases, such as those produced through the burning of fossil fuels; (2) liquids, such as the by-products of the chemical industry; and (3) solids, covering a wide variety of materials, from domestic wastes through to mine and

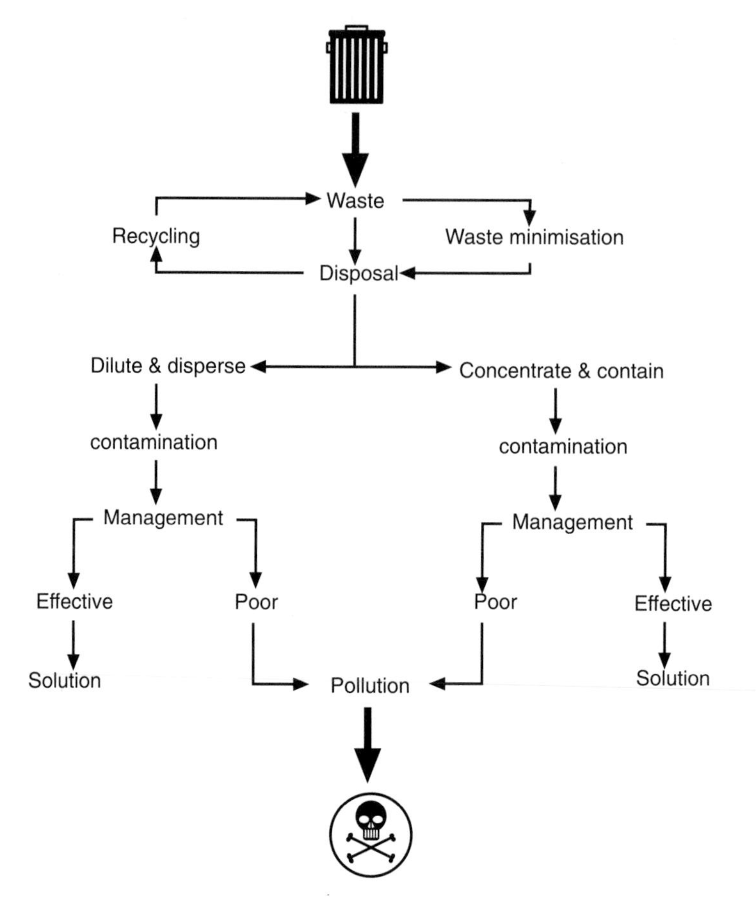

***Figure 9.2** The relationship of waste, contamination and pollution*

quarry spoil. This definition does not, however, mean that waste products necessarily represent a significant environmental problem, but simply serves to delimit that material which needs to be disposed of or recycled. It is the way in which waste is discarded, disposed of, recycled or treated which has a direct influence on the nature of its impact on the environment. In general, we can classify the treatment and disposal of waste into either managed or unmanaged wastes.

Managed waste is that which is collected, disposed of and/or recycled according to set procedures. Typically, managed wastes are those which are disposed of or treated in accordance with local and national government guidelines and policy. In some cases, this category of waste is also referred to as **controlled waste**, defined according to British legislation as any household, industrial or commercial waste which is collected and disposed of by local authorities (Table 9.1). Managed or controlled waste encompasses a wide range of substances from simple street sweepings or household domestic waste through to

Table 9.1 Current definitions of controlled waste

Waste type	Definition
Inert waste	Does not undergo significant physical, chemical or biological reactions or present a significant pollution problem e.g. masonry, bricks
Non-hazardous	No known or immediate hazard; may biodegrade, e.g. wood, paper
Difficult	Could in certain circumstances become harmful due to chemical or biological properties, e.g. batteries, mineral oils, car tyres
Hazardous	Contains a hazardous substance likely to cause death, injury, impairment to living beings, pollution of waters or unacceptable impact on the environment
Toxic	Any hazardous waste with a toxic property liable to affect living beings
Clinical	Consisting wholly or partly of human or animal tissue and/or blood; medical swabs and dressings; syringes, blades, needles and other medical instruments

the so-called **special** or **hazardous wastes**. These are managed wastes of any kind that are dangerous and difficult to keep or dispose of, and which need special provisions and/or facilities in order to do so. These categories encompass a huge range of waste products with varying impacts on the human and natural environment. For example, in Britain in 1992 it was estimated that the total waste production was 516 million tonnes per year, of which 137 million tonnes was controlled waste. Broken down further, the controlled waste fraction comprised: household (20 million tonnes); commercial (15 million tonnes); industrial (69 million tonnes); demolition and construction (32 million tonnes); and sewage sludge (one million tonnes). Of this total, 90–95% was destined for landfill.

Effluent is a term used to encompass liquid wastes, with or without suspended particles, which are managed through discharge into sewers. There are two types: **trade effluent**, derived from industrial sources, and **sewage effluent**, derived from domestic sources. Both types of effluent are controlled through legislation, and are often dispersed directly into the sea from offshore pipes.

Unmanaged waste encompasses those waste products which are discarded without reference to current guidelines or legislation, are left in an untreated state, and which often represent a hazard to public health and the natural environment in general. They also include **non-controlled wastes** which encompass mine and quarry spoil and agricultural waste which is usually dealt with in the immediate vicinity of the mine or farm, as appropriate.

9.2.2 Contamination

Contamination is the introduction or presence in the environment of alien substances or energy which may or may not be hazardous to human health or damaging to the natural environment. Contamination may occur both by design, as part of a concentrate-and-contain management strategy, and as a result of poor waste management. If contamination is not dealt with effectively it will lead to pollution.

Contamination has recently been in the public eye because of the concept of **contaminated land**, most typically, land which is derelict and which has concentrations of heavy metals, chemicals and other potentially harmful waste materials as a result of past land use. In addition, contamination may result from secondary sources, such as the leakage of gases and liquids from landfill sites. For example, there are at least 200 solid waste disposal sites in the English county of Yorkshire, of which more than half have caused a contaminated land problem through leachate and gas migration. Assessing the problem of contaminated land has become a pressing issue as cities continue to expand, and the reuse of industrial land for housing and amenity use is considered. The use of 'brown land', as it is sometimes called, is vital to avoid land-use pressure on remaining virgin or 'green land' around an urban area. Typically, assessment of contaminated land must include: (1) a detailed study of historical land use to anticipate the types of contaminant likely to be present; and (2) careful documentation of the type, concentration and distribution of contaminants present. On this basis a clean-up strategy involving either on-site treatment or removal of the contamination can be drawn up and implemented.

9.2.3 Pollution

The term **pollution** refers to the adverse environmental effects of gaseous, aqueous and solid waste generation and disposal. As legally defined, pollution is the introduction by people into the environment of substances or energy which are liable to cause hazards to human health, harm to living resources and ecosystems, damage to structure and amenity, or interference to legitimate uses of the environment. In essence, pollution is most often the result of the overloading of natural systems which may in the normal course of events be able to recycle effluents. Pollution is commonly associated with unmanaged wastes for which there are no adequate procedures for effective treatment and/or disposal, although managed wastes can also provide pollutants as a result of imperfect or inadequate management. The use of the term energy in the formal definition above demonstrates that pollution can also actively be caused by the introduction of unwanted noise, heat or visual intrusion into the natural or human environment.

Management systems for solid, liquid and gaseous wastes have traditionally followed two basic principles associated with their disposal into the natural environment: dispersal and containment. Until recently, waste management was largely carried out using the principle of **dilute-and-disperse**, which sought to minimise environmental impacts of a wide range of substances by diluting the waste through introducing it into the land, sea or atmosphere. This principle assumes that solid, liquid and gaseous wastes can most easily be dealt with by using the natural recycling processes of the environment, and that their dilution would mitigate against the concentration of harmful substances which would lead to pollution of land, sea and air. This principle has most often been used in the disposal of waste gases into the atmosphere, and the disposal of domestic sewage, industrial liquids and other aqueous wastes into the marine environment through sewage outfalls. Dumping of solid materials such as estuarine dredgings and inert building waste at sea also demonstrates the principle, as here it is hoped that the coastal and marine processes will prevent its concentration at any one place. The idea is sound provided that the rate of waste delivery to the chosen environment does not exceed the ability of that environment to recycle or redistribute the waste deposited within it. Unfortunately, the dilute-and-disperse principle frequently leads to pollution. The problem is illustrated by the creation of 'smog' in our major cities in the late nineteenth and twentieth centuries, comprising dense clouds of particulate-rich atmospheric pollution which posed significant hazards to human health. A famous example is the London 'pea-souper' of December 1952 which resulted in 4000 deaths, mostly from respiratory disease. This crisis was primarily a result of the combination of atmospheric conditions – stagnant, uncirculating air, heavy cloud cover and high humidity – with the widespread burning of fossil fuels for domestic heating, needed in the cold, foggy weather, which led to the accumulation of particulate-rich smoke and sulphur dioxide. Since then, clean-air policies preventing smoke emissions, and the decline of coal as the principal domestic fuel, have led to a dramatic improvement in atmospheric conditions in London.

Today, management policies are mostly based upon the containment of waste in 'safe' repositories from which there should be no direct source of pollution. This principle is based on the idea that wastes should be **concentrated-and-contained** in sites where there is little or no chance of pollution into the natural world. Although this is clearly desirable in principle it rarely happens in practice, since most containment barriers and liners will eventually be compromised by physical, chemical or biological action or through the ingress of ground and meteoric waters. This principle is most often associated with the management of solid and liquid waste products, the best examples being sanitary landfill operations and radioactive waste disposal repositories. The principle is less easily applied to gaseous wastes which cannot be easily contained, and are usually treated before being released into the atmosphere.

9.3 WASTE AND SOCIETY

Waste, pollution and contamination are emotive issues which engender popular interest and concern. As a consequence, there is often a social or political dimension in the selection of waste management strategies. This is discussed below in relation to waste minimisation and recycling, and in the politics of waste management.

9.3.1 Waste Minimisation and Recycling

Concern about the environment and its pollution has stimulated interest in waste minimisation through recycling and other practices, which are now increasingly seen as components of an effective waste management strategy (Figure 9.2).

Waste minimisation is a principle that is applied at the source of waste generation. This is reflected today in the reduction of packaging for many domestic items, the emphasis on reuse of containers and bags in many retail outlets, and company policies on recycling and reclamation of important items involved in industrial processes. The move towards this process reflects motives stimulated both by increased environmental awareness and by economic constraints. For example, in the United States, many companies employ so-called WRAP (Waste Reduction Always Pays) programmes which seek to formalise and emphasise waste reduction procedures in the workplace, through the integration into management of plans which allow monitoring and priority planning. Such programmes identify opportunities for waste reduction based on environmental protection and revenue issues, and identify a course of action for their implementation. For example, there is much accidental waste involved in the production of polystyrene pellets which form the raw material for a wide range of polystyrene products. Pellets are often lost in production, storage and transport. One American chemical company identified this as an unnecessary waste product providing an opportunity for waste reduction, and implemented a process of water flushing within the factory to a common drainage system which trapped the buoyant pellets. The recovered pellets were then collected and recycled. This example emphasises that waste reduction is usually economically sound and the implementation of WRAP procedures should ultimately lead to the minimisation of certain wastes targeted for disposal in containment sites or elsewhere.

Recycling is a broad term which refers to the conversion of waste products back to a useful material. Recycling is now commonly employed by industrial and commercial concerns as a way of increasing revenue from their production processes and of conserving primary mineral resources (see Section 2.2). Not all schemes are necessarily economically viable, however, although they are usually environmentally sound. This is simply because the balance of economic advantage between waste disposal and waste reclamation is often a very fine one. This balance is between the **waste disposal costs**, that is the cost of containment, transport, treatment and final disposal, and the cost of waste reclamation. The cost of waste reclamation may either be positive (i.e. **waste reclamation savings**) or negative, and this largely depends on the extent of the investment necessary to set up a reclamation scheme and the price obtained for the recycled commodity. Recently, many governments have tried to encourage recycling through official policies, and in some cases through financial incentives. This has led to the widespread development of bottle and paper reclamation and recycling schemes. For example, 10 million tonnes of paper products are consumed in Britain alone each year; three million tonnes of paper are recycled, and a further three million tonnes have potential for reuse.

9.3.2 Politics and Waste Management

Waste management is a multi-million dollar industry world-wide, and in most countries the application for disposal licences, particularly of solid wastes, is strongly reviewed and subject to stringent procedures. Granting of a licence to dump solid and liquid wastes in particular is subject to satisfactory site investigation and evaluation procedures. In most cases these have to be supported with detailed evaluations of impact and with scientific evidence of the reliability of the particular containment technique to be used. This is particularly important for the disposal of radioactive wastes, but is directly applicable in the case of domestic and commercial waste products as well. Political pressure will also be brought to bear through the application of two basic tenets, summarised in the acronyms: nimby and nimtoo. **Nimby** ('not in my back yard') reflects the strong local lobby, who, although perhaps accepting the need for the development of a particular managed waste disposal facility, would rather that it was developed closer to someone else's house or business. For most people the prospect of having a waste disposal site situated in the immediate vicinity of their home is an unattractive proposition. For example, in 1992 a 121 acre site was selected as being appropriate for waste collection from Birmingham and other large conurbations of the British Midlands. The site designated was close to the village of Hanbury in Worcestershire and was intended to receive over 200 000 tonnes of rubbish a year. The objections of the local people were based on nimby; their suggestion was that Hanbury was a scenically beautiful part of England, and that the developers should turn their attention to the more industrially 'blighted' parts of the Midlands, the so-called Black Country. **Nimtoo** ('not in my term of office') is even more fundamental, as it is a function of voter appeal; local politicians are unlikely to grant planning permission for an unpopular local waste disposal facility if there is a local or national election to be fought.

9.4 WASTES IN OPEN DUMPS

Open dumps, along with landfill sites which are discussed later, provide the principal mechanisms by which domestic, commercial and industrial waste is disposed of. The disposal of waste products in open dumps is usually unconstrained and undertaken with little regard to location and management. In the main, these dumps were associated with on-site disposal of industrial by-products, particularly colliery and mineral mine spoil, although historically most domestic waste was also handled in this way. The use of such methods in the disposal of solid wastes has contributed to concerns about contaminated land. For the most part, the philosophy of disposal has been the restriction of waste products close to source, with consequent environmental problems.

9.4.1 Mine Tips and Colliery Waste

Spoil heaps are a common feature of mining areas, reflecting the need for the disposal of overburden and unprofitable waste rock. In Britain today, these make up around 25% of the total volume of all managed wastes and are an important problem in other parts of the world, particularly where deep mining is still widely practised. In all cases, spoil and overburden are mostly dumped in close proximity to the workings, although in some cases spoil is dumped off-site with often damaging consequences. For example, in County Durham, northern England, colliery spoil was carried out to sea by long conveyor belts with the hope that it might be dispersed offshore. Although this practice has now been suspended, it has left a legacy of unsightly blackened beaches, in an otherwise scenically beautiful coastline, which are proving difficult to clean.

In general, mine and quarry wastes are disposed of in one of two open dump methods: firstly, by dumping into the void spaces created by strip mining or quarrying; and secondly, by stockpiling into spoil heaps, often associated with deep mining activities, but also characteristic of contour strip mines. The backfilling of quarries with waste material and overburden is a common consequence of quarrying and area strip mining (see Section 3.3; Figure 3.15B). It utilises the void space created by the cutting of the quarry, and provides some means of quarry restoration, usually required as a condition of planning acceptance. In most cases, this is the simplest method of spoil waste disposal, and is usually no more unsightly than the quarry itself, although it requires significant landscaping.

Stockpiled waste is a common sight in the colliery regions of northern Europe and North America and is usually composed of dumped overburden. These waste tips may be extremely large and unsightly, typical examples being the conical waste heaps of the now mostly-abandoned collieries of northern France near Lens. Here, conical spoil heaps several hundred metres in height dominate the skyline and are aesthetically displeasing. Vegetation is sparse because of the lack of topsoil, and the dereliction of the land is a large-scale problem. The stability of such dumps is also a major problem (see Section 3.4).

Stockpiled waste may also be a problem outside of colliery regions, particularly in scenic areas, such as the large-scale contour mined slate quarries of the Ffestiniog region of North Wales, on the borders of a National Park. Open waste dumps from metallic mineral mines present a particular hazard as they may contain many harmful elements which can find their way into the soil and water systems (see Section 3.4.4). Derelict mines provide a significant problem for the assessment of contaminated land, and the removal of open dumps of waste is often an important priority. For example, the gold mining district of Rodelquilar in southern Spain has a large mine tailings system which spreads out from the mine downslope (Figure 9.3). Analyses of associated soils show high concentrations of arsenic and other toxic metals associated with the process of gold extraction. No vegetation grows on the deeply dissected surface of these dumps, which have been open to the elements and surface water runoff since the mines ceased operation in the 1960s. These mine tailings represent a scenic eyesore and a potential hazard to health. However, it must be recognised that some mine dumps may have considerable archaeological, mineralogical or botanical importance, and that their wholesale removal may have scientific or other consequences.

Figure 9.3 *The mine tailing system at Rodelquilar, southeast Spain. These tailings were produced by gold extraction and are enriched in heavy metals such as arsenic [Photograph: D.S. Wray]*

9.4.2 Domestic, Commercial and Industrial Waste

The dumping of domestic, light commercial and industrial wastes in open dumps is an old practice which is now, for the most part, legislated against in most countries. As a management strategy it involves the unconstrained dumping of collected waste in open pits, river valleys, shallow depressions or other convenient repositories. Such dumps are open to the atmosphere, and are often in direct contact with permeable soils and bedrock. In river valley locations they are liable to erosion and can provide a pollution source which can be carried downriver. Active management is limited, although in some cases periodic incineration and compacting are employed. This type of dumping was, until recently, carried out for all domestic wastes collected by the town of Sorbas in Almería Province, southern Spain. Here, regular domestic waste collections were dumped on the steep slopes of the incised valley of the Rio Aguas, a river which is rarely in spate in this generally arid environment. Although the aridity of the location meant that groundwater pollution through leachates was not a problem, much of the wastes found their way as a pollution source into the river. Vermin control and volume reduction were largely carried out by periodic burning and covering by bulldozed topsoil, leaving an unstable river valley slope which was prone to slumping. Even in the most arid regions, periodic storm events have the potential to redistribute this material.

In most cases, open dumps are more typical today for unmanaged wastes, particularly where material is illegally dumped by householders or by commercial and industrial concerns. Although relatively limited in scope, these dumps can cause considerable environmental problems. Open dumps of domestic waste encourage vermin, the decay producing an unpleasant smell. Their interaction with groundwaters in river valleys, or in unlined and permeable soils, sands, gravels and well jointed rocks such as limestones, encourages groundwater contamination and pollution through the migration of unconstrained leachates. Tipping over cliff lines and down slopes is also a common but illegal practice, which contributes to slope failures and causes a decline in aesthetic appeal. Illegal tipping in this way is difficult to legislate against, but does represent a serious hazard, particularly where the waste products are chemicals or other materials derived from industrial practices.

9.5 LANDFILLING WASTE

Landfill refers to a method of disposal of solid or semiliquid wastes from domestic, commercial and industrial sources by tipping it on to the land or into excavations. There are two basic types: (1) **landfill**, which is, strictly speaking, the term used to encompass tipping into appropriate natural depressions in the landscape, such as valleys, or into ground excavations, such as quarries; and (2) **landraise**, the build-up of waste to form a hill of variable dimensions which exceeds the initial elevation of the landscape (Box 9.1). Together, these represent by far the most important of all waste disposal methods, and the operation of

BOX 9.1: LANDSCAPING WITH WASTE

In recent years there has been growing interest in the idea of landscaping with inert waste to replace or restore the original contours of the land often destroyed by resource exploitation. These ideas have been developed for the most part in Scandinavia. In Sweden, eskers – long sinuous ridges of sand and gravel – have been a locus for historic settlement, first as a strategic vantage point, and second as a source of aggregate and water (Box 5.2) for the developing urban centre. Gravel extraction completely removes the landform, often leaving a shallow depression where the gravel has been followed beneath the current land surface. Landform recreation has formed an important aspect of the restoration of some of these sites.

For example, inert demolition waste has been used to recreate the Haga and Järvakrog eskers near Stockholm, reproducing the contours of the original landform. Re-excavation of the esker has been necessary during the construction of new road schemes, which involves problems of excavation and slope stabilisation within the 'made' ground of the recreated esker. Despite this, landform recreation as both a restoration option and a positive use of inert fill is a promising idea.

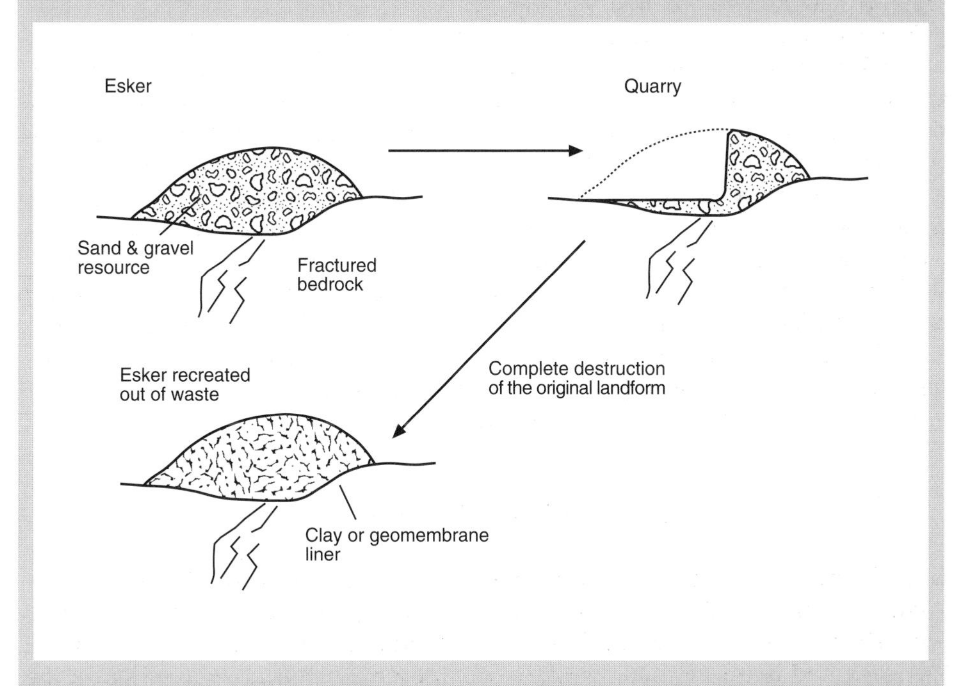

Source: Morfeldt, C.O. 1993. Landscaping with waste. *Engineering Geology*, **34**, 135–143.

landfill is a multi-million dollar industry in developed and densely populated countries.

Landfill is essentially a problem of the urban environment, as the majority of domestic wastes produced in rural settings are dealt with by simple methods of incineration or through localised dumping which is unlikely to create a large-scale environmental problem. As an example of the scale of the problem of urban wastes, the 5.8 million people of Hong Kong live in an area of just 1074 km^2 and produce 20 000 tonnes of solid waste a day, of which 14 000 tonnes are landfilled, 4000 tonnes are recycled and 2000 tonnes are incinerated. In Britain over 57% of all managed wastes from urban centres are dealt with by landfilling, and this is by far the most common method of disposal; incineration amounts to only 2% while recycling of reusable materials runs to about 2.5%. Currently, the need for void space is exceeding the rate at which old quarries and clay pits are exhausted of their mineral resources, and other methods of disposal will have to be investigated. This problem is also important in the United States, where there are more than 75 000 landfill sites, of which it has been estimated that over half will be closed within the next 10 years. New York alone produces over 25 000 tonnes of waste per day, and in order to keep pace with decreasing landfill availability, the city authorities need to dispose of at least 10 000 tonnes per day by other methods, primarily incineration.

Good landfill schemes are developed via a series of well established stages. Firstly, the site must be assessed for suitability against a series of selection criteria, such as: (1) its location with respect to main roads and conurbations; (2) the nature of its geology, and particularly its permeability; (3) the hydrogeological conditions; (4) the suitability of the site for particular waste types; and (5) the volume. Secondly, the site must then be prepared to receive waste, particularly if it requires engineering or lining by clay or geomembranes. Next, proper operation and maintenance of the site are necessary to ensure that the environmental impacts of the site are kept to the minimum. Finally, after filling, the site must be properly restored so that it can accept an appropriate afteruse.

9.5.1 Leachate and Landfill Gas: the By-products of Waste Management

The decomposition of biodegradable or **putrescible** waste is a complex process which takes place when a variety of micro-organisms react with the waste in a moist environment. There are three stages in its decomposition (Figure 9.4). Firstly, the waste is attacked by aerobic (oxygen-dependent) bacteria in a damp environment to produce simple organic compounds, carbon dioxide and water. Heat is generated as a by-product, encouraging the organisms to multiply and continue the process. Secondly, the carbon dioxide produced gradually excludes the aerobic bacteria and degradation is then taken over by organisms which can survive in an oxygen-poor environment. These organisms break down the large organic molecules present in food, paper and other biodegradable wastes into more simple compounds such as hydrogen, ammonia, carbon dioxide and organic acids. This is referred to as the **acetogenic** phase. Finally, in the

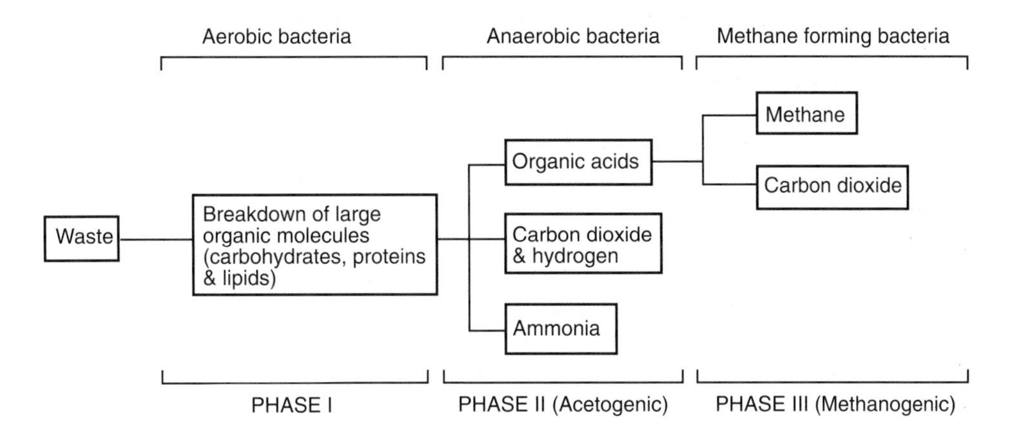

Figure 9.4 *The decomposition of domestic waste [Modified from: Attewell (1993)* Ground pollution. *E&F Spon, Fig. 10.2, p. 88]*

methanogenic phase, methane-forming bacteria become established which consume the organic compounds produced by the acetogenic phase, and produce a leachate which is high in ammonia content and is accompanied by the generation of methane-rich landfill gases.

Moisture is probably the most important factor in the breakdown of waste as it provides the basic medium for the microbial activity, as well as providing a means for the transport of nutrients and microbes through the waste pile. Optimum conditions are reached where the moisture content of the fill reaches around 50–60% saturation. The moisture content is determined by the interaction of: (1) rainfall, causing surface water infiltration; (2) the hydrogeological conditions of the landfill, allowing wetting and drying of the fill through changes in the level of the water table; and (3) water produced through the decomposition of the waste itself. In addition, pH levels of below 6 and above 8 tend to inhibit methanogenic activity. The optimum pH levels are between 6.7 and 7.2. Temperature is also important, as the optimum for anaerobic decomposition is about 30–40°C, and is often self-regulating since heat is generated by the decomposition of the waste. Shallow landfills may, however, be sensitive to climatic conditions, decay only occurring during warm summer months. Deeper landfills can have temperatures as high as 60°C, with 40–45°C being common in the first five years after landfilling.

Effective landfill is primarily a balance between the containment of the waste, and the control of the two main by-products caused by the decay of the waste: (1) **leachate**, the organically charged liquor created by the passage of mostly atmospheric waters through the body of the waste; and (2) **landfill gas**, mostly methane, an odorous and highly explosive gas which builds up in the landfill, produced by natural decay processes in the waste. Leachate is produced through the decay of all domestic waste and most industrial wastes. Leachate composition varies according to the type of waste, but tends to be consistent for

***Table 9.2** Typical composition of leachates from typical landfill sites. All results in milligrams per litre excepting pH value. BOD is the Biological Oxygen Demand. [Modified from: Department of the Environment (1990)* Landfilling wastes. A technical memorandum for the disposal of wastes on landfill sites. *Department of the Environment Waste Management Paper, No.26, HMSO, Table 3.23B, p. 25]*

Determinant	Domestic waste	Industrial (43% total waste)	Industrial (66% total waste)
pH	5.8–7.5	8.0–8.5	6.8
COD	100–62 400	850–1350	470
BOD	2–38 000	80–250	320
TOC	20–19 000	200–650	100
Volatile acids (C_1–C_6)	ND–3700	20	10
Ammoniacal N	5–1000	200–600	120
Organic N	ND–770	5–20	60
Nitrate N	0.5–5		0.04
Nitrite N	0.2–2	0.10–10	
Organic phosphate	0.02–3	0.20	0.6 (Total)
Chloride	100–3000	3400	680
Sulphate	60–460	340	30
Sodium	40–2800	2185	462
Potassium	20–2050	888	200
Magnesium	10–480	214	66
Calcium	1.0–165	88	188
Chromium	0.05–1.0	0.05	0.02
Manganese	0.3–250	0.5	
Iron	0.1–2050	10	70
Nickel	0.05–1.70	0.04	0.1
Copper	0.01–0.15	0.09	0.09
Zinc	0.05–130	0.16	0.06
Cadmium	0.005–0.01	0.02	0.0005
Lead	0.05–0.60	0.10	0.004
Monohydric phenols		0.01	
Total cyanide		0.01	
Organochlorine pesticides		0.01	
Organophosphorus pesticides		0.05	
PCBs		0.05	

domestic waste, while for industrial sources it is variable (Table 9.2). The main chemical and organic components of landfill leachates may be grouped into four classes, dependent on the type of waste: (1) major elements and ions, such as calcium, magnesium, iron, sodium, ammonia, carbonate, sulphate and chloride; (2) trace elements, such as manganese, chromium, nickel, lead and cadmium; (3) many organic compounds, usually measured as total organic carbon (TOC) or chemical oxygen demand (COD); and (4) microbiological components. Leachate is an extremely potent source of pollution, particularly where there are high concentrations of organic material and inorganic solutions of unreduced metals. When released into watercourses these will severely deplete the oxygen content of water, due to the bacterial consumption of oxygen during aerobic organic decay, leading to the death of much of the aquatic life. Often the effects of the organically charged leachate liquor and its variable pH cannot be assessed, but it is clear that they can be extremely limiting for life in surface waters. Release into groundwater also has extremely damaging effects which often remain for some time. Once groundwaters have become polluted they may become unsuitable as a source of drinkable water. The nature of the strata and the groundwater flow will control the nature and extent of the pollution. Some aquifers have purifying properties, however, and pollutants may be removed naturally if the concentrations are sufficiently low (Box 5.2).

The decomposition of organic materials within wastes leads to the production of a variety of gases collectively known as landfill gas. These gases can pose considerable hazards to human health, particularly as they are usually explosive (Box 9.2) and constitute a significant danger to areas surrounding active or completed landfill operations. Landfill gas also contributes a significant 'nuisance factor' to landfill operation as decomposition gases are usually noxious and unpleasant, making the proposed location of landfill operations close to human habitation a difficult proposition for developers. There are many factors associated with the production of landfill gas. The primary gases produced from landfill decomposition are methane, which typically makes up around 55 to 70% of the gas composition, and carbon dioxide which comprises around 30 to 40% of landfill gas. Other gases produced, in minor concentrations, are hydrogen, hydrogen sulphide and ethane. Methane is a particularly hazardous by-product as it can explode when its concentration reaches between 5 and 15% of air. Organisms can be asphyxiated by carbon dioxide when it reaches concentrations greater than 0.5%, and displacement of air by landfill gas will commonly create an asphyxiating atmosphere. As both methane and carbon dioxide are heavier than air, this can be a particular problem in smaller void spaces, trenches, underground access holes or depressions which are close to, or part of, landfill sites. Landfill gas production varies according to the age of the fill (Figure 9.5). For example, after fill placement, maximum methane gas yields will be achieved within one to two years (Figure 9.5), dependent on the prevailing conditions. Typically, yields of 30 to 200 m^3 of methane per tonne of dry refuse are produced, and yield is usually highest in the early years, when the biological activity is at its greatest, declining to low levels after 20 years of emplacement (Figure 9.5). This is of importance where the gas is to be commercially utilised as an energy source. As discussed above, the uncontrolled or unexpected escape of one or both of these by-products can contribute to significant problems of contaminated land and significant costs to the landfill operator, often on the basis of the principle that 'the polluter pays'. For example, in Britain, the Environment Protection Act of 1990 carries with it tough penalties for those who allow the escape of substances hazardous to health or the environment at any stage in the process, from waste generation to waste disposal.

The optimum conditions for leachate and landfill gas production are a function of a combination of characteristics which reflect both the method of waste management and the physical characteristics of the site such as the geology, hydrogeology and climate (Figure 9.6). The main factors are discussed below.

BOX 9.2: LESSONS FROM LOSCOE: THE DANGERS OF LANDFILL GAS

At 6.30 am on 24 March 1986, a bungalow at 51 Clarke Avenue, in the English town of Loscoe in Derbyshire, was completely destroyed by a methane gas explosion, badly injuring the three occupants. Loscoe is located 16 km north of Derby in a coal mining area. The explosion was caused by the migration of landfill gas from an adjacent landfilled brick pit. The houses were built around 1973 prior to an application for a waste disposal licence in 1977 to fill the former brick pit with domestic waste. Disposal at the site continued until 1982 when all tipping ceased. The site had been capped first with a light soil cover in 1984 and then, just prior to the accident, with a clay cap. Detailed investigation suggested that the landfill gas had migrated along fractured sandstone strata outcropping beneath Clarke Avenue. Methane built up in the space beneath the floor of these houses and was ignited beneath number 51 by the pilot light on the central heating boiler. Prior to the explosion several unexplained phenomena had provided, with hindsight, an indication of the impending problem. First, vegetation in the gardens began to show signs of distress and to die due primarily to soil 'hot spots', originally explained by the underground burning of near-surface coal seams, but now believed to be due to the bacterial oxidation of escaping methane within the soil. In addition, noxious smells were reported within these gardens during excavation work.

This example illustrates the potential dangers associated with landfill gas and the need to consider its containment carefully. In particular, identifying potential migration routes for landfill gas from landfill sites is an essential task.

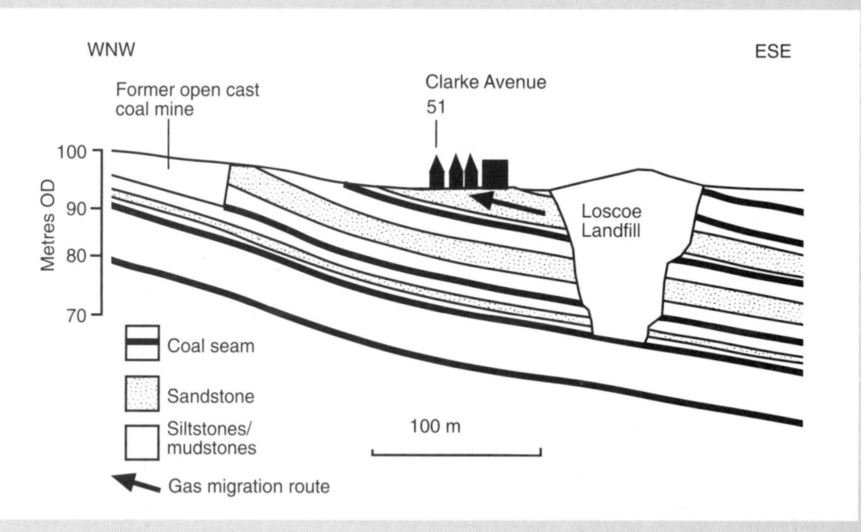

Source: Williams, G.M. & Aitkenhead, N. 1991. Lessons from Loscoe: the uncontrolled migration of landfill gas. *Quarterly Journal of Engineering Geology*, **24**, 191–207. [Diagrams modified from: Williams & Aitkenhead (1991) *Quarterly Journal of Engineering Geology*, **24**, Fig. 7, p. 200]

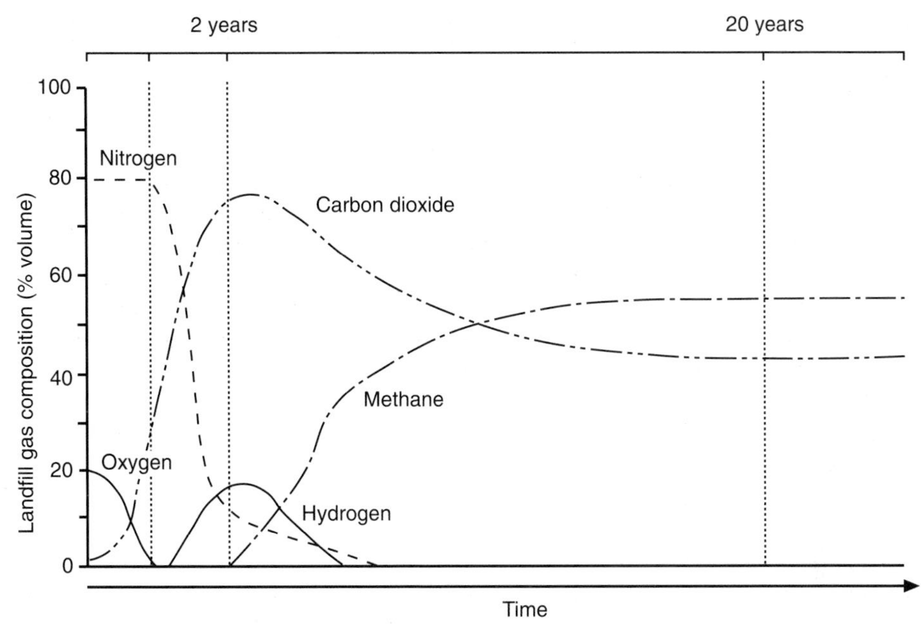

Figure 9.5 *Variation in the production of landfill gas with time [Modified from: Department of the Environment (1986)* Landfilling wastes. *Waste Management Paper No. 26, HMSO, Fig. 3.73, p. 35]*

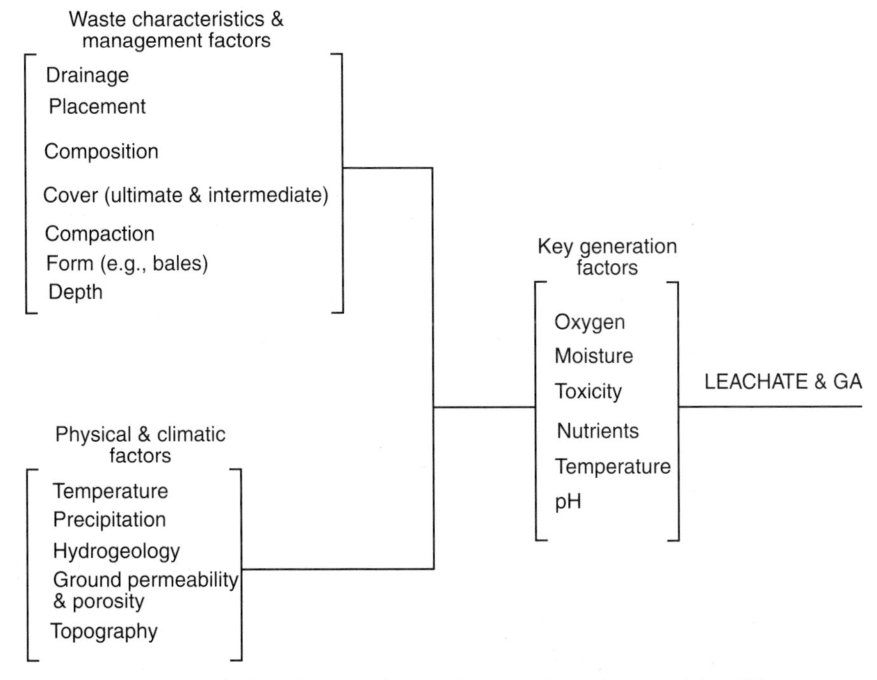

Figure 9.6 *Factors which influence the production of leachate and landfill gas*

1. **The nature and type of waste**. Waste products which are prone to degradation through bacterial or chemical decay produce both leachate and landfill gas. Wastes for disposal in landfill sites commonly comprise solids, sludges or even liquid wastes, and can be divided into three categories on the basis of their potential for pollution of groundwaters in particular. **Inert wastes** typically comprise the waste products of demolition and construction, together with commercial and industrial wastes such as glass, plastics or other inert solid waste. These waste types do not physically degrade to produce hazardous by-products and thus do not present a particular hazard to groundwaters, although they may present a barrier to groundwater flow. **Domestic and commercial wastes**, comprising waste foods and other putrescible materials from domestic, commercial and light industrial sources, are capable of breakdown to produce noxious by-products. Leachates produced from the decay of these wastes have the potential to seriously pollute groundwater sources if not dealt with through an appropriate containment strategy. Finally, **hazardous wastes** effectively contain one or more substances which represent a serious threat to the environment or to human health. Particularly troublesome are those wastes which are soluble or which can produce a mobile fraction through decay which can then lead to the pollution of groundwaters. Other hazardous wastes, such as asbestos, although harmful to health, may be insoluble and mostly immobile and are therefore easily dealt with in landfill sites without danger of leachate production.
2. **Climate**. Landfills in arid environments often produce very little leachate simply because there is little rainfall to percolate through the waste, and the water table is often very deep, preventing periodic flooding of the landfill site through rises in the water table. Leachate is much more of a problem in humid environments with high rainfall rates; in most humid environments at least some leachate will almost always be produced.
3. **Hydrogeology of the landfill site**. Five hydrogeological environments are commonly present in a given landfill site, and these directly affect its efficiency (Figure 9.7). These are: (A) the body of landfilled waste; (B) the landfill–bedrock boundary; (C) the unsaturated zone beneath the landfill–bedrock boundary; (D) the zone of water table fluctuation beneath the landfill–bedrock boundary; and (E) the saturated zone beneath the landfill–bedrock boundary. These zones are particularly important in understanding the development and migration of leachates, particularly as each one has its own combination of physical, chemical and biological processes. The most important variable in the waste itself is the degree of its saturation, as it is this which controls the processes that affect **attenuation** of the leachate, that is, the decrease in concentration of potential chemical pollutants in the liquor. In unsaturated landfill, the permeability of the waste body will be relatively low, preventing leachate migration, and oxygen is available to assist in the degradation of the waste and attenuation of the leachate; in saturated landfills, the reverse is true.

   Leachate will eventually migrate to the waste–bedrock boundary where several factors control its onwards migration into the surrounding strata.

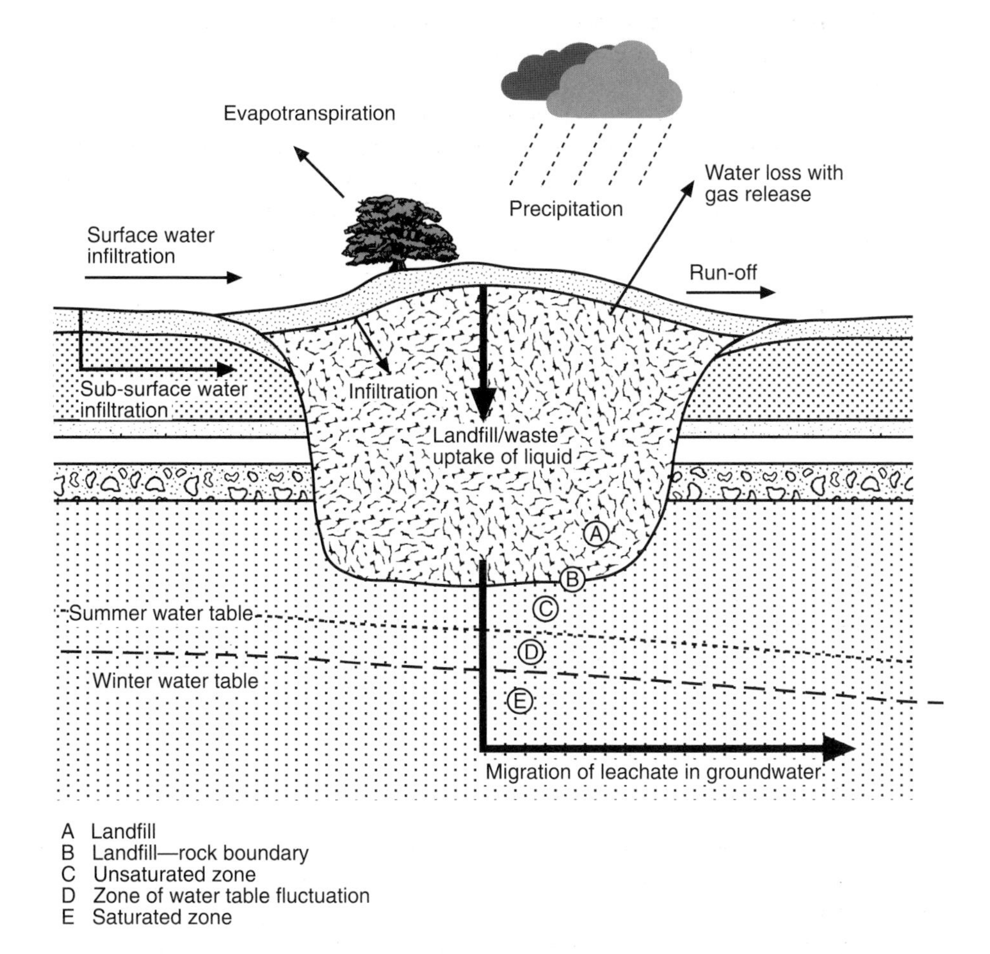

Figure 9.7 *The five hydrogeological environments present within landfill sites [Modified from: Department of the Environment (1986)* Landfilling wastes. *Waste Management Paper No. 26, HMSO, Fig. 3.33, p. 28]*

The bedrock will normally be less porous than the waste body, and will have a much smaller pore diameter, which may assist in filtering out particulate matter held in suspension in the leachate. The permeability of the bedrock is also of importance. The bedrock may also be much less permeable than the landfill, preventing direct migration and causing leachate ponding at the waste–bedrock boundary. Where the bedrock is permeable, however, the leachate will migrate vertically downwards into the unsaturated zone above the water table. Attenuation of the leachate will continue in this zone as long as there is enough oxygen to assist in its aerobic degradation. Finally, with continued vertical migration the leachate will reach the zone of water table fluctuation and ultimately the saturated zone. In the fluctuation zone attenuation of the leachate will occur through aerobic decay only when it is periodically unsaturated. In the saturated zone, leachate migration becomes horizontal, directed by groundwater flow. Continued attenuation is through dilution or by ion exchange with the bedrock; anaerobic conditions usually prevent further organic decay.

9.5.2 Types of Containment

The selection and management of suitable sites for landfill are dependent on the type of containment required and are a function of the types of waste to be disposed of, particularly in relation to the type and quantity of leachates which will be produced. Waste types are often mixed in landfill sites in order to dilute the effects of potentially hazardous materials. Typically, there are three types of waste mixture: (1) **mono-disposal**, with a single waste type, usually inert wastes; (2) **co-disposal**, with limited amounts of difficult or hazardous wastes mixed with domestic and commercial wastes in order to promote decay of the hazardous effects through the normal processes of attenuation; and (3) **multi-disposal**, in which two or more types of controlled waste are dumped together without specific regard to the co-reactions which might occur.

Effectively, landfill sites can be one of two types: (1) total containment sites, and (2) slow leachate dispersal sites, both of which are discussed below.

1. **Total containment sites**. Total containment sites are operated on the concentrate-and-contain principle in which the by-products of decomposition are retained within what is often referred to as a **sanitary landfill**, in which leachates and landfill gas are not allowed to disperse freely into either the ground or the atmosphere (Figure 9.8A). In containment sites, any attenuation of leachates is expected to take place within the body of the waste, and often noxious liquors are collected and treated before disposal. Monitoring of leachate is an important activity for all containment site operators, and is a responsibility which has to be continued long after the site has been filled, in order to ensure that complete containment has been achieved. The production of landfill gas is also carefully monitored and it is either vented in a controlled manner, or utilised to produce electrical energy.

   In containment sites it is the permeability of the bedrock geology which is of paramount importance. For total containment the rocks should have extremely low permeability, and typical examples are fine grained and compact lithologies such as clay, marl and other mudrocks, as well as unfractured igneous and metamorphic rocks. Unsatisfactory rocks include coarse permeable sandstones, limestones and any rocks with a high degree of fissuring or fracturing. These have a high permeability and are often aquifers, with a consequently greater risk of groundwater pollution (Figure 5.3). In some cases, strata which would otherwise appear to be suitable can prove to be unsatisfactory due to internal inhomogeneities caused by lateral facies changes, a good example being the sand or gravel lenses which are

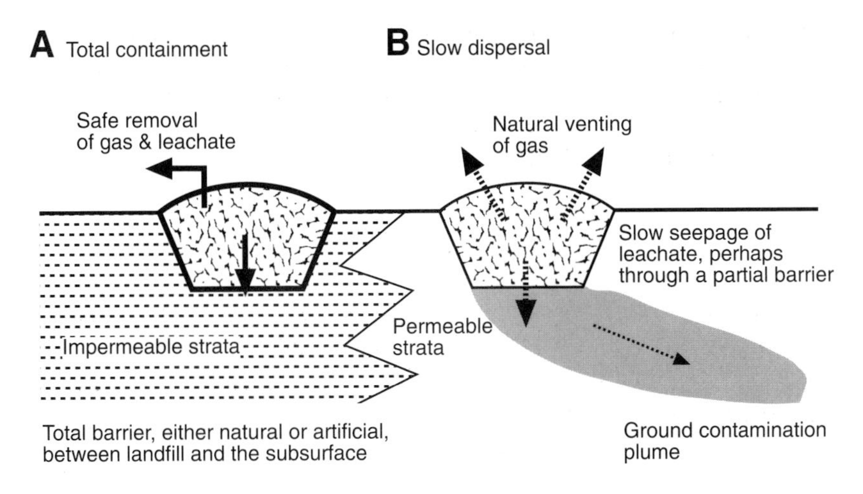

Figure 9.8 *Types of landfill site.* ***A.*** *Total containment.* ***B.*** *Slow dispersal*

commonly found in otherwise impermeable Quaternary clays (Box 9.3). Where there is pressure to utilise imperfect sites, it is possible to create containment sites through the use of barrier linings which help seal the landfill from the permeable bedrock. Typical linings utilised are puddled clay, fines from industrial processes, asphalt, and butyl rubber sheeting or other types of synthetic geomembrane. The potential exists for all of these to be compromised through time by decay or root penetration, and it is important that more than one method is employed in order to build up a multiple barrier to leachate migration.

Gas migration is perhaps the greatest problem associated with landfill gas. Like leachate, the containment of landfill gas is primarily a function of the structure and lithology of the surrounding strata. Containment is important where the gas is to be utilised commercially, but it is also important in the prevention of potentially explosive build-ups, such as that which happened at Loscoe in northern England (Box 9.2). In containment sites, landfill gas is contained by the same impermeable bottom layer as the leachate, but it is also contained by the application of a dense and impermeable capping layer which allows careful venting and/or collection of the gas for use elsewhere as a power source.

2. **Slow dispersal sites**. Slow dispersal sites allow leachates to pass through the landfill–bedrock boundary into the unsaturated zone, and landfill gas is able to disperse into both the surrounding strata and the atmosphere (Figure 9.8B). The operation of these sites, now largely out of favour because of their potential for pollution, is based on the dilute-and-disperse principle. In theory, slow dispersal of leachates into the ground provides for the further aerobic attenuation of the leachate, but this is dependent on the groundwater conditions. It also dispenses with the need for regular collection, treatment and disposal of leachates.

BOX 9.3: PERMEABILITY OF LODGEMENT TILLS: THE STORY OF HARDWICK AIR FIELD

A recently proposed development at Hardwick Air Field in Norfolk illustrates the problems associated with fissures and shear planes within lodgement tills (Gray, 1993). In 1991 Norfolk County Council applied for planning permission to build a waste hill (landraise) 10 m high to dispose of 1.5 million m^3 of domestic waste over 20 years. Crucial to their proposal was that the area was underlain by lodgement till, rich in clay, which would act as a natural impermeable barrier to the leachate generated within the decomposing waste. Normally an expensive containment liner is required to prevent contamination of the ground water by the leachate. This proposal became the subject of local debate and as a consequence the planning application was called to public planning enquiry in 1993. At this enquiry the objectors used a detailed knowledge of lodgement till to argue that it was inadequate as an impermeable barrier in its natural state. The till contained fissures, sand lenses and abundant chalk clasts. The fissures and sand lenses provide ideal flow networks for the leachate, while the chalk clasts would be dissolved by the acidic leachate increasing the permeability of the till. Consequently, in its natural state the lodgement till was not impermeable to leachate. Partly on the basis of this evidence the proposal was rejected. The example illustrates how a detailed knowledge of glacial sediments is vital to engineering within glaciated terrains.

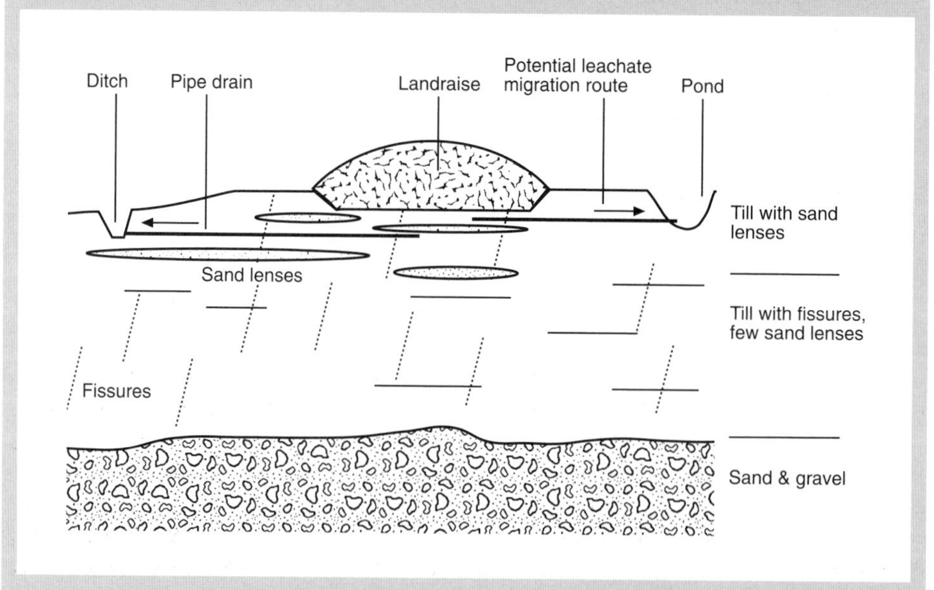

Source: Gray, J.M. 1993. Quaternary geology and waste disposal in south Norfolk, England. *Quaternary Science Reviews*, **12**, 899–912. [Diagram modified from: Gray (1993) *Quaternary Science Reviews*, **12**, Fig. 9, p. 905]

Slow dispersal of leachate into the bedrock is achieved where groundwater migration occurs through intergranular flow. This is found in permeable siltstones and fine sandstones. The most important factor in the success of these sites is the thickness of the unsaturated zone above the seasonal fluctuation of the water table (Figure 9.7). Sites with an extremely deep water table in relatively arid environments provide the best conditions for the successful operation of such sites. The unsaturated zone usually contains both a gas and a liquid phase which helps delay leachate migration through reduction in its effective permeability. This causes a delay in the further vertical migration of the leachates and this enhances the processes of attenuation through aerobic degradation, chemical reaction and soluble diffusion. It is important in such sites that the rate of infiltration of the leachate does not exceed the natural infiltration rate of the bedrock, and therefore the disposal of liquid wastes in such sites should be carefully monitored. Sites where there is little or no unsaturated zone are clearly inappropriate for this type of landfill operation, as there is no time delay in the downward migration of leachates to the groundwater. This usually causes direct pollution of the groundwater unless the waste type is strictly inert. Rapid dispersal can also take place where the bedrock is highly permeable, such as in clean sands and gravels or fissured limestones. These are often aquifers and therefore diffusion of the leachate into the bedrock will rapidly lead to pollution.

The damaging effects of gas migration into the surrounding strata are particularly acute where the site is operated on the dispersal policy, as the gas may be able to migrate considerable distances and can collect under homes and other urban areas (Figure 9.9; Box 9.2). As with leachate, it is the permeability of the strata, either as primary permeability or through the utilisation of joints and/or fissures in the rock, which has a direct impact on the site. Sites with these bedrocks should not be operated on the disperse principle, although they may be suitable for inert wastes or, by using effective liners, can be converted to containment sites.

9.5.3 Site Identification and Assessment for Landfill

Efficient site investigation is the cornerstone of good landfill practice. It has three basic components: (1) the identification of the need for landfill operations, usually considered with reference to the strategic plans of the local authority, particularly in relation to large conurbations and transport infrastructure; (2) the investigation of the site itself as a suitable repository for waste; and (3) an assessment of the likely environmental impact of the proposed waste operations. In most cases it is important for the landfill operator to submit a detailed report on these proposals in order to be granted planning permission and to receive a licence to operate, although the detailed contents of the report will differ according to the legislative requirements of the particular country or region. Operators will also have to consult widely with authorities in charge of

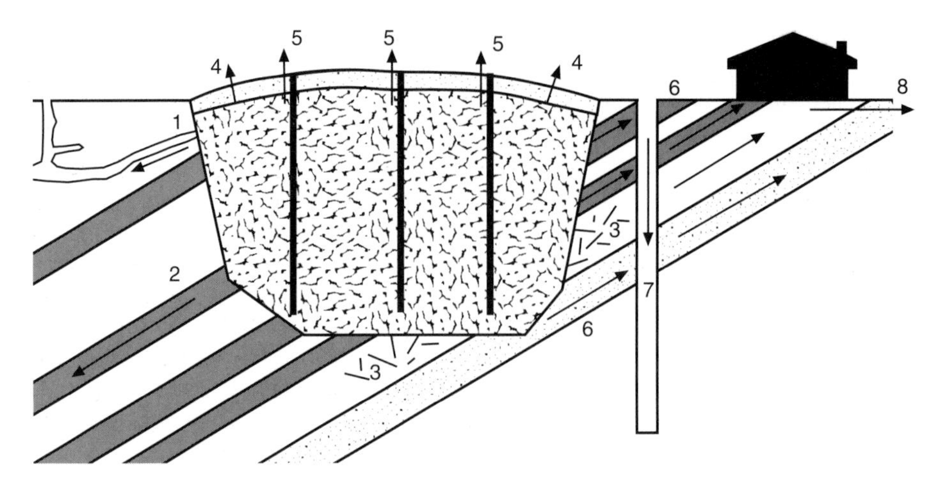

1 Through caves and other natural cavities
2 Through high permeability strata down the bedding planes
3 Through highly fissured strata, either natural or caused by explosives
4 Through desiccation cracks in the cap or around tree roots, etc.
5 Around site infrastructure providing vertical pathways; gas or leachate wells
6 Through high permeability strata up the bedding planes
7 Along mine shafts
8 Along underground services such as sewer pipes

Figure 9.9 *Schematic model of the possible routes by which landfill gas can migrate from a site [Modified from: Attewell (1993)* Ground pollution. *E&F Spon, Fig. 10.1, p. 86]*

transport infrastructure planning, environmental and water management, and aspects of nature conservation, heritage and amenity. Typically, licences to operate can be revoked at any time if the operator is found to be in contravention of any of the conditions laid down by the granting authority. The three stages to this procedure are discussed further below.

1. **Relationship to strategic plans**. In general, the perceived need for landfill sites in the vicinity of large conurbations is perhaps the most important factor in the assessment of overall suitability of a site for landfill. For example, in southern England, all quarry sites within a 40 km radius of London are viewed as potential voids for landfill operations, regardless of their geology, whereas there is less pressure to develop landfill sites in the more sparsely populated but densely quarried areas of Devon close to the city of Plymouth. Proximity to transport systems is also an important consideration as the location of sites close to motorways, rail junctions and even ports provides local authorities and operators with the potential to gain revenue through the importation of wastes. In most cases this is unpopular with the local public and environmentally aware politicians, who often view such operations through their perspectives of 'nimby' and 'nimtoo'.
2. **Site assessment**. Sites identified for waste disposal on strategic planning considerations must be capable of retaining the waste such that it does not

cause pollution of the surrounding ground or groundwaters. The most important considerations in any site assessment are: (1) the nature of the wastes to be contained; (2) the general topography and structure of the area; (3) the bedrock and surface geology; and (4) the hydrogeological regime. The design of the site will take into consideration all of these factors. The **waste type** and its quantity are of fundamental importance as these will ultimately control the type of containment (Box 9.4). Clearly, inert builders' wastes can be tipped at almost any site; putrescible and hazardous wastes needing specialist containment facilities require appropriately impermeable bedrock or a lining and careful environmental monitoring. The **topography** and **structure** of an area are of great importance as they indicate the stability of existing slopes and allow the risks from natural hazards in general to be assessed. For example, assessment of the tectonic features may indicate the presence of active earthquake or fault activity. Topographical surveys will also provide detailed knowledge of surface waters and help in determining their relationship to groundwater. This is particularly important in the determination of caves and other karst features. Evaluation of topography will also include the assessment of any particular conservation aspects, such as heritage features, protected natural areas and the like.

The nature of the **bedrock** and **surface geology** is extremely important as it is this which will ultimately determine whether the waste can be effectively contained away from the influence of groundwaters. Primary characteristics of both surface and bedrock strata to be investigated include: (1) the composition, both physical and chemical, of the lithologies present; (2) weathering, solubility and erosion resistance characteristics; (3) the porosity of individual lithologies and their permeability to water, leachate and landfill gas; (4) the vertical and lateral extent of stratigraphical units and their boundary characteristics, and the nature and spacing of joints and other fracture systems; and (5) the deformation characteristics of the rockmass.

Hydrogeological data to be assessed would include, in addition to permeability: (1) the nature of the groundwater regime and its gradient and rate of flow, inclusive of seasonal fluctuations; (2) the distribution and depth of aquifers; (3) groundwater levels and groundwater chemistry; and (4) effective rainfall, runoff rates, evaporation and groundwater recharge. In addition, the effects of groundwater abstraction should also be considered.

Once these characteristics have been collated they must be analysed with reference to the planned waste disposal operation, taking account of all of the safety and environmental impact requirements. In most cases it is the presence and suitability of natural barrier materials, the permeability and presence of groundwaters, and the potential for slope failure or other natural hazards which are of most importance in the final decisions on suitability.

3. **Environmental impact**. In most countries, particularly in Europe and the United States, it is a requirement that applications for landfill operations be accompanied by a detailed assessment of the impact the operation of the site will have on the surrounding area. Although the contents of these reports will vary from country to country, it is clear that the potential hazards and

BOX 9.4: LANDFILL DESIGN AND THE NATURE OF THE WASTE TO BE DISPOSED OF

Gray *et al.* (1974) developed a flow diagram to assist in the selection of landfill sites, placing particular emphasis on their hydrogeology. They identified three categories of waste.

Category One. Waste with potential to cause groundwater pollution through the introduction of substances hazardous to health.

Category Two. Wastes with potential to cause groundwater pollution not because of any specific toxicity they possess but by increasing the concentration of certain organic or inorganic substances to a level sufficient to render groundwaters unpotable. This includes most domestic and industrial waste.

Category Three. Wastes which are both solid and inert and from which there is no risk of groundwater pollution.

As the flow diagram below shows, those landfill sites at which it is proposed to deposit Category One waste must afford the maximum protection to groundwater supply. Total containment must be the aim. Landfills designed

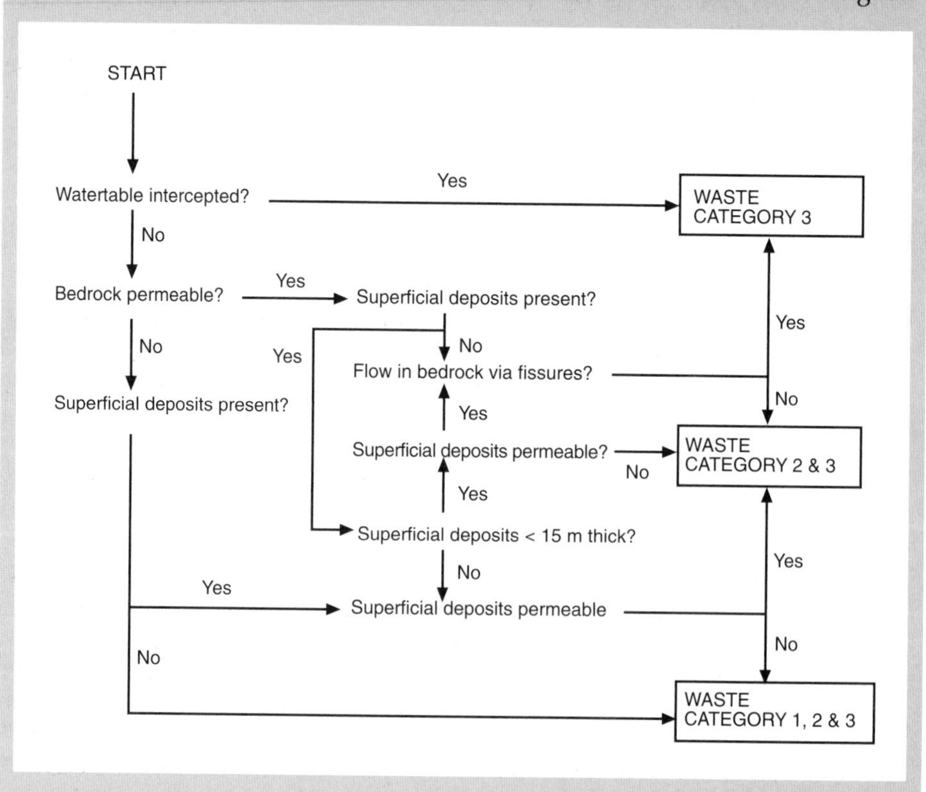

to contain Category Two waste need not necessarily involve total containment and can potentially incorporate leachate attenuation. Site selection should ensure that this attenuation process can be completed before any groundwater source is intercepted. In theory Category Three wastes can be disposed of at landfill sites which afford little or no protection to groundwater and no hydrogeological criteria need necessarily be considered in site selection. In practice the problem is to ensure that the waste mixture is uniformly of one type. More importantly, the boundaries between one category of waste and the next may not always be very clear.

Source: Gray, D.A., Mather, J.D. & Harrison, I.B. 1974. Review of groundwater pollution from waste disposal sites in England and Wales, with provisional guidelines for future site selection. *Quarterly Journal of Engineering Geology*, **7**, 181–196. [Diagram modified from: Gray *et al.* (1974) *Quarterly Journal of Engineering Geology*, **7**, Fig. 2, p. 193]

adverse effects on the environment need to be assessed. This would include plans for: (1) environmental monitoring and containment of leachates and landfill gases; (2) the control of nuisance such as noise of operations, visual intrusion, noxious odours, litter and pest control; (3) mitigation of adverse traffic loads from the coming and going of waste lorries; and (4) the protection of nature conservation, heritage and amenity areas from damage or disadvantage. In most cases, adequate provision for monitoring is of paramount importance in the forward plan.

9.5.4 Site Preparation and Operation

There are three stages to the operation of landfills once planning permission and a licence to operate have been granted. Firstly, the site must be prepared to receive waste, particularly if it requires engineering or lining. Secondly, proper operation of the site is necessary to ensure that the environmental impacts of the site are kept to the minimum. Finally, after filling the site must be properly restored so that it can accept an appropriate afteruse. These stages are discussed below.

1. **Site preparation**. Preparation of a site to receive waste is dependent on the outcome of the initial site investigation. Efficient site design is intended to minimise the amount of leachate production and reduce the risk of contamination and pollution from leachate and landfill gas migration. The design therefore needs to take into account whether a lining material is necessary before waste can be introduced to the site, and whether any other remedial engineering work is needed to prevent slope failures, or to provide leachate barriers or **bunds**. This might entail the construction of a considerable

number of earthworks, for example, or it might include the battering and clay sealing of old quarry faces. Earthworks to provide screening during operation may also be necessary.

The type and extent of site preparations will also vary according to whether the site is to be a containment or slow dispersal site (Figure 9.8). For all sites, however, it is necessary to minimise the water content of the site through: (1) reducing surface water inflow by removing standing surface water by drainage and/or pumping; and (2) by the subdivision of the disposal area into a number of **cells** which effectively minimise the area of waste which can be infiltrated by surface waters. Ideally, containment sites should be self-contained with an impermeable waste–bedrock boundary, but it is possible that contamination can be achieved through the presence of occasional permeable zones, and through the 'bath-tub' effect which allows leachate to brim over the edges of the landfill site. This is a problem only where rainfall exceeds evaporation, and here drainage and leachate collection schemes are necessary.

Many sites will require lining to achieve containment, and again, the need for this is largely dependent on the type of waste intended for the site. Where total containment of putrescible wastes is necessary, lining can be achieved through the use of imported clay materials, through the use of geomembranes and synthetic liners, or by using bitumen and aggregate lining. In most cases it is necessary to use a composite of different liner systems in order to achieve total containment (Figure 9.10A). **Clay sealing** can be achieved by the progressive importation of clay from other sites which provide a naturally impermeable layer. Clay is naturally self-sealing, a property which is a distinct advantage where there is potential for rupture of the seal from tree roots or other sources. Clay can also be banked against old quarry faces to provide a working bund (Figure 9.10B). **Geomembranes** are synthetic liners which are usually 0.5–2 mm thick with a low permeability, a long-term resistance to chemical attack, and an ability to withstand some puncturing. Most geomembranes are used in conjunction with a basal layer of compacted clay 1 m thick, and are overlain by a sand layer to provide protection for the liner material. Typical materials used include low and high density polyethylene (LDPE and HDPE, respectively), PVC and butyl rubber. Many other products have also been used. In each case, the geomembrane is supplied in rolls of up to 10 m in width, and it is necessary to weld the strips together using hot-air welding methods. In most cases it is recommended that composite liners, made of alternating geomembrane layers spaced with mineral layers, are used to seal landfill sites where there is risk of groundwater contamination (Figure 9.10A). **Bitumen and aggregate liners** provide a relatively cheap method of lining. They utilise bitumen and shotcrete layers to effectively seal unsatisfactory waste repositories prior to filling.

2. **Site operation**. Effective site operation includes: (1) the pre-treatment of wastes; (2) the method of filling and sealing; and (3) proper environmental control and monitoring. Some pre-treatment of the waste enables the more

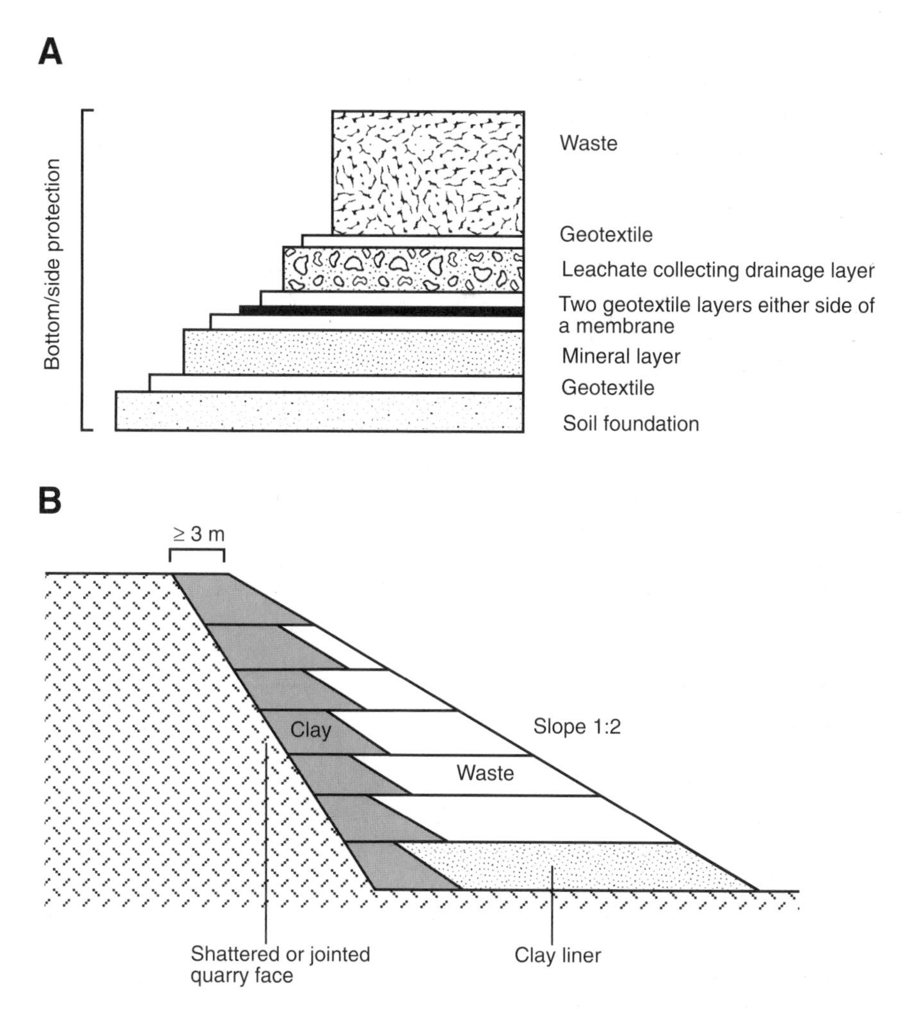

Figure 9.10 *Sealing a landfill site to achieve total containment.* ***A.*** *Use of a composite lining of landfill sites using geomembranes and mineral sealant.* ***B.*** *The use of clay to seal fractured or porous cliff faces [Modified from: Attewell (1993)* Ground pollution. *E&F Spon, Figs 10.12 & 10.15, pp. 111 & 114]*

efficient packing of wastes into the landfill, thereby optimising the effective use of the void space. Typical methods include: the compression of domestic wastes into bales; the pulverisation of wastes in a wet state to produce a homogeneous material; and the pulverising of wastes in a dry state to produce a powdered material which can be more easily disposed of.

Three types of fill method are commonly carried out in landfill sites. The **trench method** involves the excavation of a trench into which waste is deposited. The **area method** involves the deposition of waste in layers which form terraces over the whole of the available area. The area method can lead to high leachate production through the large surface area available

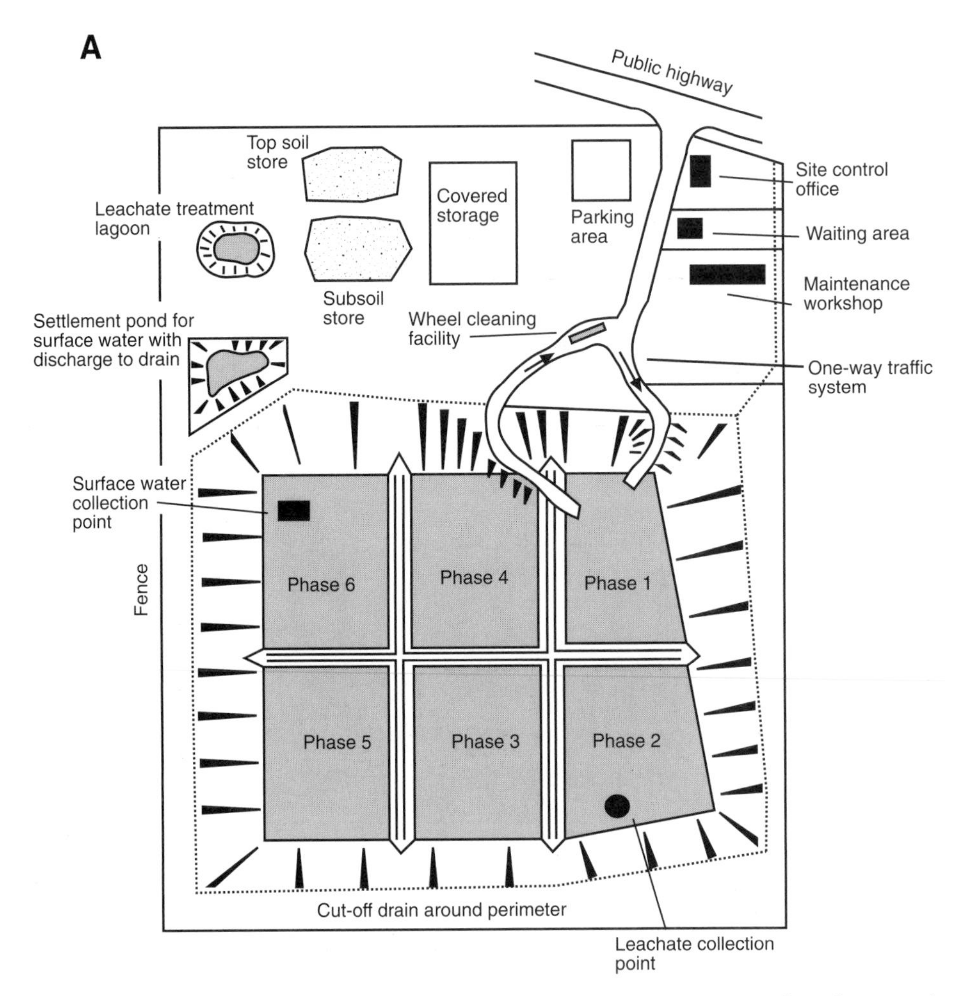

Figure 9.11 *Operation of typical landfill and landraise sites.* ***A.*** *Site plan showing the landfill cells, each of which is filled during a successive phase of operation.* ***B.*** *Section through a typical landfill site using the principle of cells.* ***C.*** *Section through a typical landraise site using the principle of cells [A Modified from: Department of the Environment (1986)* Landfilling wastes. *Waste Management Paper No. 26, HMSO, Fig. 5.16b, p. 64; B Modified from: Department of the Environment (1986)* Landfilling wastes. *Waste Management Paper No. 26, HMSO, Fig. 6.5, p. 74; C Modified from: Attewell (1993)* Ground pollution. *E&F Spon, Fig. 10.10b, p. 108]*

for rainwater penetration. The **cell method** is most commonly used and involves subdivision of the available area into clay-bunded cells which allow progressive landfilling of a site and which provide extra protection from leachate migration (Figure 9.11). In all cases, the waste material is deposited over a gently sloping (usually 1 in 12) face encouraging lateral build-out and allowing the necessary compaction to take place. At the end of a working

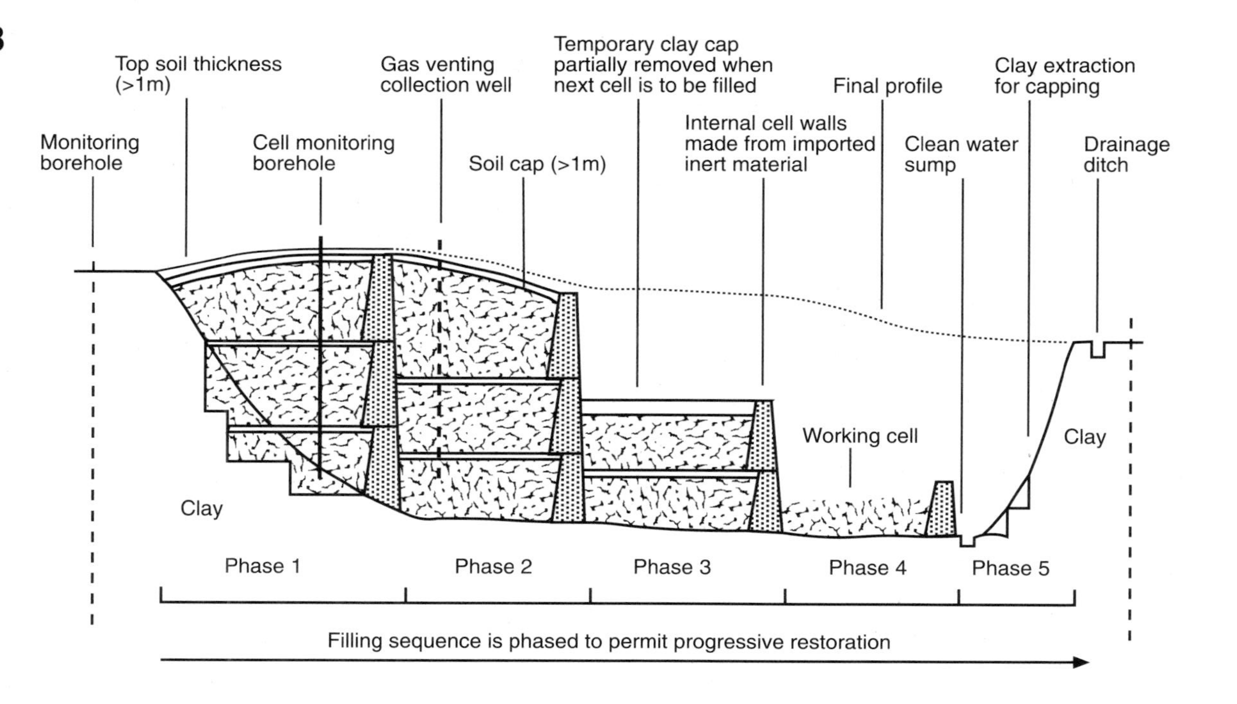

continues overleaf

C

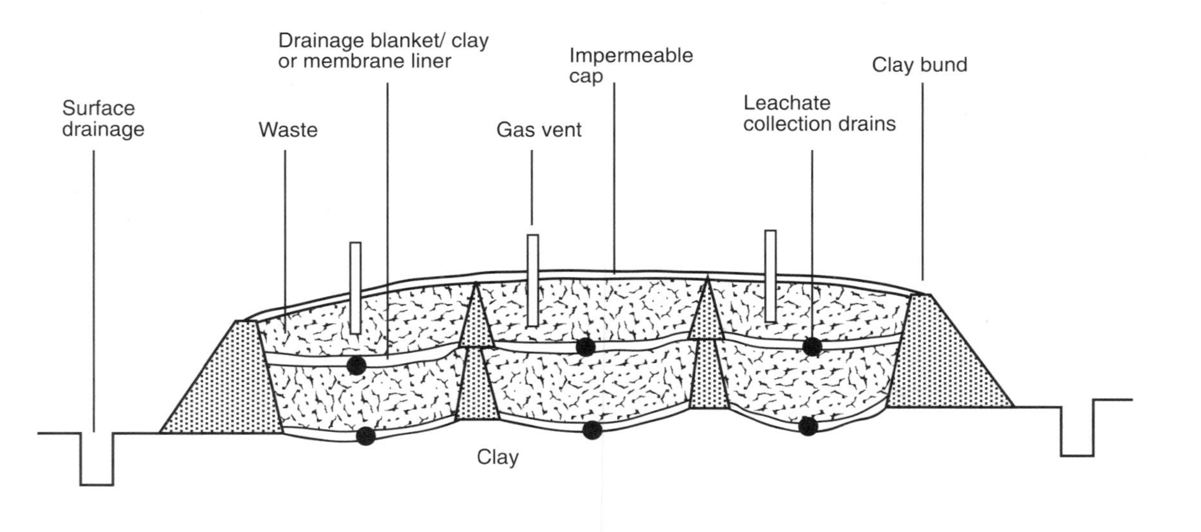

Figure 9.11 *(continued)*

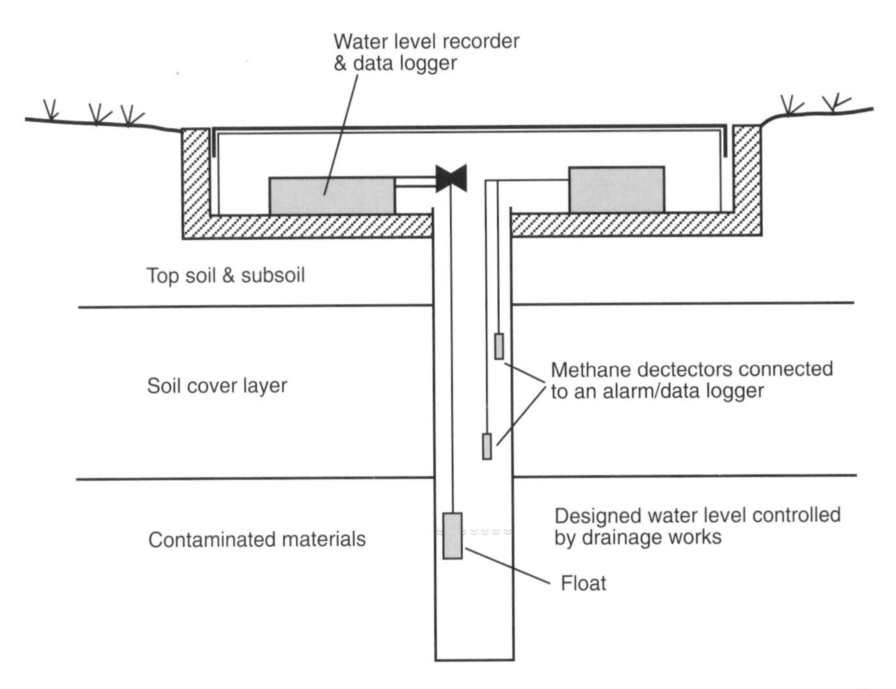

Figure 9.12 *An example of a monitoring device to record water levels and gas emission from a landfill site [Modified from: Attewell (1993)* Ground pollution. *E&F Spon, Fig. 10.5, p. 91]*

day, the deposited waste is covered with a suitably inert material, which helps to inhibit infiltration of rainwater, wind blow of waste, odours and infestation by pests. Typical cover materials can be clays and waste materials such as pulverised fly ash from power stations.

Environmental control is an important aspect of waste site operations and includes noise abatement, prevention of visual intrusion, litter control, bird and pest control, the prevention of odours and fires, and the control of the adverse effects of excess traffic. Monitoring of such aspects is particularly important to effective landfill operations, through both the operational and aftercare phases of development. Monitoring is particularly important for the mitigation of leachate and landfill gas hazards, and is usually a requirement of any landfill operation planning application (Figure 9.12).

Noise is due to the nature of the machinery used at landfill sites, and although screening with trees may help, often the most important factor is the restriction of working hours. Visual intrusion is an important factor to many people sensitive to property values, and screening with earthworks and/or tree lines is often effective. Litter control is primarily at the mercy of the prevailing wind direction. Bunds and netting have some success, but the daily placing of a seal over the tipped waste is the most effective solution. Birds are a hazard as they may contaminate adjacent water supplies with their excreta and provide a nuisance, especially if close to airports and airfields. Nets, bird scarers and falcons provide useful methods of control. Effective control of pest infestation and the emission of noxious odours is achieved through good site management, and in particular the daily coverage and sealing of landfill cells and working areas. Fires should be avoided as they can be hazardous to health and the smoke emissions are a serious source of atmospheric pollution. Fires can be caused by the casual or accidental ignition of dry wastes and are exacerbated by the emission of landfill gases. Traffic management, particularly the size and frequency of loads, the damage to roads, bridges and other infrastructure, and the choice of routes by drivers, is an important part of environmental management, and all of these factors require careful monitoring.

3. **Site restoration**. Landfill provides one of the most important routes for the restoration of old quarries and pits back to the original land surface contours, and it provides a mechanism for the creation of new land areas for agricultural, domestic or industrial developments. Capping of the site with an impermeable layer prevents the migration of landfill gas and leachates, but adequate monitoring and removal of residual fluids is necessary. Monitoring of compaction rates is also important. It should be noted that landfill sites pose a significant problem to future land use, particularly for large civil engineering projects (Box 9.5).

9.6 THE DISPOSAL OF RADIOACTIVE WASTES

The disposal of radioactive wastes represents one of the most emotive environmental issues and presents a challenge for any environmental geologist. Radioactive wastes may be produced as by-products of industrial processes such as the creation of uranium fuel rods for nuclear power stations, or as a result of military and medical applications of radioactive materials. In almost all cases radioactive wastes are hazardous to human health, although the risk is dependent on the type of radioactive particles which are emitted, and the length of time of exposure to the radiation. On this basis, there are effectively two currently recognised types of radioactive wastes: high level and low level wastes.

High level waste is effectively derived from the operation of nuclear reactors and is characterised by a large percentage of radioactive nuclides which are emitting alpha, beta and gamma particles. Typical wastes are the spent fuel rods from nuclear power stations, and the waste liquids produced from reprocessing of these rods. Nuclear reactors rely upon the fission of the atoms of the ^{235}U isotope when bombarded by neutrons. The most common fuel used in nuclear reactors is uranium dioxide in the form of pellets (Figure 9.13; see Section 3.1.1; Figure 3.5). These are placed in metal tubes in clusters in water, graphite or deuterium oxide which acts as a neutron modulator and prevents the fission chain reaction from producing a nuclear explosion. The process of fission produces a wide variety of isotopes including selenium, cobalt, strontium and

BOX 9.5: THE PROBLEM OF LANDFILL AND FUTURE SITE DEVELOPMENT

Filling a quarry with waste effectively seals and thereby sterilises any remaining mineral resource. Equally, landfill sites pose a significant problem to future civil engineering. Construction on top of old landfill sites is possible where problems of landfill gas and leachate do not occur or are effectively managed, but careful design of foundations is required to prevent problems of settlement. Major civil engineering projects may require the removal of part or all of a landfill site. These problems were encountered during the construction of the London orbital motorway, the M25.

Most of the sand and gravel pits around London have been historically filled with landfill. The planned route for parts of the M25 dissected several major landfill sites. Excavation through these old landfill sites was necessary in several places. Where the material was inert this was relatively simple, although careful design of cut slopes was required due to the heterogeneous character of the waste. However, where the road cut through domestic or industrial waste sites, resealing of the cut slopes and investigation of the potential impacts of such activity on the landfill containment was required. Cut slopes were resealed using geomembranes keyed into the base of the embankment as shown in the diagram below. The essential aim is to ensure that the leachate from the waste tip does not enter the storm drainage system for the motorway.

In addition to the problem of resealing opened landfill sites, one of the major difficulties is the lack of adequate records of the contents of old landfill sites. Historically, quarries have been infilled with all kinds of waste prior to modern environmental control; documentation of the contents of an old landfill site may be very poor, adding to the construction problems.

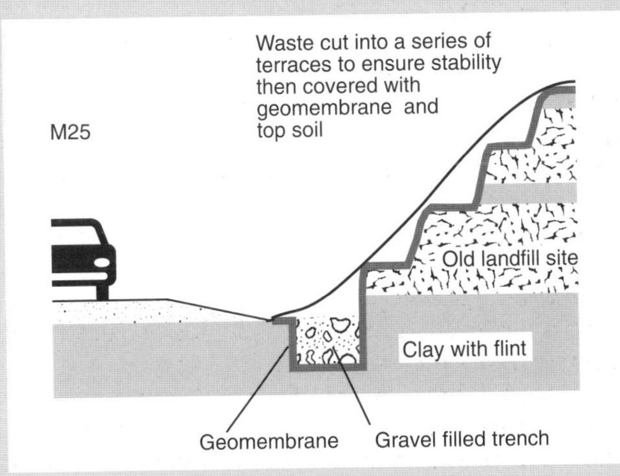

Source: Winney, M. 1985. Unsupported trench collapse kills two. *New Civil Engineer*, 5 September, 8–9.

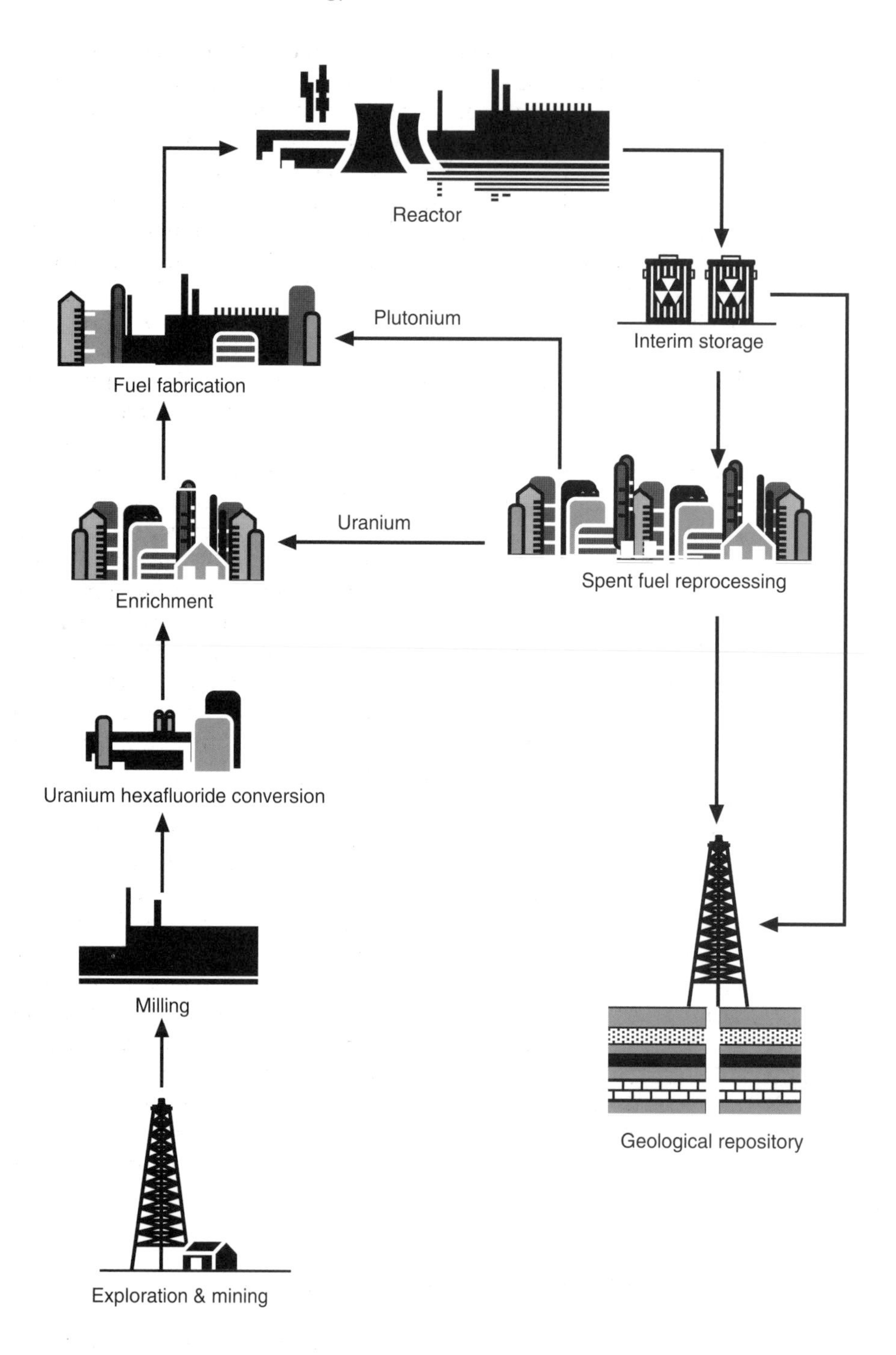

Figure 9.13 *The commercial nuclear fuel cycle [Modified from: Krauskopf (1988)* Radioactive waste disposal and geology. *Chapman & Hall, Fig. 2.6, p. 20]*

caesium. These accumulate in the fuel rods and gradually decrease their efficiency by absorbing and deflecting the neutrons which are essential to the fission process. Fuel rods have to be replaced approximately once a year. The spent rods are removed and are stored in large tanks of circulating water.

The spent fuel rods have commonly been **reprocessed** in order to produce plutonium for nuclear weaponry (Figure 9.13). The bulk of the fuel rods is composed of the heavy isotope of uranium, ^{238}U, which does not undergo fission but actually adds neutrons to the nuclei of some of its atoms. These form atoms of the isotope ^{239}U which are themselves unstable, and which, through a combination of the addition of further neutrons and the process of radioactive decay, transform into atoms of other elements with even higher atomic masses. These are called the **transuranic elements** and include plutonium, neptunium, americium and curium, all of which are radioactive and emit alpha particles, some of which have long half-lives. Plutonium is used in the creation of nuclear weapons as, like ^{235}U, ^{239}Pu atoms give out considerable energy when struck by neutrons. The plutonium is recovered from the fuel rods through a process of nitric acid digestion of the remaining uranium dioxide and the fission products, the plutonium being separated out by the addition of an organic chemical (Figure 9.13). The leftover nitric acid solution, formally referred to as **reprocessing waste**, is another source of high level nuclear waste, and it is retained in large tanks awaiting disposal. It is generally expected that reprocessing waste will have to be solidified in some way, possibly as a glass, before final disposal can take place. Both types of high level waste are characterised by radioactive elements with long half-lives, such as the isotopes of uranium and the transuranic elements (Table 9.3). The disposal of these wastes needs a long-term strategy on an almost geological timescale.

Low level waste is a general term for all radioactive wastes other than those produced from reactors. The majority of low level wastes contain very little radioactivity, often at a level which is below the ambient levels of the environment. Typical low level wastes are derived from medical sources, laboratory debris, and the metal tubes of fuel rods once the fuel has been dissolved. They can also include mine waste from uranium mining which, in common with most mine waste, is usually left in open dumps at the mine site. In some cases an additional classification of **medium** or **intermediate level** waste may be used for materials containing a significant concentration of radioactive materials, such as that which is produced during the production and handling of plutonium for military purposes. Much low to intermediate level waste contains radioisotopes with half-lives on a human timescale, of a few hundred years or more (Table 9.3).

9.6.1 Disposal of Radioactive Wastes

The disposal of radioactive wastes is a difficult problem, not least in that it evokes the greatest 'nimby' reactions from local people and 'nimtoo' reactions from politicians. This is especially true when it comes to the selection of

Table 9.3 *Half-lives of some typical radionuclides [Modified from: Krauskopf (1988)* Radioactive waste disposal and geology. *Chapman & Hall, Table 1.2, p. 8]*

Isotope	Emitted particle	Half-life
Americium-241	alpha	433 years
Americium-243	alpha	7370 years
Bismuth-210	beta	5.0 days
Cobalt-60	beta	5.27 years
Caesium-135	beta	3×10^6 years
Caesium-137	beta	30.17 years
Curium-245	alpha	8500 years
Lead-210	beta	22.3 years
Neptunium-237	alpha	2.14×10^6 years
Polonium-210	alpha	138.4 days
Plutonium-239	alpha	24 000 years
Plutonium-240	alpha	6470 years
Radium-226	alpha	1630 years
Radon-222	alpha	3.82 days
Strontium-90	beta	28.8 years
Thorium-230	beta	80 000 years
Uranium-234	alpha	2.45×10^5 years
Uranium-235	alpha	7.04×10^8 years
Uranium-238	alpha	4.47×10^9 years

repositories for high level radioactive wastes, as the integrity of such sites will have to remain in perpetuity, on a geological timescale of millions of years. Low level wastes are often mixed with domestic wastes and landfilled on the basis that exposure to radiation from such sources is only harmful where people have been continuously exposed to the source for a long period of time. However, it is rapidly approaching a time when the storage of high level reactor and reprocessing wastes from power stations will exceed the facility for stockpiling, and it is clearly a necessity to design appropriate methods of disposal for this extremely hazardous waste. In the early days of nuclear power, when it was considered to be the cure for all the world's power production problems, little regard was given to any future needs for waste disposal. Most decision-makers expected that the science and technology of the future would hold the answer to the problem, but 50 years later the stockpile continues to grow and the problems remain to be solved. These are questions which are taxing the minds of governments and scientists alike, and so far no definite decisions beyond the first stages of investigation have been made by any government.

9.6.2 Disposal Options for High Level Wastes

A number of disposal options for high level nuclear waste have been suggested, all of which operate on the principle of isolation from human activity. These are discussed in outline below.

1. **Disposal in a deep mined repository**. Here the waste is disposed of in an underground repository within a suitable rockbody (Figure 9.14). The rockbody itself should be the final barrier to the migration of the nuclear waste away from the repository, such that any breaching of waste containment systems can be mitigated. It is most unlikely that inert metals such as gold and platinum, resistant to corrosion, would be used to encapsulate the waste, so it is important to plan for the eventual breaching of other barrier metals which are open to corrosion. The most important geological factors to consider are therefore: geological structure and seismic/tectonic history; lateral and vertical extent of the rockbody; porosity and permeability of the rock type; and the hydrogeological regime. Suggested repositories in Britain and the United States include deep mines in volcanic tuffs, salt deposits and igneous rocks, all of which are naturally impervious and impermeable.
2. **Disposal in a surface repository**. In this type of containment facility the waste would be maintained in a repository close to the surface in order to provide greater facility for monitoring. This is essentially an extension of the present storage arrangements, although a greater degree of shielding is necessary, and this could be attained through the construction of a large monolithic structure to contain the waste, or through the cutting of shallow tunnels into mountains.
3. **Disposal in deep wells**. This would involve the drilling of extremely deep boreholes, in order that the radioactive material could be assimilated through melting of the surrounding rock through the combined heat of both geothermal and radioactive processes. This proposition is highly theoretical with too many variables and unknowns, and is not being seriously considered at present.
4. **Disposal in ice sheets**. It has been suggested that waste could be disposed of within the Antarctic Ice Sheet. Accumulation and glacier flow are sufficiently slow for the waste, if placed in the ice sheet centre, to take several hundred thousand years to pass through the ice sheet before re-emerging at the margin, during which time it would become relatively harmless. Several problems exist with this solution: not least is the possibility of unstable ice flow which could accelerate the throughput of the waste; more importantly, widespread dispersal of the waste could occur if its containment cells were compromised. The remoteness of Antarctica does, however, make the concept appealing to some politicians at least.
5. **Ocean floor dumping**. Dumping of waste at sea has been considered because it has been thought possible that the waste could ultimately be carried into the mantle through subduction processes. Dumping of wastes into the deep ocean trenches could theoretically achieve this goal, but not enough is known about the deformation of trench sediments to be sure that the waste would not simply be thrust out of the trench as part of an accretionary prism.
6. **Disposal in space**. The idea of putting the radioactive material in a rocket and blasting it into space has some attraction as it removes the problem

Figure 9.14 *The Äspö hard rock larboratory in Sweden. This test site is designed to provide an experimental environment at depths similar to those required by a high level nuclear waste repository. Rock properties and groundwater flow are being monitored as part of the scientific feasibility study of deep waste repositories. A similar laboratory has been suggested for Sellafield in Britain prior to the development of a repository, although this has been currently shelved [Photograph: A.F. Bennett]*

completely from the Earth. This option is, however, unrealistic, as the whole process would be dangerous – especially if the rocket exploded on take-off – and costly, and on too small a scale to be of any value.

The majority of these management options have been considered by the nuclear industry, although it is now clear that the preferred option, in the USA and the UK at least, is that of containment in a deep mined underground repository.

There are many factors which need to be considered in the construction of such a facility, some of which are discussed below (Figure 9.15).

1. **Principle of containment**. The first and most important principle to be considered in design of a suitable repository for high level waste is the application of the concentrate-and-contain principle, in which it is important to completely isolate the hazardous wastes from the environment and from interference, accidental or otherwise, by humans. Unlike the expected attenuation of hazardous leachates and the decline in gas production in normal landfill operations, which take place over a human timescale, the isolation of high level nuclear waste has to be on a geological timescale, and therefore the containment must take into account the potential for

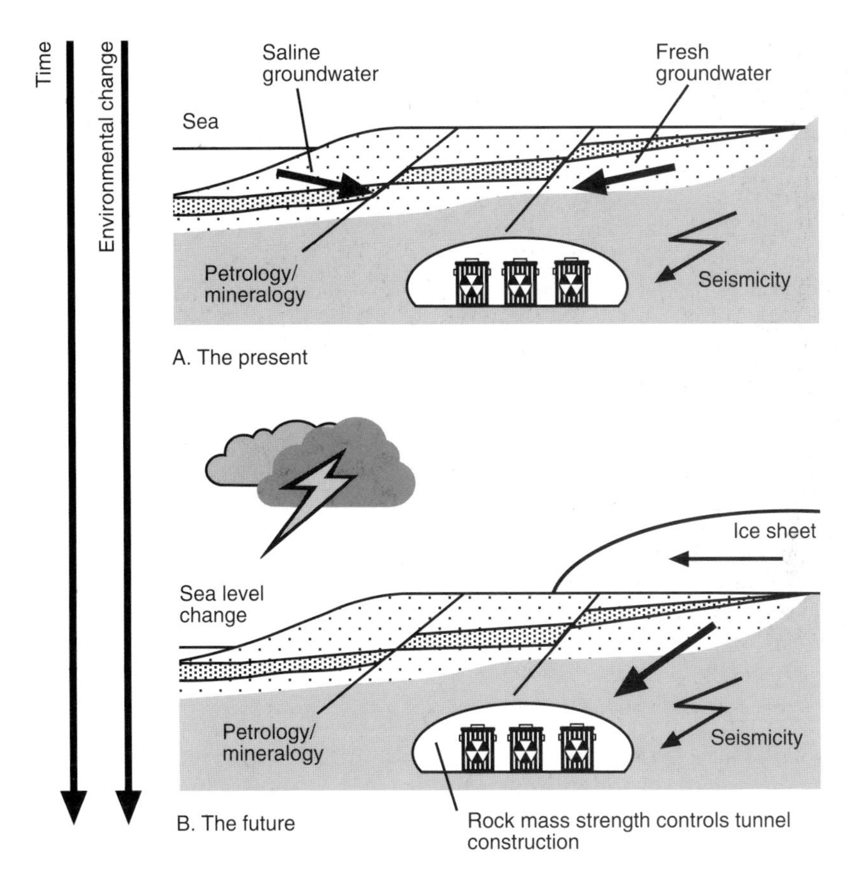

Figure 9.15 *Factors to be considered in the construction of a deep repository for high level nuclear waste*

breaching of any barriers. This means that the containment principle must operate a multi-barrier system. For example, plans for deep repositories have at least three barriers: the first is the sealed metal capsules, capable of shielding the effects of radiation in which the waste is stored; the second is the medium in which the metal capsules are protected, likely to be concrete; and the final barrier is the surrounding geology.

2. **Predicting environmental change**. Given the geological time range during which some of the radioactive isotopes decay, with many millions of years required for the isotopes of uranium and plutonium to become harmless, it is necessary to design a containment site which will survive, for the sake of argument, the growth of ice sheets over the repository during some future glacial cycle (Box 1.1). This is of paramount importance as the growth of ice sheets would not only cause problems for surface containment systems, but would also have a dramatic effect on the groundwater systems. Sea level rise is another important factor which could lead to flooding of repository sites and result in adverse changes in saline groundwater systems. For under-

ground sites it is clearly necessary to understand not only the fluctuations in saline and fresh groundwaters controlled by climatic effects, but also to model the potential effects of major global climate shifts.

3. **Predicting natural hazards**. Clearly it would be nonsensical to locate high level waste facilities in areas of known seismicity, volcanicity or any of the other natural hazards discussed in Chapters 11 and 12. Once again, however, it is important to remember that the safe containment of nuclear waste must be on a very long timescale and therefore it is necessary to plan, predict and model future changes in global tectonics which are likely to take place on a timescale of millions of years and may affect the distribution of geological hazards.
4. **Understanding material properties**. The long-term integrity of the materials used in high level waste sites is also of great importance. Understanding the reaction of containers, of natural or human origin, is essential. For example, in underground geological repositories it is important to consider not only the metallurgy of the manufactured capsules, but also the reaction of the surrounding lithology to bombardment by nuclear particles, as this may lead to physical weathering of the rock, thereby weakening its potential to contain the waste. Similarly, the ability of the rock to support the engineered caverns, and its porosity and permeability, are all pertinent features in the design of a repository.

The disposal of low level nuclear wastes is less of a problem, as they are more usually capable of being treated using the dilute-and-disperse principle, in which dilution is achieved through mixing with other wastes before landfill disposal. These wastes are the subject to the usual problems of leachate development and migration as for normal landfill wastes. In most cases, however, it is more appropriate to maintain specific landfill sites for the containment of low level radioactive waste alone, in order that they can be carefully monitored. These sites would necessarily have geological and hydrogeological conditions, such as limited permeability, which would limit the migration of radioactive waste into the surrounding strata.

9.6.3 Deep Mine Disposal Sites: Site Assessment and Selection

The majority view, at present, is that disposal in deep mines represents the most appropriate option for a long-term waste disposal facility. This is the case in both the USA and Britain where it has been a matter of government policy to seek out and investigate the most appropriate deep mine site for high level waste disposal. In both countries there has been a decision to fully investigate the suitability of certain sites for waste disposal: three in the USA (Hanford Reservation, Washington, in basalts; Yucca Mountain, Nevada, in tuffs; and Deaf Smith County, Texas, in salt); and one in Britain (Sellafield, Cumbria, in volcanic rocks) (Figures 9.16 & 9.17). Currently, much detailed research is being carried out in order to characterise the material properties, hydrology, hazards and future climatic conditions of each of the potential repositories. Final decisions will await

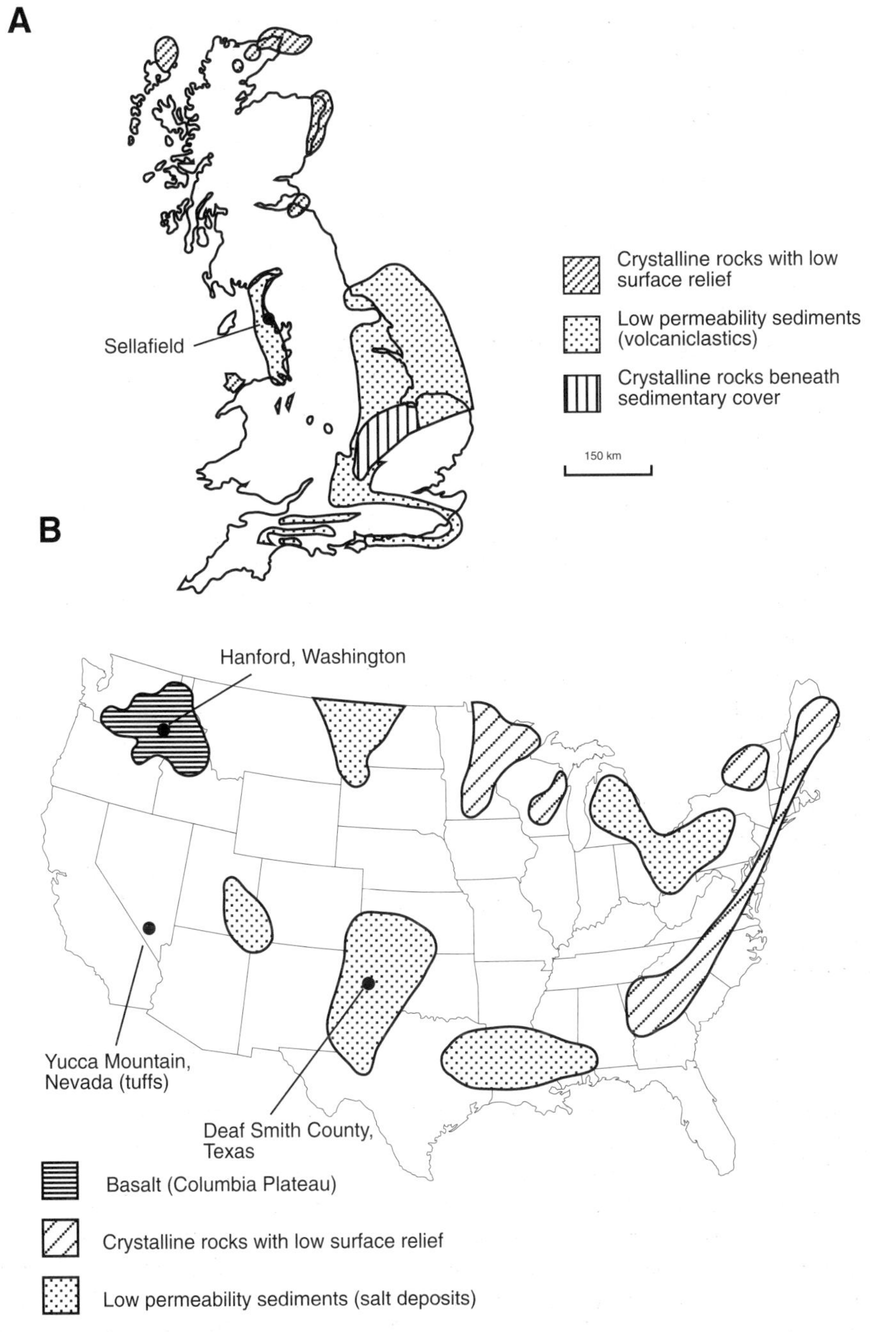

Figure 9.16 *Maps of proposed nuclear waste facilities in the UK and USA [A Modified from: Woodcock (1994)* Geology and environment in Britain and Ireland. *UCL Press, Fig. 16.8, p. 132; B Modified from: Baillieul (1987)* Bulletin of the Association of Engineering Geologists, **24**, *Figs 1 & 3, pp. 209 & 211]*

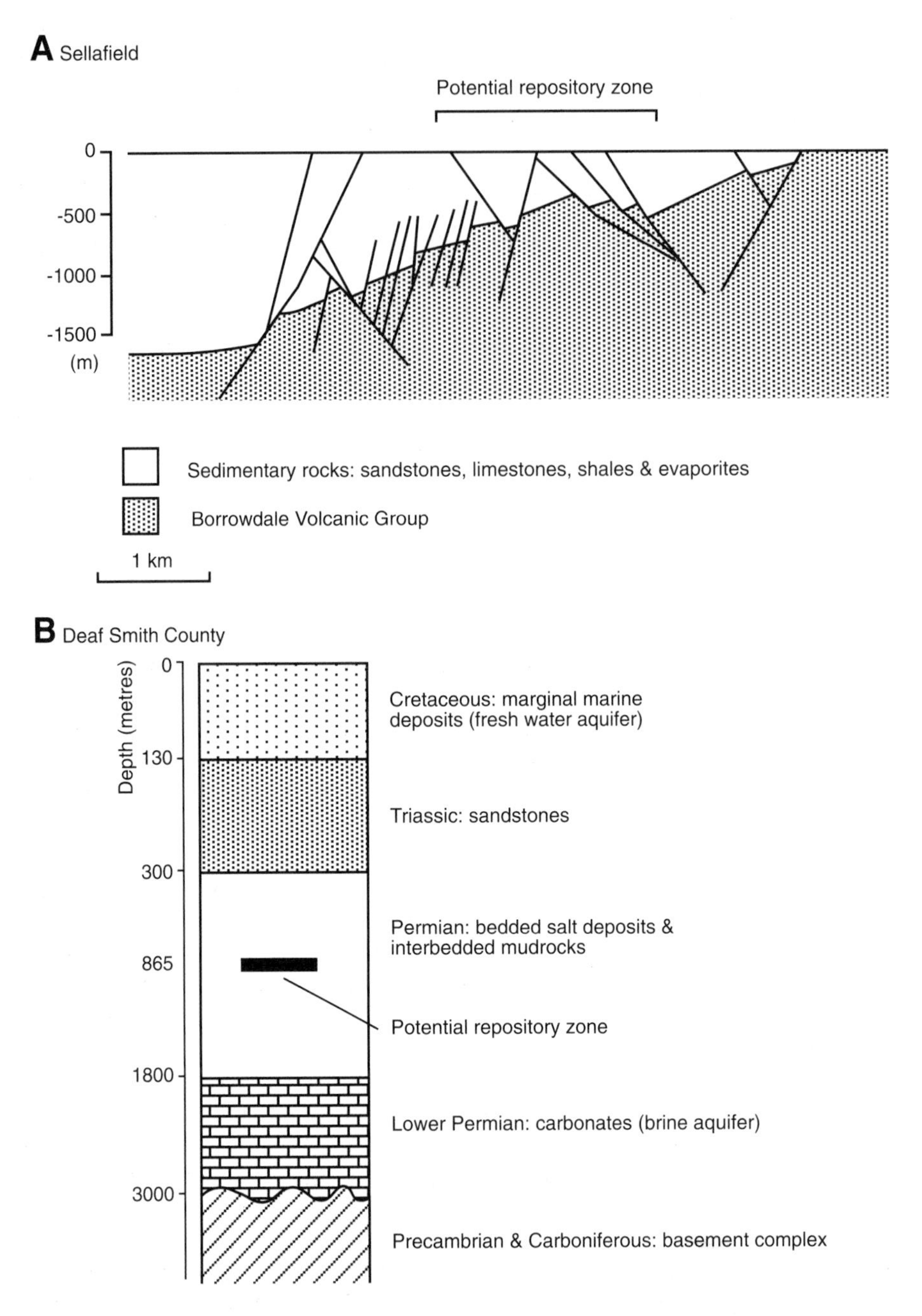

Figure 9.17 *The geology of two possible nuclear waste sites.* ***A.*** *Simplified geology of the Sellafield tuff repository in Britain.* ***B.*** *Simplified geology of the Deaf Smith County salt repository in Texas, USA [A Modified from: Chaplow (1996)* Quarterly Journal of Engineering Geology, **29**, *Fig. 4, p. S5. B Modified from: Krauskopf (1988)* Radioactive waste disposal and geology. *Chapman & Hall, Fig. 3.10, p. 49; Peck (1987)* Bulletin of the Association of Engineering Geologists, **24**, *Fig. 2, p. 218]*

the outcome of this painstaking work by many scientists. The range of factors which must be considered in choosing a site and in its design are summarised conceptually in Figure 9.15. Taking the Deaf Smith County and Sellafield sites as examples, the following characteristics have been studied so far.

1. **Future environmental change**. This area is the most difficult to predict, although some modelling has been carried out to investigate the likely effects of isostatic loading and changes in groundwater regime at Sellafield associated with a future ice sheet. Work has also been done on the effects of sea level fluctuations on the groundwater regime.
2. **Lithology**. Lithology is important as it ultimately controls the integrity of the containment. Typical lithologies are impermeable and impervious to prevent intergranular fluid migration, and include salt, basalt and volcanic tuffs. The best lithologies are capable of withstanding considerable heat generation and bombardment from radioactive particles. In the USA, rock salt is a preferred lithology, although tuff and basalt have also been investigated. The salt deposits at the Deaf Smith County site in Texas are bedded salt deposits at a depth of 850 m, which are of Permian age (Figure 9.17). Salt is particularly important as it is widespread, easy to mine, plastic at depth, which allows self-healing, and can withstand radioactive bombardment. It is unsatisfactory in so far as it is liable to solution, and deforms when loaded. This will also be a problem in the construction of the containment site, particularly in its long-term stability, as it has long been understood that mine workings become unstable with age, leading to surface subsidence and the attendant environmental problems. Tuff is the preferred lithology at Sellafield in Cumbria. Here the suggested repository is in the Ordovician Borrowdale Volcanic Group at a depth of 400–600 m (Figure 9.17). The majority of the levels selected as potential repository sites consist of tuffs which are dense, with low porosity and permeability, and are capable of accepting a high thermal loading. There is some possibility of mineralogical change caused by bombardment by nuclear particles, but the risk is thought to be slight. This material is also structurally strong and capable of maintaining the structural integrity of the containment facility.
3. **Geological structure**. Geological structure is indicative of both tectonic activity and the potential for groundwater flow along joints and fractures rather than between grain boundaries. The Deaf Smith County site consists primarily of bedded salts and related rocks which have not been subject to intense folding, and there are few faults within the immediate vicinity of the proposed repository. This is clearly valuable in inhibiting groundwater flow. At Sellafield, however, the structure is much more complex, the Borrowdale Volcanic Group having undergone several phases of tectonic evolution to produce a well faulted basin (Figure 9.17). This clearly influences groundwater flow at the site, and detailed hydrogeological modelling is currently being carried out to investigate how this will affect the overall permeability of the site. This has identified severe drawbacks with the Sellafield site and it has been temporarily shelved as a consequence.

4. **Hydrogeology**. Understanding the groundwater flow regime at any given deep repository is of paramount importance in maintaining total containment of the radioactive waste. For example, at Sellafield, groundwater flow is predominantly through the fracture systems. A detailed analysis of the way in which the flow system operates is underway, and this will have a very important bearing on the viability of the site as a waste repository. Current knowledge suggests that flow is controlled by the density mixing of gravity-driven fresh and saline waters. These flow at different rates, but the low hydraulic conductivity of the Sellafield tuffs suggests that any flow will be slight, and that these conditions are likely to prevail for some time. Notwithstanding this, modelling the future climate change regime is a clear challenge to ensure that resulting effects of groundwater flow will not be adverse.
5. **Seismicity**. Seismic activity could contribute to the rupture of the containment barrier systems, and is clearly a hazard to safe waste management. Neither Deaf Smith County nor Sellafield have a record of intense seismicity which could provide a problem, but it is clearly necessary to consider the future potential of this activity in order to ensure containment.

9.7 EFFLUENT TREATMENT AND DISPOSAL

The disposal and treatment of domestic and industrial water waste is an important part of water resource management, not least because treated water may be recycled for human use downstream (Figure 5.4). In Britain, trade and domestic effluent is dealt with through dispersal directly into the sea, with little or no direct treatment, using the principle of dilute and disperse. It is, however, illegal to discharge trade effluents known to contain chemical substances which are hazardous to humans or the environment directly into a range of water bodies without effective treatment. The disposal of domestic and trade effluents is a contentious issue because of its potential impacts on the natural environment, particularly its impact upon water resources for human consumption and recreational use, and on the ecology of river and coastal systems. The review that follows is written from the perspective of the developed world, in particular Britain, although it should be recognised that treatments employed across the world vary according to climate (see Section 8.1.2) and resource availability.

9.7.1 Sewage Effluent Treatment

In Britain, typical sewage flows are 99.5% water, while the rest consists of minorgenic particles (700 ppm), organic solids (100–350 ppm) and dissolved organic and inorganic solids (200–1000 ppm). Minerorgenic particles are particles derived from soil and are produced by storm runoff if foul-water sewers also carry storm water. This is commonest in older developments and consequently the minorgenic component of this type of runoff is relatively high. In most modern towns, however, there are separate sewer systems for storm water, and domestic and trade sewage. In these modern sewage systems the minorgenic component is relatively low and is primarily derived from waste disposal units, from vegetable washing, and from broken crockery.

The biggest component of sewage flows is organic solids, primarily faeces and food fragments which are rich in bacteria, protozoans and a variety of parasites. Dissolved organic matter is primarily derived from urine and from fluid within faeces, as well as from cooking liquors, detergents, washing water and a wide range of commercial inputs from trade effluents. Dissolved inorganic components are derived from the original water supply, but may also be enriched in cooking salt and in a variety of industrial chemicals. The range of industrial components present will depend on the industries involved and on the regulations concerning their waste disposal. The principal inputs come from small commercial units; larger units have special disposal provision. The water temperature of sewage is usually in the range of 10 to 20°C and is commonly warmer than the domestic supply water due to inputs of hot water and heat exchange from warm buildings.

The aim of sewage effluent treatment is first to separate the insoluble solids, and then to remove the dissolved components, thereby producing clean water and solids for disposal. The basic processes are: screening to catch large objects; sedimentation to remove the bulk of the solid particles; biochemical reaction to remove dissolved organic components; filtration; and precipitation of the inorganic dissolved components before the final water is sterilised to remove any bacterial components. This is organised into three stages: primary, secondary and tertiary treatments, as shown in Figure 9.18 and discussed below.

1. **Primary treatment**. In this process, water is passed through a screen of metal bars set between 50 and 150 mm apart to remove all floating objects. It is then passed into a small tank (**grit chamber**), and its flow is slowed sufficiently for dense inorganic grit, but not the organic solids, to settle out. At this stage the sewage is often stored in a holding tank to even out the fluctuations inherent in the daily introduction of new waters, and to ensure a continuous supply to the subsequent processes. These holding tanks are stirred and aerated to prevent the production of objectionable odours, most of which are produced by anaerobic decay. The first significant stage in water quality improvement follows as water is passed to the **primary settling tank**. Here, water is held virtually stationary for several hours in a tank 0.5 to 3 m deep to allow all visible solids to settle out of suspension. At the end of the primary treatment stage, the water will have lost 50 to 60% of its suspended solids.
2. **Secondary treatment**. The aim of the secondary treatment process is to reduce the volume of dissolved organic components. This is achieved by the aeration of the water to encourage growth of aerobic bacteria. Such bacteria use the dissolved organic material and convert it to carbon dioxide, water and biomass which can be settled out. This whole process may be achieved by three different techniques. The first method is the use of a **trickling filter**.

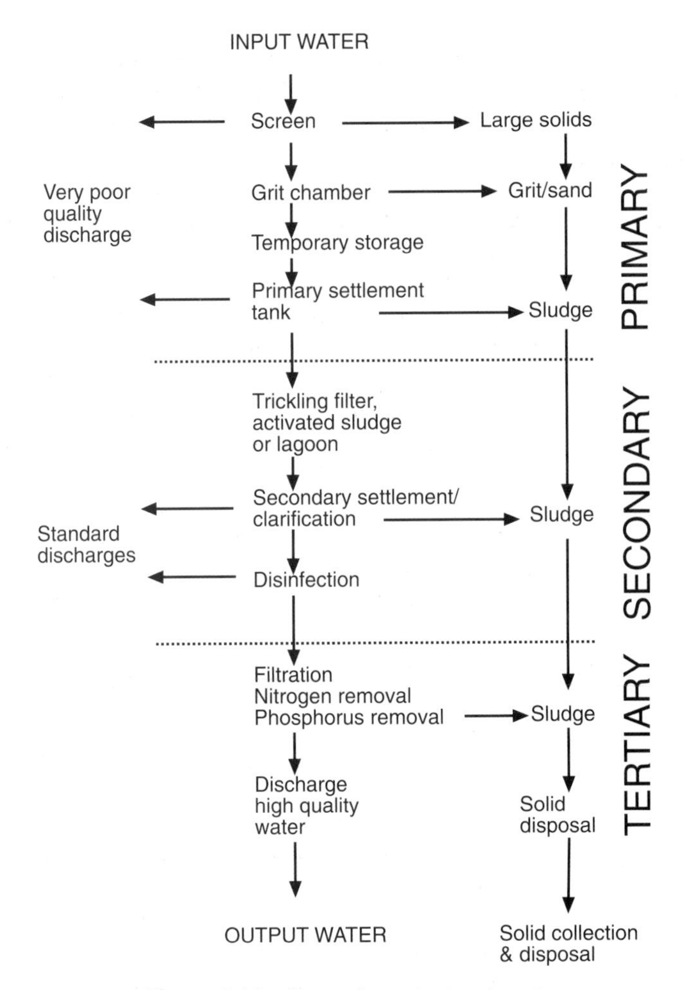

Figure 9.18 *Stages in water treatment*

These are large tanks, 5–20 m in diameter, standing 1 to 2 m above the ground which are filled with a porous and permeable mass of crushed rock or plastic rings. This filling provides a large fixed surface area that is colonised by mats of aerobic bacteria. Water is sprinkled from a rotating boom at a rate just sufficient to maintain a film of surface water on the fill. The bacterial mats 'feed' on the organic-rich water, improving its quality. Having passed through the trickling filter the water is then placed in settling tanks (**clarifiers**), where any remaining solids or biomass is removed from suspension.

The second method is through a process known as **activated sludge treatment**. In this treatment, sewage is mixed with a bacterial sludge in a tank, and is aerated and stirred for six to eight hours. Aeration is achieved by bubbling air through the mixture and more than 8 m^3 of air is required to treat 1 m^3 of sewage. This process causes a very rapid growth of bacteria, producing a large amount of biomass. Stirring helps aggregate the biomass into large flocks. The treated water is then removed, and about 70% of the sludge is also removed. The rest is retained to inoculate the next batch of sewage with bacteria.

The third method involves shallow (1.5 m deep) **oxidation lagoons** in which water is stored for up to six months. The water becomes stratified, with an upper oxygenated layer in which aerobic decay occurs, and a lower anoxic zone. The high nutrient content of the water stimulates algal growth. Aerobic bacteria feed on the dissolved organic material removing it as biomass, carbon dioxide and water. Decomposition occurs anaerobically at depth causing acid fermentation and the production of methane. The oxygenated zone gradually migrates downwards as the aerobic bacteria consume the available food, thereby causing the water to clear. However, evaporation losses may be considerable and careful disposal of the sludge is required owing to its anoxic state and consequently noxious smell.

Whichever method is used, the final stage of secondary treatment is disinfection to kill pathogenic bacteria and other pathogenic micro-organisms. At this stage almost all of the dissolved inorganic material has been removed.

3. **Tertiary treatment**. At this stage the waste water contains small amounts of resistant organic compounds, very fine suspended organic particles, inorganic sludges such as nitrogen, nitrates, ammonia, phosphates and a variety of metal salts. Filtration, using multi-media granular filters, is used to remove all the remaining organic suspended material. These filters are composed of beds of garnet sand (lowest layer), quartz sand (middle layer) and ground coal (top layer) and the water is simply passed through them. Resistant organic compounds in a dissolved state are removed by passing over freshly heated charcoal or coke. The organic compounds become attached to the charcoal chemically as the water passes over the carbon. Periodic heating of the carbon is required to drive off the adsorbed carbon compounds and therefore reactivate the adsorption properties of the charcoal. Phosphate ions are removed by adding iron chloride, aluminium sulphate or calcium hydroxide causing them to be precipitated as iron, aluminium or calcium phosphates. The chemicals are added to the water which is then held in a vessel while the precipitates settle out. Finally, nitrogen may be removed using denitrifying bacteria which thrive in anoxic conditions. An anoxic reaction chamber is therefore required along with some organic food, usually methanol, to promote bacterial growth.

The level of treatment sewage receives varies dramatically from location to location, dependent on the requirements of local legislation. For example, in Britain, sewage is simply screened and pumped out to sea at many coastal towns, causing concern over safe bathing facilities. In general, sewage is rarely treated beyond a secondary stage. With increasing environmental concerns, however, the tertiary treatment process is of growing importance.

Where an expensive infrastructure is not available, secondary and tertiary treatment can be achieved by spreading sewage effluent onto dedicated land, a process which has been practised at sewage farms for many years. There are three main approaches. The first involves the slow irrigation of water, at a rate equivalent to rainfall at 3 to 10 mm per day, to crops or grass on permeable soils. Flow rates should be low enough to prevent ground saturation. Aerobic biological degradation of organic matter takes place in the soil while the growing plants remove some of the phosphorus and nitrogen. Free drainage is essential to prevent anoxic conditions developing within the soil. The second method involves saturating a sloping soil surface to such an extent that a surface film of water is maintained, which then moves downslope. A large surface area of water per unit depth is presented to the atmosphere and therefore the water remains oxygenated, allowing aerobic decay. Impermeable soils may assist this process. The final method involves the use of reed swamps and lagoons. Here, cultivated reeds trap organic matter, consume phosphorus and nitrogen and also adsorb or precipitate some metals. The water is then passed into lagoons when oxygenation by algal photosynthesis further reduces the organic content of the water.

9.7.2 Disposal of Treated Effluents

Two products result from all types of effluent treatment, whether primary, secondary or tertiary, and are formed at all stages of the process: (1) clean water, and (2) sludge. **Clean water** is an obvious end product of effluent treatment systems, and it is usually discharged into river systems or pumped out to sea. The level of impact which it has on the fluvial or coastal system depends on the degree to which it has been processed and on the degree of dilution by the receiving flow. If poorly cleansed, concentrated bacterial decay may cause a reduction in the oxygen content of the receiving water, causing adverse effects on its ecology. This reduction does not reach its maximum value at the point of inflow but at a point downstream, since the bacterial decay of its organic content, which consumes oxygen, takes time. Beyond this critical point the level of oxygen will rise as the organic content is removed (Figure 9.19). The impact of sewage outflow will vary seasonally with river discharge. Sewage outflow is more or less continuous and at low flows the level of dilution by river water may be minimal, causing severe problems for river ecology. Coastal discharges, particularly in Britain where the level of processing is often minimal, may cause severe problems during calm conditions. Sewage water is fresh and therefore floats on seawater and will rise to the surface beyond the outfall pipe. Wave action gradually mixes the two water bodies, but in calm weather slicks of sewage may form on the surface. Tidal currents acting alongshore may transport such slicks to neighbouring areas.

Sludge produced by water treatment also requires careful disposal and some form of processing is usually required. Initially the sludges contain between 2 and 10% solids, and this percentage needs to be increased prior to processing.

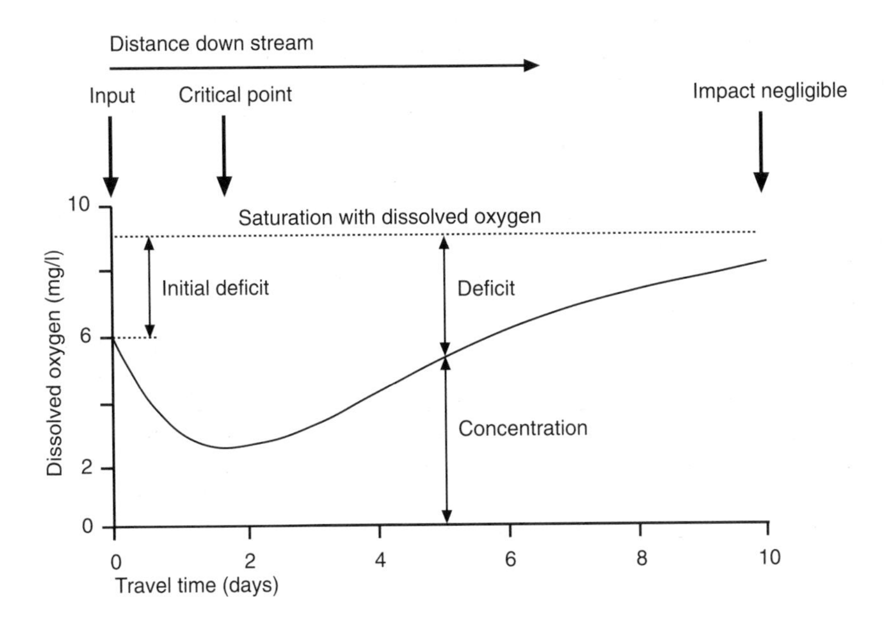

Figure 9.19 *Changes in the oxygen demand downstream of a sewage input*

Water treatment sludges are difficult to handle as they are both fluid and prone to rapid anaerobic deterioration, thereby giving off offensive odours. The basic processes of sludge treatment include: (1) thickening by gravity drainage to increase the percentage of solids; (2) stabilisation or digestion through the utilisation of aerobic and anaerobic bacteria to reduce the organic content of the sludge; (3) conditioning, which involves treatment of the sludge with chemicals and heat to release water; (4) dewatering, using vacuum pumps or evaporation, and; (5) the disposal or incineration of the remaining sludge.

Of these processes, the most important involves the digestion or biochemical decomposition of the sludge. Both aerobic and anaerobic bacteria are used. Aerobic reactors are similar to activated sludge tanks except that residence times are higher so that carbon dioxide and water are the dominant products. Anaerobic reactors utilise a class of bacteria that obtain all their metabolic energy by the breakdown of complex carbon molecules to give methane and water. These bacteria only live in oxygen-free environments. This process takes place in enclosed oxygen-free reactors which are heated to between 20 and 35°C to maximise the bacterial growth. Methane produced as a by-product is then collected and used as fuel. Some components may not be decomposed and residual sludges may need removal for disposal in landfill sites.

The dewatered and digested sludge remains still have to be disposed of in some way. This material is relatively inert and may be placed in landfills or used as a base for horticultural composts. Dewatered sludges which have not been subject to bacterial digestion are sometimes incinerated and have been used directly as a fertiliser on agricultural fields, although in recent years there has been increasing concern about the presence of heavy metal contaminants. London's sewage sludge has been dumped in the Thames Estuary for a number of years.

9.8 WASTE GASES AND THE ATMOSPHERE

9.8.1 Waste Gases and Particulates

Waste gases are produced by a wide variety of industrial, commercial and domestic processes, and are commonly released into the atmosphere along with particulate matter (smoke), which is mostly produced by burning of substances such as fossil fuels. The most important atmospheric waste producers are coal- and oil-fired power stations, industrial processes and road traffic, although in the past the widespread domestic use of solid, hydrocarbon-based fuels was particularly important in the production of the smoke and particulate-laden atmospheres characteristic of the industrialised nations during the late nineteenth and early twentieth centuries.

Management of such potential pollutants has mostly been on the dilute-and-disperse principle, and this has led to a widespread and clearly recognisable overload of the natural system which has set its mark on the post-Industrial Revolution world. Although some gases, particularly from power stations, are condensed in order to reduce the emission of particulate matter into the atmosphere, the majority of emissions are unchecked. Up until recently, this has been particularly true of emissions from motor vehicle exhausts, although all new cars are now made with catalytic converters as standard, which is expected to cut down on the emissions of certain gases.

There are five primary waste gas and particulate emissions into the atmosphere: (1) **carbon monoxide**, a common, naturally occurring gas which is also derived from fires and car exhausts; (2) **nitrogen oxides**, and particularly **nitrogen dioxide**, an important component of smog, nearly all of which are derived from the burning of fossil fuels and from car emissions; (3) **sulphur dioxide**, a gas derived from coal burning and industrial processes such as oil refining; (4) **volatile organic compounds** (**VOC**), particulates derived from car exhausts and industrial solvents; and (5) **smoke**, particulates of mostly carbonaceous composition derived from the burning of coal and other hydrocarbons. All of these materials contribute to the pollution of the atmosphere through both primary and secondary means. Primary effects are a direct result of the emission, while secondary effects are a result of the reaction of the pollutants with atmospheric gases. The emissions of smoke and sulphur dioxide in many of the northern industrial nations have been falling in response to efficient 'clean air' legislation which has sought to restrict the effects of the dilute-and-disperse emissions from power stations and industrial plants, and which have led to the introduction of catalytic converters to check the unmanaged emissions from car exhausts. Such activities have, however, led to an increase in the emission of carbon dioxide, now recognised as an extremely important 'greenhouse gas', and this is becoming one of the most important concerns of atmospheric pollution.

Over the last 20 years, there has been a growing realisation that this area of human activity is likely to cause significantly enhanced global warming, which in turn may lead to climate change and ultimately sea level fluctuations. This phenomenon is of concern on a human timescale as it is likely to cause marine inundation of coastal regions, with a consequent need for increased coastal management. These factors are discussed in detail below.

9.8.2 Atmospheric Pollution and Climate Change

The greenhouse effect is well known as a natural phenomenon of the Earth's atmosphere, which causes modulation of surface air temperature, and which ultimately has been implicated as a natural cause of global climate warming. This effect results from the action of the atmosphere as an insulating blanket which is more transparent to short wave radiation from the sun than it is to long wave radiation emitted by the cooler Earth. In this way a significant proportion of the heat obtained from the sun is retained and maintains a stable Earth surface temperature. The temperature of the Earth's surface is controlled by the insulating property of the atmosphere which is a function primarily of the amount of moisture, carbon dioxide and methane present, and these so-called 'greenhouse gases' have varied naturally through geological time. For example, an increased period of volcanic activity may lead to an increase in atmospheric carbon dioxide, while the weathering of fresh crystalline rock following major mountain-building episodes may actually reduce atmospheric carbon dioxide. Natural fluctuations in the amount of carbon dioxide within the atmosphere may have been the primary cause of the onset of global warming or cooling in the geological past. Atmospheric concentrations of carbon dioxide are controlled by the carbon cycle (Figure 9.20) which involves the flux of carbon between different reservoirs. This system has a natural equilibrium, which is increasingly disrupted by the release of carbon dioxide into the atmosphere through the burning of fossil fuels (Figure 9.20).

However, carbon dioxide is not the only greenhouse gas. People have introduced into the atmosphere a range of halocarbons, in particular chlorofluorocarbons, which are especially effective as greenhouse gases (Figure 9.21). The release of methane into the atmosphere is also of importance, particularly so since global warming in high latitude regions may cause the release of frozen biogenic methane, otherwise known as methane clathrates, which are held in the permafrost of arctic regions. The release of this natural methane into the atmosphere as a result of even a small amount of warming could dramatically accelerate, through a positive feedback loop, the rate of global warming (Box 8.1).

The impact of the human-induced release of greenhouse gases into the atmosphere has been a rise in the global surface temperature of between 0.3 and 0.6°C over the last 100 years. Documenting this subtle rise is difficult and

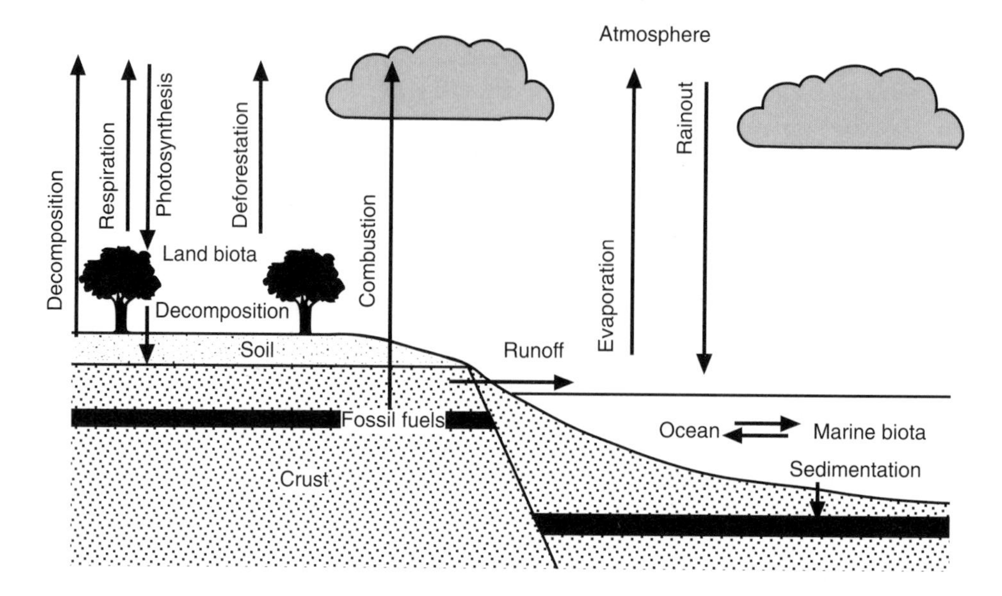

Figure 9.20 Conceptual diagram of the main components of the carbon cycle [Modified from: Woodcock (1994) Geology and environment in Britain and Ireland. UCL Press, Fig. 17.7, p. 138]

predicting future increases is even harder, particularly in the face of sceptical public and political opinion. Figure 9.22 shows a series of projections of future warming, should the emission rates continue to increase at their current levels.

9.8.3 Atmospheric Pollution and Sea Level Change

During Earth history, global sea level has fluctuated dramatically as plate tectonics has changed the volume of the ocean basins, and as ice sheets have grown and decayed. Should ice sheets return to the mid-latitudes in the future, which seems likely given the regularity of the glacial–interglacial cycles characteristic of the Cenozoic Ice Age, then sea level will fall, revealing much of the continental shelf as it did during the last glacial maximum, just 20 000 years ago. Despite this, most concern today is focused on the environmental effects of an increase in sea level which may result from global warming, particularly since it has been estimated that approximately 75% of the world's population live within 1.5 km of the coast. Global warming may affect sea level in two ways: (1) through a decrease in global ice cover; and (2) through the thermal expansion of the oceans.

If the Antarctic and Greenland Ice Sheets, which contain 99% of ice on Earth, were to melt, the resultant global sea level rise would be in the order of 70 m – the Antarctic Ice Sheet alone would contribute 65 m of this rise. Melting the remaining 1% of the Earth's ice, contained in mountain ice fields, ice caps and

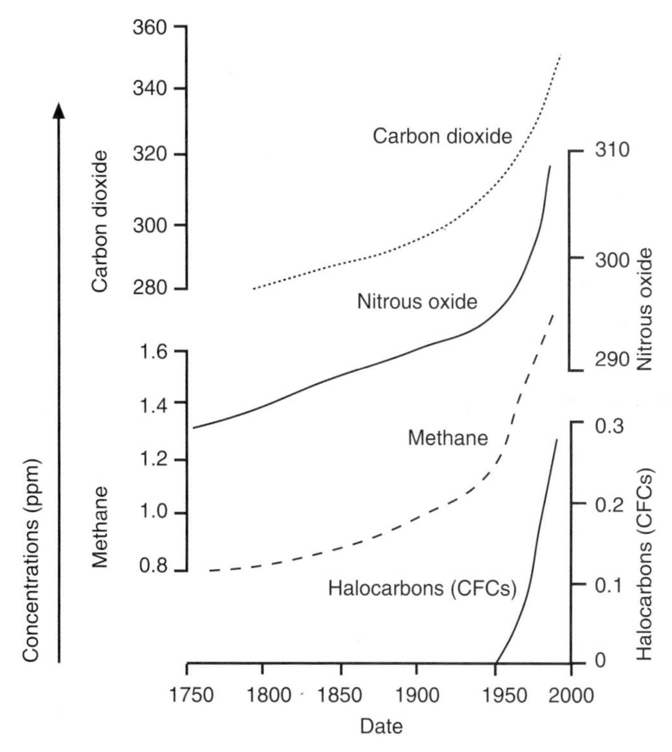

Figure 9.21 *Concentrations of some important greenhouse gases through time [Modified from: Woodcock (1994)* Geology and environment in Britain and Ireland. *UCL Press, Fig. 17.5, p. 137]*

valley glaciers, would lead to an increase of the order of 0.3 to 0.6 m. It is an aspect of popular conception that there is a simple causal link between global warming, ice sheet melting and sea level rise. Such views are apparently supported by the rapid retreat of most mountain ice fields and valley glaciers since the Little Ice Age of medieval times – a time of widespread glacial expansion. In practice, however, glacier–climate interaction is complex, and no simple causal linkage between global warming and glacier volume can be established. Global warming may in fact cause Antarctica to grow rather than decay, at least in the short term, by increasing the snowfall it receives.

Global warming may cause sea level to rise directly through the thermal expansion of ocean water. The density of seawater, and therefore its volume, can change significantly with temperature. At its present salinity, around 35‰, seawater has a maximum density at close to 0°C. Therefore as temperature increases, density decreases and seawater expands. The rate of expansion is controlled by the thermal expansion coefficient which varies from 114×10^{-4} per °C at a temperature of 5°C, typical of high latitudes, to 297×10^{-6} per °C in the tropics (25°C), assuming a salinity of 35‰. This would correspond to a 0.57 or 1.46 m rise in sea level if a 500 m deep layer of the ocean was warmed

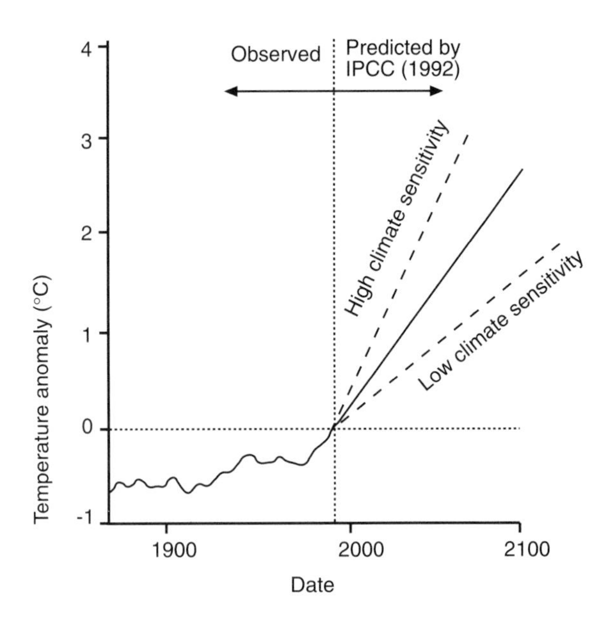

Figure 9.22 *Observed and predicted trends of temperature change caused by global warming*

uniformly. In practice, the effect of ocean expansion may be very complex. The rate of mixing between surface and deeper ocean water is critical. High rates of mixing will slow the rate of global cooling owing to the thermal inertia of the oceans. However, efficient mixing will increase thermal expansion and sea level rise as more heat is dispersed throughout the ocean depths.

Obtaining reliable data on global (eustatic) sea level is difficult owing to the variety of regional factors which may affect relative sea level, such as tectonic subsidence, uplift, or glacio-isostatic rebound. However, as Figure 9.23 shows, there is some evidence to suggest that there has been an increase in global or eustatic sea level during the last 100 years. This is probably due to a combination of thermal expansion of ocean water and the melting of small mountain glaciers.

A variety of sea level projections have been made using different computer simulations, some of which are shown in Table 9.4. They range from zero to just over 3 m during the next 100 years, although most estimates seem to suggest a rise of the order of 0.5 m.

The magnitude of a sea level rise may vary from one location to the next depending upon the nature of the geological environment. It is not sufficient to select a particular scenario of, say, a 50 cm rise, then select an appropriate contour and simply assume that all land below this will be flooded. For example, in a deltaic region a rise in global sea level may be offset by increased sedimentary deposition, effectively increasing the land level. The Bangladesh (Ganges–Brahmaputra–Meghna) Delta, for example, may not be at risk from marine flooding but from inland flooding, which could have a significant

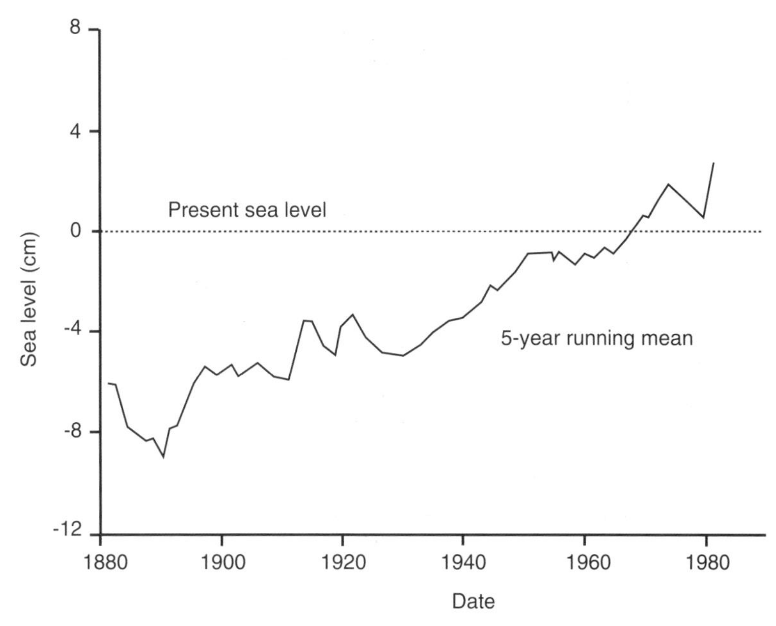

Figure 9.23 *Sea level variation during the twentieth century: evidence for sea level rise*

Table 9.4 *Estimates of future global sea level rise (cm) [Modified from: Warrick (1993) In: Warrick* et al. *(Eds)* Climate and sea level change. *CUP, Table 1.1, p. 13]*

	Contributing factors					
Estimate	Thermal expansion	Alpine glaciers	Greenland icesheet	Antarctic icesheet	Total rise (cm)	By (year)
1	20	—— 20 (combined) ——			40	2050
2	30	12	13		71	2080
3	28–115	—— 28–230 (combined) ——			56–345	2100
4		10–30	10–30	−10–100	10–160	2100
5	28–83	12–37	6–27	12–220	58–367	2100
6	28–83	14–35	9–45	13–80	100	2100
7	4–18	2–19	1–4	−2–3	21	2030
8	31	15	0	0	61	2087

adverse effect on the agricultural productivity of this region. The rivers of this delta complex are dynamic and a sea level rise is likely to be matched by a rise in the river levees and by coastal sedimentation. Relative sea level would be maintained, but inland regions behind the river levees would remain at a lower elevation, increasing the risk from inland flood waters which cannot drain into the raised rivers and coastal areas. The design of effective defence measures against this flooding is difficult owing to the need for pumping of flood water. In other regions the extent of the flood problem associated with rising sea level may be amplified by regional conditions. For example, in the Salado Basin of

Argentina, raised sea levels are likely to cause significant increases in groundwater levels, impeding the drainage of inland flood water and thereby increasing the area likely to be inundated by flood water during storm events. Biological response to rising sea level may reduce its impact. For example, rising sea level may not threaten low-lying tropical islands, since land loss may be compensated, up to a point, by the growth of coral reefs and by sediment entrapment within mangrove swamps. In summary, therefore, the impact of a global sea level rise may vary dramatically given the nature of the geological environment, and consequently understanding the dynamics of a particular coastal system is vital in predicting the likely impact of a sea level rise.

Human action within coastal regions may amplify the impact of rising sea level. For example, in Hong Kong, widespread urbanisation on reclaimed land has caused subsidence which has increased the risk of inundation by rising sea level and by storm surges. Equally, the pollution and mining of coral as a construction material in the Republic of the Maldives threatens to disrupt the natural capacity of these islands to respond to rising sea level.

It is possible to identify five major impacts of raised sea levels, each of which is discussed below.

1. **Land loss**. This may be of considerable importance in areas of extreme or expanding population density.
2. **Increased coastal erosion**. Deeper coastal waters will allow the onshore penetration of larger storm waves. At present around 70% of the world's sandy beaches are experiencing erosion, a problem which it is widely suggested will be exacerbated by continued rises in sea level, although relatively little is currently understood about onshore wave-induced sediment transport. Changes in coastal water depth will also change tidal distributions and associated currents. The retreat of unconsolidated coastal cliffs is also likely to accelerate under conditions of rising sea level owing to the increased wave activity made possible by deeper coastal waters.
3. **Increased flooding hazard of lowland regions**. Increases in sea level will increase the area prone to coastal flooding and will necessitate the upgrading of coastal flood defence works to accommodate water levels which may exceed the design criteria used in their construction. The impact of storm surges is likely to be more extensive, since it has been suggested that global warming may increase the frequency of tropical storms and the areas subject to them, although this has yet to be proven. In areas such as the Bay of Bengal, where tropical storm surges are a significant coastal flood hazard already, the potential for an increase in the scale of natural disaster due to increasing the potential inundation area is considerable.
4. **Salinisation of coastal aquifers**. Rising sea level is likely to increase coastal penetration of saline groundwater. On some sandy coastal islands, such as those on the Dutch coast, this may pose a significant threat to the water supply. An increased water table may also have an adverse effect on the drainage of inland flood waters, exacerbating fluvial flooding in coastal regions.

5. **Inland migration and potential loss of coastal ecosystems.** Rising sea level will cause the inland migration of sedimentary and ecological coastal habitats. This is particularly important for wetland ecosystems on low-lying coasts. Where sufficient space is available these ecosystems and depositional environments will simply move inland. However, where space is restricted, either by high land, embankments or coastal defence works, such environments may simply disappear. Not only is this of ecological significance, but it may also adversely affect coastal defences. Salt marshes, mud flats and other similar environments are important coastal sediment sinks and their loss from the coastal system may lead to sediment depletion within the coastal cell. Equally, such environments often provide a natural coastal buffer between the sea and coastal defence works which may not have been designed to cope with direct wave attack.

9.9 SUMMARY OF KEY POINTS

- Waste refers to any substance which constitutes a scrap material, effluent or unwanted surplus material arising from the application of a process, or any substance or article which requires disposal through being broken, worn-out, contaminated or otherwise spoilt. It may be either managed or unmanaged. Unmanaged waste is a frequent source of pollution.
- Pollution is the introduction by people into the environment of a substance or energy which is liable to cause hazards to human health, harm to living resources and ecosystems, damage to structure and amenity, or interference to legitimate uses of the environment.
- Contamination is the introduction or presence in the environment of alien substances or energy which may or may not be hazardous to human health or damaging to the natural environment.
- Waste disposal may be managed either by the principle of dilute-and-disperse or by concentrate-and-containment.
- Waste may be disposed of either in open dumps or in some form of landfill site of which there are two types: (1) landfill, where the waste is placed in a void within the ground; and (2) landraise, where the waste is used to create a positive feature in the landscape.
- Landfill sites may be designed on the basis of: total containment, in which all the noxious by-products and pollutants are retained; or slow dispersal, where noxious substances are released slowly to the environment, ideally at a rate less than that at which it is able to sanitise the released material.
- The selection of a landfill site involves three stages or levels of decision-making: (1) identification of the need for a landfill site within a given area, which is usually in line with a strategic waste disposal plan for the region; (2) selection of a suitable site from the range of possible options within the chosen search area; and (3) assessment of the likely environmental impacts associated with the landfill of the chosen site.

- Disposal of radioactive waste is based on the principle of concentration and total containment. The preferred option at present is disposal within deep underground mines or caverns. There are three important considerations in site selection: (1) the prediction of likely patterns and consequences of environmental change, due to the very long timescales over which the integrity of the site must be maintained; (2) prediction of natural hazards such as seismicity, which might threaten site containment; and (3) the material properties of the site, in particular groundwater flows and lithological characteristics.
- Effluent, liquid waste, is dominated by sewage disposal. Sewage treatment involves a process of filtration and bacterial decomposition.
- In recent years there has been growing concern about atmospheric pollutants due to their role in global warming. The principal geological impact of global warming is sea level rise due to the ablation of glaciers and the thermal expansion of oceans. There are five main consequences of rising sea level: (1) land loss; (2) accelerated coastal erosion; (3) increased vulnerability of coastal areas to marine flooding; (4) salinisation of coastal aquifers; and (4) inland migration and loss of ecosystems.

9.10 SUGGESTED READING

There are several good working practice guides to waste management. These are technically presented, but provide a good overview of current working practice; typical examples include Anon (1991), Daly & Wright (1992), Attewell (1993), Department of the Environment (1990, 1992). The selection of landfill sites is covered by Gray *et al.* (1974) and Cartright (1982). Good case histories on landfill are provided by Raybould & Anderson (1987), Williams & Aitkenhead (1991), Eyles *et al.* (1992) and Andriashek *et al.* (1993). The problem of contaminated land is covered by Cairney (1993) and Bell *et al.* (1996). Several papers which deal with contamination and pollution control can be found in the volumes edited by Culshaw *et al.* (1987) and by Lumsden (1994). Radioactive waste disposal is covered by Testa (1994), Baillieul (1987), Williams & St Ivany (1987), Krauskopf (1988), Arnould *et al.* (1993) and NIREX (1996). A very readable, if a little dated, account of climate change is provided by Gribbin (1978), while the definitive work of the IPCC is summarised by Houghton *et al.* (1990, 1992). The problems of sea level rise are covered in a series of papers within the volume edited by Warrick *et al.* (1993).

Anon, 1991. *Croner's waste management.* Croner Publications, Kingston upon Thames.

Andriashek, L.D., Thomson, D.G. & Jackson, R. 1993. Edmonton landfill site investigation: a case study applying hydraulic information to interpret a glacially disturbed site. *Geoscience Canada*, **20**, 157–164.

Arnould, M., Furuichi, T. & Koide, H. (Eds) 1993. *Management of hazardous and radioactive waste disposal sites.* Special issue, *Engineering Geology*, **34**(3/4).

Attewell, P. 1993. *Ground pollution. Environment, geology, engineering and law.* E & FN Spon, London.

Baillieul, T.A. 1987. Disposal of high-level nuclear waste in America. *Bulletin of the Association of Engineering Geologists*, **24**, 207–216.

Bell, F.G., Duane, M.J., Bell, A.W. & Hytiris, N. 1996. Contaminated land: the British position and some case histories. *Environmental & Engineering Geoscience*, **2**, 355–368.

Cairney, T. (Ed.) 1993. *Contaminated land. Problems and solutions.* Blackie, London and Lewis Publishers, Boca Raton.

Cartright, K. 1982. Selection of waste disposal sites. *Bulletin of the Association of Engineering Geologists*, **19**, 197–201.

Culshaw, M.G., Bell, F.G., Cripps, J.C. & O'Hara, M. (Eds) 1987. *Planning and engineering geology*. Geological Society, London.

Daly, D. & Wright, G.R. 1992. *Waste disposal sites. Geotechnical guidelines for their selection, design and management*. Geological Survey of Ireland Information Circular 82/1, Geological Survey of Ireland, Dublin.

Department of the Environment, 1990. *Landfilling wastes. A technical memorandum for the disposal of wastes on landfill sites*. Department of the Environment Waste Management Paper, No.26, HMSO, London.

Department of the Environment, 1992. *Landfill gas. A technical memorandum providing guidance on the monitoring and control of landfill gas*. Department of the Environment Waste Management Paper, No.27, HMSO, London.

Eyles, N., Boyce, J.I. & Hibbert, J.W. 1992. The geology of garbage in southern Ontario. *Geoscience Canada*, **19**, 50–62.

Gray, D.A., Mather, J.D. & Harrison, I.B. 1974. Review of groundwater pollution from waste disposal sites in England and Wales, with provisional guidelines for future site selection. *Quarterly Journal of Engineering Geology*, **7**, 181–196.

Gribbin, J. 1978. *Climate change*. Cambridge University Press, Cambridge.

Houghton, J.T., Jenkins, G.J. & Ephraums, J.J. 1990. *Climate change: the IPCC scientific assessment*. Cambridge University Press, Cambridge.

Houghton, J.T., Callander, B.A. & Varney, S.K. 1992. *Climate change 1992: the IPCC scientific assessment*. Cambridge University Press, Cambridge.

Krauskopf, K.B. 1988. *Radioactive waste disposal and geology*. Chapman & Hall, London.

Lumsden, G.I. (Ed.) 1994. *Geology and the environment in Western Europe*. Oxford University Press, Oxford.

NIREX, 1996. *Geology and hydrogeology of the Sellafield area*. Proceedings of the NIREX seminar, 11 May 1994. *Quarterly Journal of Engineering Geology*, **29**, Supplement 1.

Raybould, J.G. & Anderson, D.J. 1987. Migration of landfill gas and its control by grouting – a case history. *Quarterly Journal of Engineering Geology*, **20**, 75–83.

Testa, S.M. 1994. *Geological aspects of hazardous waste management*. Lewis Publishers, Boca Raton.

Warrick, R.A., Barrow, E.M. & Wigley, T.M.L. 1993. *Climate and sea level change*. Cambridge University Press, Cambridge.

Williams, G.M. & Aitkenhead, N. 1991. Lessons from Loscoe: the uncontrolled migration of landfill gas. *Quarterly Journal of Engineering Geology*, **24**, 191–207.

Williams, J.H. & St Ivany, G. 1987. Geologic aspects of hazardous waste site investigations in Missouri. *Bulletin of the International Association of Engineering Geologists*, **24**, 43–51.

10
The Geology of Natural Hazards

The natural environment kills and disrupts human infrastructure, and it does so frequently. Natural hazards range from high magnitude, low frequency events, such as earthquakes and volcanic eruptions which cause major loss of life, to low magnitude, high frequency events such as soil, fluvial and coastal erosion which rarely threaten life, but which cause widespread economic loss. It has been estimated that in the last 25 years in excess of three million people have been killed and over $1000 billion worth of damage has been caused by natural hazards. Any discussion of such hazards tends to be dominated by the dramatic life-threatening events, and yet it is the more sedate, cumulative hazards that rarely threaten life which have the greatest global impact. For example, it has been estimated that natural hazards affect the lives of over 100 million people each year in some respect, either through the disruption of transport links or by the loss of property.

A **geological hazard** can be defined as any geological or geomorphological event which has an adverse socio-economic impact on the human use system. Geological hazards are naturally occurring phenomena which only become hazardous due to the presence of human infrastructure. For example, coastal erosion is a natural process which is vital to replenish nearshore sediment, and only becomes a hazard when property is at risk. Not only is the presence of human infrastructure of importance in the consideration of hazards, but also its vulnerability to destruction. The distribution and impact of natural disasters is unequal, with the greatest loss of life being concentrated in less developed and developing countries (Table 10.1). This is not due to a greater hazard frequency, but simply a greater vulnerability; richer, developed countries can respond more efficiently, minimising loss of life, although the economic loss may be much greater owing to the greater level of structural investment.

The frequency of major disasters appears to have increased since 1960. The number of people killed each year has increased at a rate of 6% and the amount of economic loss has also risen steadily (Figure 10.1). This has occurred against a

Table 10.1 *Distribution of natural disasters as a function of the economic development of countries [Modified from: Lumsden (1994)* Geology and the environment in Western Europe. *Oxford University Press, Table 4.3, p. 141]*

GNP of country	Number of disasters	Number of victims	Percentage of world total
Low	329	1 090 900	76
Medium	451	335 000	23.3
High	79	10 000	0.7
Total	859	1 435 900	100

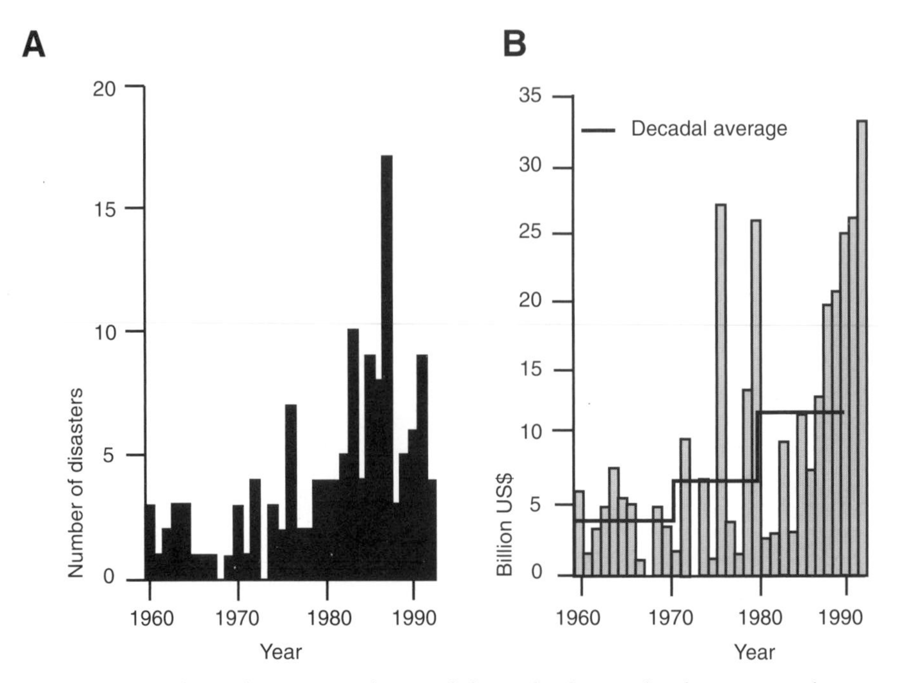

Figure 10.1 *Trends in frequency of natural hazards during the late twentieth century [Modified from: Rosenfeld (1994)* Geomorphology, **10**, *Figs 1 & 2, pp. 28 & 29]*

background of increasing hazard awareness and management. For example, in Japan storm-triggered landslides destroyed more than 130 000 homes and killed in excess of 500 people in 1938, a figure reduced to 2000 homes and 125 lives in 1976 as a consequence of a comprehensive management strategy (Figure 10.2). The increase in the frequency of natural hazards can be assigned to four basic factors: (1) increasing reporting as news networks extend globally and enter more remote regions; (2) increasing world population, which has doubled since 1960; (3) the increasing concentration of people into smaller and smaller areas – it is estimated, for example, that by the year 2000, 50% of the world's population will live on less than 0.4% of the Earth's surface; and (4) increasing natural degradation – since the second world war, for example, about 40% of the world's

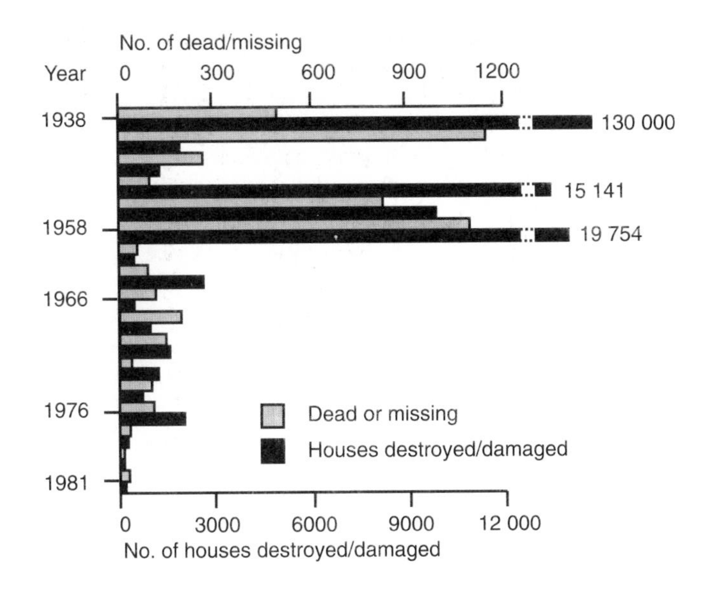

Figure 10.2 *Impact of mass movement hazard management in reducing the number of deaths and level of economic loss in Japan [Modified from: Smith (1992)* Environmental hazards. *First edition. Routledge, Fig. 8.3, p. 171]*

tropical rainforest has been destroyed leading to soil erosion and increased flooding. Natural hazards are therefore of considerable global importance and will continue to be of increasing concern in the future.

Three broad groups of natural hazards can be identified: (1) **atmospheric hazards** caused by atmospheric processes such as tropical storms, hail storms, hurricanes and droughts; (2) **exogenic hazards** caused by the operation of natural Earth surface processes, such as flooding, coastal erosion, mass movement and soil erosion; and (3) **endogenic hazards** which result from internal Earth processes, such as volcanoes and earthquakes. The overlap between these different events is somewhat arbitrary since many floods result from atmospheric storms, and earthquakes may trigger a wide range of exogenic processes. However, it forms a crude basis on which to separate geological hazards (exogenic and endogenic) from atmospheric or meteorological events.

Different hazard types can be profiled against seven basic criteria, namely: (1) **event magnitude**, referring to the amount of energy involved; (2) **frequency of occurrence**, which provides information on how often an event of a given magnitude occurs in a given length of time (Box 10.1); (3) **duration**, the length of time the event lasts; (4) **areal extent**, referring to the physical space affected; (5) **speed of onset**, that is, the amount of time from the first appearance to the maximum intensity of an event; (6) **spatial dispersion**, referring to the space likely to be affected by all hazards of a given type; and (7) **temporal spacing**, given that hazards may occur at regular (i.e. seasonally or in a cycle) or random intervals. Using these criteria, Figure 10.3 shows characteristic profiles for some of

BOX 10.1: HAZARD RECURRENCE INTERVALS: EXTREME EVENT ANALYSIS

Knowledge of the probability of a hazard with a given magnitude occurring in a given time is vital if effective hazard management is to take place. This information is usually expressed as the recurrence interval of a particular event. For example, the one in 50 year flood means that a flood of similar magnitude to the event under discussion will occur once in 50 years or has a 2% probability of occurring in any given year. Recurrence intervals are used to design engineering works, such as flood banks, which are designed to withstand a one in 100 year event. It does not make sense to defend, at increasing cost, against events which are so infrequent that they may not occur within the life span of a building or structure.

Recurrence intervals are usually calculated using extreme event analysis. A temporal record of all the recorded events with their measured or estimated magnitude is assembled and ranked. For example data, usually discharge, on all the flood events on a river in the last 50 years are assembled. These data are then ranked, with a rank of one being given to the large event. For each event the return period is then calculated using the following formula:

$$\text{Return period} = \frac{\text{Number of years of record} + 1}{\text{The rank of a given event}}$$

Return periods can then be plotted, as shown below, against their magnitude. This method is good where long records are available. However, where they are not it is necessary to extrapolate the data set. For example, if we only have 30 years of records we could extrapolate the relationship between recurrence and magnitude to determine the magnitude of the 50 year event. It may be necessary, as shown below, to normalise the data in order to facilitate extrapolation. Extrapolating the data in this way can introduce serious problems into the data set and assumes that the processes generating the event remain constant, which off course they may not.

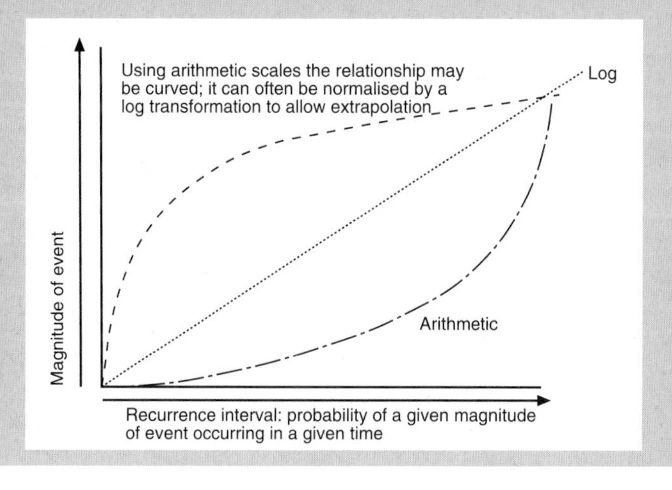

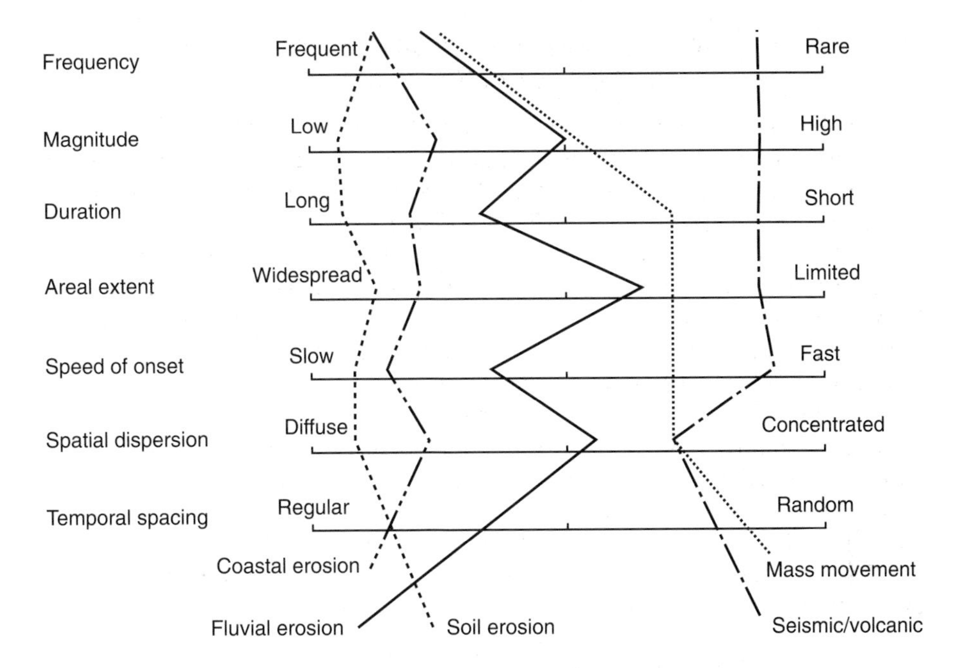

Figure 10.3 *Characteristic profiles for selected natural hazards [Based on data in: Gares* et al. *(1994)* Geomorphology, ***10****, Table 7, p. 14]*

the hazards described in the following chapters. Exogenic hazards are typically of low magnitude and high frequency, long in duration and slow in onset, affect a variable area both singularly and collectively, and often have regular or repetitive occurrence. In contrast, the dramatic endogenic seismic and volcanic hazards are generally of high magnitude but low frequency, of short duration, and affect a relatively small area both singularly and collectively. Figure 10.3 emphasises the importance of the 'quiet' exogenic hazards which tend to affect a large area at regular intervals (e.g. fluvial and coastal erosion) and therefore in many respects achieve a greater human impact than the more dramatic, infrequent and spatially more restricted events (e.g. volcanic and seismic).

10.1 HAZARD AND RISK MANAGEMENT

Hazards are intimately linked to the concept of risk. In addition to the definition already given, a hazard may be defined as the probability of a change in the geological environment of a given magnitude occurring within a specified time period in a given area, while the associated risk is the consequent damage or loss of life, property and services. The risk posed by a hazard is primarily a function of the socio-economic status and vulnerability of the country or region (Table 10.1). There are four main aspects of risk to be considered in dealing with geological hazards.

1. **Risk assessment**. This involves the precise description of the nature of a hazard and the extent and type of exposure to the hazard in various populations. Risk is the product of the probability of occurrence of a natural hazard and its socio-economic consequences, and there are two elements to this: (1) **hazard mapping/zoning** and prediction; and (2) **assessing vulnerability** or degree of loss. Risk assessment can be summarised in the expression: $R_x = P_x X D_x$, where R_x is the risk, P_x is the probability of a hazard reaching a level X which is dangerous during a given period (hazard), and D_x is the probability of a given damage when level X is reached (vulnerability). The value D_x will grow as a town grows and the number of dwellings, infrastructure, production capacity, and population increases.
2. **Risk perception**. Community perception of the hazard is a vital part of the equation. In simple terms, the level of perception is represented by whether the local population feel 'safe' or 'unsafe'. This has a major control on the decision-making process and both the type and success of the response to the hazard. For example, local residents in coastal communities often feel 'safer' after the construction of a concrete sea wall rather than after the simple replenishment or nourishment of the beach.
3. **Risk communication**. This involves social dialogue about the risk between the various stakeholders and it may often be driven by conflict between those who wish to see action and those who do not.
4. **Risk management**. The final component is managing the problems and finding a solution which is acceptable to all parties and which incorporates risk assessment, perception and communication.

Environmental geologists have an important role in all these aspects, particularly in the development of hazard assessments associated with remedial works. The development of risk control or management plans for a given hazard is as important economically as it is socially, and the decision to put in place some form of hazard management is usually controlled by a consideration of **cost-benefit analysis**.

The concept of cost-benefit analysis is illustrated schematically in Figure 10.4. Faced with a hazard, it is possible to compare for each situation the amount of predicted damage C_d with the cost of the damage limitation work C_p which is necessary to limit or reduce the cost to C_d'. Then the benefit (B) to cost (C) ratio of such investment is given by: $B/C = (C_d - C_d')/C_p$. In simple terms, when the cost exceeds the benefit, value for money is not obtained. This may occur either when the amount of damage is minor or where the level of investment required to reduce the hazard is simply too large (Figure 10.4). A cost-benefit analysis for various hazards within California is shown in Table 10.2; the higher the benefit to cost ratio, the greater the financial sense in undertaking remedial work. Cost-benefit analysis is fundamental to hazard planning, but it should be noted that it is very hard to place financial values on many of the elements involved. For example, one can place a financial value on a house, but one cannot place a figure on its value or significance to the owner, especially if their family history is intimately woven into its fabric.

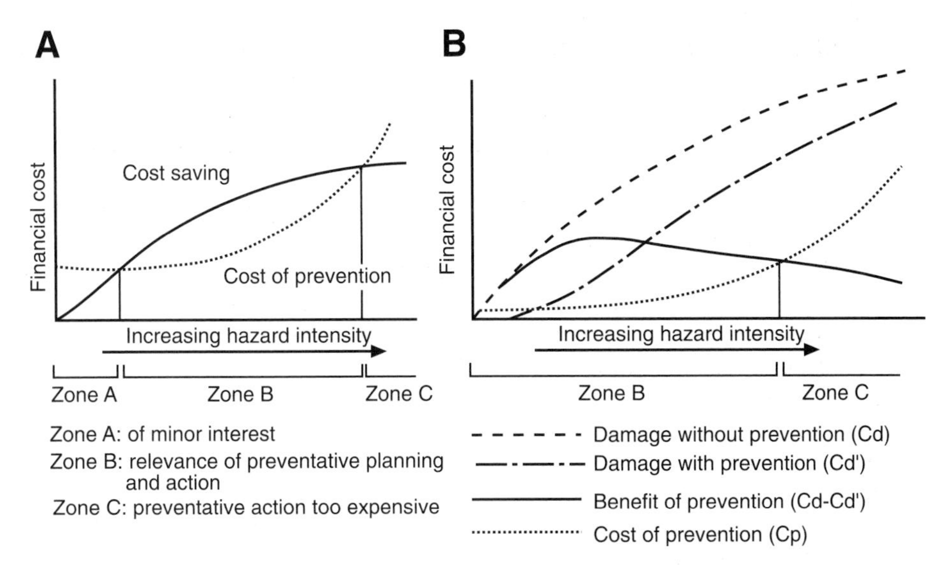

Figure 10.4 *The principles of cost-benefit analysis [Modified from: Lumsden (1992)* Geology and the environment in Western Europe. *Oxford University Press, Figs 4.31 & 4.32, pp. 190 & 191]*

Figure 10.5 shows the various generic types of response or management options to an actual or perceived natural hazard. These are: (1) **defence**, such as forecasting, early warning, evacuation and rescue planning; (2) **prevention**, such as land-use planning to avoid hazardous areas, development of building codes, remedial or defensive works; and (3) **accept and adjust** to the loss. In the development of any response or management plan there are six main stages. Stage One: risk assessment to determine the hazard intensity and vulnerability. Stage Two: evaluation of the management options in dialogue with the stakeholders such as the local residents or other interested parties. Stage Three: choice of solution and design. Stage Four: communication of the solution to the stakeholders and education/promotion of its benefits. Stage Five: implementation of the chosen solution. Stage Six: monitoring of its success.

These concepts will be explored further with regard to the specific hazards discussed in the next two chapters. The first of these chapters looks at 'exogenic hazards' or the subaerial geomorphological hazards, while the second considers 'endogenic hazards' or those from within the Earth. Each one will describe the main hazards and consider their mitigation.

10.2 SUMMARY OF KEY POINTS

- Natural hazards pose a significant threat to lives and human infrastructure.
- Losses due to natural hazards are increasing due to increasing population concentration, and to the degradation of the natural world.

Table 10.2 *Projected losses due to geological hazards in California, for the period 1970 to 2000; costs are in millions of dollars at 1970 value [Modified from: Lumsden (1994)* Geology and the environment in Western Europe. *Oxford University Press, Table 4.16, p. 191]*

Hazard	Projected total losses, without preventative policy (C_d)	Possible total reduction with prevention as a %	Possible total reduction with prevention	Estimated total cost of prevention as a % of total loss	Estimated total cost of prevention	Benefit/cost ratio
Earthquakes	21 035	50	10 517.5	10	2103.5	5
Mass movements	9850	90	8865	10.3	1018	8.7
Flooding	6532	52.5	3432	41.4	2703	1.3
Erosion	656	66	377	45.7	250	1.5
Fault displacement	76	17	12.6	10	7.5	1.7
Volcano	49.38	16.5	8.135	3.5	1.655	4.9
Tsunami	40.8	95	37.76	63	25.7	1.5

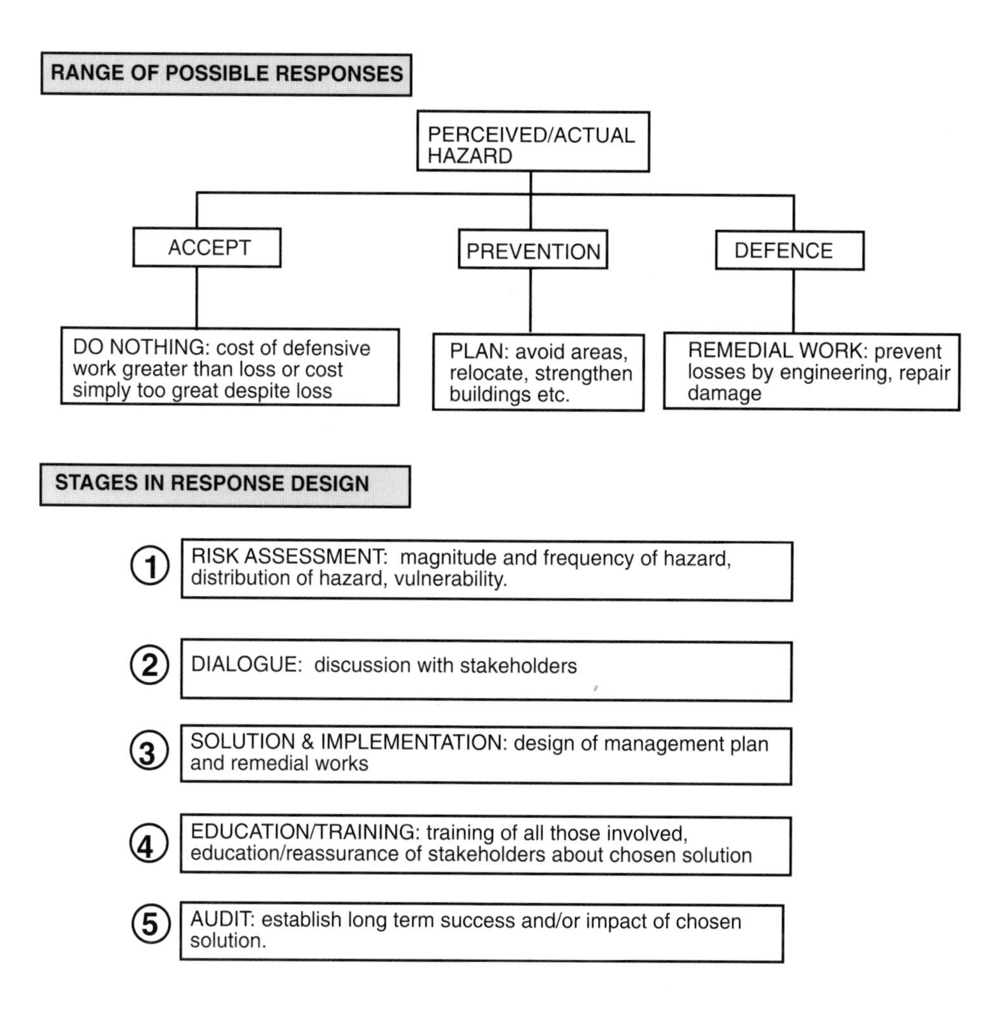

Figure 10.5 *Range of possible responses and generic stages in the design of a response to natural hazards*

- Hazard and risk are intimately linked. A hazard can be defined as the probability of a change in the geological environment of a given magnitude occurring within a specified time period in a given area, while the associated risk is the consequent damage or loss of life, property and services.
- Key aspects of risk management include risk assessment, risk perception, risk communication, risk management.
- An important tool in risk management is cost-benefit analysis in which the cost of damage due to a hazard is compared to the cost of the investment in hazard mitigation.
- Six stages in hazard management exist: Stage One, risk assessment; Stage Two, evaluation of the management options (defend, prevent, accept); Stage

Three, choice of solution and design; Stage Four, communication of the solution; Stage Five, implementation; and Stage Six, monitoring.
- Two types of geological hazard are identified: exogenic and endogenic hazards.

10.3 SUGGESTED READING

In recent years there has been a proliferation of textbooks on natural hazards, including Cooke & Doornkamp (1990), Bryant (1991), Burton *et al.* (1993) and Smith (1996) as well as the edited volumes by McCall *et al.* (1992) and Slaymaker (1996). Gares *et al.* (1994) provide a useful account of the history of hazard research and the role of geomorphology within it. The chapter on geological hazards within the volume edited by Lumsden (1994) is particularly good and gives a clear account of the mechanism of cost-benefit analysis.

Bryant, E.A. 1991. *Natural hazards*. Cambridge University Press, Cambridge.
Burton, I., Kates, R.W. & White, G.F. 1993. *The environment as a hazard*. The Guildford Press, New York.
Cooke, R.U. & Doornkamp, J.C. 1990. *Geomorphology in environmental management*. Oxford University Press, Oxford.
Gares, P.A., Sherman, D.J. & Nordstrom, K.F. 1994. Geomorphology and natural hazards. *Geomorphology*, **10**, 1–18
Lumsden, G.I. (Ed.) 1994. *Geology and the environment in Western Europe*. Oxford University Press, Oxford.
McCall, G.J.H., Laming, D.J.C. & Scott, S.C. (Eds) 1992. *Geohazards: natural and man-made*. Chapman & Hall, London.
Slaymaker, O. (Ed.) 1996. *Geomorphic hazards*. John Wiley & Sons, Chichester.
Smith, K. 1996. *Environmental hazards. Assessing risk and reducing disaster*. Second edition. Routledge, London.

11
Exogenic Hazards

In this chapter we examine those hazards which occur as a result of the operation of natural Earth surface processes. Many of these exogenic hazards are low magnitude, high frequency events which are primarily a product of geomorphology, and which have a wide spatial distribution (Figure 10.3). Their impact is therefore not primarily associated with dramatic local events involving large loss of life (although they can be), but is more pervasive, causing widespread economic loss. In this chapter, we focus on just four exogenic hazards: fluvial, coastal, mass movement and glacial. For each of the chosen hazards we examine its nature, impact and magnitude before reviewing methods of risk assessment, and strategies for hazard mitigation.

11.1 FLUVIAL HAZARDS

Fluvial hazards are among the most common of all exogenic problems. The reason lies in the widespread geographical occurrence of river valleys and their attraction for human occupation. There are two types of fluvial hazard: (1) flooding, and (2) erosion/sedimentation along fluvial channels. Both types have important beneficial properties, such as fertilisation and irrigation of agricultural land, or the regular rejuvenation of river bank habitats, as well as being hazardous.

Flooding is of great importance to many countries. For example, in the USA nearly two-thirds of all the federally declared disasters between 1965 and 1985 were floods, and it has been estimated that 10% of the population in the USA live within flood-prone areas. The problem is perhaps greatest in southeast Asia, where the population is often concentrated in low-lying deltaic areas, living in dwellings which are easily damaged by flood waters. About four million hectares of land and crops are affected by flood water and the lives of over 17 million people are blighted each year within this region. In Bangladesh, for example, 110 million people live mostly unprotected on the floodplain of

Figure 11.1 *Flood hazards*

southern Asia's most flood-prone river system of the Ganges–Brahmaputra–Megna. Monsoon-driven floods frequently cover between 20 and 30% of the country and in 1988, 46% of the country was flooded, killing an estimated 1500 people. In China, over five million people lost their lives to flooding between 1860 and 1960. The economic cost of flooding is also considerable, in terms of both damage and inconvenience to transport or commerce (Figure 11.1). For example, in 1985 the insured losses due to flooding were of the order of $20 million dollars in New Zealand and were in excess of $28 billion dollars in the USA for the decade 1976 to 1985.

A flood can be defined either as a relatively high flow which exceeds the capacity of the natural or artificial channel provided for runoff, or alternatively as a body of water which rises to overflow land which is not normally submerged. In a river, flood levels are recorded by a hydrograph, which is a plot of changes in discharge over time. **Discharge** is the amount of water which passes a point in a given time and is usually measured in cubic metres per second, a unit commonly referred to as the **cumsec**. Discharge is measured on rivers at gauging stations which monitor water level through a measured cross-section which can be used, via rating curves, to provide data on variation in discharge through time. Figure 11.2A shows a typical flood hydrograph, and consists of a rising flood limb as discharge increases to its peak, followed by a falling limb back to the base flow or normal flow level. The area beneath this curve corresponds to the total volume of water in the flood. The shape of the hydrograph is determined by the rate at which water drains from adjacent areas and is a function of the properties of the drainage basin, as set out in Figure 5.1, and those of the channel system. This flood wave will pass downstream and change in shape with the addition of flood water from tributaries (Figure 11.2B). This action may either increase the magnitude of the flood peak or decrease it to become attenuated (Figure 11.2D). A flood is therefore best visualised as a wave of water progressively moving down a river system.

At any point on a given river, the flood hydrograph will always have the same shape, whatever the volume of flood water, provided that the characteristics of the drainage basin and channel system remain constant. This follows since if the components of the drainage basin (Figure 5.1) and channel system remain constant, then the processes and rate with which water drains into the channel and moves downstream should be the same. This concept is known as the **unit hydrograph**. The shape of a unit hydrograph can, however, be altered by changing the characteristics of the drainage basin or channel system. For example, deforestation, drainage or urbanisation will all change the rate of water throughput and storage within a drainage basin (Figure 5.1). Urbanisation has perhaps the most dramatic effect on the flood hydrograph by increasing the drainage density, reducing infiltration and thereby generating runoff more quickly, and by reducing storage capacity within the basin; concrete car parks and pavements are much more effective in generating runoff than grassy fields. Figure 11.2C shows the increasing flood peak on Cannon's Brook near the new town of Harlow, northeast of London, since it was constructed in the late 1950s and early 1960s.

Flood-prone areas are not only associated with low-lying parts of active floodplains and estuaries but are also encountered within small drainage basins in arid and semiarid areas which are subject to flash floods, and in areas below unsafe dams. Small drainage basins in arid and semiarid regions often contain steep topography, little vegetation and are subject to intense rainfall events of short duration which combine to give rise to **flash floods** that occur rapidly and with little warning. Flash floods are now the main cause of weather-related deaths in the USA and in one event in Colorado in 1976, 139 people were

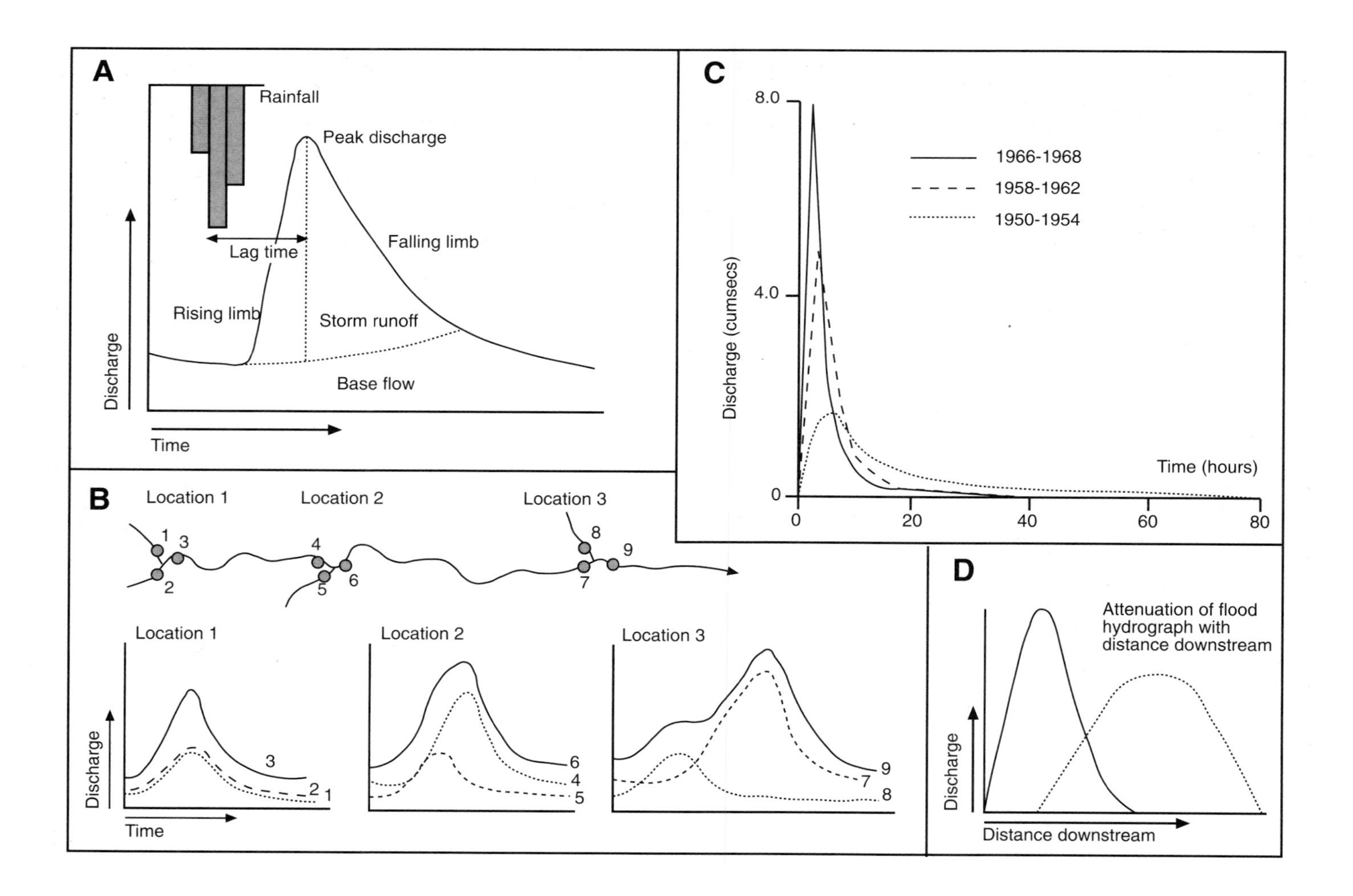
A
Rainfall
Peak discharge
Lag time
Falling limb
Rising limb
Storm runoff
Base flow
Discharge
Time
B
Location 1
Location 2
Location 3
1
2
3
4
5
6
7
8
9
Discharge
Time
C
8.0
4.0
0
Discharge (cumsecs)
1966-1968
1958-1962
1950-1954
Time (hours)
0
20
40
60
80
D
Attenuation of flood hydrograph with distance downstream
Discharge
Distance downstream

drowned in Big Thompson Canyon. Similarly, over 60 people were killed and 189 injured by a flash flood in the Pyrenees in August 1995 which inundated a campsite where over 634 holidaymakers were staying. Floods as a consequence of dam failure are also of considerable importance and in the USA alone over 30 000 dams are considered unsafe, threatening over 2000 communities.

High river flows may be caused by such things as intense rainfall, snow melt and dam failure; however, it is the presence of intensifying or amplifying factors which turns high discharges into floods. The River Conway in North Wales illustrates how flood amplification can occur. Here, extensive flooding is associated with the coincidence of high tides in the estuary and high river discharge. The rising tidal water within the estuary essentially blocks the outflowing discharge from the River Conway, causing it to overtop its banks. At low tide the equivalent discharge would not be sufficient to cause flooding. In this case the amplification is natural, but in many cases it is a consequence of human action, specifically due to such things as land drainage, urbanisation and deforestation.

Flooding as an exogenic problem appears to be on the increase due to three main factors: (1) increasing urbanisation of river catchments, increasing flood peaks (Figure 11.2C); (2) continued development and encroachment into flood-prone areas such as floodplains (Figure 11.3); (3) riverside development and the construction of bridges leading to reduction in channel capacity; and (4) the adverse effects of some flood protection works. It is also important to emphasise that the impact of flooding is not simply restricted to water damage. Flood waters may scour building foundations leading to collapse, and more significantly may deposit large quantities of sediment, which frequently contains sewage and industrial effluents.

Fluvial hazards are not just restricted to floods, but **bank erosion** and **channel metamorphosis** are also important problems. River channels may be cut either in sediment (alluvial channels), in bedrock or in a combination of both. The cross-sectional and channel pattern geometry of alluvial channels is adjusted to, and in equilibrium with, the dominant discharge and rate of sediment supply. Changes in sediment and water discharge cause changes in channel geometry through deposition and erosion (see Section 5.2.1; Figure 5.10). Channel patterns may also change and evolve. Figure 11.4 shows the general relationship between sediment load within a river, channel stability and channel form. Channels with fine sediment moving in suspension and resistant bank material are more stable and follow straight or sinuous courses. Mixed load streams, with more mobile

---

Figure 11.2 *(opposite) The flood hydrograph.* **A.** *Typical storm hydrograph, also illustrating the principle of the unit hydrograph.* **B.** *Downstream changes in the unit hydrograph during a storm event due to the addition of asynchronous flood peaks from tributaries.* **C.** *Impact of urbanisation on peak discharge; changes in the flood hydrograph of the Cannon Brook following the construction of the new town of Harlow to the northeast of London.* **D.** *Downstream attenuation of a flood hydrograph, due to floodplain storage [A Modified from: Petts & Foster (1985)* Rivers and landscapes, *Arnold, Fig. 2.12A, p. 40; D. Modified from: Petts & Foster (1985)* Rivers and landscapes. *Arnold, Fig. 2.13D, p. 45]*

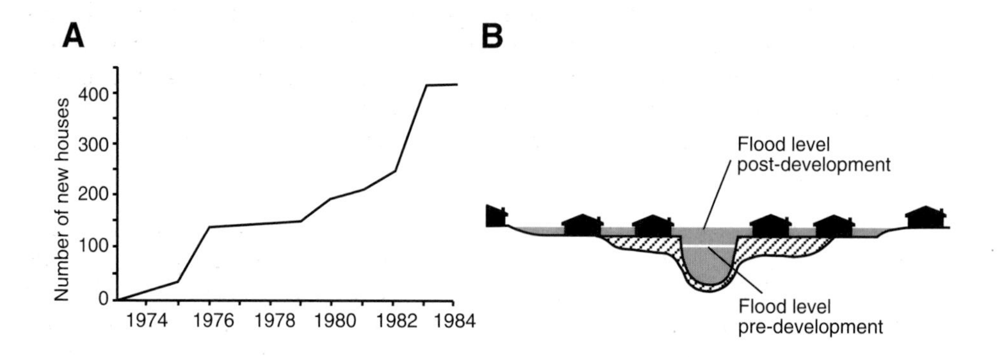

Figure 11.3 *Problems of floodplain encroachment.* **A.** *Encroachment of new residential housing onto the flood plain of the River Thames at Datchet, England, between 1974 and 1984.* **B.** *Changes in flood height as a result of floodplain encroachment [A Modified from: Smith (1992)* Environmental hazards. *Routledge, Fig. 11.4, p. 232; B Modified from: Keller (1990)* Environmental geology. *Merill, Fig. 6.20, p. 105]*

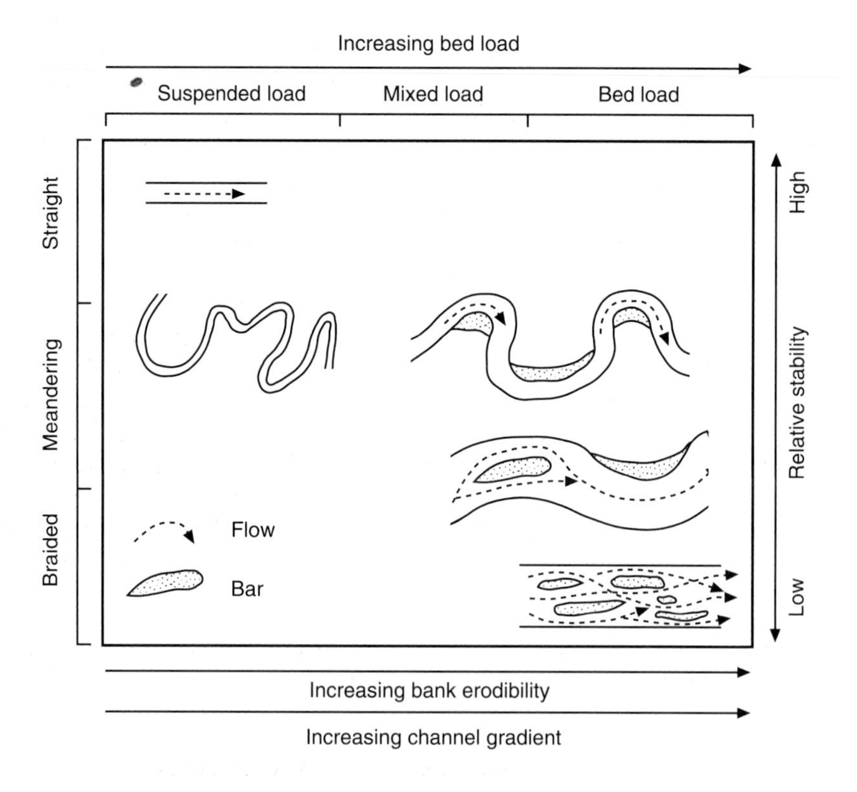

Figure 11.4 *Qualitative relationship between sediment load, channel stability and channel form [Modified from: Thorne* et al. *(1992) In: Stevens* et al. *(Eds)* Conserving our landscape. *English Nature, Fig. 2, p. 138]*

bed material and less resistant banks, adopt shifting, meandering courses. Rivers with coarse sediment loads and easily eroded bank material, which is transported as mobile bed load, tend to have wide braided channels. Meandering and braided channels evolve through time in response to changes in discharge and sediment supply (Figure 11.4). There are two major hazards associated with changing channel patterns.

1. **Bank erosion**. This is usually associated with the lateral migration of natural channels across floodplains, particularly meandering channels, posing threats to river bank infrastructure (Figure 11.5A&B). Similarly, where channels have been straightened or confined and are therefore out of phase with the water and sediment discharge regime, then deposition within a channel and bank erosion may occur as the channel attempts to adopt a more natural form. Table 11.1 provides data on the rate of bank erosion on some temperate rivers, with a maximum of around 10 m a^{-1}. However, in tropical regions rates may be an order of magnitude greater. For example, the Brahmaputra River is known to have migrated laterally at a rate of between 200 and 400 m a^{-1} between 1975 and 1981. As Figure 11.5C shows, the rate of bank erosion increases rapidly with drainage area (discharge). Large rivers are likely to be associated with more bank erosion owing to their greater discharge. Bank erosion may cause a variety of channel pattern modifications as shown in Figure 11.5A&F. For example, banks erode through small mass failures which may be developed as either rotational slips or as some form of cantilever failure associated with undercutting (Figure 11.5B&E). Bank erosion is particularly common on meandering rivers, where erosion is frequently concentrated on the outside bank of meanders. Erosion of these outside banks is facilitated by the presence of **secondary currents** within the river. Secondary currents are flows of water which occur perpendicular to the primary downstream current. The presence of a small secondary current on the outside flank of a river substantially increases water shear at the base of the bank, thereby accelerating erosion (Figure 11.5D).
2. **Channel pattern metamorphosis**. Shifts in the type of channel pattern adopted by a river may occur either naturally or as a result of human interference. For example, Box 11.1 shows how the dumping of mine tailings within a river caused a dramatic shift in channel form from a single meandering thread to a complex braided system. Similarly, Box 11.2 describes how reservoir construction and mining from a channel bed has caused major channel modification.

In summary, therefore, rivers provide a hazard not only through flooding but also by bank erosion and channel pattern metamorphosis.

11.1.1 Risk Assessment of Fluvial Hazards

In the previous section we identified two main types of fluvial hazard: flooding and bank erosion. Here we examine methods of risk assessment for each.

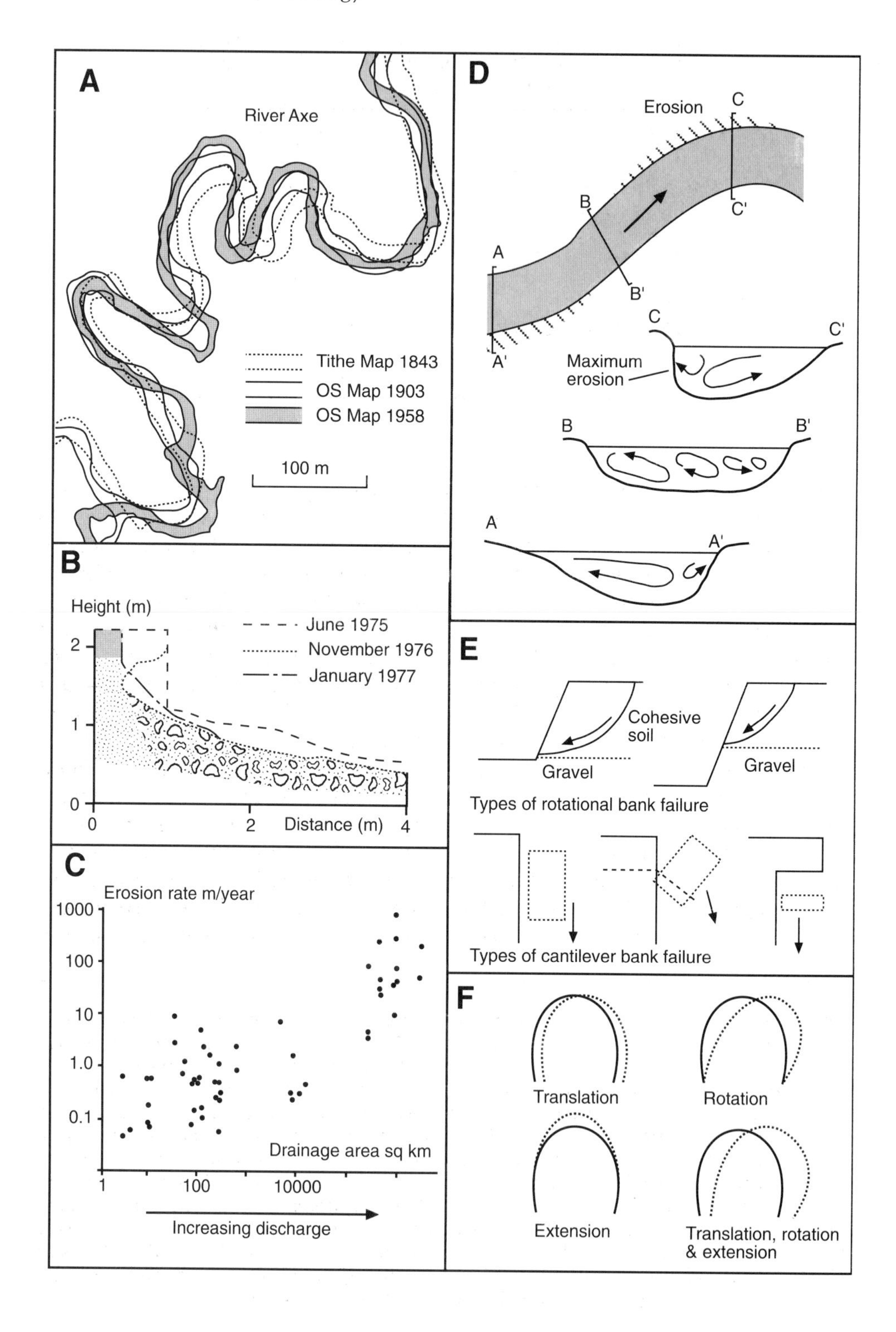
A
River Axe
Tithe Map 1843
OS Map 1903
OS Map 1958
100 m
B
Height (m)
June 1975
November 1976
January 1977
2
1
0
0
2
Distance (m)
4
C
Erosion rate m/year
1000
100
10
1.0
0.1
Drainage area sq km
1
100
10000
Increasing discharge
D
Erosion
C
C'
B
B'
A
A'
Maximum erosion
E
Cohesive soil
Gravel
Gravel
Types of rotational bank failure
Types of cantilever bank failure
F
Translation
Rotation
Extension
Translation, rotation & extension

Table 11.1 *Selected rates of bank erosion [Modified from: Knighton (1984)* Fluvial forms and processes. *Arnold, Table 3.5, p. 65]*

River	Average rate of retreat ($m\ a^{-1}$)
Axe, Britain	0.15–0.46
Bollin-Dean, Britain	0–0.9
Cound, Britain	0.64
Crawford, N. Ireland	0–0.5
Exe, Britain	0.6–1.2
Mississippi, USA	4.5
Watts Branch, USA	0.5–0.6
Wisloka, Poland	8–11
Torrens, Australia	0.58

Flood risk assessment involves two elements: firstly, a prediction of the extent and probability of flooding; and secondly, a determination of the vulnerability to that flooding. Predictions are most commonly a question of the availability of good time series data on the following: (1) variation in river and sediment discharge; (2) the extent of flood water and damage for each flood event; and (3) knowledge of the meteorological conditions or other factors which precipitated the flood. In many countries these data are simply not available. In Britain, for example, there are 750 river gauging stations, but the average length of record is only of the order of 23 years. This restricts the reliability of the hazard assessments that can be made. There are two fundamental tools in flood documentation: (1) the flood frequency curve; and (2) the unit flood hydrograph. **Flood frequency curves** are constructed using extreme event analysis (Box 10.1) which plots the **recurrence interval** or probability of a given flood against its magnitude or discharge (Figure 11.6). By examining the hydrographs for floods of a given frequency, for example once in 50 years, it is possible to construct a unit hydrograph. By combining this information with records of flood extent and damage obtained from previous events of a similar magnitude, it is possible to

Figure 11.5 *(opposite) Channel migration and bank erosion.* ***A.*** *The course of the River Axe, near Axminster, Devon between 1843 and 1958.* ***B.*** *The pattern of bank erosion via cantilever failure on the River Severn in England.* ***C.*** *Relationship between bank erosion and drainage area.* ***D.*** *Secondary currents within meandering rivers; the small secondary current cell on the outside of each meander is responsible for increasing fluid shear at the base of the bank and accelerating bank erosion.* ***E.*** *Mechanisms of river bank failure.* ***F.*** *Models of meander evolution. [A Modified from: Hooke & Kain (1982)* Historical change in the physical environment. *Butterworths, Fig. 4.17, p. 126; B Modified from Petts & Foster (1985)* Rivers and landscapes. *Arnold, Fig. 4.9B, p. 113; C Modified from: Hooke (1980)* Earth Surface Processes, ***5****, Fig. 2, p. 150; D Modified from: Hooke (1992)* In: *Stevens* et al. *(Eds)* Conserving our landscape. *English Nature, Fig. 6, p. 114; E Modified from: Petts & Foster (1985)* Rivers and landscapes. *Arnold, Fig. 4.9A, p. 112; F Modified from: Bathurst* et al. *(1977)* Nature, ***269****, Figs 1–3, pp. 505 & 506]*

BOX 11.1: CHANNEL METAMORPHOSIS CAUSED BY RIVER BED DISPOSAL OF MINE TAILINGS

In North Yorkshire close to the town of Alston there are numerous abandoned lead and zinc mines. This mining activity reached a peak in the 1880s and 1890s. Mining was undertaken for the most part through shallow adits cut horizontally into the hillside from the valley floor. Waste rock and mine tailings were commonly disposed of by direct dumping into adjacent rivers.

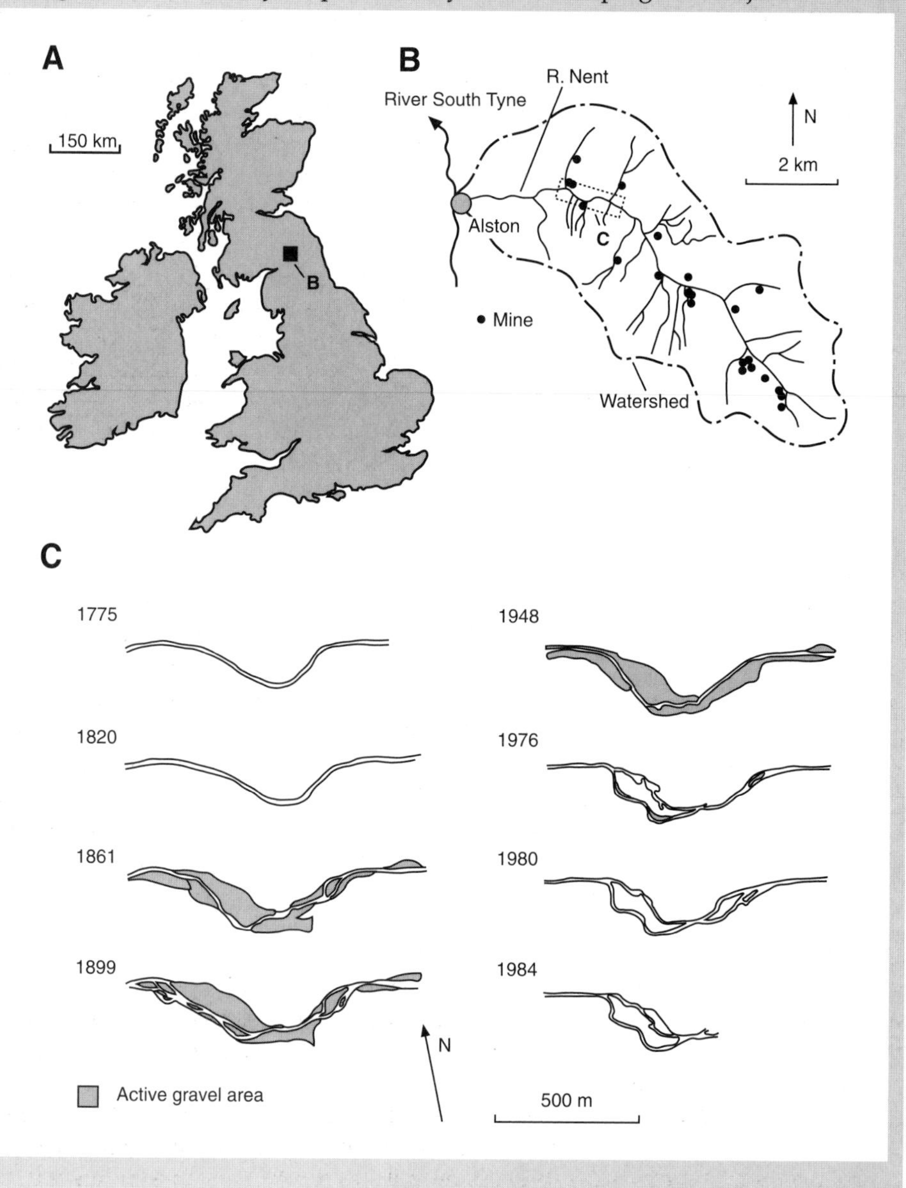

This not only spread the toxic metals within the tailings downstream but caused dramatic river metamorphosis. The mine tailings increased river bedload and caused them to adopt a braided course. The floodplain sediments deposited by these rivers as they metamorphosed were enriched in lead, zinc and other toxic metals. Today vegetation on these floodplains is restricted to metal-tolerant species and the floodplain sediments can be dated in relation to the extent of mining activity by the degree of metal contamination within them. Since mining activity declined early in the twentieth century, the river channel patterns have gradually begun to return to their original form.

This channel metamorphosis is particularly well illustrated by the River Nent. The diagrams opposite show how the river metamorphosed from a meandering course into a braided river due to the input of mine tailings. Since mining stopped in the 1930s the river has gradually begun to return to a meandering course as the excess bedload is removed from the system.

Source: Macklin, M.G. & Rose, J. 1986. *Quaternary river landforms and sediments in the Northern Pennines: field guide.* British Geomorphological Research Group/Quaternary Research Association, Leighton Buzzard. [Diagrams modified from: Macklin (1986) In: Macklin & Rose (Eds) *Quaternary river landforms and sediments in the Northern Pennines: field guide.* British Geomorphological Research Group/Quaternary Research Association, Figs 1 & 3, pp. 20 & 23]

define flood hazard zones. For events with recurrence intervals in excess of the length of the record it is necessary to extrapolate the flood frequency curve to obtain the flood magnitude. The hydrograph and peak discharge for this flood can then be obtained by using the principle of the unit hydrograph which states that all floods, whatever their magnitude, will have similarly shaped hydrographs. From this information it should be possible to predict the likely elevation of the water above the banks and obtain an idea of the area at risk from flooding. Such extrapolation can be problematic and therefore hazard assessments constructed in this way should be treated with caution. In areas without adequate discharge records, determining recurrence intervals and therefore predicting the frequency of the different flood events observed historically is difficult.

Assessing vulnerability to flood damage is also of importance. This is determined by the type of land use and nature of the building stock which will be affected by the flooding. Typically, vulnerability is greatest where a flood has the potential to destroy not only housing but also the agricultural and industrial productivity of a region.

Assessing the hazard potential of bank erosion and channel metamorphosis depends either on direct observation or on the use of historical records. Historical records, such as old maps, air photographs and other documentary evidence,

BOX 11.2: CHANNEL METAMORPHOSIS DUE TO HUMAN IMPACT

Kondolf & Swanson (1993) describe the response of Stony Creek, a tributary of the Sacramento River in California, to both reservoir construction and gravel extraction from its channel. Both have caused major changes in the river's planform which has undermined bridges necessitating repairs costing in excess of $1.4 million.

Stony Creek drains a 1920 km^2 drainage basin near Hamilton City, about 200 km north of San Francisco, California. The river emerges from a mountainous catchment just downstream of the Black Butte Reservoir and originally formed a braided reach due to the rapid deposition of bedload as the channel gradient was reduced. The catchment now contains three reservoirs, of which Black Butte Reservoir is the most downstream, and was constructed to reduce flood peaks and provide irrigation in 1963. It not only reduces flood peaks but traps over 100 000 m^3 of sediment each year. Gravel is extracted directly from the river channel at several locations downstream. In one reach, between 230 000 and 580 000 m^3 of gravel is removed each year. The channel response to these impacts has been dramatic and was monitored at 64 surveyed cross-sections in 1967 and 1990, and through the use of serial air photographs. The channel changes are described in relation to four reaches, successively downstream, below the Black Butte Dam.

Reach One. Originally the river had a broad braided planform. On closure of the dam in 1963 bedload was reduced completely by sedimentation in the reservoir. The sediment-free water, often referred to as 'hungry water', caused rapid channel incision and replaced the braided channel system with a single-thread meandering channel. Rapid lateral meander migration has caused the loss of over 90 ha of land since 1963. The old braided floodplain has become vegetated.

Reach Two. Here the channel has also changed from a braided to meandering form. However, lateral migration of the channel is restricted by three bridges (Interstate Highways 5 and 99, plus Southern Pacific Railroad (SPR)). Little channel incision has occurred. The channel pattern has been adversely influenced by borrow pits on the floodplain, used to obtain gravel for Interstate Highway 5. These pits have been flooded by channel migration and now influence the channel planform. Water is now directed obliquely at the bridge for the Interstate Highway 5 and scour of the northern end of the bridge has occurred.

Reach Three. Channel incision of over 5 m has occurred in this reach since 1973 due to two large gravel quarries located immediately upstream and downstream of the bridge for Highway 32. Since its construction in 1973, incision has occurred undermining five of the eastern piers. Remedial work is likely to cost over $1.4 million. Again the channel pattern has changed from being braided to meandering. Channel adjustment to this gravel extraction is likely to continue long after it has ceased.

Reach Four. In this reach the impact of both the mining and reservoir are less apparent, although the active channel has narrowed and become more sinuous.

The impact of both reservoir construction and gravel extraction has been profound on Stony Creek, causing major change in its planform. It illustrates the potential of river morphogenesis to cause problems for environmental management.

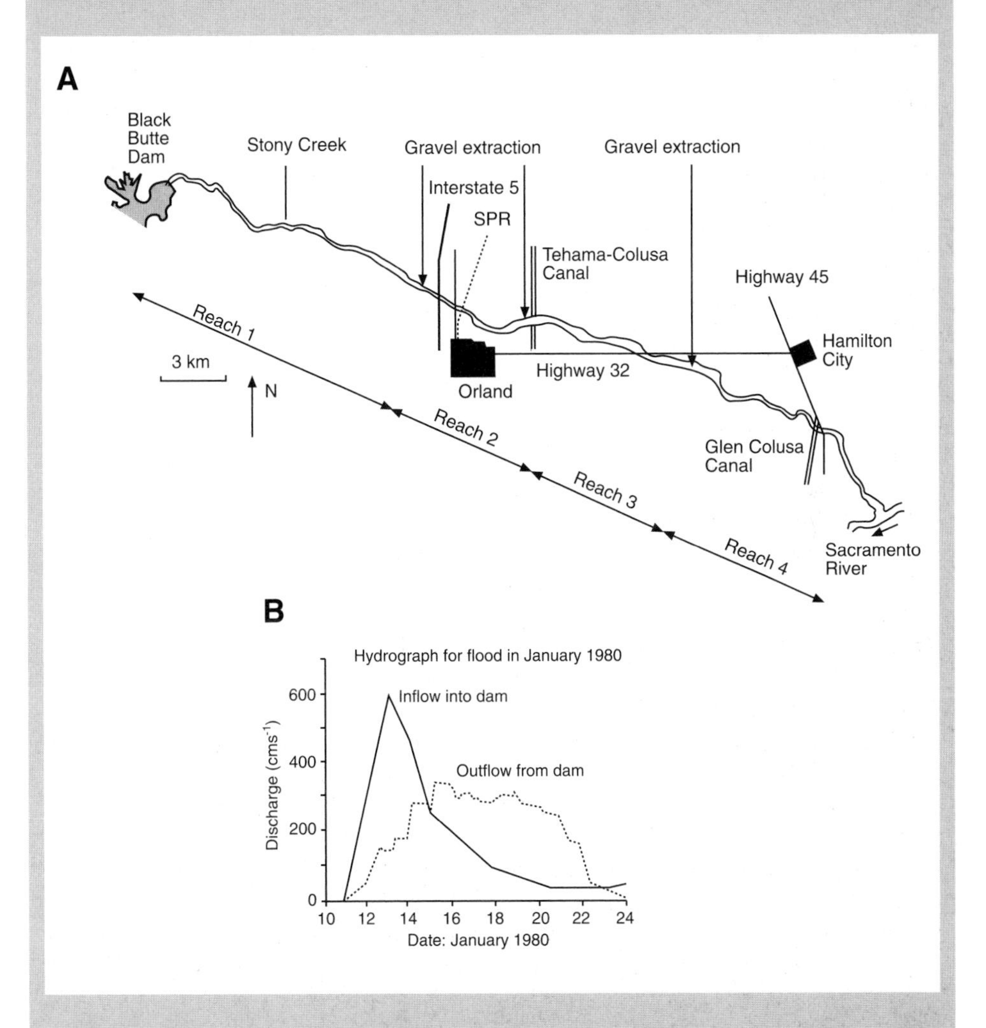

continues overleaf

BOX 11.2: *(continued)*

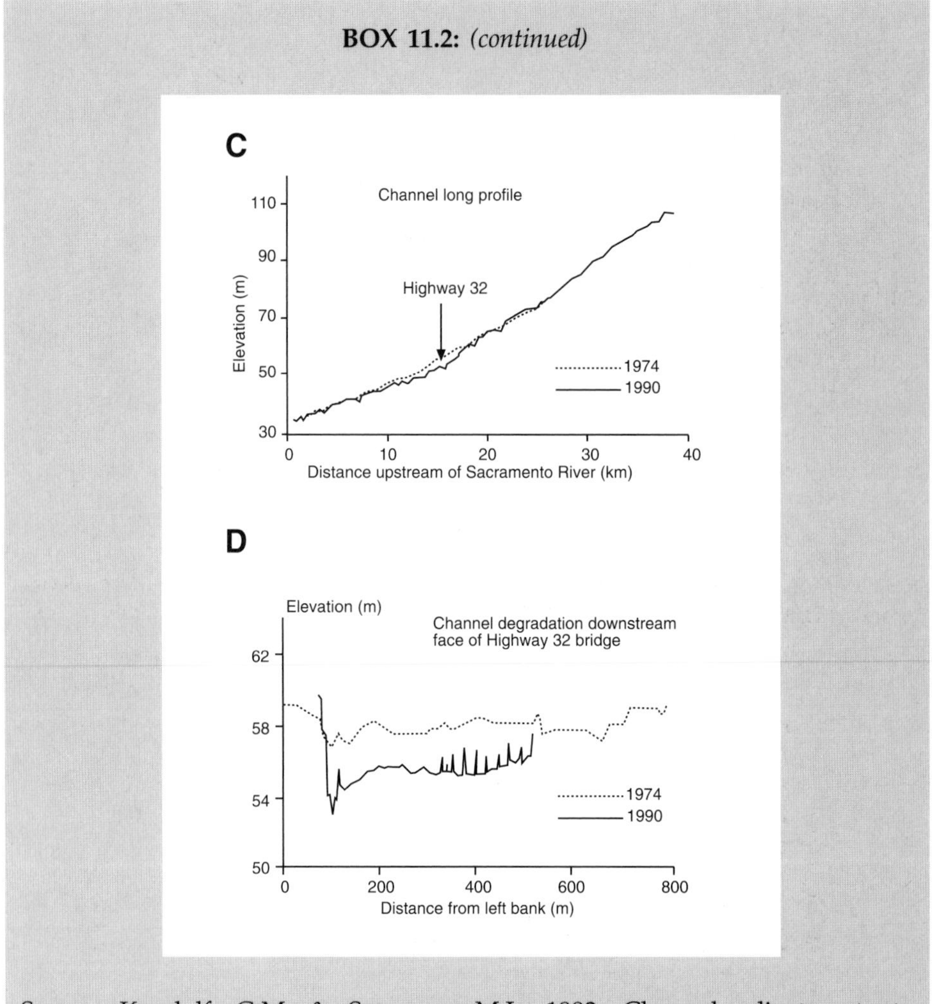

Source: Kondolf, G.M. & Swanson, M.L. 1993. Channel adjustments to reservoir construction and gravel extraction along Stony Creek, California. *Environmental Geology*, **21**, 256–269

can be employed to establish the characteristics and rate of bank erosion or channel pattern change in the past, and thereby can be used to predict future change (Figure 11.5A). Where such records are not available, geomorphological and sedimentological evidence may be obtained from the floodplain or alternatively through the direct monitoring of erosion by the use of repetitive surveys or erosion pins. In this way it is possible to predict future channel pattern changes. Such predictions rely upon the assumption of a steady state within the river system; that is, erosion will continue at the same rate and in the same direction as at present. However, fluvial systems often contain thresholds or

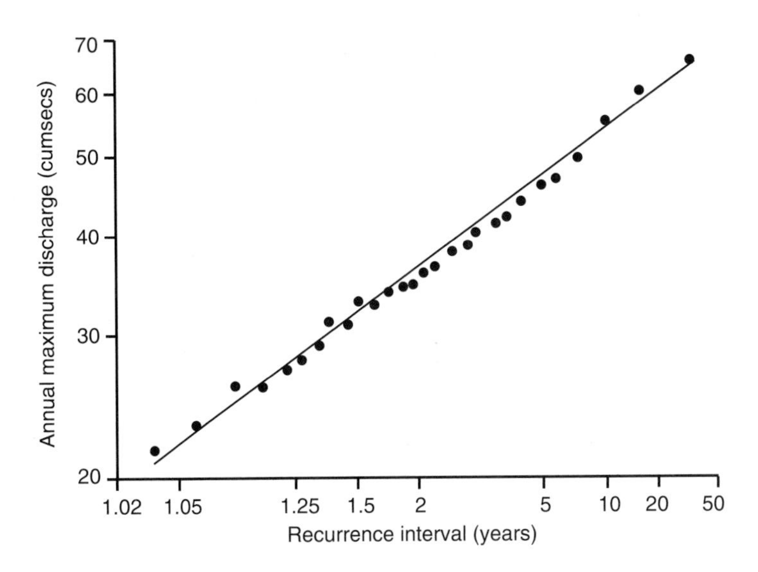

Figure 11.6 *Flood recurrence intervals plotted against discharge for the River Bollin, Cheshire, England [Modified from: Knighton (1984)* Fluvial forms and processes. *Arnold, Fig. 3.2B, p. 46]*

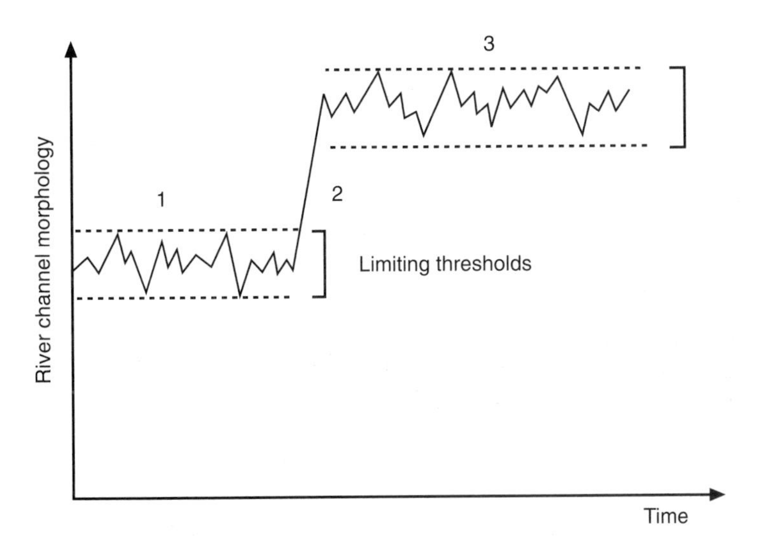

Figure 11.7 *Thresholds within the fluvial system*

boundaries, either internal or external, at which their behaviour may change radically and in a way not necessarily predicted by the record. Figure 11.7 shows how channel patterns or other aspects of a river's morphology may be controlled by process thresholds. Three zones can be identified on this diagram. Zones 1 and 3 are those in which the river response is stable, although oscillating around a mean or between well defined limits. An example of this oscillation is the equilibrium between the size of a channel's cross-section and the dominant discharge; if the channel becomes too wide, discharge falls, since discharge is the sum of velocity and cross-sectional area, and as a consequence deposition occurs within the channel, reducing its size. Equally, if too much deposition occurs this will reduce the channel capacity thereby increasing the velocity and causing erosion. In this way the channel cross-section oscillates around a mean size or between two defined limits determined by the dominant discharge regime. In contrast, Zone 2 identifies sensitive behaviour in which a river, in response to an external event or change (Boxes 11.1 & 11.2), crosses a threshold to a new process regime, destroying the old fluvial landforms and creating new ones. Identifying these thresholds and the sensitivity of the fluvial system to them – essentially, estimating how big a change or event must be to cause the fluvial system to cross the threshold – is critical in predicting potential fluvial hazards associated with channel metamorphosis. This type of information is specific to a given fluvial system and can only be obtained by detailed knowledge of past events and their character. Significantly, fluvial sensitivity to a particular event may remain undiscovered until it is too late.

11.1.2 Mitigation and Management of Fluvial Hazards

In dealing with flood hazards there are three principal approaches: (1) control and limitation; (2) abatement; and (3) adjustment.

The **control and limitation** strategy involves control by land defence through the containment of flood water. Containment may be achieved through the use of reservoirs, overflow or by-pass channels, artificial levees or embankments (Figure 11.8). The construction of reservoirs upstream of a town or city subject to flooding is an expensive but highly effective solution. In this option the reservoir is kept partially empty in order to store flood water which is then released slowly over a period of days or weeks (Figure 11.8A). It is a highly effective solution but its use is limited by the following factors.

1. **Cost**. Dam construction is expensive.
2. **Availability of suitable sites**. The beneficial impact of a dam in flood abatement declines downstream with the addition of each undammed tributary. Consequently, there need to be potential reservoir sites close to the settlement to be protected. In flat terrain the size of dams required would be prohibitive, and therefore the presence of some topographic relief and suitable dam sites is essential.

Figure 11.8 *(opposite) Approaches to flood protection and management.* ***A.*** *Dams and reservoirs.* ***B.*** *By-pass channels.* ***C.*** *Flood embankments.* ***D.*** *Flood ponds. [A Modified from: Petts & Foster (1985)* Rivers and landscapes. *Arnold, Fig. 2.14A, p. 46]*

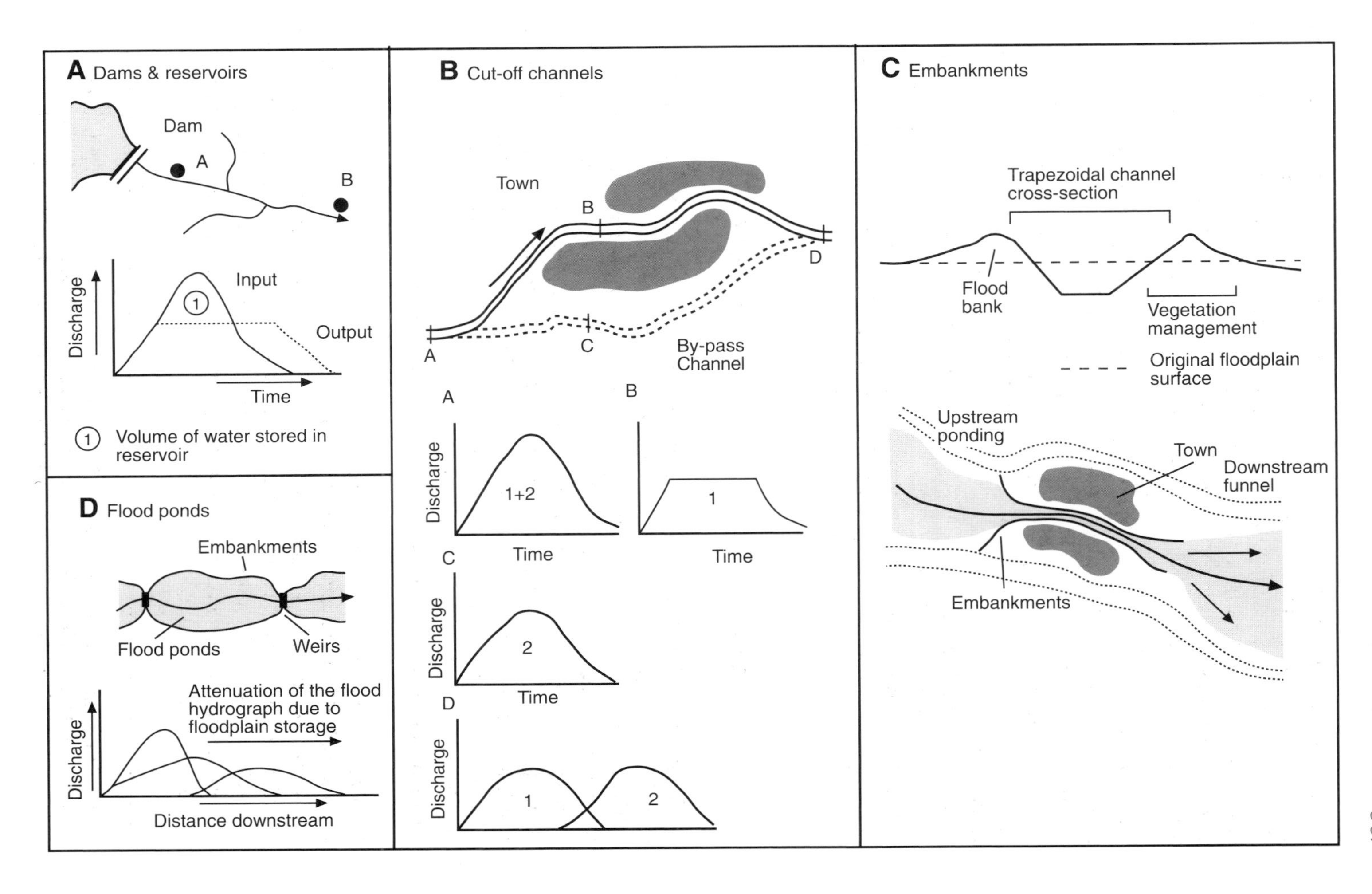
A Dams & reservoirs
Dam
A
B
Discharge
Input
Output
Time
1 Volume of water stored in reservoir
B Cut-off channels
Town
A
B
C
D
By-pass Channel
Discharge
1+2
1
2
Time
C Embankments
Trapezoidal channel cross-section
Flood bank
Vegetation management
Original floodplain surface
Upstream ponding
Town
Downstream funnel
Embankments
D Flood ponds
Embankments
Flood ponds
Weirs
Attenuation of the flood hydrograph due to floodplain storage
Distance downstream

3. **Aesthetics**. The water level within the reservoir needs to be maintained, at least seasonally, at a reduced level to provide the necessary flood storage capacity. This may not be aesthetically pleasing and may restrict the recreational use of the reservoir.
4. **Sedimentation**. The reservoir not only stores flood water but will also trap flood sediment and may fill rapidly, limiting its long-term effectiveness.

An alternative approach is the construction of by-pass or overflow channels which are designed to provide additional channel capacity during flood events and consequently to reduce the chance of overtopping of channel banks (Figure 11.8B). If well designed, water entering the by-pass channel (cut-off) is moved downstream to rejoin the main channel either in front of the flood wave or after it has passed, thereby attenuating the flood hydrograph (Figure 11.8B). This type of approach is being used on the River Thames in southern England to alleviate flooding in the towns of Windsor and Eton. It is only possible where sufficient land adjacent to the river is available and can be acquired cheaply.

The traditional solution is the construction of artificial levees or embankments to contain flood water within the existing channel (Figure 11.8C). This solution has been used throughout history and is still the most commonly used flood control method. The channel bed and adjacent floodplain are dredged up to form the embankments and then reinforced, where necessary, with stone and concrete revetments. In some cases natural levees may be enhanced, and channel enlargement to increase capacity may also be undertaken. Artificial levee construction provides a simple, tested and relatively inexpensive solution to flood hazards. Over 4500 km of the Mississippi River in the USA is embanked in this way. More elaborate systems include flood ponds or flood corridors to provide storage of flood water and these consequently attenuate the flood hydrograph through storage and slow release (Figure 11.8D). There are, however, significant disadvantages with the construction of flood embankments.

1. **Adverse downstream impacts**. If the scheme does not include significant flood storage capacity then the flood wave is simply moved downstream, increasing the potential flood problem. In many cases this has led to progressive downstream proliferation of flood embankments (Figure 11.8C).
2. **Adverse upstream impacts**. Due to the funnelling effect of the levees, flood water may be ponded upstream exacerbating the flooding problem there and leading to demands for further flood protection work (Figure 11.8C).
3. **Channel scour or sedimentation**. During flood events, very high flow velocities will occur within the restricted channel and this may cause scouring of the embankments and channel floor, in some cases undermining them and leading to the collapse of the embankments. Equally, any sedimentation during waning or normal flows will be restricted to the channel bed causing it to rise above the adjacent floodplain. With time, the embankments may be required to contain not only flood flows but also more normal discharge levels (Figure 11.9).
4. **Catastrophic flooding due to ponding and embankment collapse**. If flood embankments fail, the extent of flood damage is likely to be much greater

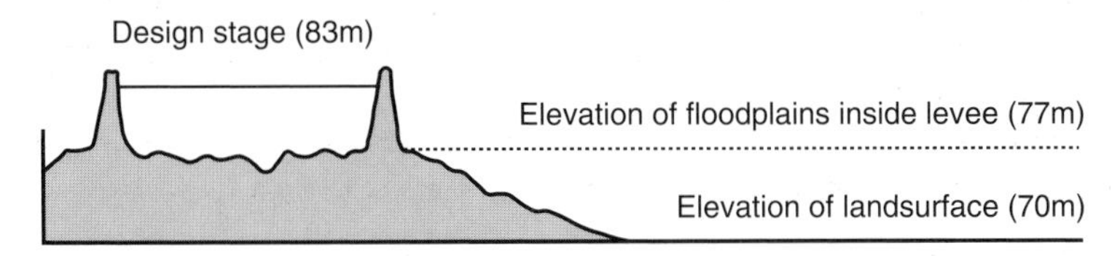

Figure 11.9 *The perched nature of the Yellow River is due to sedimentation on the channel bed between the two levees. The river is now elevated above the adjacent floodplain. Should the flood bank fail, even during normal flows, the flood damage will be considerable [Modified from: Shu & Finlayson (1993)* Sedimentary Geology, ***85****, Fig. 6, p. 290]*

due to escape of ponded water. Much of the severe winter flooding in northern Europe in 1995 was due to the collapse of protective works.

5. **Adverse ecological effects**. The construction of flood embankments and their subsequent maintenance may destroy wildlife habitats in the river bank. Such construction is important not only for conservation, but also for fisheries and for recreation. The impact on river banks may be minimised by the construction of flood corridors in which the embankments are placed some distance back from the normal river edge.

The second type of generic approach to flood management is the use of **flood abatement strategies**. The aim here is to prevent the flood occurring in the first place or at the very least to limit its magnitude through catchment management, changing the characteristics within the drainage basin system (Figure 5.1). To be effective, management must be applied to the whole catchment and must succeed in reducing throughput of water within the system. Typical strategies in rural areas include: (1) reforestation or reseeding of sparsely vegetated areas to increase interception and evaporation losses; (2) contour ploughing and terracing to reduce runoff and sediment supply; (3) vegetation protection against fires, overgrazing, deforestation or other activities likely to increase runoff and therefore flood discharge; (4) maintenance and creation of ponds, holding pools and wetlands to preserve or increase natural storage capacity; and (5) restriction of drainage work and channel modifications such as dredging or straightening. Within urban areas, increased water storage can be achieved through the creation of ponds, lakes and parkland. The effectiveness of such strategies depends primarily on the magnitude of the flood problem and the level of planning and land-use control which can be imposed.

The third type of approach to flood management is **adjustment**. At its most radical this may involve the relocation of settlements, but in most cases it involves: (1) the introduction of planning control based on hazard maps and insurance incentives; (2) the introduction of control through building codes; (3) the use of flood forecasting and warning; and (4) contingency and disaster aid planning.

Floodplains may be zoned to show those areas likely to be inundated by different magnitudes of flood event, such as the one in 50 or one in 100 year

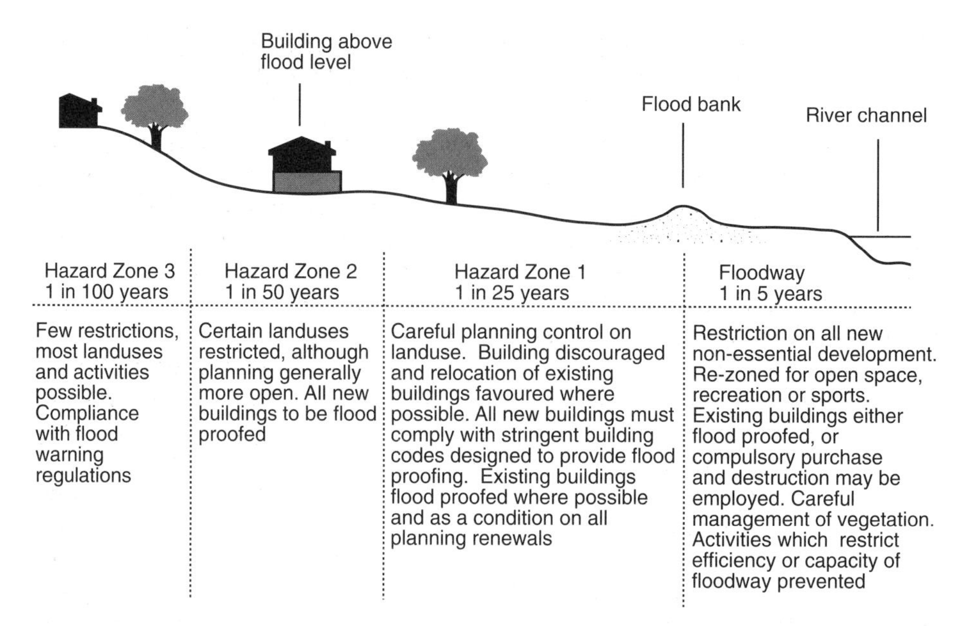

Figure 11.10 *Hypothetical example of flood hazard zones*

event. Development or, where it has already taken place, redevelopment can be prevented in the floodway adjacent to the river and controlled in adjacent flood zones (Figure 11.10). For example, building codes can be used to ensure that structures are flood-proofed to varying degrees depending on the probability of flooding. In the USA the adoption of planning constraints based on flood hazard zonation is encouraged and enforced through a National Flood Insurance Program which makes floodplain management a requirement of insurance provision. The development of flood zone planning schemes can be a sensitive issue, particularly in residential areas, due to the potential impact on the value of property. Research in New Zealand, however, has rather surprisingly shown that this is not the case, due primarily to poor hazard perception by residents (Box 11.3). Flood-proofing strategies include the elevation of buildings above predicted flood water levels, foundation protection, the fitting of water-tight doors to subways and other low-level buildings, and the careful design of sewer/drainage systems to avoid contamination of flood water by sewage.

Flood warning schemes also have an important role in flood management. A combination of meteorological data and information about rising discharge in upstream areas allows a series of graded warnings to be issued. This relies on the recognition of a series of threshold events, based on past empirical experience, such as rainfall events of a given intensity or a given rate of increase in upstream discharge. These threshold events trigger a series of flood warnings and civil defence actions. The idea is to put in place temporary flood-proofing and evacuation measures, a basic premise being that the earlier a warning is given the more time there is for effective countermeasures. Critical in the recognition of

BOX 11.3: HAZARD AREA DISCLOSURE AND PROPERTY VALUE IN NEW ZEALAND

The publication of hazard maps for existing towns as a guide to future planning is an increasingly important tool in hazard management, but is often unpopular amongst local residents and the business community. Clearly, if a building is located in a high hazard zone its value may suffer. Research in New Zealand has challenged this widely held assumption and also illustrates the poor public perception of threat from natural hazards (Montz, 1993). The property values in the town of Te Aroha before and after the disclosure of hazard zonations were monitored. The hazard zones and associated planning restrictions are outlined in Table A. The future development or redevelopment of buildings in high hazard zones is clearly restricted and one would expect an adverse effect on property value. The results suggest that hazard zone disclosure had no adverse effect on property value (Table B), although the occurrence of major flood/landslide events did temporarily affect property values in the whole town. The potential hazard of a location was not a factor in house prices, above such factors as size and availability of car parking or similar facilities. This demonstrates the lack of hazard awareness of most citizens; if you cannot see the hazard it does not exist. From a planning point of view the lack of an adverse reaction to hazard disclosure in the work by Montz

Table A

Category	Description	Planning restrictions
A	Primary floodway of Waihou River	No development or redevelopment allowed
B	Secondary areas subject to flooding from Waihou River	Development allowed if 80% of section is 200 mm over a 100 year flood level: floor must be 500 mm above 100 year flood level
C	Areas subject to inundation from tributary streams	No development or redevelopment
D	Areas possibly subject to overland flows from tributary streams	Development must meet certain criteria such as waterproofing external walls and locating large openings away from flow paths
E	Roadway areas along which overland flow paths should be created	
F	All other areas	No restrictions

continues overleaf

BOX 11.3: *(continued)*

(1993) is good, otherwise people might be looking for compensation on disclosure; however, it did illustrate the absence of hazard awareness amongst the general population.

Table B

Area	Selling price		Difference
	Before designation	After designation	
Entire town	35 480	36 255	775
Non-flood prone	36 197	36 994	797
Flood prone	19 710	19 995	285

The price of the same house is compared before and after designation, average values in NZ$

Source: Montz, B.E. 1993. Hazard area disclosure in New Zealand: the impacts on residential property value in two communities. *Applied Geography*, **13**, 225–242. [Diagrams modified from: Montz (1993) *Applied Geography*, **13**, Table 2, p. 231.]

the threshold or trigger events is the balance between achieving the earliest possible warning and being confident that the event will actually occur. The size and character of the catchment determines the amount of time available for flood warning; the more responsive or 'flashy' the hydrograph to a storm event (i.e. the shorter the lag time; Figure 11.2), the less time is available. In Britain, a combination of short rivers and highly urbanised basins ensures very limited warning times. As a result, a network of integrated radar weather stations was built during the 1980s to recognise intense rainfall events likely to give rise to flooding. Despite this, over 50% of unprotected dwellings in Britain still have less than six hours' warning. In less developed countries, the problem of flood warning is hindered by the lack of communication. For example, most of the floods in Bangladesh start well upstream in India, but the authorities receive no more than 72 hours' warning of floods on the Ganges and Brahmaputra rivers, yet if they received information from further upstream they could have up to eight days' warning. In practice, flood warning schemes rarely fulfil their potential; this is primarily due to difficulties in disseminating the flood warning and in accurately predicting flood events, causing many false alerts.

11.2 COASTAL HAZARDS

Coastal hazards can be divided into those which are essentially 'quiet' and progressive, such as coastal erosion or the gradual rise in global sea level (see

Section 9.8.3), and those that are 'dramatic' events such as coastal flooding or the impact of tsunamis.

The world's coastline is highly variable, ranging from cliffed coasts to sandy or gravel beaches, and from barrier beaches to large intertidal mudflats and deltas. Many of these environments are sensitive to change and have long been desirable locations for human development. Approximately 75% of the world's population live within 1.5 km of the coast and there has been rapid urban expansion in coastal areas both in developed and in less developed countries. For example, in the USA the coastal population doubled between 1940 and 1980. Not only is the coastal zone densely populated but in many countries it is a focus for tourism. For example, more than 20 million people invade just 2000 km of French coast each summer. Over 51% of the French coastline is urbanised and 13 000 new holiday homes are constructed in coastal areas each year. Tourism on the French coast has an annual turnover of 50 billion francs, which amounts to 12 times the turnover of its commercial ports, and eight times that of its fishing industry.

Coastal infrastructure is threatened every year by coastal erosion and coastal flooding (Figure 11.11; Table 11.2). It has been estimated that over 95% of the world's coastline is currently eroding. Coastal recession is a natural process and is vital in replenishing sediment within the littoral system, although in recent years it has been exacerbated by coastal development and poor management.

Coastal erosion is most marked on cliffed coasts where land loss and property damage are most visible. The rate of recession is a function of coastal exposure (i.e. wave climate) and the nature of the cliff. If the cliff is composed of 'soft' rocks and is low, recession is likely to be continuous and therefore fast (Figure 11.12A). However, as cliff height increases recession will tend to become the product of large mass movements, driven by coastal erosion at the cliff base. In this case cliff recession will be episodic, since the debris from each episode of mass failure will act as a natural coastal barrier to coastal erosion (Figure 11.12B). Although the magnitude of cliff loss may be larger, it only occurs episodically and consequently overall recession rates may be lower. Cliffs composed of 'hard' rock will erode or fail less frequently, but are also affected by this balance between the size of cliff failure and the episodic nature of cliff recession (Figure 11.12A).

Whatever the mechanism and rate of cliff recession, and of coastal erosion in general, human action has played an important role in exacerbating the problem. In particular, urbanisation and development within the coastal zone have accelerated this problem in many areas through depletion of beaches, which otherwise provide an effective dynamic and natural coastal defence. Beach depletion occurs through such activities as: (1) direct mining of beach and dune sand; (2) coastal defence reducing cliff erosion and sediment supply; (3) offshore dredging; and (4) the reduction of longshore sediment flux due to breakwaters, groynes and other structures which interrupt longshore drift (Figure 11.13). For example, Figure 11.11 shows the rapid rate of coastal retreat at Mar Chiquita beach, Mar del Plata in Argentina, between 1957 and 1979. The population of Mar del Plata doubles in the summer to in excess of one million, and beach depletion and consequent erosion are the product of: (1) sand mining of the

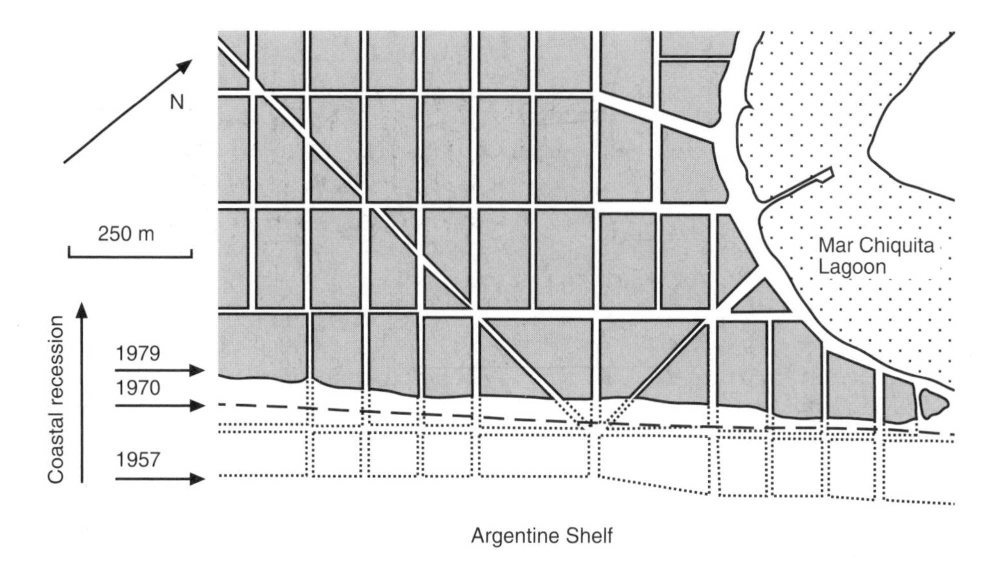

Figure 11.11 *Coastal retreat at Mar Chiquita beach, Mar del Plata, Argentina, based on air photographs and field surveys [Modified from: Schnack (1993) In: Warrick* et al. *(Eds)* Climate and sea level change: observation, projections and implications. *Cambridge University Press, Fig. 21.7, p. 344]*

Table 11.2 *Erosion rates around the coast of the European Community [Modified from: Lumsden (1994)* Geology and the environment in western Europe. *Oxford University Press, Table 4.12, p. 181]*

Member state	Location	Geology	Erosion rate (m a^{-1})
United Kingdom	Humberside	Glacial sediments	0.3–3.3
	Norfolk	Glacial sediments	0.2–5.7
	Kent	London Clay	0.7–3.4
	Yorkshire	Liassic claystone	0.02–0.04
France	Somme	Chalk	0.08–0.37
	Landes	–	1–3
	Seine Maritime	Chalk	0–0.4
Germany	Helgoland	Sandstone	1
	Baltic Coast	Glacial deposits	0.6–2.0
Italy	Romagna	Sand dunes	<9
	Calabria	Alluvium	1.2–3.3
	Calabria	Sand & gravel	8–11
	Gulf of Tarento	–	4

beach to the south at a rate of 250 000 m^3 a^{-1}, which is approximately half the northward littoral drift; (2) the extensive use of groynes to retain beach sand causing down-drift sediment starvation; and (3) the fact that the beach is effectively a fossil sand body and is not being replenished by cliff erosion, which now only contributes occasional pebbles or fine silts.

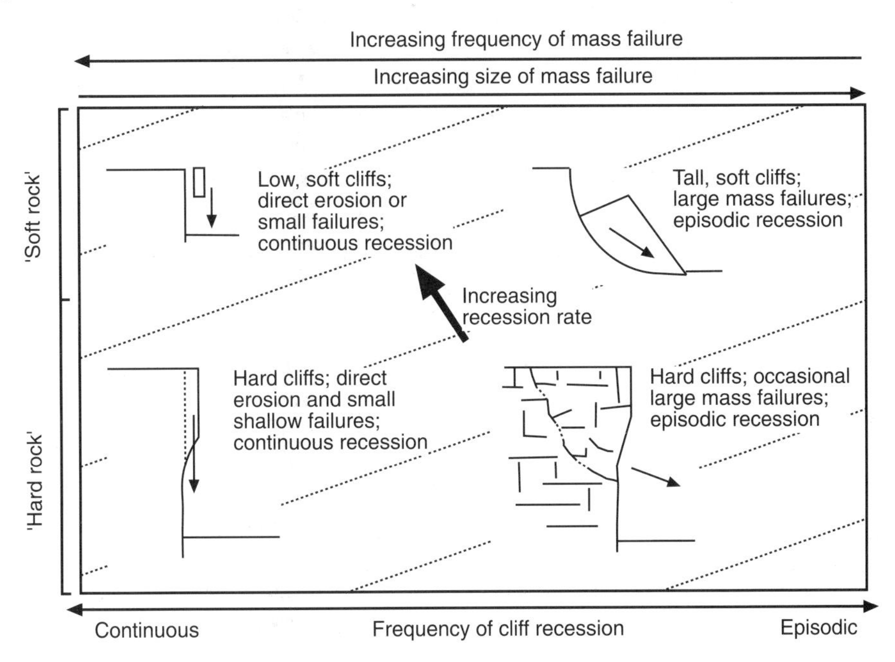

Figure 11.12 *Coastal cliff reccession.* ***A.*** *A model of coastal cliff recession based on the size of mass movement caused by coastal erosion at the cliff base. Large mass failures lead to dramatic but episodic coastal recession, while small continuous failure is less dramatic but is usually associated with higher recession rates.* ***B.*** *An example of the episodic nature of coastal recession via large mass failure*

This type of problem is typical of the coastal pressure leading to erosion. The impact of sea defences, such as breakwaters and groynes, which interrupt longshore sediment supply is illustrated in Figure 11.14. Here, construction of a pier on Point Lonsdale at the entrance to Port Philip Bay in Australia has caused progressive beach depletion in a down-drift direction, matching the progressive construction of sea defences (Figure 11.14). Coastal erosion causes the loss not only of agricultural land but also of buildings and transport infrastructure. The threat of coastal erosion is poorly perceived, especially during the purchase of property, and most residents consider it their right to expect their property to be defended from the sea at whatever cost. Throughout the twentieth century the proportion of coast, in developed countries such as Britain, which has been defended in some way has increased dramatically and this has often accelerated the problem elsewhere by reducing sediment supply or hindering longshore drift. The problem of cliff recession illustrates the many complex and conflicting issues associated with the coast and the difficulties facing decision-makers (Figure 6.9; Box 6.6). In Britain this problem has been exacerbated in the past by the sheer number of bodies responsible for different stretches of the coast. For example, no less than 10 separate authorities have responsibility for coastal management on

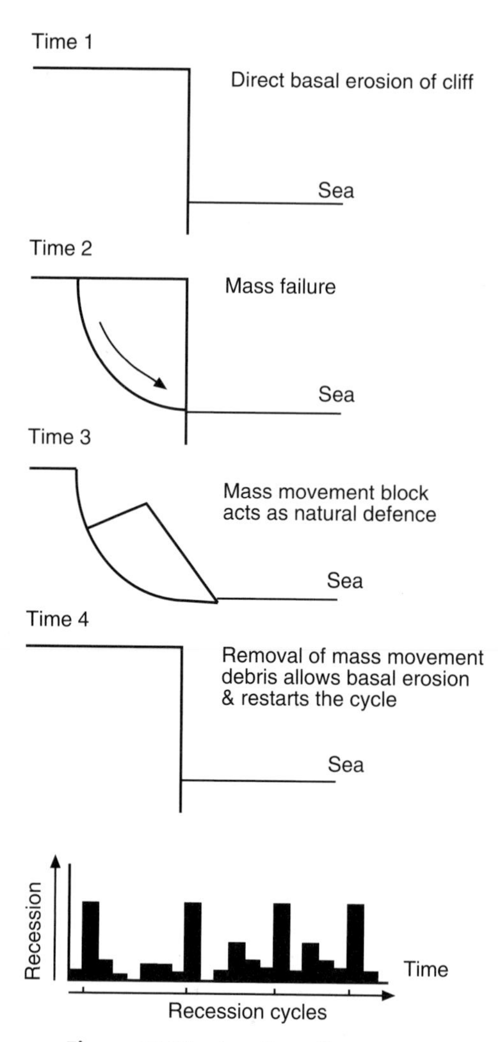

Figure 11.12 *(continued)*

one 64 km stretch of the English coast (Figure 11.15). As a consequence, coastal management has often proceeded within stretches of coast defined by political boundaries as opposed to geomorphological ones. This has led to natural coastal cells being interrupted and disrupted by coastal works, with the long-term effect of passing on the problem to neighbouring authorities.

Coastal flooding is associated with abnormally high tides or storm surges which cause the overtopping of natural or artificial coastal defences. The flooding of low-lying land by saline water contaminates fresh water bodies and causes water damage. In Britain, the worst episode of coastal flooding in recorded history occurred during the 1953 storm surge when large parts of eastern Britain were flooded.

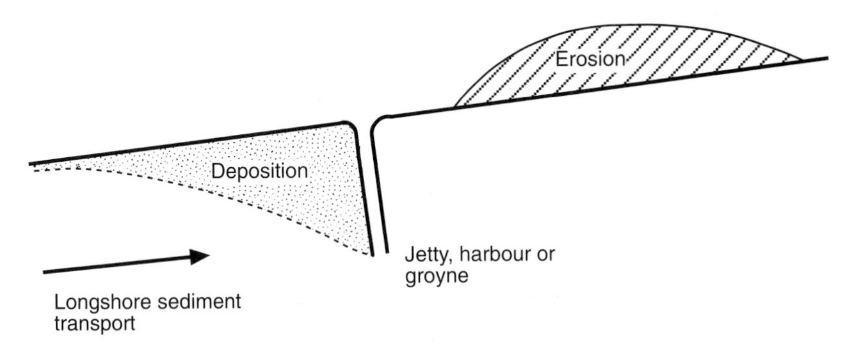

Figure 11.13 *Erosion and deposition associated with the construction of a pier, long groyne or impermeable jetty which interrupts longshore sediment transport*

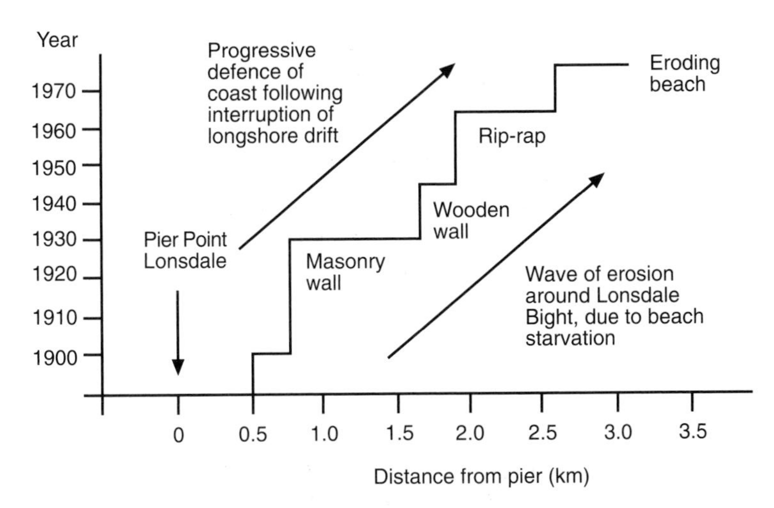

Figure 11.14 *Sequence of sea wall construction at Point Lonsdale, at the entrance to Port Philip Bay, Melbourne, Australia, as a consequence of the construction of a pier interrupting longshore sediment transport [Based on data in: Cooke & Doornkamp (1990)* Geomorphology in environmental management. *Oxford University Press]*

Storm surges are exceptionally high tidal events induced by synoptic weather patterns. In a European context, storm surges occurred in the North Sea in the years 1897 and 1953. The 1953 event took place between 30 January and 1 February. On 29 January a depression of considerable intensity was situated northwest of Scotland, while at the same time a ridge of high pressure was developing to the west of Ireland. Normally depressions in this area move to the east over Scandinavia, but in this case the depression moved southwards across Denmark and into Germany. A very steep pressure gradient developed between the high over Ireland and the low in the North Sea. As a consequence, very strong winds blew down the east coast of Britain, which resulted in storm waves up to 6 m high and a southward transport of the water mass. Water levels were also

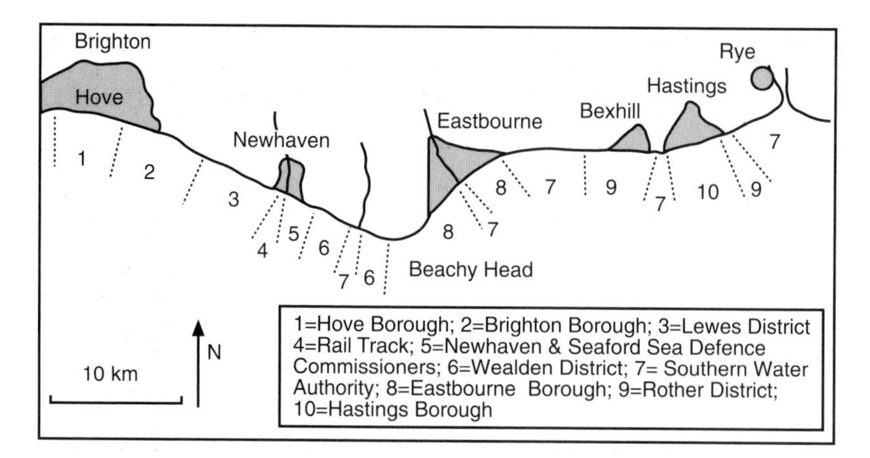

Figure 11.15 *A 64 km stretch of the East Sussex coast showing the number of different authorities responsible for coast management [Modified from: Hansom (1988) Coasts. Cambridge University Press, Fig. 8.13, p. 92]*

raised by between 2 and 3 m due to the intense low pressure. The combination of these events in conjunction with spring tides caused a major flooding event around the margins of the North Sea. Not only did this flood cause severe property damage, leading to the evacuation of over 30 000 people in Britain, but it also claimed 264 lives. The losses were even greater in Holland where 1350 people died. With rising sea levels (see Section 9.8.3) the impact of such events is likely to increase. This is particularly worrying as large concentrations of people in southeast Asia live on exposed deltas and coastal areas which are frequently threatened by flooding from tropical storm surges.

Perhaps the most dramatic of all coastal hazards are **tsunamis**. Since 1900 over 350 people have died in the USA due to tsunamis and over $500 million worth of property damage has been recorded. These large waves of water are the product of a wide range of phenomena which cause a large-scale displacement of water, including submarine landsliding, volcano-induced mass movements, and earthquakes. The size of the wave is a function of the magnitude of the event producing it. For example, Table 11.3 shows the relationship between tsunami run-up height and the magnitude of the earthquake producing the wave. In recent years they have been most frequently associated with the Pacific Ocean (Figure 11.16). However, tsunamis are recorded in the Holocene coastal record of Britain (Box 11.4) and tsunamis from volcano-induced mass movements on the Canary Islands have the potential to threaten the whole of the Atlantic fringe.

11.2.1 Risk Assessment of Coastal Hazards

The production of hazard maps for coastal areas involves the following tasks: (1) identifying the range of coastal hazards present; (2) assessing the area likely to

Table 11.3 *Earthquake magnitude versus tsunami run-up height, in Japan. Only earthquakes of magnitude seven or greater are responsible for significant tsunami waters with run-up heights in excess of 1 m [Modified from: Bryant (1991)* Natural hazards. *Cambridge University Press, Table 11.3, p. 211]*

Earthquake magnitude: Richter Scale	Tsunami magnitude	Maximum run-up (m)
6	−2	<0.3
6.5	−1	0.5–0.75
7	0	1.0–1.5
7.5	1	2.0–3.0
8	2	4.0–6.0
8.25	3	8.0–12.0
8.5	4	16.0–24.0
8.75	5	>32.0

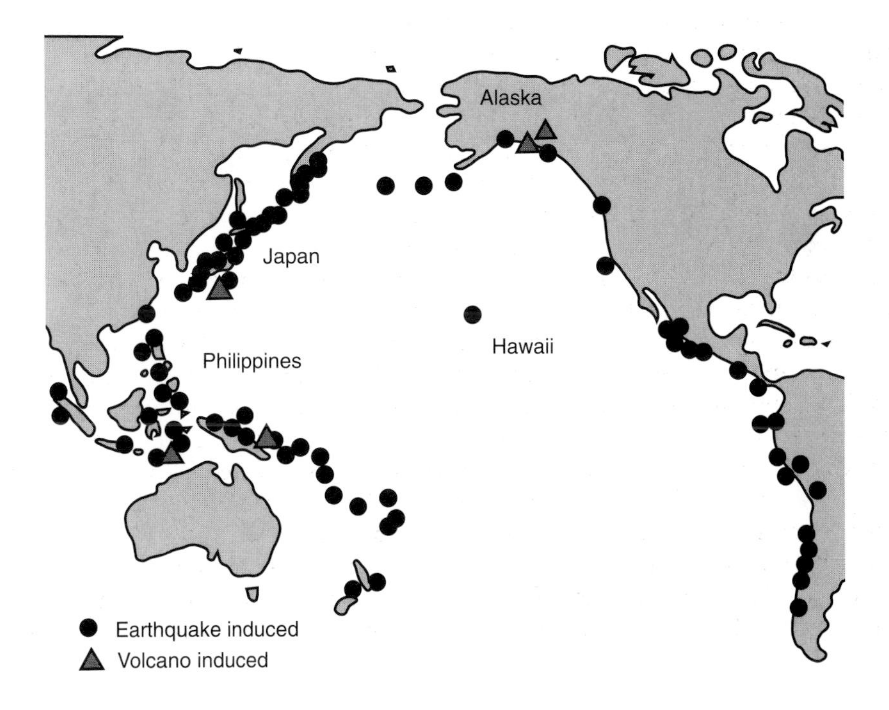

Figure 11.16 *Source areas for tsunami in the Pacific region generating run-up elevation greater than 2.5 m [Modified from: Bryant (1991)* Natural hazards. *Cambridge University Press, Fig. 11.11, p. 207]*

BOX 11.4: A TSUNAMI IN THE NORTH SEA

A layer of sand up to 1 m thick has been reported from raised coastal sediments of Holocene age in eastern Scotland. Where the sand layer overlies peat it has been dated using radiocarbon techniques, falling in the range from 7140 ± 120 to 6850 ± 75 years BP. Originally the sand layer was interpreted as a storm surge deposit; however, it appears to be laterally very extensive and is found in locations all along the east coast of Scotland so that this interpretation now seems unlikely. It was reinterpreted during the late 1980s as a deposit produced by a tsunami generated by a huge submarine landslide off the Norwegian continental slope. This submarine landslide is one of the largest so far found in coastal waters and is known as the Storegga Slide. This slide comprises three events, the first and largest of which involved 3880 km^3 of material and occurred 30 000 years BP. A second and a smaller third slide occurred between 8000 and 5000 years BP and involved 1700 km^3, and were probably triggered by an earthquake with a magnitude of at least 7 on the Richter Scale. If this entire block of material had failed simultaneously, the water displacement would have generated a wave with a height of 350 m. In practice the wave would have been much smaller but still sufficient to deposit a sand layer along much of the eastern coast of Scotland up to 4 m above sea level at that time.

Following this discovery, concern about future tsunami hazards in the North Sea has been expressed. Subsidence associated with the extraction of

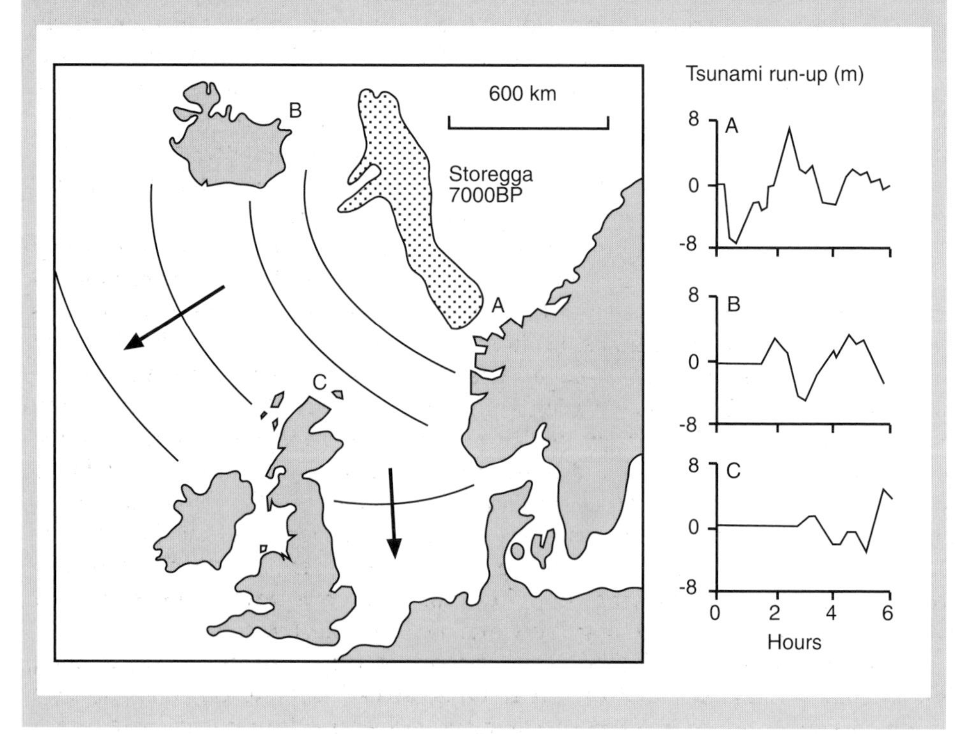

North Sea oil and gas causes small earthquakes which may have the potential to precipitate further submarine slides. Tsunami hazards are not normally considered on European coasts when siting sensitive facilities such as toxic waste dumps or nuclear power stations, but perhaps in the future they should be.

Sources: Long, D., Smith, D.E. & Dawson, A.G. 1989. A Holocene tsunami deposit in eastern Scotland. *Journal of Quaternary Science*, **4**, 61–66; Bugge, T., Belderson, R.H. & Kenyon, N.H. 1988. The Storegga Slide. *Philosophical Transactions of the Royal Society of London*, **A325**, 357–388. [Diagrams modified from: Dawson *et al.* (1991) *Science of Tsunami Hazards*, **9**, Fig. 1, p. 75]

be affected by each hazard on the basis of past experience or direct monitoring; and (3) assessing the probability of that hazard occurring and affecting a given land parcel. This information can be obtained by a careful review of historical data and records for coastal areas (see Section 1.4.3).

The rate of coastal recession can be obtained either by using time-sequential maps, air photographs and oblique photographs or alternatively by direct measurement. The usefulness of direct measurements is usually limited by the availability of time. More importantly, short records may not sample the full range of magnitude and frequency of events operating on a coast, particularly where cliff recession occurs episodically (Figure 11.12B). Long records are, therefore, essential to ensure that high magnitude, low frequency events are detected, or that the chance occurrence of such an event within the data obtained does not bias the observations. Once the rate and style of coastal cliff recession has been established, these rates can be used to predict future coastal geometry after different time periods and thereby define hazard zones.

Determining the extent of coastal flooding is more problematic. Unfortunately it is not simply a task of determining storm water levels and identifying an equivalent land contour to define the flood area. In practice, the flood area is determined by such variables as: (1) height of average storm water level and of the storm waves above that average in relation to any coastal embankments or protective works; (2) the point of ingress, that is, the most likely location or low point along a coastal reach at which water may transgress inland; (3) the geometry of both the natural and built topography, that is where water is likely to flow, collect and pond; (4) where water can subsequently drain offshore, a factor which may be restricted by the presence of sea defences; and (5) how the coastal water will affect inland drainage. In practice, defining flood zones may therefore be complex, involving careful examination of historical flood events, as well as in some cases modelling the water flow within the area.

Predicting tsunamis and their likely impact is very difficult owing to their infrequency and the fact that they may be generated in a variety of ways, each of which may produce a different magnitude of event. Run-up distances are a function of such variables as: (1) the magnitude of the event and distance from the source; (2) coastal orientation to the incident wave; and (3) coastal topography and building geometry which control the way in which water is funnelled onshore, raising run-up distance locally. In Japan, historical records and extreme event analysis (Box 10.1) are used to define typical run-up distances for events with known return periods.

Vulnerability to coastal hazards is a function of the land use, population concentration and building infrastructure located in the hazard zone. Vulnerability of some areas is increasing as a consequence of sea level rise (see Section 9.8.3) and also land reclamation. This is often critical in determining the need for coastal defence. For example, in Britain coastal defence or protection works are normally only funded when it can be shown that the cost-benefit ratio (see Section 10.1) for a given defensive scheme is greater than one; that is, the cost of the damage and its probable occurrence exceeds the cost of the proposed works. This is a financial equation which does not allow consideration of more subjective but equally important issues. For example, it is very difficult to put a financial value to an aesthetic or scientific coastal resource which would be destroyed by a coastal defence structure (Box 6.6). Equally, the financial cost of a house does not include the emotional attachment that the occupants might have to it as their home. It is for this reason that issues of coastal defence can be extremely emotive.

11.2.2 Mitigation and Management of Coastal Hazards

Two distinct historical phases can be identified in the management of coastal hazards: (1) the 'structural era' when civil engineering or 'hard' coastal defence options were routinely employed; and (2) the recent move towards 'soft' coastal defences with a geomorphologically focused emphasis on the design of defensive works. This transition from structural to geomorphological orientated solutions has proceeded at different rates in different countries, and Britain, for example, is only just emerging from the structural era. This shift in approach can be summarised as a shift from conflict with the natural system to an attempt to work with the natural system wherever possible. For example, throughout much of the twentieth century the emphasis in Britain has been placed on defending the coast from attack through concrete sea walls and similar defensive works, and these have traditionally been applied to the coast on the basis of political boundaries. The limits of district and borough councils, each with responsibility for its own section of coast, have formed the end point of the construction of coastal defence schemes, often with little consideration for the adverse effects that such structures might have on a neighbouring authority's coast (Figure 11.15). This has led to exacerbation of coastal problems by defensive works. In recent years, emphasis has been placed on recognising natural geomorphological coastal sediment cells along the coast, which form the natural division for coastal management. Attempts have been made to get groups of authorities to manage their cost in an integrated fashion, but these are yet to have a major impact.

In many countries such as Britain, a legal distinction is made between coastal protection and sea defence. Coastal protection involves land defence against erosion while sea defence is concerned with coastal flooding of low lying land. This is, however, an arbitrary distinction. There are five principal types of coastal protection or defence strategy, each of which is discussed below.

1. **Do nothing**. This involves non-interference with the shoreline sediment budget allowing the coast to continue to evolve naturally. It has the advantages of being cheap, although this can depend on the amount of compensation which may need to be paid or on the cost of relocating essential infrastructure. Retreat may also outflank adjacent sections of coast which have been previously defended. Popular dissent is likely, especially if the politico-legal infrastructure for compensation is not well established. This technique involves rehousing, relocation of transport infrastructure and compensation.
2. **Fall-back**. Here the aim is to consolidate behind the existing shoreline or behind some predetermined fall-back line. This type of strategy has been used in Britain recently whereby agricultural land is allowed to flood and revert to tidal flats or salt marshes thereby creating new habitats and, perhaps more importantly, creating a natural coastal buffer between the sea and improved coastal defence works onshore. Coastal set-back like this requires compensation for the land lost and investment in coastal defence along the chosen fall-back line.
3. **Hold-on**. This method tries to maintain the status quo. Traditionally this has involved the use of hard engineering to prevent any erosion along a stretch of coast. Strong public support is usually received since property is protected and often amenity and access are enhanced. However, it destroys the natural integrity of the foreshore and backshore environment and may exacerbate coastal erosion in down-drift areas by restricting sediment supply to the coast and its longshore transport. This requires large capital investment as well as recurrent maintenance costs.
4. **Build forward**. The aim here is to create a new coastal buffer zone between the coastal infrastructure and the sea, either by creating new beaches or by reclaiming land which can be used sacrificially if required. Beach nourishment is one of the most effective sympathetic coastal defence strategies. However, the sediment has to be 'borrowed' from somewhere and may be associated with environmental degradation in these places. Changes in coastal geometry, particularly through land reclamation, may have an adverse effect on the natural coastal system and sediment transfer. This type of scheme has the advantage of creating new coastal resources and most particularly space in the case of land reclamation. However, in part of Hong Kong, land reclamation has simply increased the problem by encouraging coastal development on reclaimed land, thereby increasing vulnerability.
5. **Build-off**. This employs offshore breakwaters and islands. Such structures create areas of low wave energy suitable for sedimentation and beach accretion, and reduce coastal cliff recession. Their construction may increase

the safety of beaches and coastal water for recreation. However, this type of scheme provides only local protection and usually interferes with coastal sediment flow, particularly with nearshore circulation. Capital costs are high and maintenance costs are recurrent.

The different types of coastal engineering which can be used are listed, along with their relative advantages and disadvantages, in Table 11.4. One of the commonest responses is the construction of some form of sea wall with which to reduce coastal erosion and associated flooding (Figure 11.17). Beach nourishment schemes, however, provide a more dynamic solution, which does not prevent erosion but effectively puts the erosion clock back (Figure 11.18). Beaches are natural energy sinks, reshaping and absorbing energy in response to changing wave conditions. In contrast, a sea wall is static and therefore unable to adapt to changing wave conditions. This contrast represents the difference between what are often referred to as 'hard engineering' (e.g. sea walls, groynes etc.) and 'soft engineering' solutions (e.g. beach nourishment, crenulated beach models, offshore breakwaters etc.).

The choice of a given coastal defence option depends on a complex range of often very local factors (Box 6.6). These considerations can be broadly summarised into five main categories. Firstly, the severity of the problem and the value of the threatened infrastructure is an important consideration. Expensive sea walls are more likely where the problem is severe and the potential losses are high. Secondly, the amenity use of the beach will need to be maintained. Coastal defence schemes involving armourstone and rip-rap, for example, pose a hazard to amenity use and may not always be aesthetically pleasing. **Sea walls** with broad promenades, on the other hand, are more popular due to the beach access they provide. The third consideration may be financial, particularly if protective works are only barely justified by the cost-benefit analysis. The fourth consideration is public perception. Residents often feel safer if there is a visible defence such as a sea wall; beach nourishment may be more effective in some cases but does not provide the same sense of action and security. The final consideration is the background, training and experience of the local coastal defence managers and engineers responsible for drawing up or commissioning a particular scheme. For example, in Britain local officials responsible for coastal defence often have backgrounds in civil engineering and may have little appreciation of the natural coastal system or of geomorphological processes. As a consequence, solutions have often focused on 'hard engineering'. Past experience obtained by trial and error on a particular section of coast will also colour the local coastal engineer's perspective. With increased training and awareness of coastal geomorphological processes, these problems become less apparent.

Some of the structures – sea walls and embankments – outlined in Table 11.4 can be used in flood protection (sea defence in Britain). In addition, tidal barrages, such as the Thames Barrier in London, can be built. The Thames Barrier was planned and commissioned partly as a response to the 1953 storm surge which caused the worst coastal flooding in Europe this century. Approximately 116 km^2

Table 11.4 *Types of coastal defence structure*

Structure	Aim	Design characteristics	Advantages	Disadvantages
A. Individual structures				
Groynes	To trap sediment in longshore transport and thereby retain and build beach	Permeable or impermeable barrier normal to the beach. Construction materials highly variable ranging from wooden panels, to rip-rap, sheet metal and gabion mesh baskets. Wide range of different designs have been tried including; L-shaped, T-shaped and zigzag structures in plan. Variation in permeability to sediment and water has also been tried. Design parameters include length and spacing. These are usually adjusted by experience to ensure that: (1) sediment fillets between groynes remain full preventing beach scour around the landward end of the groyne; and (2) the maximum amount of sediment is trapped.	Simple, relatively cheap coastal defence, large amounts of empirical observations and experience available	Starves sediment from down-drift areas; may simply transfer down-drift the problem of beach depletion
Beach nourishment	Restore the beach to previous levels thereby putting the erosional clock back (Figure 11.18)	Critical design problem is obtaining a source of sediment within an economically viable transport cost distance. The source sediment must be of comparable grain size to the original beach; too coarse and the sediment will be immobile on the beach, too fine and it will be lost rapidly to sea; spreading methods on the receiving beach include pumping and use of earth-moving equipment	A natural beach is the best type of coastal defence	Needs to be periodically repeated as beach erosion continues; may have poor public perception

continues overleaf

Table 11.4 *(continued)*

Structure	Aim	Design characteristics	Advantages	Disadvantages
Revetments	To protect the base of a sea wall or seaward face of an embankment	They may be smooth concrete or stone ramps or alternatively stepped in order to help dissipate wave energy. Permeability may be an issue; some revetments are made from wood to act as wave baffles, but to be permeable to sediment recharge of beach	Effective means of toe protection	May adversely affect the run-up characteristics of breaking waves if poorly designed
Sea walls	To stabilise shoreline position, prevent basal erosion, prevent overtopping	A wide range of designs exist from recurve sea walls which aim to reflect wave energy back to sea to vertical walls exist (Figure 11.17). Usually constructed of concrete and require some form of toe protection via a revetment, armour or rip-rap. Critical issues are outflanking at the lateral end of a sea wall, basal toe erosion and the build-up of water behind the sea wall	Effective sea defence, visible impact may improve access and amenity value of beach	Immobile; need for constant maintenance prevents sediment supply to beach
Embankments	Usually designed to prevent flooding	Earth or hard-core embankments which may be faced on the seaward side by a revetment or protected by armourstone or rip-rap; design cross-sections for earth embankments depend on the solid properties of the material used	Cheap and effective	May cause serious coastal flooding during high tides if they fail; impede inland drainage
Dunes	Dunes provide natural flood protection, beach sediment reservoir	Stabilise existing dunes and encourage initiation of new ones; use of sand fences, surface mulches and planting; control of recreational use may be an important part of any dune stabilisation scheme	Enhances the natural beach-dune system	

Toe protection	To prevent erosion at the base of a coastal slope	Rip-rap or armourstone units are placed at the base of slope to prevent erosion (Figure 11.17)	May be relatively inexpensive and simple to design and deploy	Aesthetically displeasing; dangerous to recreational users
Offshore breakwaters	Reduce wave energy reaching a coast	Constructed from either armourstone blocks or rip-rap; may either remain submerged at low tide or be visible depending on tidal range	Reduces wave erosion and may allow continued coastal slope erosion, but on a reduced scale	Expensive; hazard to shipping and recreational use; may be visually displeasing
B. Composite schemes				
Coastal slope stabilisation	Reduce coastal cliff recession	Involves slope battering and drainage to produce a stable slope. Toe erosion is then prevented either by armourstone or rip-rap barrier or by a sea wall (Figure 11.25). Beach nourishment or groyne construction may also be used to protect the toe structure.	Reduced cliff erosion; improves beach access; tangible and comforting sign of action for local residents	Expensive; destroys natural geological and biological habitats found along eroding cliffs; prevents natural beach recharge
Crenulated bay model	To produce a stable coastal geometry	A series of 'hard points' made of rip-rap are introduced along a coast and extended as coastal erosion bites into the intervening cliff line; wave erosion decreases as natural embayments between hard points are created; a stable shoreline beach and coastline is created (Box 6.6)	Reduces coastal erosion but maintains natural geological and biological coastal habitats	Expensive and largely untried; potentially aesthetically displeasing; lack of tangible sign of protection

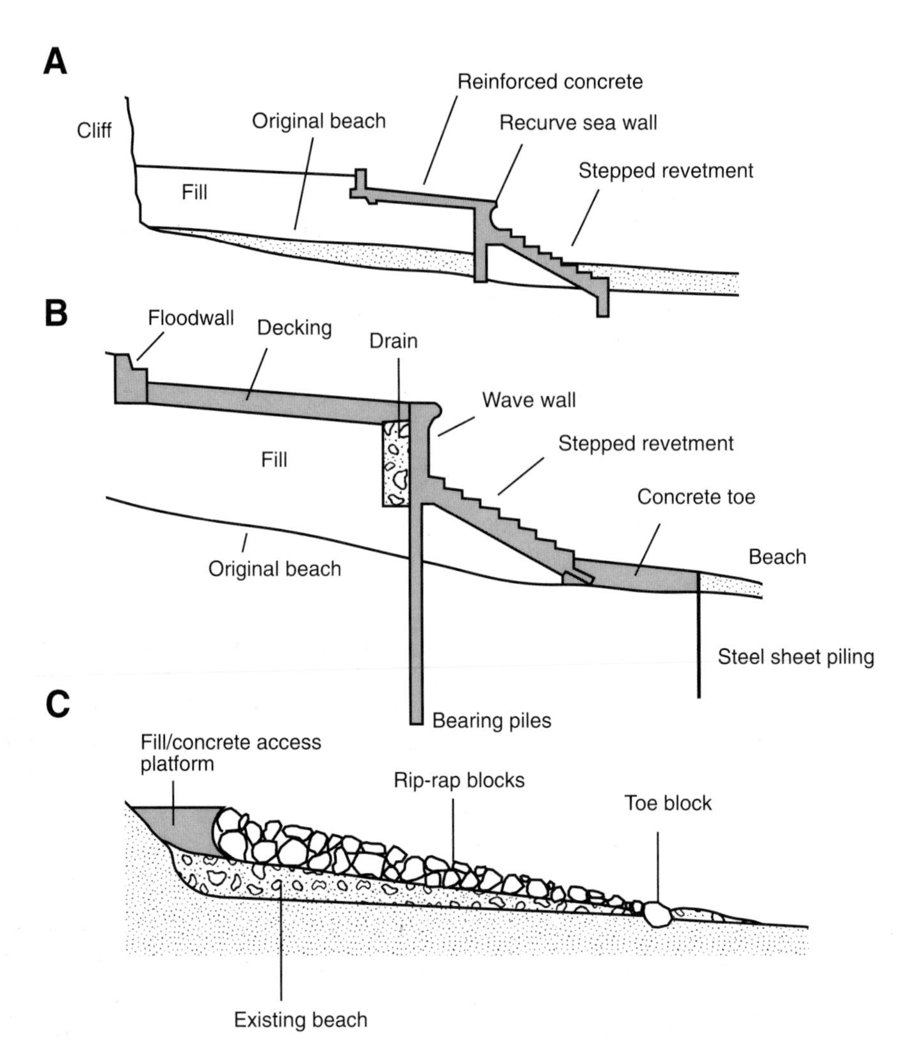

Figure 11.17 *Examples of different types of sea wall from large concrete structures (**A**&**B**) to simpler ones constructed from armourstone and/or rip-rap blocks (**C**)*

of London and over a million people are at risk of flooding if a similar storm surge were to recur. The Thames Barrier was completed in 1983 at Woolwich and provides protection against this. The barrier consists of a series of arcuate gates which can be raised during storm surges to prevent the inland penetration of flood water. The barrier is linked to an improved network of coastal embankments around the outer Thames Estuary to provide complete protection for the capital of Britain.

The management of tsunami hazards is difficult, not least because in most cases they occur too infrequently to justify economically the defensive measures required, despite the catastrophic damage that they can cause. Japan is one of

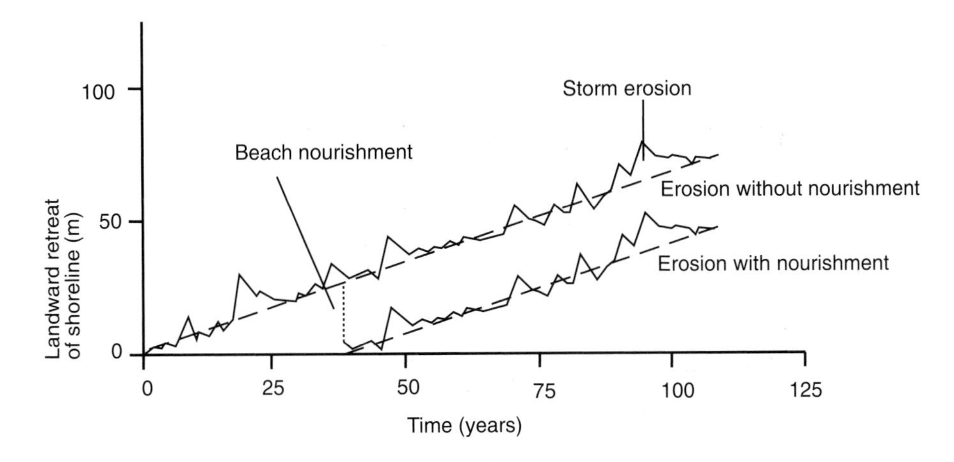

Figure 11.18 *Conceptual model of a beach nourishment scheme. Beach nourishment does not necessarily prevent erosion, but simply restarts the erosion clock*

the few countries where defensive measures have been put into place, due primarily to the extremely high population densities threatened and the frequency of occurrence. These measures consist of two main defensive strategies (Figure 11.19). The first is coastal relocation and redevelopment within the inland limit of the one in 100 year tsunami. Where buildings are located in this zone or on its perimeter they are extensively flood-proofed and raised above the likely flood level. This redevelopment is associated with a new coastal road system designed with rapid evacuation in mind. The second approach is to plant trees in the run-up zone so as to dissipate wave energy. Similarly, in some cases very large offshore breakwaters have been constructed or are proposed. In most places, however, tsunami management simply involves the identification of a wave and rapid coastal evacuation. In the Pacific Ocean, for example, waves can be identified and tracked for a considerable time prior to impact, allowing coastal evacuation.

11.3 MASS MOVEMENT HAZARDS

Mass movements can be defined as the downslope movement of soil, sediment and rock. They range in magnitude from small continuous movements associated with soil creep through to rapid and catastrophic landslides and rock avalanches. On a global scale, mass movements are responsible for only a small proportion of all the deaths caused by natural disasters. For example, between 1947 and 1988, only 29 of the 1062 natural catastrophes involving over 100 deaths were due to mass movement, although Table 11.5 demonstrates that the number of deaths from individual catastrophes can be significant. However, by focusing on loss of life, the true impact of mass movements, which is primarily

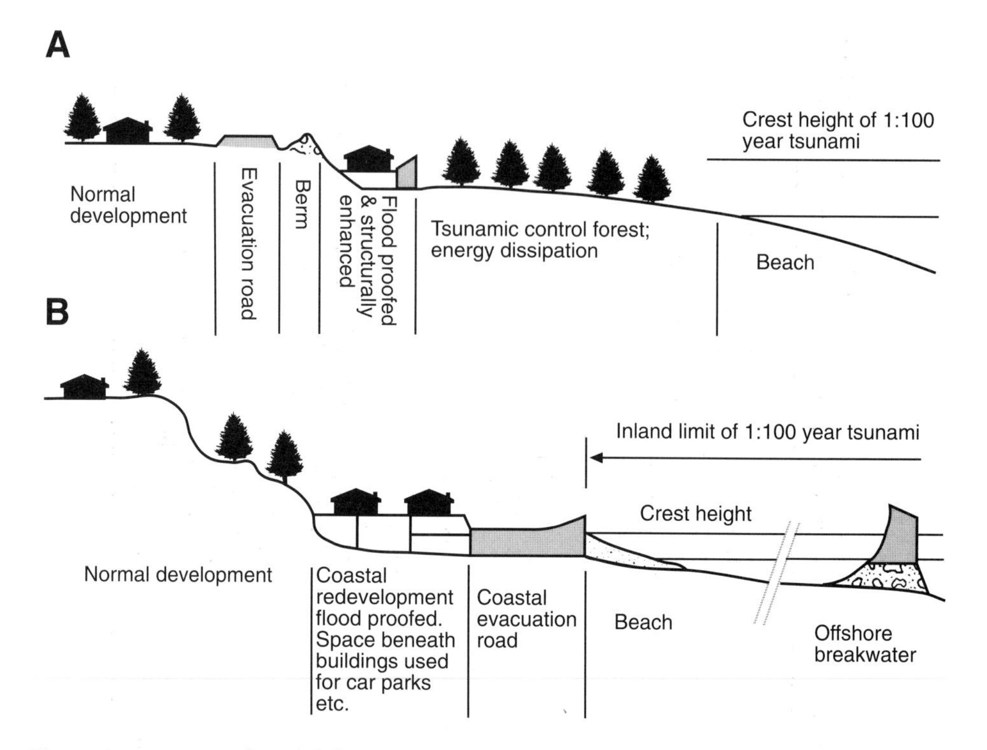

Figure 11.19 *Examples of defensive structures aimed at reducing the impact of tsunamis in Japan [Modified from: Smith (1992)* Environmental hazards. *First edition. Routledge, Figs 6.7 & 6.10, pp. 126 & 134]*

economic, is neglected. Landslides cause widespread and frequent destruction and disruption to human infrastructure and transportation, exacting a considerable economic cost (Box 11.5; Figure 11.20; Table 11.6). The total cost of landslide damage is unknown, but it was estimated in the late 1970s in the USA at over $1000 million, while annual losses in Italy are of the order of $1140 million. Within Britain alone there are over 8365 recorded active and inactive mass movements. The importance of landslides as a constraint on human activity is not well documented, but they are a ubiquitous and underrecognised problem, not least because of the diversity of processes and types of movement which can occur.

A slope becomes unstable and therefore prone to failure by landsliding whenever the forces driving failure exceed those resisting downslope movement. This balance is known as the **factor of safety (F)**, and can be expressed as:

$$F = \frac{\text{Sum of the resisting forces}}{\text{Sum of the driving forces}} = \frac{\text{Strength}}{\text{Strain}}$$

A slope with a factor of safety of less than one is unstable and therefore in a condition for failure, while a slope with a factor of safety of greater than one is

Table 11.5 *Some of the most important mass movement disasters of the twentieth century [Modified from: Jones (1995)* Proceedings of Royal Academy of Engineering Conference on Landslide hazard mitigation. *Royal Academy of Engineering, London, Table 3, p. 24]*

Place	Date	Type	Impact
Java	1919	Debris flow	5100 killed, 140 villages destroyed
Kansu, China	1920	Loess flow	c. 200 000 killed
California, USA	1934	Debris flow	40 killed, 400 houses destroyed
Kure, Japan	1945		1154 killed
Tokyo, Japan	1958		1100 killed
Ranrachirca, Peru	1962	Ice & rock avalanche	3500+ killed
Vaiont, Italy	1963	Rockslide into reservoir	c. 2600 killed
Aberfan, Britain	1966		144 killed
Rio de Janeiro, Brazil	1966		1000 killed
Rio de Janeiro, Brazil	1967		1700 killed
Virginia, USA	1969	Derbis flow	150 killed
Yungay, Peru	1970	Earthquake-triggered debris avalanche	25 000 killed
Chungar	1971		259 killed
Hong Kong	1972		138 killed
Kamijima, Japan	1972		112 killed
S. Italy	1972–73		100 villages abandoned, affecting 200 000 people
Mayunmarca, Peru	1974	Derbis flow	Town destroyed, 451 killed
Mantaro Valley, Peru	1974		450 killed
Mt Semeru	1981		500 killed
Yacitan, Peru	1983		233+ killed
W. Nepal	1983		186 killed
Dongziang, China	1983		4 villages destroyed, 227 killed
Armero, Colombia	1985	Lahar	c. 22 000 killed
Çatak, Turkey	1988		66 killed

BOX 11.5: THE MAM TOR LANDSLIP: AN EXAMPLE OF A LARGE COMPOSITE MASS MOVEMENT

The Mam Tor landslip is located approximately 2 km northwest of Castleton in Derbyshire, northern England. This large composite failure occurred some 3600 years ago, but is still active today. The failure consists of a large arcuate scar beneath which there is a debris lobe of over a kilometre in length. Geologically the area consists of Carboniferous shales with thin sandstone interbeds. The shales are particularly susceptible to weathering, which rapidly reduces their strength. A combination of glacial oversteepening during the last glacial cycle and rapid weathering is believed to be responsible for the failure. The debris lobe remains mobile, probably as a consequence of secondary failures within it. These appear to be precipitated by meteorologically induced fluctuations in the water table.

In 1810 an important road linking Manchester and Sheffield (A625) was built across the debris lobe. Owing to the presence of a hairpin bend the road traversed the landslip twice. Ever since its construction, constant repair and

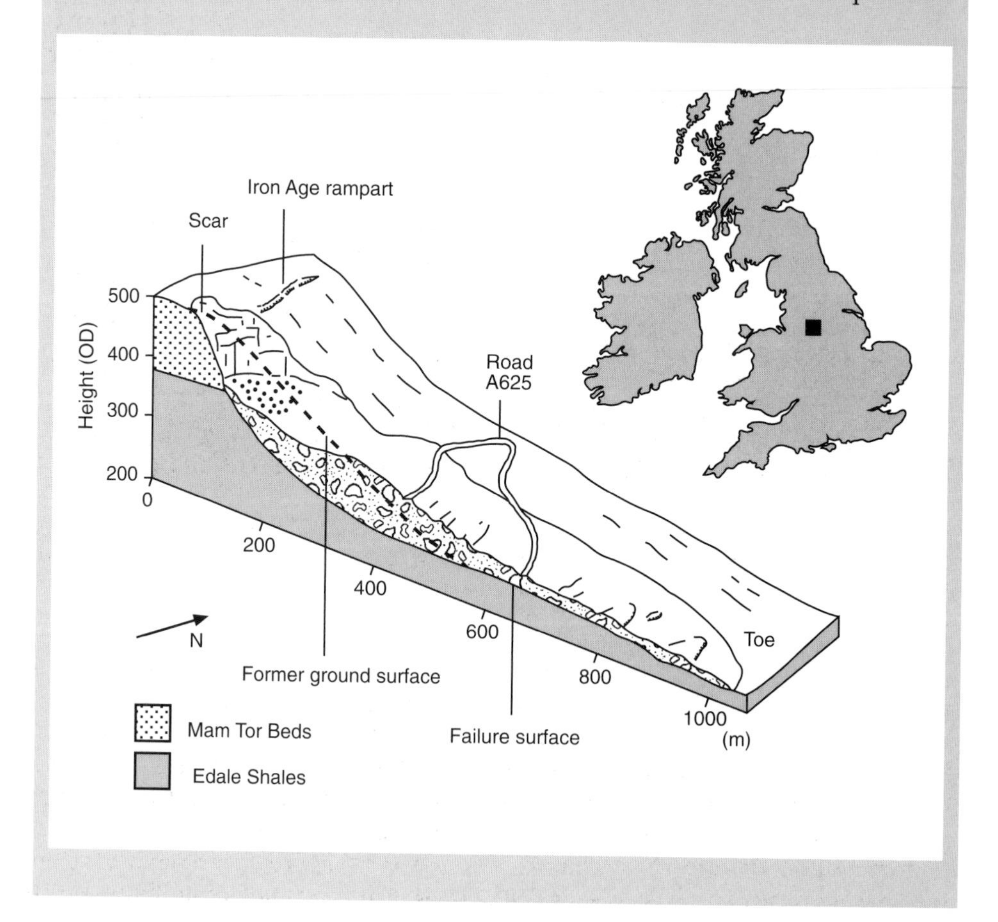

maintenance have been required due to ground movement and in 1977 a major failure reduced the road to a single lane. In 1979 the road was closed, the cost of remedial work being considered too expensive for the available funds. It is one of the few examples where a major trunk road has been closed permanently in Britain as a consequence of mass movement.

Source: Skempton, A.W., Leadbeater, A.D. & Chandler, R.J. 1989. The Mam Tor landslide, north Derbyshire. *Philosophical Transactions of the Royal Society of London*, **A329**, 503–547. [Diagrams modified from: Cripps & Hird (1992) *Geoscientist*, **2**, Fig. 3, p. 24]

Figure 11.20 *The Mam Tor landslide which permanently closed the A625 road between Manchester and Sheffield in England, causing disruption to the transport network within this area*

stable. A slope with a factor of safety of one is in critical equilibrium. Most slopes are stable and mass failure is caused by a progressive change in the relative balance between the resisting and driving forces (Figure 11.21A&B). This change is caused by preparatory factors which move the slope to a critical state with a factor of safety close to one (marginal stability), at which point a landslide may be caused by some triggering factor such as an earthquake, storm or similar event (Figure 11.21B). Figure 11.21C shows some of the common preparatory factors which can operate to cause slope failure. These can be grouped into those which

Table 11.6 *Cost of recent mass movement damage [Modified from Gares (1994) Geomorphology,* **10**, *Table 4, p. 9]*

Location	Year	Damage (millions of US$)
San Jose, California	1968–70	1.76
San Francisco Bay Area	1968–69	25.18
Los Angeles, California	1969	6.3
Los Angles, California	1978	>60
Seattle, Washington	1971–72	0.46
Hamilton Co., Ohio	1974–76	17.44

reduce slope strength and those which increase the stress acting upon it. Slope strength can be reduced by weathering and variation in porewater content, while stress may be increased by such things as loading, removal of toe support, and by transitory events such as traffic vibration or earthquakes. Six types of landslide-prone terrains can be identified, although they are not mutually exclusive.

1. **Seismically active regions**. Earthquakes are important triggers for landslides, both by temporarily increasing shear stress and by causing liquefaction of fine grained sediment. For example, the 1976 Guatemala City earthquake caused 10 000 mass movements each involving more than 15 000 m^3 of material.
2. **Mountainous environments**. Areas of high relative relief and oversteepened slopes are commonly due to recent glaciation or intense downcutting by tectonic uplift. These high energy environments are prone to widespread rockfalls, while the resulting abundant debris may give rise to significant debris flows. The most important types of event within these areas are catastrophic rock avalanches involving huge failures (5–100 million m^3 of material), high velocities (90–400 km per hour) and long transport distances. Of these events, the 1970 rock and ice Huascaran failure in Peru killed over 25 000 people.
3. **Land degradation**. In regions of moderate relief, friable soils and severe land degradation as a consequence of human action, landslides are common. In southern Italy, for example, historically important hill-top villages are being threatened and in some cases abandoned, owing to mass failure on surrounding slopes caused by poor land management. Deforestation in tropical regions has also created large areas of terrain prone to such failures.
4. **Areas covered by thick loess sheets**. These are prone to slope failure, especially where deeply eroded and subject to earthquake shaking or prolonged rainfall. The massive death tolls produced by the 1920 Kansu earthquakes in Japan, in which over 200 000 people died, were largely due to widespread failure of loess slopes.
5. **Adverse meteorology**. Areas subject to frequent periods of intense rainfall, such as those with a monsoonal climate, are often associated with mass movement problems.

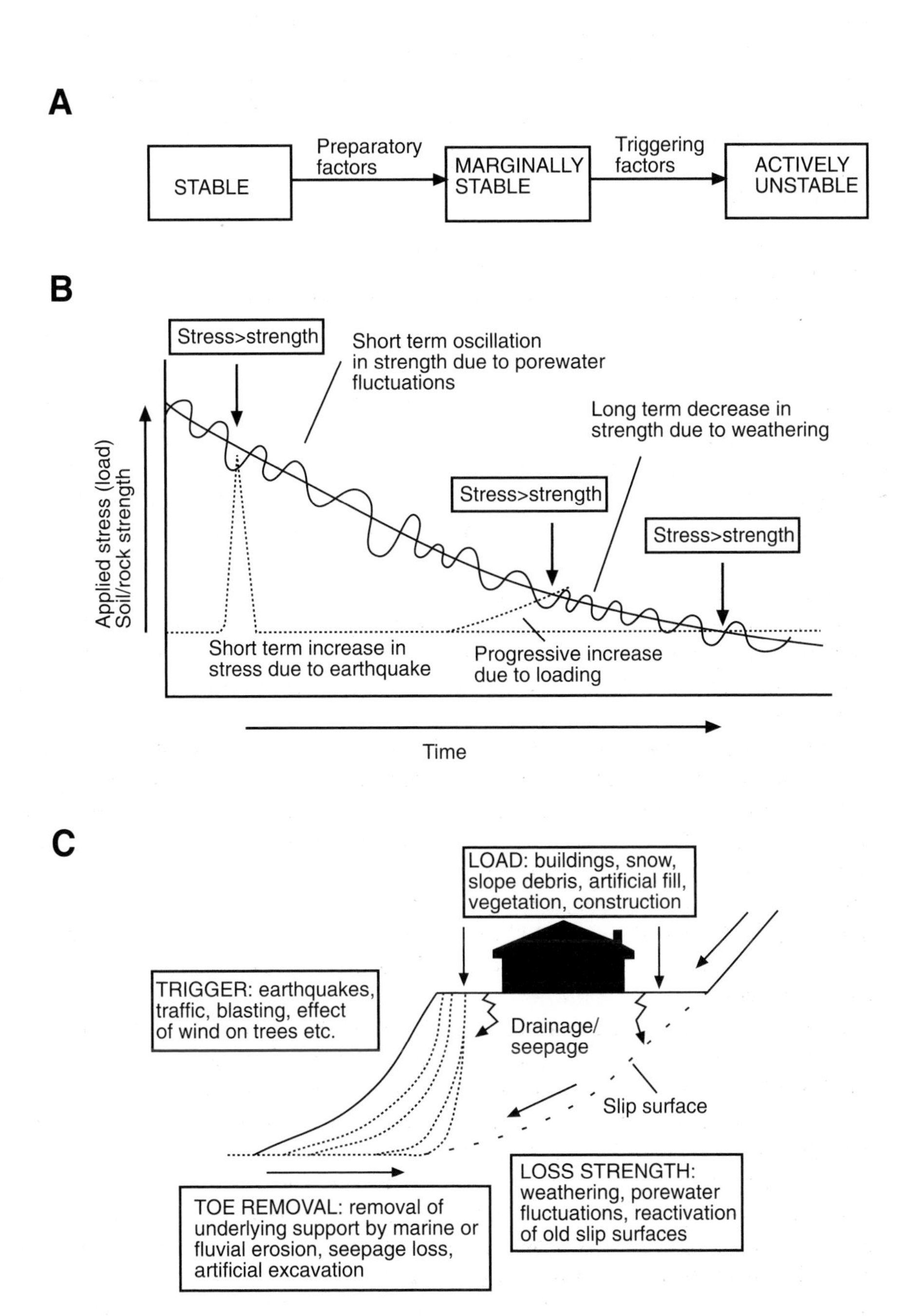

Figure 11.21 *The causes of mass movement.* ***A.*** *Stages in the development of a mass movement.* ***B.*** *Progressive change in slope strength and shear stress, leading to slope failure when stress exceeds strength.* ***C.*** *Mechanisms by which the factor of safety of a slope can be reduced, either by decreasing slope strength, or increasing the stress on a slope*

6. **Areas subject to rapid development**. Some regions may be subject to major changes in stability due to human development. These include slope remodelling, vegetation clearance, irrigation, poor maintenance of water infrastructure, and urbanisation on unstable slopes.

Mass movements come in many shapes and sizes. They can be classified on the basis of: the materials involved, whether soft sediments or competent rocks; the type of movement, whether flows, slides or heave-type motions; and the speed of movement (Figure 11.22). Common mass movements include soil creep, rotational slumps, mud flows, rockfalls and avalanches, and block slides (Figure 11.22). In practice, however, as the scale of mass movements increases they become composite in nature, involving many different styles of failure and motion.

11.3.1 Risk Assessment of Mass Movement Hazards

In the context of mass movements, risk assessment has largely focused on the production of **hazard maps** depicting the relative stability of slopes, although it must be remembered that assessment of the vulnerability to damage should also be considered. In practice, mass movement hazard assessment can be divided into: (1) prediction, concerned with the likelihood of a given slope failing; and (2) forecasting, involving identification of the location, timing and magnitude of an individual failure.

1. **Prediction**. This operates at two very different scales: (A) determining the relative instability of a chosen site; and (B) determining the regional variation of mass movement hazard potential. The first is largely the preserve of the civil engineer or engineering geologist and is often an important part of a site investigation. It involves the use of the factor of safety, in which all the shear stresses acting on a slope are calculated and compared to the total shear strength of the slope. Detailed site investigations are therefore necessary in order to obtain data on the strengths of the materials composing the slope and to determine their likely mode of failure, as well as to identify the presence of existing shear surfaces or other lines of weakness. The stability of a given slope in a range of conditions is then expressed in terms of the factor of safety. In most cases this involves a detailed understanding of soil mechanics.

   Of more importance to regional planning is the production of regional hazard maps depicting susceptibility to slope failure. There is a wide range of different and frequently overlapping approaches in the production of mass movement hazard maps, but we can identify three main approaches to the problem: (A) mass movement inventory mapping; (B) qualitative slope classification; and (C) numerical scoring systems.

   Mass movement inventory mapping involves mapping and recording all active and relict mass movements within an area using a combination of air photographs and field mapping. In this way those areas prone to failure,

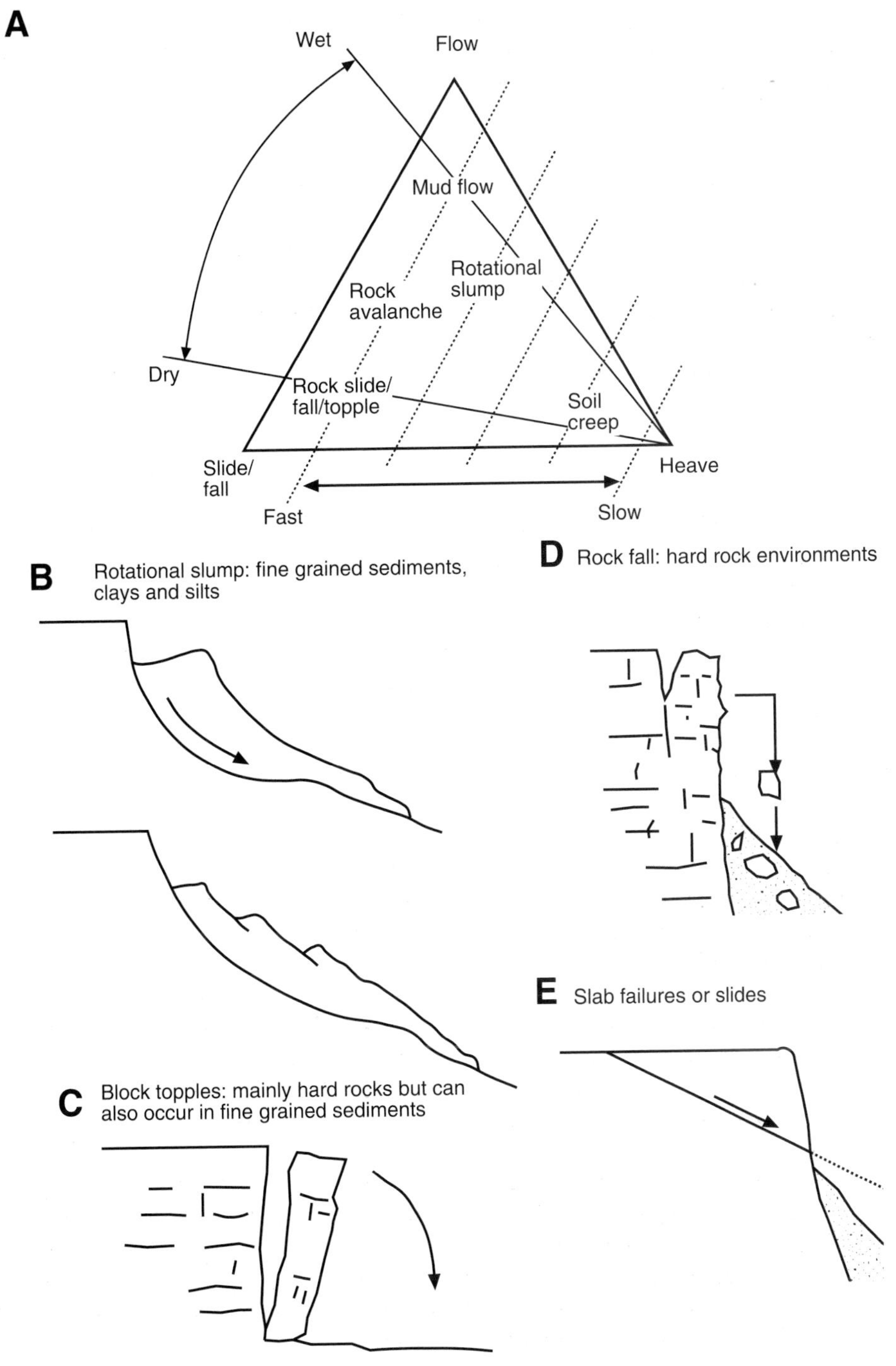

Figure 11.22 *Classification and characteristics of common types of mass movement [A Modifed from: Carson & Kirby (1972)* Hillslope form and process. *Cambridge University Press, Fig. 5.2, p. 100]*

F Translational slide: blocks moving over a weaker shear zone or layer

G Rock avalanche: rocks and sediments

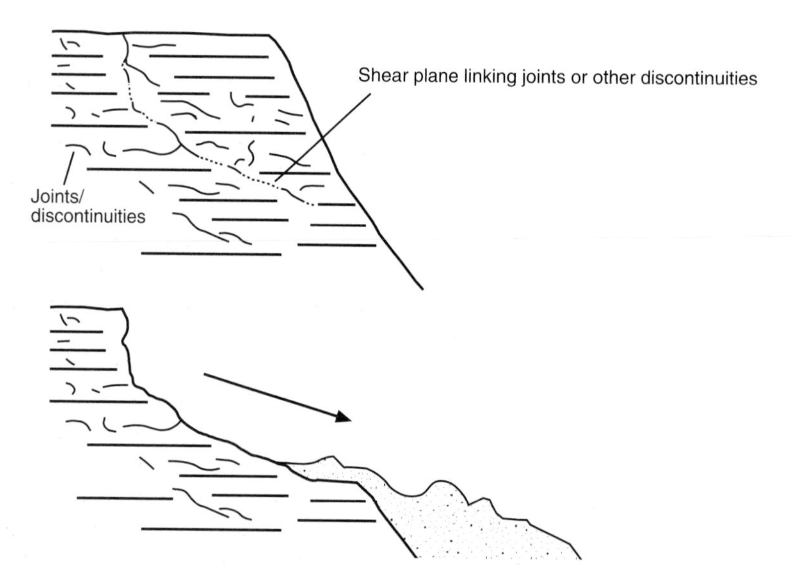

H Mudflow: shallow failure in fine grained soils and sediments

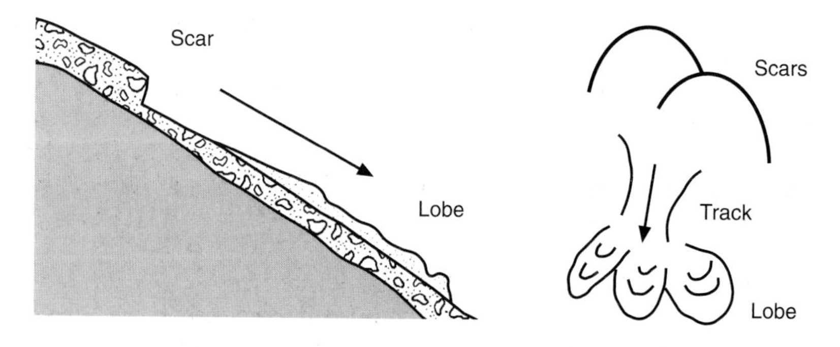

Figure 11.22 *(continued)*

either in the past or at present, are identified as problem areas. The quality of maps or databases derived from this process depends largely on the ability and the experience of the survey team as well as the criteria or definitions used to identify each landslide. In addition, this technique will commonly fail to identify those areas which are unstable at present, but have no history of failure.

The second approach involves the use of **qualitative slope classification** such as that shown in Table 11.7 in which individual parcels of land are classified. In this case the field geologist is asked to assess the stability of a slope from visible evidence; that is, to exercise judgement about how close a slope is to failure. This is both difficult and a serious responsibility with liability concerns, since incorrect judgements may lead to engineering problems if a mistake is made. This approach is totally dependent on the experience and skill of the geologist concerned. It is also important to realise that simply erring on the side of caution may lead to unneeded and expensive site investigation work, causing an escalation of project cost. More importantly, it is labour intensive and the quality of such surveys is difficult to audit. As a consequence, in recent years greater emphasis has been placed on the use of numerical scoring systems which incorporate all the variables likely to affect the stability of a slope.

Numerical scoring systems are based on an approach in which all the available data relevant to regional slope stability are collated and then cross-tabulated in order to derive a numerical value of hazard susceptibly. The approach is illustrated generically in Figure 11.23. A list of all the variables likely to affect the stability of a slope is drawn up – lithology, structural geology, slope angle, soil/sediment cover, land use, hydrology – and maps are produced for each variable. These maps are then combined or cross-tabulated in some way in order to produce a hazard map in which the susceptibility to failure is reduced to three or five classes. There are two approaches to this cross-tabulation. The first is based on geological judgement. Using past experience, either personal or drawn from published literature, an assessment is made of the relative importance in determining slope stability of each variable and a scoring system is drawn up. Using this scoring system, each land parcel is given a value for hazard susceptibility, and a hazard map can be drawn up based on these values.

An alternative method introduces a mass movement inventory map and removes the subjectivity introduced by the scoring system. In this method the mass movement inventory map is compared statistically with all the other variables in order to determine which combinations of variables result in the most mass movements. This is then used predictively in the production of a hazard map.

There are a wide range of strategies which have been used in the production of hazard maps. The importance of numerical scoring systems is that they allow an external audit of a map's quality or reliability to be undertaken. They do not, however, necessarily remove all the subjectivity involved in hazard assessment. Whatever the approach used in the produc-

Table 11.7 *Qualitative landslide probability classification [Modified from: Crozier (1986)* Landslides: causes, consequences and environment. *Croom Helm, Table 6.3, p. 212]*

Class	Description
I	Slopes which show no evidence of previous landslide activity and which by (1) stress analysis, (2) analogy with other slopes, or (3) analysis of stability factors, are considered highly unlikely to develop landslides in the foreseeable future
II	Slopes which show no evidence of previous landslide activity but which are considered likely to develop landslides in the future. Landslide potential indicated by stress analysis, analogy with other slopes, or by analysis of stability factors
III	Slopes with evidence of previous landslide activity but which have not undergone movement in the previous 100 years
IV	Slopes infrequently subject to new or renewed landslide activity. Triggering of landslides results from events with recurrence intervals greater than five years
V	Slopes frequently subject to new or renewed landslide activity. Triggering of landslides results from events with recurrence intervals of up to five years
VI	Slopes with active landslides. Material is continually moving, and landslide forms are fresh and well defined. Movement may be continuous or seasonal

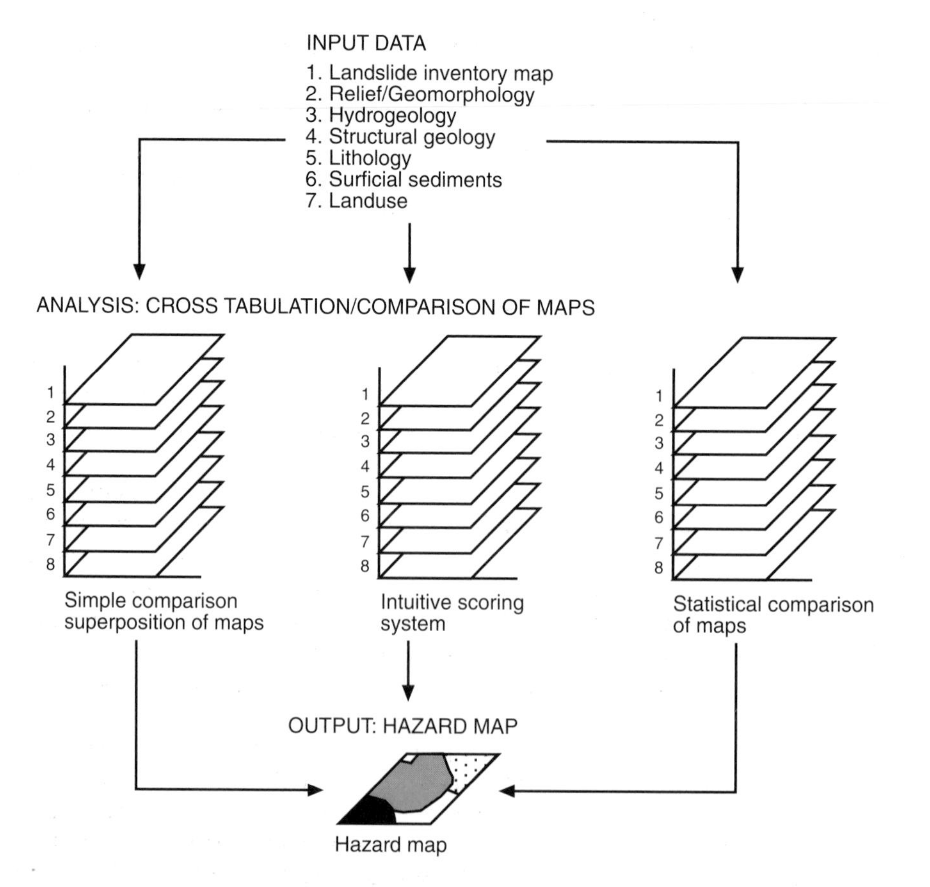

Figure 11.23 *Generic model for the production of a mass movement hazard map*

Table 11.8 Hazard categories in the Grindelwald area of the Swiss Alps [Modified from: Kienholz (1978) Arctic and Alpine Research, **10**, p. 178]

Hazard category	Description
3	Houses are destroyed and people are in danger from landsliding
2	Houses in little danger, but areas between houses may experience some landslides, hence people may be in danger
1	Houses in very little danger, slight but infrequent danger to people outside the houses
0	No known danger

tion of a hazard map it is essential that it is presented in such a way that the non-expert can make use of it. This is well illustrated by the mass movement class descriptions used on a hazard map of the Grindelwald area of the Swiss Alps (Table 11.8). Hazard maps are not static and may date quickly if land management and land use changes.

2. **Forecasting**. This is concerned with establishing the location, character, magnitude and timing of specific future slope failure phenomena, especially failures that could result in detrimental effects on human activity. The aim is either to undertake emergency remedial action or, more usually, to evacuate and reduce hazard loss. It involves careful monitoring of slope characteristics in order to recognise change in the slope stability over time. Also of importance is the identification, using the historical record, of critical triggers such as earthquakes or storm events within an area.

11.3.2 Mitigation and Management of Mass Movement Hazards

The choice of management options for mass movement hazards is similar to all others in this chapter and includes (Figure 10.5): (1) do nothing and accept the loss; (2) remove the problem; (3) avoid the site; and (4) mitigation works and careful building design.

The 'do nothing' solution is only a realistic option where the problem either cannot be avoided and the cost of stabilisation work is prohibitive, or alternatively is of a trivial nature. Where the problem is significant landslide forecasting and a civil defence programme are essential to ensure that hazard loss is minimised. In some cases the problem can be removed. This is particularly useful in hard rock cliffs where loose block may be removed prior to construction, either by hand using simple tools or by blasting; this process is known as **scaling**.

The best solution, however, is to avoid problem sites and in this context mass movement inventory maps and hazard maps are essential. The use of such maps needs to be an integral part of the planning process to ensure that planning/building permission is not granted in areas of severe mass movement hazard, or at least the potential future impacts are fully accommodated into the design. In Britain such hazard maps may enter the planning system first as advice, then as

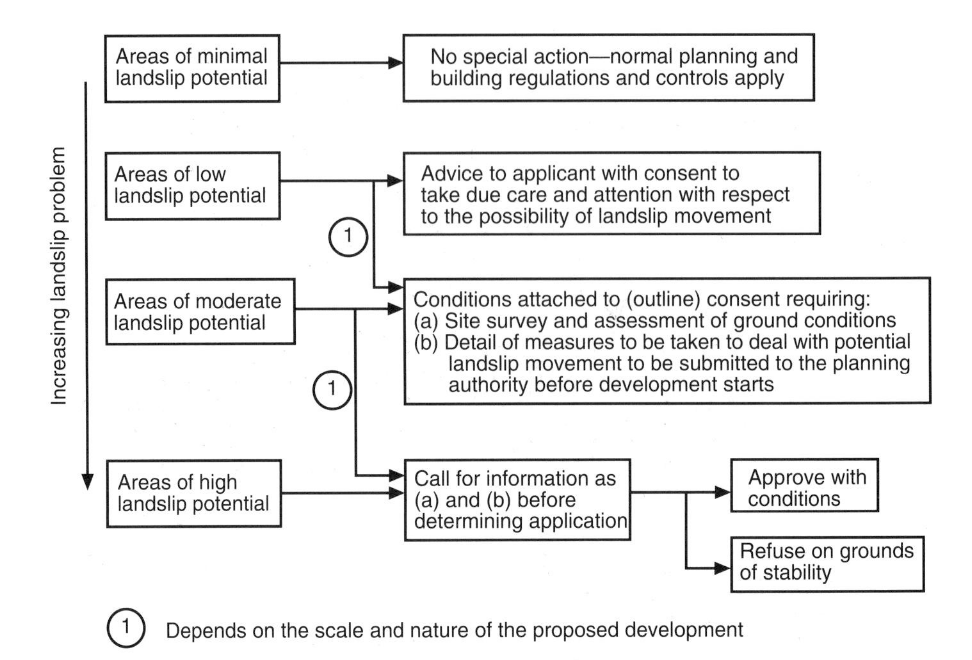

Figure 11.24 *Potential relationship between hazard maps and the planning process [Modified from: Jones (1995) In: McGregor & Thompson (Eds)* Geomorphology and land management in a changing environment. *John Wiley & Sons, Fig. 2.5, p. 29]*

a series of planning conditions, and ultimately by refusing planning consent as the magnitude of the landslide problem increases (Figure 11.24). Avoidance is not always possible, particularly when dealing with areas which have only slight stability problems. In this case, careful ground stability analysis and careful design are essential to ensure that any structure increases slope stability as opposed to decreasing it. This may involve the installation of efficient drainage systems, the reduction of slope angles or the careful design of filled areas to reduce any increase in the weight acting on a slope.

Where construction has already taken place in an unstable location or where a choice of location is not possible, mitigation work and careful site management are essential. Figure 11.25 illustrates some of the ways in which slopes may be stabilised. In dealing with slopes composed of 'soft rocks', such as clays and other sediments which are prone to slumping and flow, there are three basic principles (Figure 11.25A): (1) reduce the slope angle (i.e. increase the batter); (2) install some form of toe protection to prevent basal erosion resteepening the slope; and (3) drain the slope so as to increase its strength. In dealing with 'hard rock' slopes the following techniques are commonly used (Figure 11.25B&C):

Figure 11.25 *(opposite) Methods of slope stabilisation.* ***A.*** *'Soft rock' or sediment cliffs.* ***B&C.*** *'Hard rock' slopes*

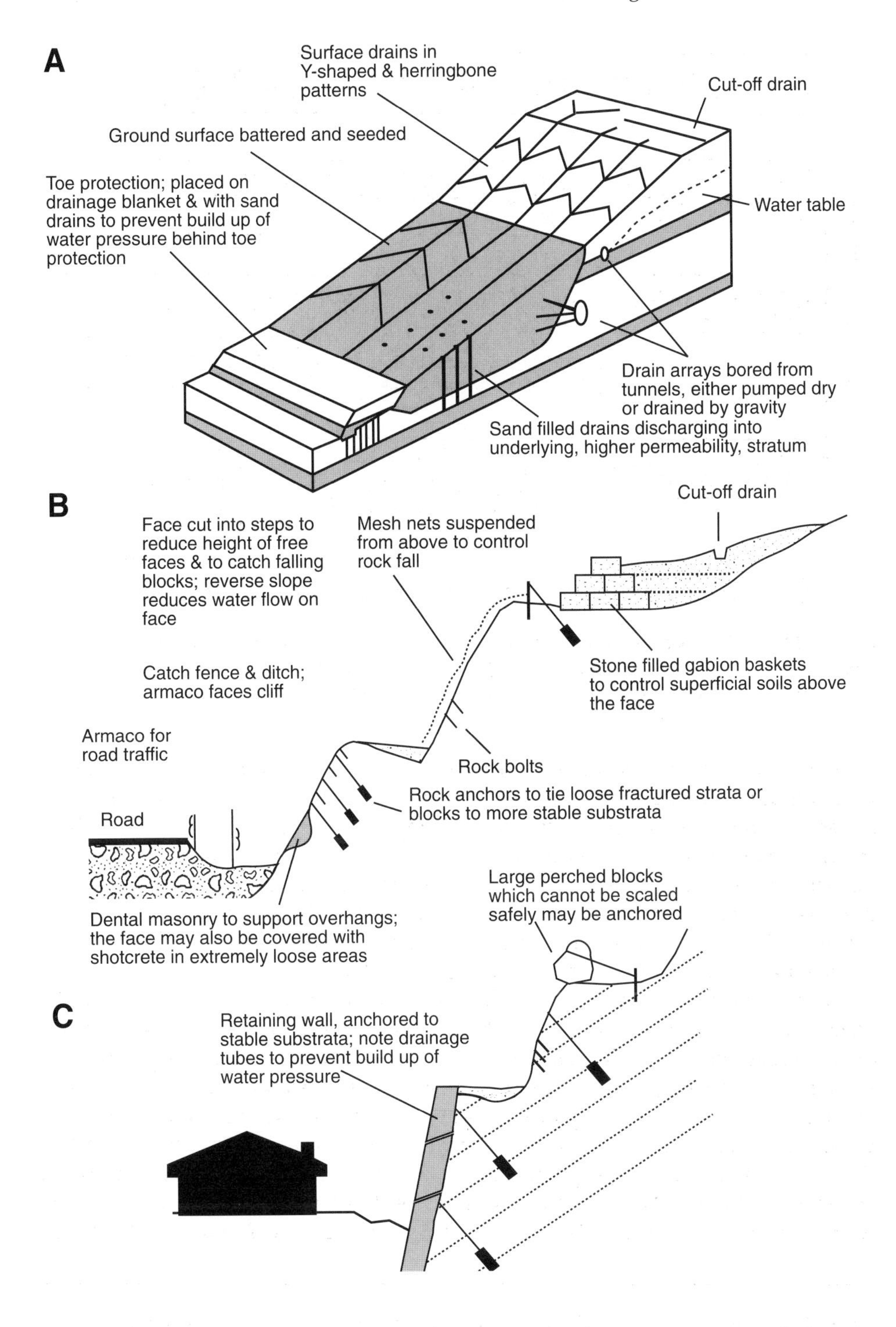
A
Surface drains in Y-shaped & herringbone patterns
Cut-off drain
Ground surface battered and seeded
Toe protection; placed on drainage blanket & with sand drains to prevent build up of water pressure behind toe protection
Water table
Drain arrays bored from tunnels, either pumped dry or drained by gravity
Sand filled drains discharging into underlying, higher permeability, stratum
B
Cut-off drain
Face cut into steps to reduce height of free faces & to catch falling blocks; reverse slope reduces water flow on face
Mesh nets suspended from above to control rock fall
Stone filled gabion baskets to control superficial soils above the face
Catch fence & ditch; armaco faces cliff
Armaco for road traffic
Rock bolts
Rock anchors to tie loose fractured strata or blocks to more stable substrata
Road
Dental masonry to support overhangs; the face may also be covered with shotcrete in extremely loose areas
Large perched blocks which cannot be scaled safely may be anchored
C
Retaining wall, anchored to stable substrata; note drainage tubes to prevent build up of water pressure

Figure 11.26 *Use of stone-filled gabion baskets to control scree slopes encroaching onto houses in Bleanau, Ffestininog in North Wales*

(1) reduction of slope height and angle through the use of benches; (2) the use of retaining walls tied via rock anchors to stable substrata; (3) the use of gabion baskets to retain loose material (Figure 11.26); (4) the use of rock bolts and rock anchors to secure loose or fractured rock; and (5) the use of mesh, shotcrete or catch fences to prevent damage when rockfall is unavoidable.

Effective site management also has an important role. For example, the maintenance of ditches, drains, sewers, ponds and other water bodies is essential to ensure that uncontrolled drainage is prevented; saturated slopes have lower strength characteristics. Careful control of garden works and of the construction of extensions, garages and sheds in order to prevent unsafe loading or slope steepening may be required. Education of the owner/occupiers to ensure effective management is therefore essential. In order to follow and take an interest in

Figure 11.26 *(continued)*

the advice provided, local residents will require education about the problems, and how such steps will help in their mitigation.

The implementation of such management strategies, particularly in the avoidance of unstable ground, can be achieved through educating people about the formal planning process and the use of building codes. This, however, requires basic information about the magnitude and importance of the landslide problem, and it is in the provision of this information, in the form of hazard maps, that the environmental geologist has the greatest role to play. The impact of effective landslide hazard management is well illustrated by Japan. Major landslides triggered by heavy typhoon rains were responsible for the loss of more than 130 000 homes and 500 lives in 1938, a figure reduced to 2000 homes and 125 lives by 1976 due to an effective hazard management scheme (Figure 10.2).

11.4 GLACIAL HAZARDS

Although having a somewhat restricted distribution at present, glaciers have the potential to destroy and disrupt human infrastructure. Their potential was particularly clear during the Little Ice Age of the late seventeenth and early eighteenth centuries, when alpine glaciers advanced to ruin farmland, destroy villages and cut transport links, leaving a striking imprint in the historical record. Although most glaciers have been decreasing in size since the Little Ice Age, and are likely to continue to do so, their potential as an agent of destruction is clear.

The tourist industry has constructed numerous attractions and infrastructure – ski-lifts, cable cars, hotels, restaurants and railways – close to the margins of many glaciers which would be threatened by even a small readvance. The Karakorum Highway in Pakistan, for example, crosses the snout of three glaciers, all of which would threaten it if they were to advance. Similarly, the Copper River and Northwestern Rail Road in Alaska illustrates the problems very well. Constructed between 1906 and 1910 to link the Kennicott Copper Mines with the Port of Cordova, it passes in front of six glacier snouts in one 50 km section, two of which advanced rapidly during construction. By 1911 one of these glaciers had advanced to within 450 m of the line, necessitating the construction of a bridge at a cost of over $1.5 million.

In many cases, changes in the geometry of a glacier snout can be predicted from a knowledge of the glacier's mass balance; that is, the balance between the accumulation of snow and its ablation (melting plus iceberg calving) over the glacier surface. Typically, glaciers tend to advance when air temperature falls or snowfall increases. However, the geometry of certain glaciers is not predictable owing to periods of unstable flow. Certain glaciers, known as **surge-type glaciers**, are prone to periodic episodes of fast flow several orders of magnitude greater than normal. In these glaciers, a wave of fast-flowing ice passes through the glacier and may cause its snout to advance rapidly. Such periods of flow are rarely sustained for long, but have the potential to cause serious damage to anything in the way.

Infrastructure is threatened not only by glacier advances but also by glacier decay. Glaciers provide support for valley sides which have often been over-steepened by glacial erosion. Debuttressing of slopes by the removal of glacier support as they melt may lead to major mass movements, particularly where the debris becomes mixed with snow and ice (Box 11.6). The recent rock avalanches involving 5–10 million m^3 of debris on Mount Fletcher, above the Maud Glacier in the Southern Alps of New Zealand, provide an excellent example. Glacier thinning of over 250 m since the Little Ice Age maximum exposed a steeper toe to the slope along which failure occurred (Figure 11.27A). Deep-seated deformation of slopes above the Affliction Glacier in British Columbia is also associated with glacier thinning (>100 m). Tension cracks, grabens and collapse pits have developed along the flank of the glacier and the factor of safety has gradually decreased as the glacier has thinned, debuttressing the slope (Figure 11.27B). Thinning of the glacier has decreased the factor of safety of the slope by 15% since the Little Ice Age.

Glacial meltwater may also provide a significant hazard. The braided meltwater streams, outwash or sandur plains, in front of glaciers are some of the most dynamic of all fluvial environments; they shift courses, deposit large amounts of coarse sediment, and have rapidly changing discharges. These active outwash plains pose significant hazards to construction. The Karakorum Highway (Pakistan), for example, crosses the outwash plain of the Batura Glacier. In 1973 a major shift in the pattern of meltwater led to damage to bridges and the removal of embankments necessitating the reconstruction of over 3 km of road. Similarly, the highway passes in front of the Ghulkin Glacier. One of the bridges originally

BOX 11.6: ROCK/ICE AVALANCHES: TWO EXAMPLES FROM NEVADOS HUASCARAN, PERU

Two of the most destructive rock/ice avalanches that have ever occurred took place in 1962 and 1970 on Nevados Huascaran in Peru. They were caused by a combination of factors including rapid glacier thinning on the northern side of the mountain. In 1962 approximately 13 million m^3 of rock and glacier ice broke free from the northern flank of Huascaran and travelled for 16 km

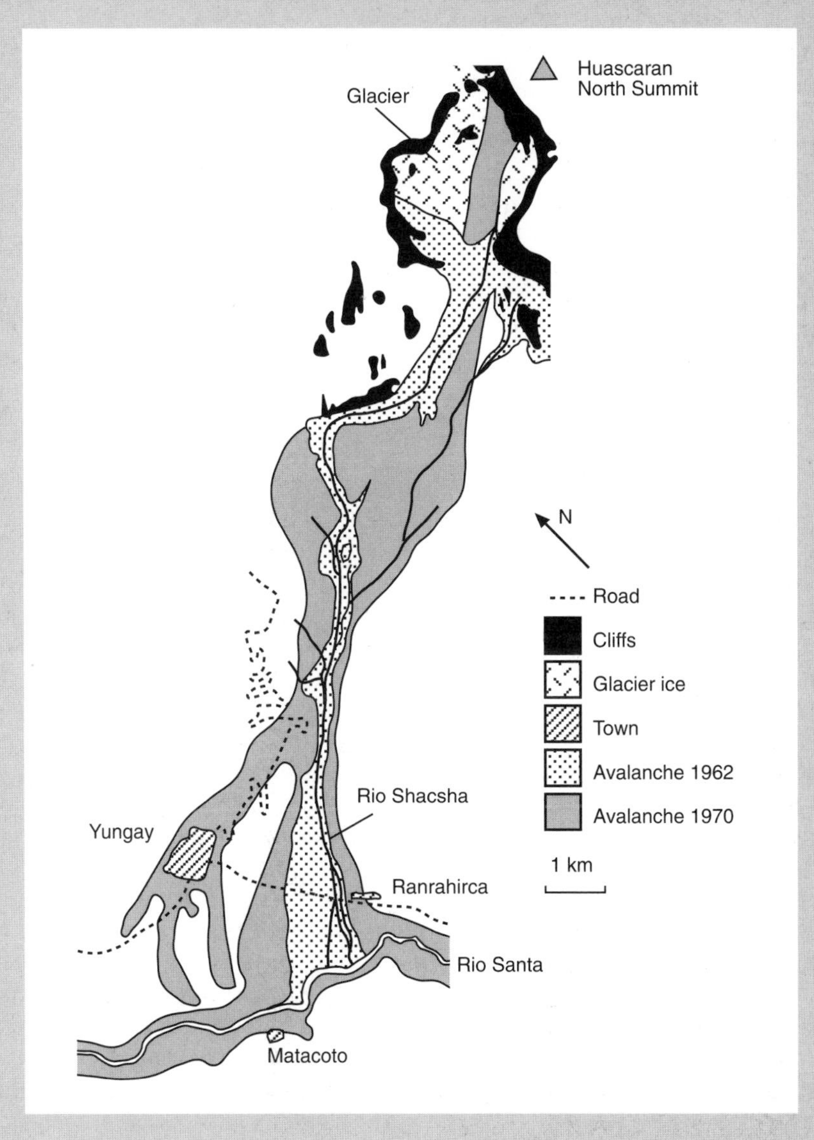

continues overleaf

BOX 11.6: *(continued)*

down the Rio Shacsha at an average velocity of 47 m s^{-1}. This fluid debris avalanche killed over 4000 people, destroyed six villages, partly buried three more, ruined 10 flour mills, four bridges and covered over 600 ha of agricultural land, killing thousands of sheep, cows, pigs and poultry. Parts of the debris flow continued for over 52 km from the source. This was followed just 8 years later by an even more devastating event in 1970. Following a large earthquake (magnitude 7.7) centred 130 km to the west, a second block of ice and rock detached from the northern side of the mountain. Some estimates suggest that it may have contained over 100 million m^3 of material. The debris was incredibly mobile in the upper reaches of the flow. Boulders were hurled into ballistic trajectory and impacted over 4 km from their launch site. The debris travelled a vertical distance of 4200 m over a horizontal distance of 16 km at an average speed of 75 m s^{-1}. The debris avalanche completely buried the town of Yungay killing an estimated 18 000 people and causing widespread destruction of communication links within the Rio Santa valley. These two events demonstrate the incredibly destructive potential of rock and ice avalanches.

Sources: Browing, J.M. 1973. Catastrophic rock slide, Mount Huascaran, north-central Peru, May 31, 1970. *American Association of Petroleum Geologists Bulletin*, **57**, 1335–1341; Plafker, G. & Ericksen, G.E. 1978. Nevados Huascaran avalanches, Peru. In: Voight, B. (Ed.) *Rock slides and avalanches*. Elsevier, Amsterdam, 277–314. [Diagrams modified from: Tufnell (1984) *Glacier hazards*. Longman, Fig. 5.2, p. 54]

constructed to cross the meltstreams was destroyed and over 300 m of road was buried under coarse gravel by shifts in the meltwater stream in 1980.

Catastrophic floods, known as **jökulhlaups**, may also be associated with glaciers where they: (1) dam up lakes (Figure 11.28A); (2) contain internal water pockets; or (3) occur in volcanic terrains. The shape of the flood hydrograph may vary depending on the nature of the event (Figure 11.28B&C), but flows are normally several orders of magnitude greater than normal flows. Some of the biggest jökulhlaup events this century have occurred in the Karakoram Himalayas. For example, the damming of the Upper Shyok River by Chong Kumdan Glacier formed a lake with an estimated volume of 1400 million m^3. A sudden outburst from this lake in 1929 produced a flood wave which travelled down the Shyok River into the Indus River and raised water levels by over 8 m at Attock, which is over 740 km away from the ice dam. A flood in 1935 from the Orba Glacier in Italy involved 22.5 million m^3 of water with discharge rates of up to 13 000 m^3 s^{-1}. The eruption of the volcano Grímsvötn beneath the Vatnajökull ice cap in Iceland in the autumn of 1996 was associated with a massive

A

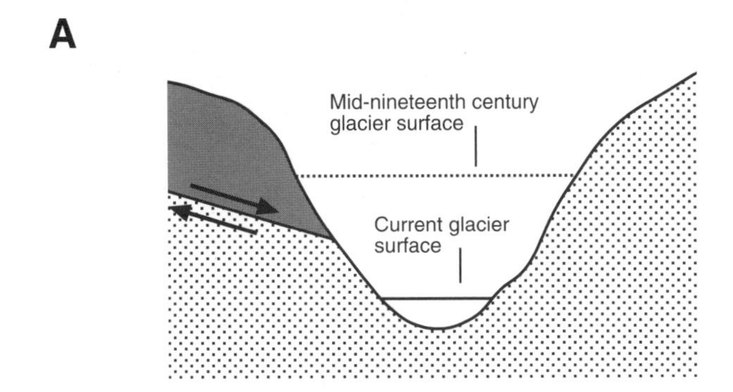

B

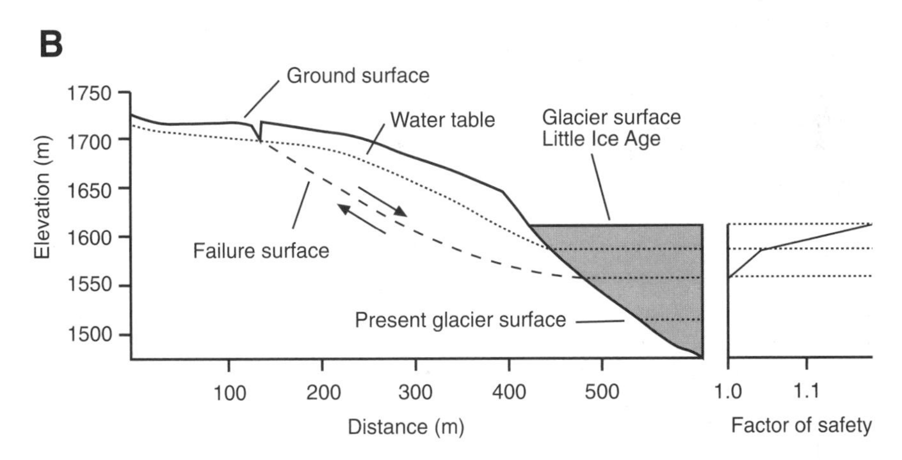

Figure 11.27 *Role of glacier thinning in debuttressing slopes causing rock failure.* ***A.*** *Profiles of the Mount Fletcher rock avalanche, New Zealand.* ***B.*** *Cross-section of slope movement at Affliction Creek and variation in the factor of safety with glacier thinning [Modified from: Evans & Claque (1994)* Geomorphology, ***10****, Figs 5 & 9, pp. 111 & 113]*

jökulhlaup event causing widespread destruction to transport infrastructure in front of the glacier.

Jökulhlaups are not just associated with dramatic floods but may trigger large mud flows, particularly since unusually large amounts of loose glacial sediment are available in these regions. A particularly good example is the debris flows triggered by drainage of ephemeral lakes on the margins of the Cathedral Glacier in British Columbia. Drainage of these lakes into steep sediment-filled ravines in front of the glacier has generated large debris flows (maximum of 100 000 m^3) several times this century, which have flowed up to 3 km and blocked the Canadian Pacific Highway and Trans-Canadian Pacific Railway. The Trans-Canadian Pacific Railway is particularly vulnerable at this point owing to two large hairpin bends in the line which meant that the tracks could be crossed by the debris flow in four areas. The problem was reduced in 1985 when the railway

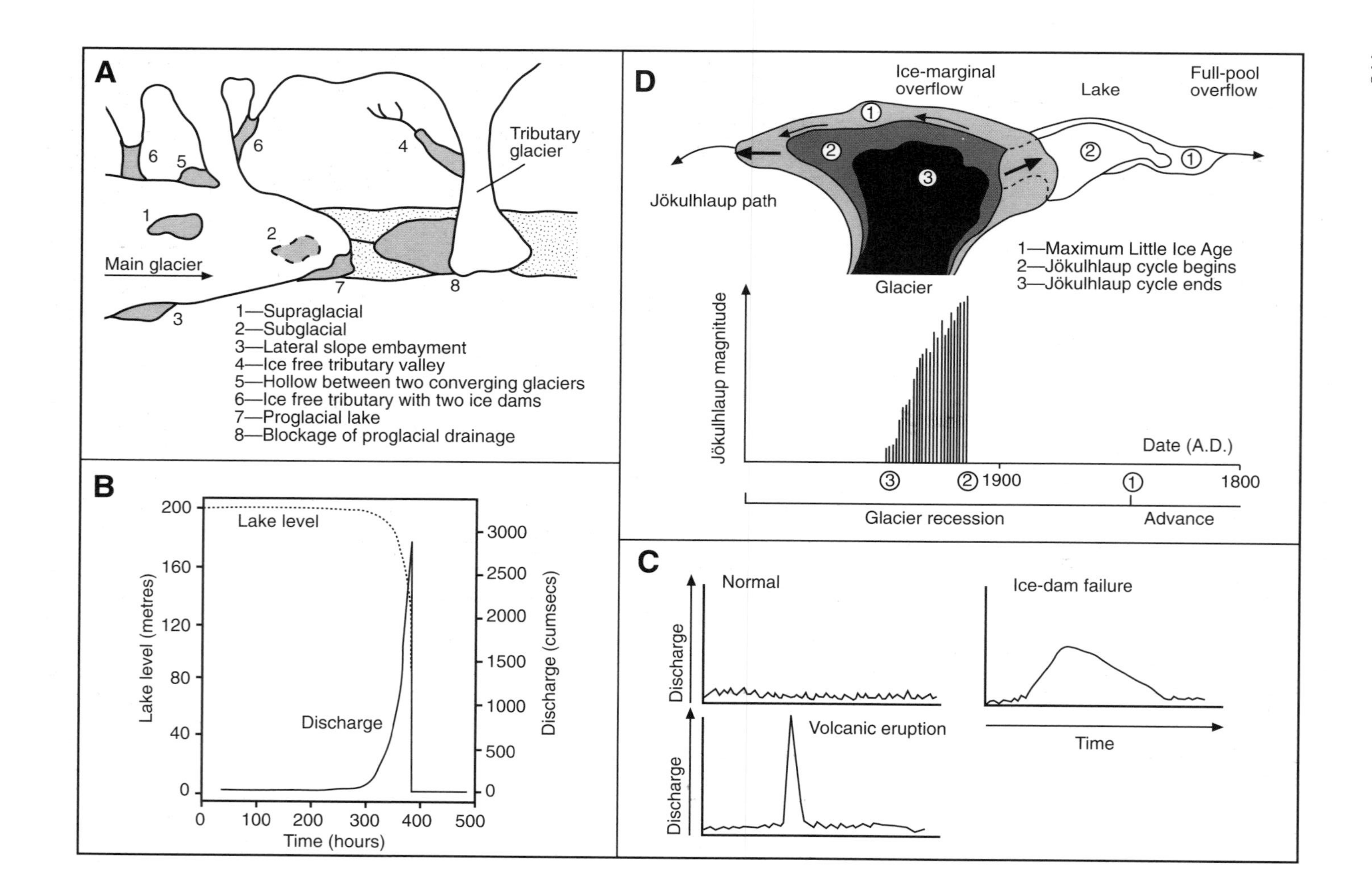

A
Main glacier
Tributary glacier
1—Supraglacial
2—Subglacial
3—Lateral slope embayment
4—Ice free tributary valley
5—Hollow between two converging glaciers
6—Ice free tributary with two ice dams
7—Proglacial lake
8—Blockage of proglacial drainage
B
Lake level (metres)
Discharge (cumsecs)
Lake level
Discharge
Time (hours)
0
100
200
300
400
500
0
40
80
120
160
200
0
500
1000
1500
2000
2500
3000
D
Ice-marginal overflow
Lake
Full-pool overflow
Jökulhlaup path
Glacier
1—Maximum Little Ice Age
2—Jökulhlaup cycle begins
3—Jökulhlaup cycle ends
Jökulhlaup magnitude
Date (A.D.)
1900
1800
Glacier recession
Advance
C
Normal
Ice-dam failure
Discharge
Volcanic eruption
Discharge
Time

company started to pump water from the Cathedral Glacier in an attempt to reduce the problem.

The most common type of jökulhlaup is associated with the failure of **ice-dammed lakes**. Failure of the ice dam occurs for several reasons. Water pressure within the lake increases with depth until it may be possible to literally float the ice dam leading to drainage; alternatively the ice dam may fail due to a seismic event, or due to the development of drainage channels beneath the ice. A cycle of jökulhlaup activity can be recognised (Figure 11.28D). As glaciers advance as they did during the Little Ice Age, the ice dam will be solid and drainage of the lake restricted to overflow via a col in the surrounding valley sides; in such cases jökulhlaups will be very rare. As the glacier thins and retreats from this maximum, the ice dam will become less stable and drainage will occur periodically via regular jökulhlaups. The periodicity between events is usually controlled by the time taken for the lake to refill to a critical depth when the water pressures within it are sufficient to float the ice dam or generate a subglacial drainage system. As the glacier thins, the size of the lake will decrease and therefore the magnitude of the jökulhlaup events will fall. Ultimately, the water in the lake will find its way around the glacier and the lake will cease to exist. Most valley glaciers in the world have experienced sustained retreat and decay since the Little Ice Age, a phenomenon which is perhaps being accelerated by global warming (see Section 9.8.2); as a consequence, the threat of jökulhlaups will gradually decline over the next century. Catastrophic floods may also be associated with the drainage of lakes dammed by ice-cored moraines, which become unstable as the ice core melts (Box 11.7).

Avalanches of snow and ice also provide significant hazards in regions of abundant snowfall. **Snow avalanches** are common in most mountain areas after snowfall. They vary from small, localised powder avalanches to large slab failures in snow, which have the potential not only to kill but to uproot trees and destroy or bury buildings and other structures under both snow and debris. In central Europe there are about 40 casualties due to avalanches each year, a figure which is steadily increasing with the growing popularity of winter sports. Of greater significance are **ice avalanches** which involve the downslope movement of blocks of glacier ice. These involve the failure of the snouts of hanging glaciers, which break off due to increased crevassing or differential melting. Ice avalanches are most marked where glaciers retreat up very steep slopes. For example, in 1965 a one million m^3 block of ice broke from the snout of the Allalingletscher in the Saas Valley in Switzerland and slid into the Mattmark Dam construction site killing 88 people. Between 1901 and 1983, 124 people were killed by ice avalanches in the Swiss Alps alone.

Figure 11.28 *(opposite) Jökulhlaups.* **A.** *Typical discharge curves from different types of jöklhlaup.* **B.** *Variation in lake level and jökulhlaup hydrograph at Summit Lake (September 1967), British Columbia.* **C.** *Patterns of jökulhlaup activity accompanying glacier retreat.* **D.** *Model of jökulhlaup activity associated with glacier thinning since the Little Ice Age [B&D Modified from: Evans & Claque (1994)* Geomorphology, **10**, *Figs 19 & 21, pp. 123 & 125; C Modified from: Maizels & Russel (1992)* Quaternary Proceedings, **3**, *Fig. 7, p. 142]*

BOX 11.7: CATASTROPHIC OUTBURSTS FROM MORAINE-DAMMED LAKES

Moraine-dammed lakes are found in high mountains close to existing glaciers. Most of the moraines impounding these lakes were built during the Little Ice Age or similar Neoglacial ice advances. Since these ice maxima the glaciers have retreated leaving behind a closed basin filled with water. Failure of these moraine dams may occur for several reasons, including (Diagrams A–C): (1) melting of ice cores within moraines; (2) displacement of water over the top of the moraine by rock/ice avalanches into the lake which results in rapid erosion of the moraine; (3) piping or seepage of water through the moraine may occur slowly at first and then accelerate as pipes develop; and (4) failure due to a seismic shock.

Moraine dam failures are particularly well reported, owing to the devastation they have caused, in the Cordillera Blanca of Peru. One of the most devastating occurred in 1941 when the moraine-dammed lake, Lake Palcacocha, drained. The moraine rapidly breached and the flood water

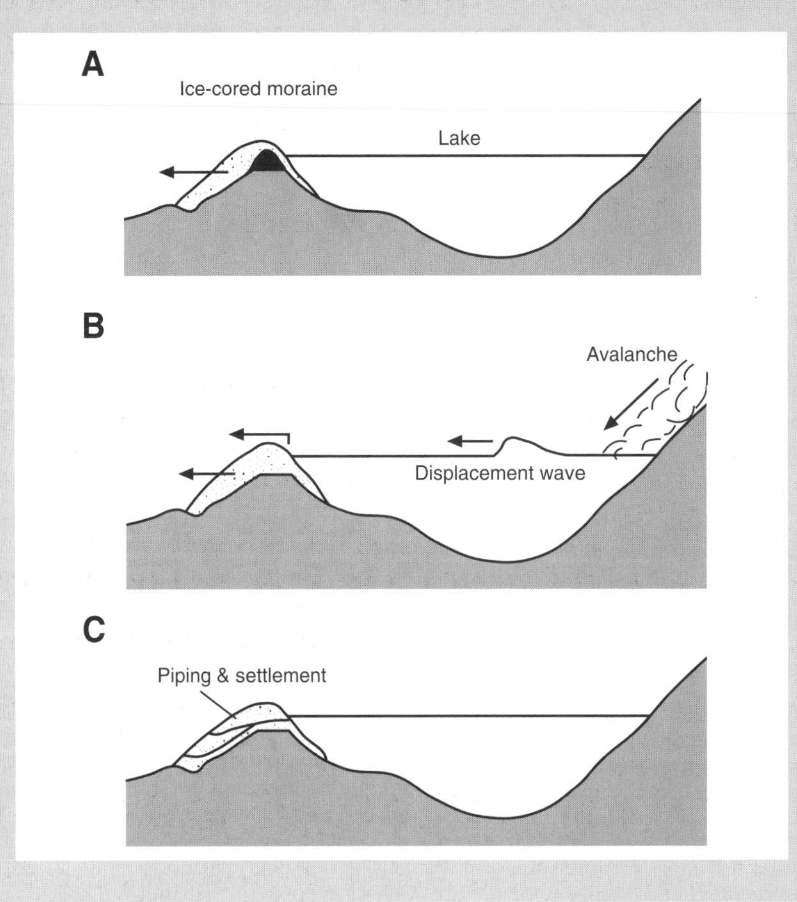

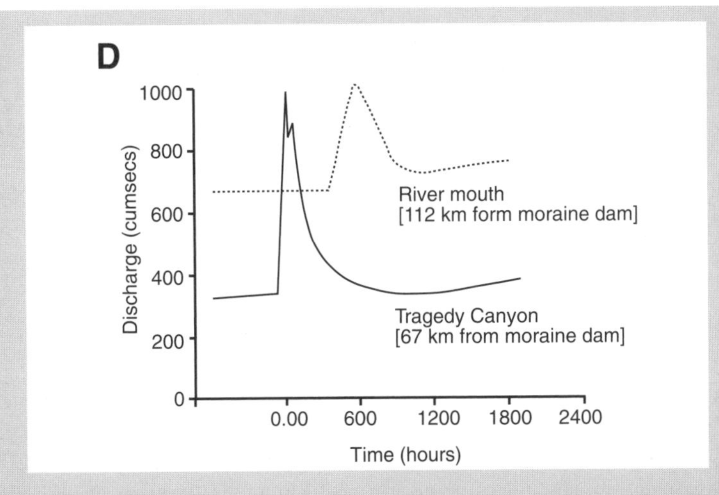

drained into a lower lake causing it to drain as well. The resulting flood water combined with glacial debris to form a debris flow over 8 million m^3 in volume which destroyed 30% of the City of Huaraz killing 6000 people. An equally spectacular but less devastating example, due to its occurrence in a poorly populated region, occurred in July 1983 in the Canadian Cordillera. Here Lake Nosteetuko, 230 km north of Vancouver, drained as a consequence of an ice avalanche into the lake. Within just four hours the moraine had been incised to a depth of over 40 m and 6 million m^3 of water had been released. This produced a destructive flood wave which swept 115 km down the Nostetuko and Homathko valleys to the sea. As the hydrographs show (Diagram D), at one site 67 km downstream discharge increased from 330 m s^{-1} to over 900 m s^{-1} in less than one hour. The hydrograph was more attenuated by the time the water reached the sea. The examples given are just two of the many possible examples of outbursts from moraine-dammed lakes which have occurred in the mountains of the world in recent times. With time, the number of moraine-dammed lakes capable of failure should decline as all those capable of failing do so. In several instances in Peru, direct drainage of dangerous lakes has been undertaken by drilling drainage holes through bedrock or the moraine to allow controlled drainage.

Sources: Reynolds, J.M. 1992. The identification and mitigation of glacier-related hazards: examples from the Cordillera Blanca, Peru. In: McCall, G.J.H., Laming, D.J.C. & Scott, S.C. (Eds) *Geohazards: natural and man-made.* Chapman & Hall, London, 143–157; Evans, S.G. & Claque, J.J. 1994. Recent climatic change and catastrophic geomorphic processes in mountain environments. *Geomorphology*, **10**, 107–128. [Diagrams A & C Modified from: Evans & Claque (1994) *Geomorphology*, **10**, Figs 11 & 14, pp. 115 & 118. Diagram B Modified from: Reynolds (1992) In: McCall *et al.* (Eds) *Geohazards: natural and man-made*. Chapman & Hall, Fig. 10, p. 152]

Finally, icebergs are produced when glaciers terminate and calve into water. They not only threaten shipping, but may also scour the sea floor disrupting cables as well as oil/gas installations. Icebergs may be derived either from the break-up of sea ice or by glacier calving. They are primarily confined to polar regions although drifting bergs frequently threaten shipping routes, such as those in the Atlantic. The most famous iceberg-related event was the sinking of the *Titanic* in 1912.

11.4.1 Risk Assessment of Glacial Hazards

Hazard mapping can be used to delineate the extent of some glacial hazards, although most are by nature unpredictable and dependent on future climatic trends. Hazard assessment should involve the following tasks. First, the identification of all areas at risk from ice and snow avalanches given information on the location of seasonal snow accumulation and potentially unstable hanging glacier snouts. Second, the identification of all unstable water bodies associated with a glacier or its moraine system and the likely discharge routes given the current glacier geometry and surrounding topography. Third, the examination of historical glacier behaviour in order to identify potential instabilities in flow, such as a tendency to surge. Fourth, through the use of direct monitoring and historical records, predictive models should be developed which link glacier fluctuations to changes in regional or local climate. In many cases this is not possible, although the geometry of some maritime glaciers can be closely linked to changes in air temperature. The final component is that of continual monitoring and modification of the hazard assessment as the geometry of a glacier changes and its fluvial system evolves. In practice, few integrated hazard maps for glacial regions have been produced, although avalanche hazard maps are widely available for most alpine regions.

11.4.2 Mitigation and Management of Glacial Hazards

Many glacial hazards, such as advancing glaciers, cannot be dealt with effectively and the best policy it to recognise hazard potential and simply avoid the site. Flood protection works may help constrain regular glacial floods or jökulhlaups. For example, the volcanic lake Grímsvötn, under the Vatnajökull ice cap in Iceland, regularly floods across the outwash plain of Skeirdarajokull threatening the campsite and associated infrastructure at Skafatfell as well as the main road. A series of embankments and holding pools have been constructed to attempt to control the flood water.

The problem of icebergs as shipping hazards has led to considerable theoretical research into methods of constraining them. For example, in 1989 the supertanker *Exxon Valdez*, carrying more than 1.2 million barrels of crude oil valued at £15 million, hit a reef 25 miles from the Port of Valdez in Alaska, while trying to avoid icebergs from the Columbia Glacier. The Port of Valdez is at the southern end of the Trans-Alaskan Pipeline (Figure 8.14). The potential threat of ice from the Columbia Glacier was identified 12 years earlier when proposals were put forward to restrain the icebergs with giant nylon cables. It was estimated at the time that up to 13 cables would be required, each around 30 cm in diameter, to restrain a single berg at an estimated cost of $30 million. Such proposals never got off the drawing board owing to the difficulties involved in the scheme.

The problems of snow and ice avalanches have been tackled by direct action such as blasting to dislodge snow and ice in a controlled fashion. More commonly, planning control based on hazard mapping is used to locate or relocate buildings, but where this is not possible protective works may include: (1) snow fences in the starting zones in order to trap and hold snow away from avalanche slopes; (2) structures in the starting zone of avalanches to help support the weight of the snow and tie it into the hillside; (3) deflection and retarding structures in the track or runout zone of the avalanche, including snow sheds over roads and railways, walls, wedge-shaped structures to deflect snow around obstacles, and earth mounds or small dams to bring avalanches to a stop in the runout zone. In the long term, reforestation of avalanche-prone slopes may improve slope stability.

11.5 SUMMARY OF KEY POINTS

- Exogenic or geomorphological hazards cause widespread economic loss and threaten life.
- Fluvial hazards consist of flooding and changes in channel pattern associated with bank erosion.
- Understanding the hydrological system and the thresholds within it is critical in hazard assessment.
- A variety of flood control measures exist which can be grouped into three categories: (1) control and limitation; (2) abatement; and (3) adjustment.
- Coastal hazards comprise the familiar problems of coastal erosion and coastal flooding as well as the impact of dramatic phenomena such as tsunamis. Approximately 95% of the world's coastline is currently eroding owing to a combination of increasing pressure on the coastal zone, rising sea levels and poor coastal management.
- Risk assessment of coastal hazards involves the use of historical events as a guide to future ones. Increasingly sophisticated computer simulation of the coastal zone is assisting this process.
- A range of coastal defence solutions exists to deal with problems of coastal recession and flooding. These can be broadly grouped into traditional 'hard engineering' solutions, and more progressive 'soft engineering' which attempts to utilise the natural coastal system. The choice of solution is very site-specific and is usually controlled by a range of local variables or considerations.

- Mass movements are variable in nature and are particularly noteworthy for the amount of economic loss which they cause.
- Mapping of mass movement hazards is well developed and a variety of approaches have evolved to produce hazard maps. These range from qualitative site assessment by experienced geologists and inventory maps of all known landslides, to more sophisticated maps based on intuitive or statistical scoring systems which synthesise regional data from variables likely to control slope stability.
- Various approaches are available to stabilise slopes, which focus on either increasing slope strength or reducing the stress leading to slope failure.
- A wide range of glacial hazards can be recognised. These include loss due to glacier advances, slope stability problems associated with glacier decay, glacier meltwater, drainage of ice-dammed lakes, ice and snow avalanches and the threat from icebergs.
- Hazard assessment is very site-specific and generic approaches cannot be identified. Careful monitoring of the glacial environment is essential. The scope for hazard management is also somewhat limited.

11.6 SUGGESTED READING

The literature on exogenic or geomorphological hazards is vast, and what follows is a list of some key texts, review papers and papers on the specific examples referred to in the text; it is by no means an exhaustive list. The subject is well covered in a number of edited volumes and books; of particular note are Cooke (1984), Cooke & Doornkamp (1990), Bryant (1991), McCall *et al.* (1992), McGregor & Thompson (1995), Smith (1996) and Slaymaker (1996). A thematic issue of *Geomorphology* (1994, Volume 10) contains a wide range of papers on geomorphological hazards and numerous specific examples. In addition, the paper by McCall (1996) provides an excellent review of the subject.

Fluvial hazards have been discussed widely in the literature. The problems of bank erosion are covered in numerous papers but attention is drawn here to the work of Hooke (1979, 1980) and Haque & Hossanin (1988). Abam (1992) discusses the potential of bank erosion to destroy natural flood defences and increase flood problems in the Niger Delta. A broad-ranging review by Brookes (1985) covers many of the impacts which channel works can have. Shu & Finlayson (1993) provide an excellent example of flood control problems from the Yellow River in China.

The available literature on **coastal management** is considerable. Jolliffe (1983) and Jolliffe & Patman (1985) provide good overviews of the hazards and management options. Dolan *et al.* (1992) examine methods of predicting shoreline recession, while Nordstrom & Renwick (1984) consider some management options for eroding coasts. Numerous examples and specialist papers are to be found within the pages of the *Journal of Shoreline Management* and *Journal of Coastal Research*. The subject of tsunami hazards is covered in the specialist journal *Science of Tsunami Hazards* published by the International Tsunami

Society, which contains numerous papers relevant to this subject. Monge & Mendoza (1993) provide a detailed case study of the assessment of tsunami hazards on the northern Chile coast.

The topic of **mass movement** is covered in a wide range of publications, of particular note are a series of authoritative review papers by Jones (1992, 1993, 1995a). The theory of mass movements and their management is covered in contributions within several volumes of edited papers; of particular note are the volumes edited by Coates (1977) and Brunsden & Prior (1984). The role of humans in causing mass movements is discussed by Slosson & Larson (1995) and the impact of landsliding within urban areas is covered by Alexander (1989). Methods of landslide hazard assessment are discussed in Carrara & Merenda (1976), Carrara *et al.* (1991), Anbalagan (1992), Pachauri & Pant (1992), Maharaj (1993) and Jones (1995b). Landslide countermeasures are discussed by Hutchinson (1984), Kockelman (1986), and Broomhead (1995).

Glacial hazards are discussed in detail in the book by Tufnell (1984) and are well covered in the review papers by Grove (1987), Reynolds (1992) and Evans & Clague (1994). The prediction of jökulhlaups is covered in the work of Maizels & Russell (1992) while Lliboutry *et al.* (1977) review catastrophic floods from the failure of moraine dams. The threat of ice avalanches is covered by Rothlisberger (1977) and Alean (1985), while conventional avalanches are covered in numerous papers, of particular note a thematic set of papers in the *Journal of Glaciology* (1980, Volume 26). The threat of icebergs to Alaskan oil shipping is reported by Dickson (1978). Problems of glacial hazard assessment are considered in Haeberli *et al.* (1989)

Abam, T.K.S. 1992. Geomorphic processes and the threat of increased flooding in the Niger Delta. *Journal of African Earth Sciences*, **15**, 59–63.

Alean, J. 1985. Ice avalanches: some empirical information about their formation and reach. *Journal of Glaciology*, **31**, 324–333.

Alexander, D. 1989. Urban landslides. *Progress in Physical Geography*, **13**, 157–191.

Anbalagan, R. 1992. Landslide hazard evaluation and zonation mapping in mountainous terrain. *Engineering Geology*, **32**, 269–277.

Brookes, A. 1985. River channelization: traditional engineering methods, physical consequences and alternative practices. *Progress in Physical Geography*, **9**, 44–73.

Broomhead, E.N. 1995. Water and landslides. *Proceedings of Royal Academy of Engineering Conference on Landslide hazard mitigation*. Royal Academy of Engineering, London, 42–55.

Brunsden, D. & Prior, D.B. (Eds) 1984. *Slope instability*. John Wiley & Sons, Chichester.

Bryant, E.A. 1991. *Natural hazards*. Cambridge University Press, Cambridge.

Carrara, A. & Merenda, L. 1976. Landslide inventory in northern Calabria, southern Italy. *Geological Society of America Bulletin*, **87**, 1153–1162.

Carrara, R., Cardinali, M., Detti, R., Guzetti, F., Pasqui, V. & Reichenbach, P. 1991. GIS techniques and statistical models in evaluating landslide hazard. *Earth Surface Processes and Landforms*, **16**, 427–445.

Coates, D.R. (Ed.) 1977. *Landslides*. Geological Society of America Reviews in Engineering Geology, 3.

Cooke, R.U. 1984. *Geomorphological hazards in Los Angeles*. Allen & Unwin, London.

Cooke, R.U. & Doornkamp, J.C. 1990. *Geomorphology in environmental management*. Oxford University Press, Oxford.

Dickson, D. 1978. Glacier retreat threatens Alaskan oil tanker route. *Nature*, **273**, 88–89.

Dolan, R., Fenster, M.S. & Holme, S.J. 1992. Spatial analysis of shoreline recession and accretion. *Journal of Coastal Research*, **8**, 263–285.
Evans, S.G. & Clague, J.J. 1994. Recent climatic change and catastrophic geomorphic processes in mountain environments. *Geomorphology*, **10**, 107–128.
Grove, J.M. 1987. Glacier fluctuations and hazards. *Geographical Journal*, **153**, 351–369.
Haeberli, W., Allean, J-C., Muller, P. & Funk, M. 1989. Assessing risks from glacier hazard in high mountain regions: some experiences in the Swiss Alps. *Annals of Glaciology*, **13**, 96–102.
Haque, C.E. & Hossanin, Z.M.D. 1988. Riverbank erosion in Bangladesh. *Geographical Review*, **78**, 20–31.
Hooke, J.M. 1979. An analysis of the processes of river bank erosion. *Journal of Hydrology*, **42**, 39–62.
Hooke, J.M. 1980. Magnitude and distribution of rates of river bank erosion. *Earth Surface Processes*, **5**, 143–157.
Hutchinson, J.N. 1984. Landslides in Britain and their countermeasures. *Journal of Japan Landslide Society*, **21**, 1–24.
Jolliffe, I.P. 1983. III. Coastal erosion and flood abatement: what are the options? *Geographical Journal*, **149**, 62–71.
Jolliffe, I.P. & Patman, C.R. 1985. The coastal zone: the challenge. *Journal of Shoreline Management*, **1**, 3–36.
Jones, D.K.C. 1992. Landslide hazard assessment in the context of development. In: McCall, G.J.H., Laming, D.J.C. & Scott, S.C. (Eds) *Geohazards: natural and man-made*. Chapman & Hall, London, 117–141.
Jones, D.K.C. 1993. Landsliding as a hazard. *Geography*, **78**, 185–190.
Jones, D.K.C. 1995a. The relevance of landslide hazard to the International Decade for Natural Disaster Reduction. *Proceedings of Royal Academy of Engineering Conference on Landslide hazard mitigation*. Royal Academy of Engineering, London, 19–33.
Jones, D.K.C. 1995b. Landslide hazard assessment. *Proceedings of Royal Academy of Engineering Conference on Landslide hazard mitigation*. Royal Academy of Engineering, London, 19–33.
Kockelman, W.J. 1986. Some techniques for reducing landslide hazards. *Bulletin of the Association of Engineering Geologists*, **23**, 29–52.
Lliboutry, L., Morales, A.B., Pautre, A. & Schneider B. 1977. Glaciological problems set by the control of dangerous lakes in Cordilera Blanca, Peru. I: Historic failures of morainic dam, their causes and prevention. *Journal of Glaciology*, **18**, 239–254.
Maharaj, R.J. 1993. Landslide processes and landslide susceptibility analysis from an upland watershed: a case study from St. Andrew, Jamaica, West Indies. *Engineering Geology*, **34**, 53–79.
Maizels, J. & Russell, A. 1992. Quaternary perspectives on jökulhlaup prediction. *Quaternary Proceedings*, **2**, 133–152.
McCall, G.J.H. 1996. Natural hazards. In: McCall, G.J.H., De Mulder, E.F.J. & Marker, B.R. (Eds) *Urban geoscience*. A.A. Balkema, Rotterdam, 81–125.
McCall, G.J.H., Laming, D.J.C. & Scott, S.C. (Eds) 1992. *Geohazards: natural and man-made*. Chapman & Hall, London.
McGregor, D.F.M. & Thompson, D.A. 1995. *Geomorphology and land management in a changing environment*. John Wiley & Sons, Chichester.
Monge, J. & Mendoza, J. 1993. Study of the effects of tsunami on the coastal cities of the region of Tarapacá, north Chile. *Tectonophysics*, **218**, 237–246.
Nordstrom, K.F. & Renwick, W.H. 1984. A coastal cliff management district for protection of eroding high relief coasts. *Environmental Management*, **8**, 197–202.
Pachauri, A.K. & Pant, M. 1992. Landslide hazard mapping based on geological attributes. *Engineering Geology*, **32**, 81–100.
Reynolds, J.M. 1992. The identification and mitigation of glacier-related hazards: examples from the Cordillera Blanca, Peru. In: McCall, G.J.H., Laming, D.J.C. & Scott, S.C. (Eds) *Geohazards: natural and man-made*. Chapman & Hall, London, 143–157.

Rothlisberger, H. 1977. Ice avalanches. *Journal of Glaciology*, **19**, 669–671.
Shu, L. & Finlayson, B. 1993. Flood management on the lower Yellow River: hydrological and geomorphological perspectives. *Sedimentary Geology*, **85**, 285–296.
Slaymaker, O. 1996. *Geomorphic hazards*. John Wiley & Sons, Chichester.
Slosson, J.E. & Larson, R.A. 1995. Slope failures in Southern California: rainfall, threshold, prediction and human causes. *Environmental & Engineering Geoscience*, **1**, 393–401.
Smith, K. 1996. *Environmental hazards: assessing risk and reducing disaster*. Second edition. Routledge, London.
Tufnell, L. 1984. *Glacier hazards*. Longman, Harlow.

12
Endogenic Hazards

Endogenic hazards are defined as those which derive their energy from within the Earth (Figure 12.1). In this chapter we identify three principal types: (1) seismic; (2) volcanic; and (3) radon emissions. Two of these, seismic and volcanic hazards, normally give rise to composite events by stimulating a wide range of exogenic processes such as mass movements, tidal waves and floods. The greatest loss of life is often associated with these secondary hazards. As in the previous chapter, we will first outline the nature of each hazard before examining methods of risk assessment and hazard management.

12.1 SEISMIC HAZARDS

Seismic hazards affect over 35 countries and together they kill more people each year than any other natural hazard. For example, the Tangshan earthquake in China in 1976 killed over 750 000. Death and destruction are caused not only by the collapse of buildings, but also by fire and a host of secondary hazards such as landslides and liquefaction of unconsolidated sediment. For example, more than 80% of the property damage caused by the 1906 San Francisco earthquake was due to fire. Similarly, the majority of the 160 000 deaths in the Tokyo and Yokohama earthquake in 1923 were also due to fire. It is also important to recognise that the impacts of an earthquake are not simply short term, but that problems associated with the destruction of homes and sanitation may have a significant impact on the death toll. For example, the San Salvador earthquake of 1986 made 250 000 people homeless and caused 1500 deaths, 10 000 injuries.

The distribution of earthquake zones is controlled by plate tectonics (Figure 12.2) and results from the relative displacement of lithospheric plates. The greatest concentration of plate boundaries, and therefore earthquake zones, is in southeast Asia, which is reflected in the fact that all the earthquakes killing over

Figure 12.1 *Endogenic hazards: rock fracture in Iceland [Photograph: A.F. Bennett]*

100 000 people have been concentrated in this region. Earthquakes also occur occasionally within plate interiors due to stress release and movement along faults. For example, a number of small earthquakes have occurred in Britain during the twentieth century.

The primary feature of an earthquake is ground shaking, while secondary phenomena include displacement along faults, ground subsidence, soil liquefaction, mass movement, snow and ice avalanches, floods due to dam failure, and fire within urban areas. The magnitude of **ground shaking** is controlled by two sets of variables. Firstly, the magnitude, duration and frequency of the shock waves is extremely important. This is determined by the nature of the event and also by the distance from it. Secondly, a range of factors may amplify this ground shaking, such as ground conditions and topography. Shock waves displace the ground both vertically and horizontally, and commonly radiate out with decreasing magnitude from earthquake foci. Buildings are particularly sensitive to horizontal displacement. Vertical forces are less important since buildings are strengthened in this direction to resist the effects of gravity. The duration and frequency of shaking events is also important. Buildings vibrate by small amounts naturally and have a distinctive frequency of vibration depending on their geometry and method of construction. If the earthquake shock waves are in phase with this natural vibration it may become amplified, thereby increasing the scale of damage. For example, high frequency shock waves tend to cause less damage to tall narrow buildings than to low ones. In contrast, low frequency shock waves are more hazardous to tall buildings. The duration of

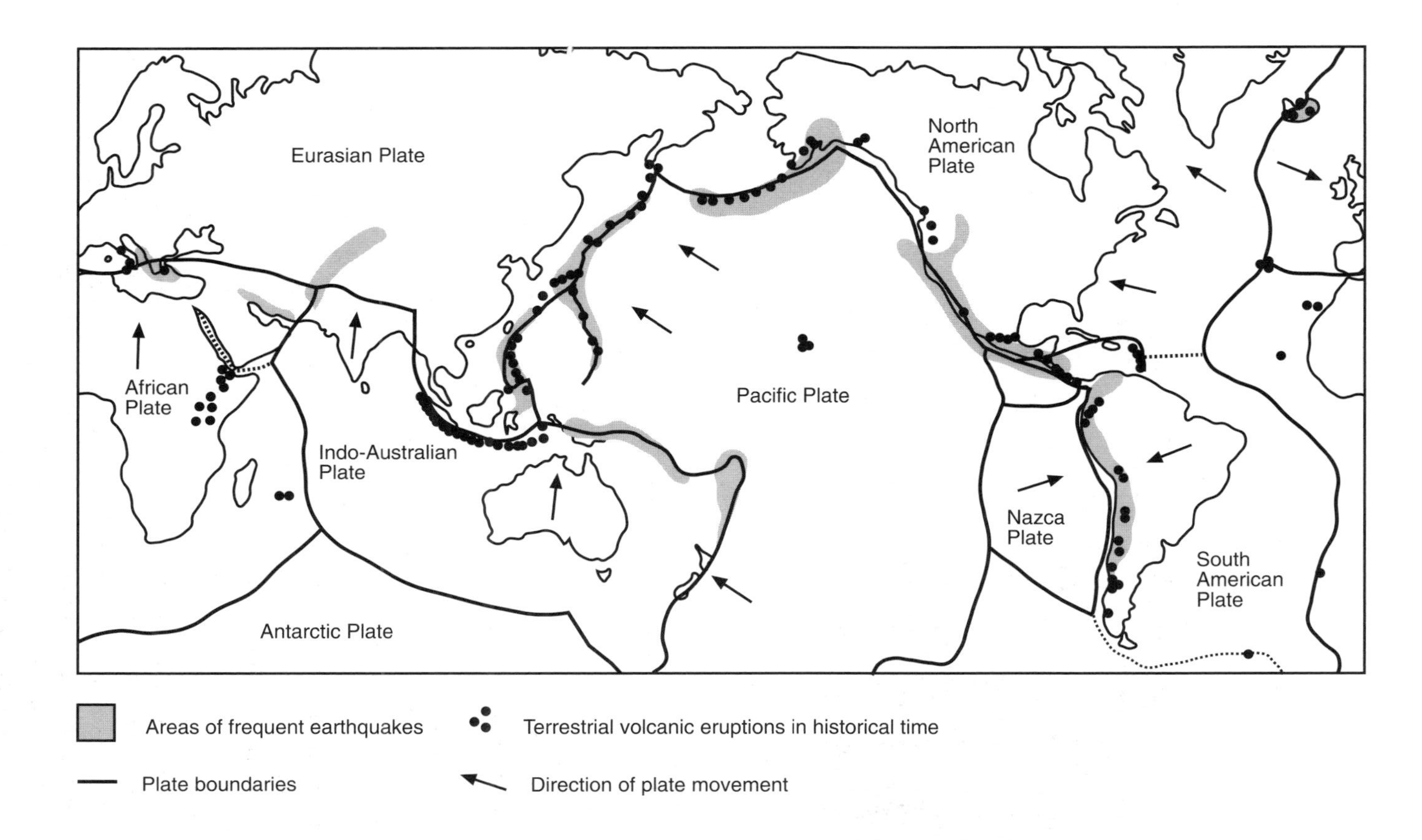

Figure 12.2 *The distribution of plate boundaries, intense earthquakes, and historical land-based volcanic eruptions [Modified from: Bryant (1991)* Natural hazards. *Cambridge University Press, Fig. 10.1, p. 174]*

ground shaking is also of importance. For example, as the Loma Prieta earthquake which hit northern California in 1989 lasted only 6 to 10 seconds it was not sufficient to cause widespread liquefaction and therefore the scale of damage was minimal.

Shock waves may also be amplified by ground conditions. In particular the presence of soft sediment causes significant amplification of the shock wave. This was of particular importance during the Mexican earthquake in 1985 and the extent of building damage was closely linked to the distribution of ancient lake sediments (Box 12.1). Topography also has a role in the amplification of shock waves. For example, steep topography and in particular sharp ridge crests have been linked with enhanced ground shaking.

The vulnerability of a town or city to earthquake damage is also critical in determining the amount of destruction caused and is a function of such variables as the nature of the building stock (Table 12.1), the population density, and the time of day. All unreinforced masonry structures are at risk from earthquakes, but most at risk are those constructed from **adobe** or sunbaked clay bricks. Adobe is the indigenous building material in arid and semiarid regions where other materials are often scarce. It consists of clay bricks bound together by straw or dung and provides fire-resistant and thermally equable brick buildings. This type of building material is very common in the Middle East and in parts of South America. In Peru for example, two-thirds of all rural dwellings and over one-third of urban buildings are made from this type of material. Such buildings are, however, very earthquake prone. In the Peruvian earthquake of 1970, 60 000 such dwellings collapsed killing at least 50 000 people and injuring a further 150 000.

In summary, therefore, the amount of destruction caused by an earthquake depends on the magnitude, frequency and duration of the shock waves as well as the distribution of rock types, topography and building types within an area. As a result an earthquake of similar magnitude can have very different impacts in different places. In practice, however, it is often the secondary hazards triggered by earthquakes which have the greatest potential to cause destruction and loss of life. One of the most important secondary hazards is soil liquefaction. **Liquefaction** occurs because water-saturated sediment loses its strength when vibrated and becomes 'quick-sand'. The most sensitive sediments are cohesionless, granular sediments with high water content. There are four possible consequences of liquefaction: (1) loss of bearing capacity; (2) ground oscillation; (3) lateral spreading; and (4) flow. **Loss of bearing strength** occurs when soil liquefies under a building. Soil movement away from a building causes differential settlement and tilting. The other three types of liquefaction phenomena form a continuum with increasing slope angle. On horizontal surfaces **ground oscillations** occur. This is often observed as a travelling ground wave, the passage of which may be accompanied by the opening and closing of fissures which are often over 1 m wide and 10 m deep. On slightly more inclined surfaces (less than 3°), **lateral spreading** occurs. Here, intact terrain blocks may move laterally on liquefied sublayers causing problems for shallow foundations, pipelines and other buried structures. On slopes greater than 3° liquefaction

BOX 12.1: THE ROLE OF SOIL CONDITIONS IN AMPLIFYING GROUND SHAKING: THE MEXICAN EARTHQUAKE IN 1985

The 1985 Mexican earthquake ranks amongst the worst natural disasters of the twentieth century; 10 000 people died, 50 000 were injured, 250 000 were made homeless and over $4000 million worth of damage was done. Much of this damage was concentrated in Mexico City, 370 km away from the epicentre in the Pacific. Within the city 7400 buildings were damaged of which 770 were lost and 1665 severely damaged. The distribution of this damage illustrates the

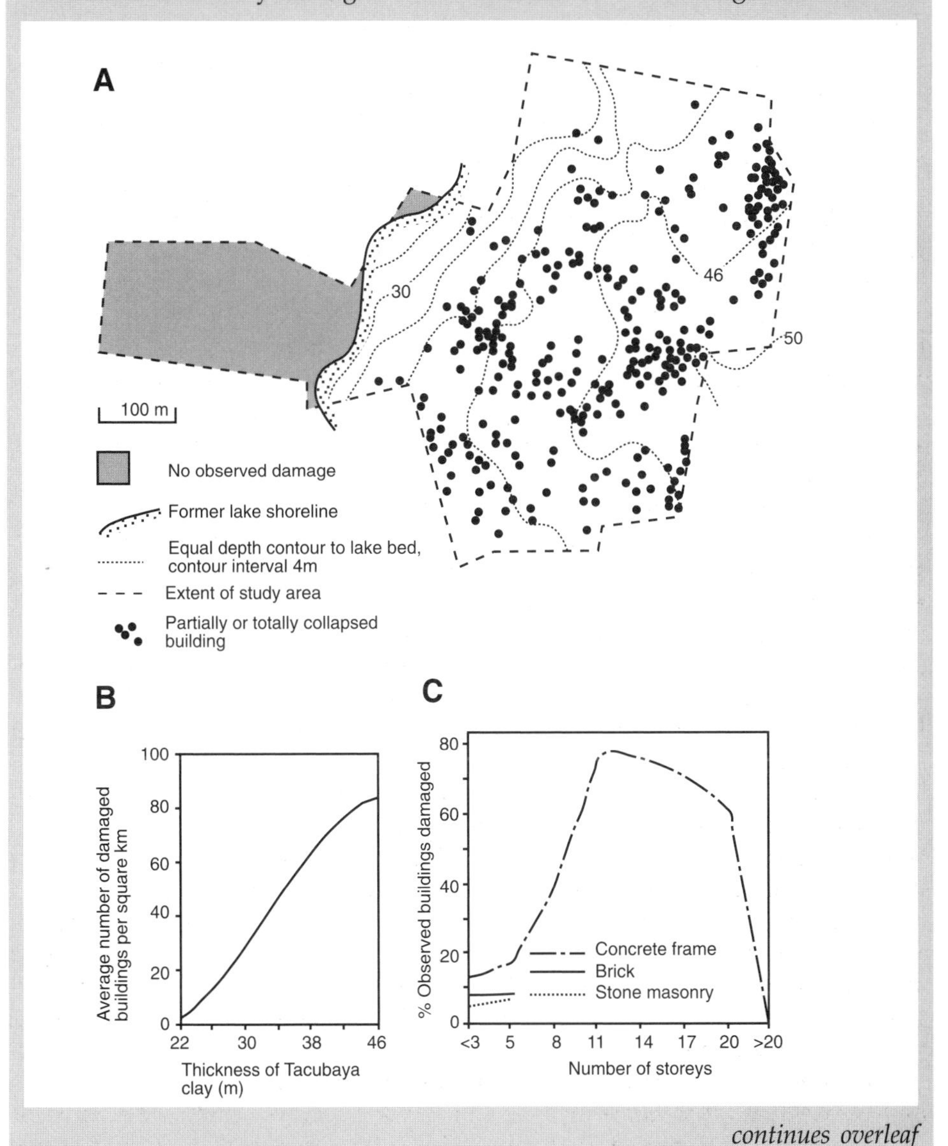

continues overleaf

BOX 12.1: *(continued)*

role of ground conditions in amplifying ground shaking. Much of Mexico City is constructed on former lake beds composed of saturated clays. As Diagram A shows, the maximum damage within the city was concentrated where these clays were thickest and least damage on the adjacent bedrock areas. Construction methods were also important in determining the degree of damage. On the lake bed rigid structures – masonry buildings – generally performed better than flexible ones made from reinforced concrete. The height of buildings was also critical with those between six and 20 storeys being worst affected. Medium- to high-rise buildings tend to have low natural frequencies of vibration – all buildings have a natural frequency of vibration in the same way that tuning forks have – compared to low-rise buildings. This low natural frequency was more 'in tune' with the low frequency of the ground motion during the earthquake causing them to resonate more, thus amplifying and prolonging the shaking. If the ground motions had been of higher frequency, fewer high-rise buildings would have been damaged. This case study demonstrates the importance of local ground conditions and building characteristics in determining the degree of earthquake damage.

Source: Degg, M.R. 1992. Some implications of the 1985 Mexican earthquake for hazard assessment. In: McCall, G.J.H., Laming, D.J.C. & Scott, S.C. (Eds) *Geohazards*. Chapman & Hall, London, 105–114. [Diagrams modified from: Degg (1992) In: McCall, G.J.H., *et al.* (Eds) *Geohazards*. Chapman & Hall, Figs 9, 10 & 11, pp. 109 & 110]

Table 12.1 *Earthquake loss susceptibility for different types of construction [Modified from: Degg (1992) In: McCall* et al. *(eds)* Geohazards: natural and man-made. *Chapman & Hall, Table 1, p. 99]*

	Average damage (%) on the Modified Mercalli Intensity Scale				
	VI	VII	VIII	IX	X
A. Non-seismic design					
Adobe	8	22	50	100	100
Unreinforced masonry	3.5	14	40	80	100
Reinforced concrete frame	2.5	11	33	70	100
Steel frames	1.8	6	18	40	60
B. Seismic design					
Reinforced masonry	0.3	1.5	5	13	25
Reinforced concrete frame	0.9	4	13	33	58
Steel frames	0.4	2	7	20	40

causes large **fluid flows** to form. For example, during the 1964 Alaskan earthquake, a flow over 2 km long and 30 m wide developed on the sandy soils of the Turnagain Heights which dislodged over 70 buildings.

Other secondary phenomena include the initiation of mass movements, snow or ice avalanches and tsunamis. The importance of these phenomena in causing earthquake deaths cannot be underestimated. For example, it has been estimated that more than half the earthquake-related deaths during the Japanese Kobayashi earthquake in 1964 were caused by landslides. The catastrophic rock and ice failure at Huascaran in Peru in 1970, in which over 4000 people died, was also initiated by an earthquake (Box 11.6). Similarly, in the past 100 years there have been over 50 000 deaths due to earthquake-induced tsunamis, affecting over 22 countries. Since 1900, more than 350 people have died in the USA due to seismic tsunamis and over $500 million worth of property damage has been recorded. These secondary hazards have been discussed already in Chapter 11.

12.1.1 Risk Assessment for Seismic Hazards

Risk assessment involves two components: (1) determining the hazard magnitude; and (2) determining the vulnerability to that hazard.

Earthquake magnitude is determined using the Richter Scale while earthquake intensity or degree of destruction is measured with the Modified Mercalli Intensity Scale. The **Richter Scale** is a complex logarithmic scale which measures the vibrational energy of an event. The scale starts at zero and has no theoretical upper limits, although very few events greater than 9 on the Richter Scale have ever been recorded. In fact, it has been argued that rockbodies would shatter before sufficient energy could be pent-up within a fault to release an event equivalent to 10 on the Richter Scale. The logarithmic scale means that for each unit increase in the scale: (1) the amplitude of the seismic wave increases 10-fold; (2) there are 10 times fewer events of that magnitude; and (3) there is a nearly 30-fold increase in energy release. Most evidence suggests that an event of at least 5.5 on the Richter Scale is required to cause a major disaster, but as we have shown in the previous section, the impact of an earthquake varies dramatically from location to location, depending on the amplifying factors and vulnerability of the building stock. The **Modified Mercalli Intensity Scale** is, in contrast, a measure of the degree of damage caused. It is a qualitative measure of seismic intensity based on subjective analysis of the observed damage (Table 12.2). It is important to emphasise that two earthquake events with different magnitudes on the Richter Scale may cause the same amount of destruction in different locations and therefore have the same Modified Mercalli Intensity score.

In the context of land-use planning, insurance and civil defence the production of earthquake hazard maps is an essential task. However, the production of such maps is a complex problem due to the number of variables involved and the need to consider not just the effects of primary shaking but also the damage caused by secondary hazards. Where good seismic records exist, from a widely

Table 12.2 *Modified Mercalli Intensity Scale for assessing earthquake damage*

I	Imperceptible
II	Felt on upper floors
III	Objects swing
IV	Felt by everybody, creaking noises
V	Felt by everybody, objects move
VI	Cracks appear in buildings
VII	Fissures in roads, difficult to stand, large bells ring
VIII	Steering on cars affected, chimneys and towers fall
IX	Landslides, underground pipes break
X	Most masonry and frame structures destroyed, serious damage to large structures such as dams and dykes
XI	Serious damage to large structures and buildings
XII	Total destruction, distortion of the Earth's surface

spaced network of seismographs with a good temporal span, it is possible to produce maps of the probability of different ground shaking events. This has been done in the USA and Australia. Ground shaking is usually expressed as **ground acceleration** – the change in velocity of the ground caused by the passage of a shock wave. By plotting the peak acceleration of shock waves and their rate of decay away from the epicentre for all recorded earthquakes within a region it is possible to estimate the probability of different ground accelerations within a fixed time period, usually 50 years, by the method of extreme event analysis (Box 10.1). The choice of 50 years reflects the average life expectancy of many urban structures. This type of map can only be constructed if a good spatial and temporal coverage of seismic records exists, and its accuracy is usually limited by the length of the record available. Figure 12.3A shows a seismic risk map for parts of North America constructed in this way. It depicts areas likely to experience a given value of ground acceleration within a 50 year period. The probability threshold is 10%; that is, there is a 10% probability that within 50 years the specified ground acceleration rate will be exceeded. The map identifies three zones of seismic hazard: one in California, another on the eastern foothills of the Rockies and a third in the St Lawrence and Quebec area. This type of national survey provides valuable planning data which can be used to define zones in which different building codes apply to ensure earthquake resistance. However, such maps provide little information about the degree of

---

Figure 12.3 *(opposite) Various types of seismic hazard map.* **A.** *Seismic hazard map for the USA based on the probability of a given magnitude of ground shaking expressed as horizontal acceleration with a probability of occurrence of 10% in 50 years.* **B.** *Seismic risk map for the USA based on the amount of damage which might occur within a region based on past experience and the presence of active fault systems; the probability of such damage occurring is not given (MM = Modified Mercalli Intensity Scale) [A Modified from: Cooke & Doornkamp (1990)* Geomorphology in environmental management. *Oxford University Press, Fig. 13.13, p. 357; B Modified from: Keller (1990)* Environmental geology. *Merrill Publishing Company, Fig. 8.4, p. 157]*

A

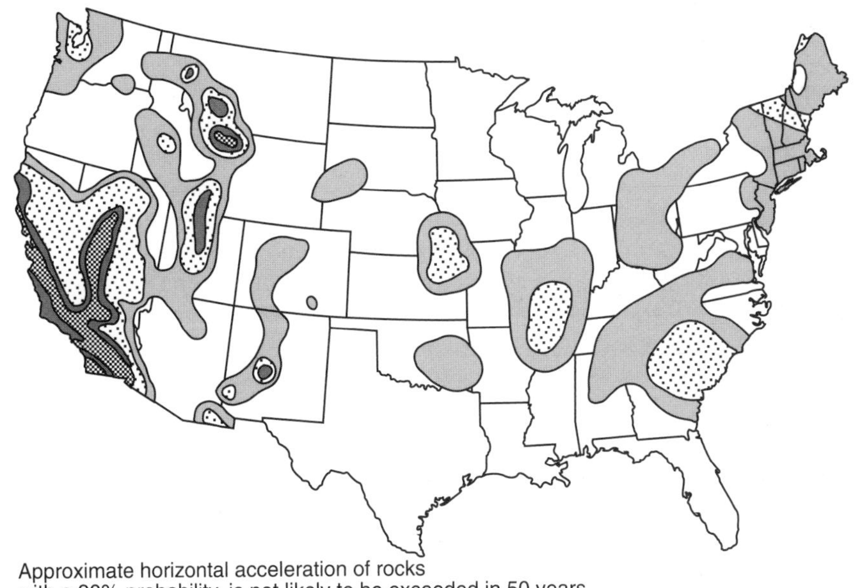
Approximate horizontal acceleration of rocks
with a 90% probability, is not likely to be exceeded in 50 years

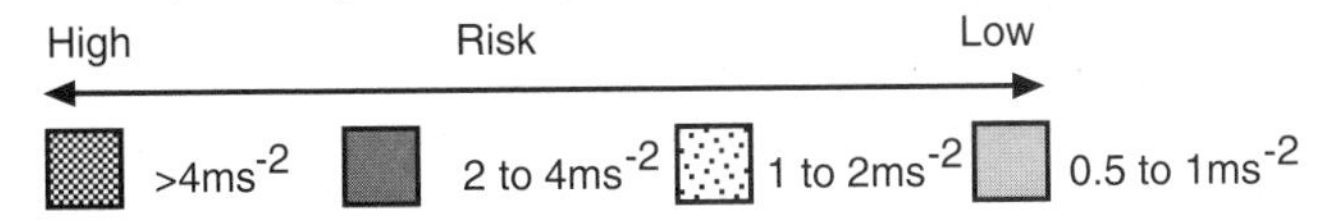
High
Risk
Low
>4ms-2
2 to 4ms-2
1 to 2ms-2
0.5 to 1ms-2

B

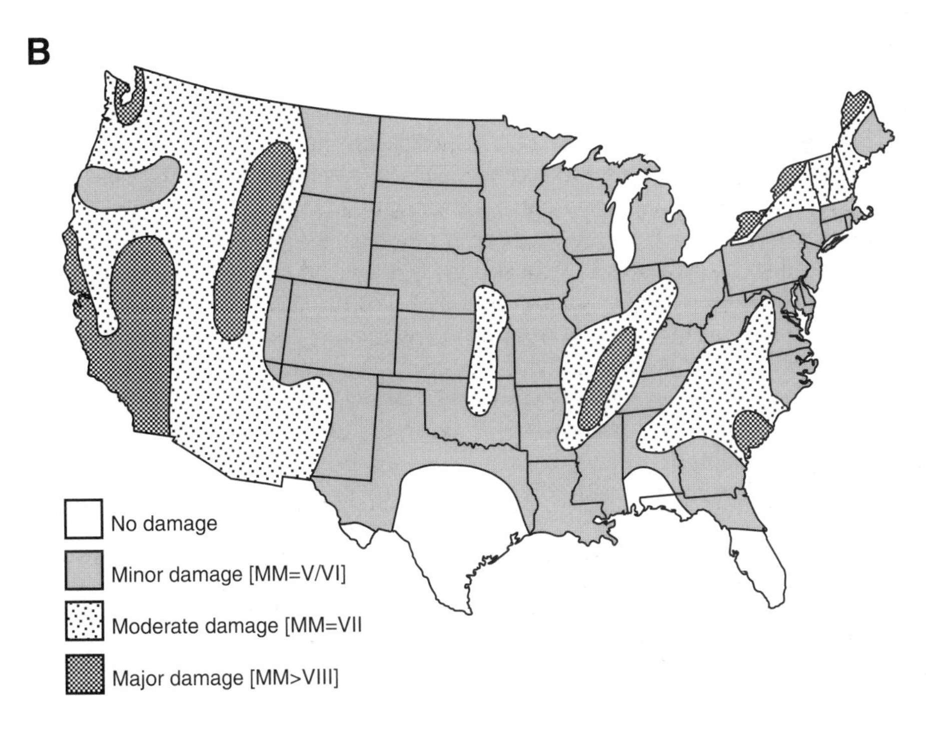
No damage
Minor damage [MM=V/VI]
Moderate damage [MM=VII
Major damage [MM>VIII]

damage which might occur to existing building stock and infrastructure. Figure 12.3B shows a hazard map which provides this information. It is based on a review of all known (historically) earthquakes and the damage caused. It identifies areas subject to different degrees of destruction in the past and on this basis predicts the likely future pattern of damage. Although widely used, such maps do not contain information about the likelihood of damage occurring, but simply records that it could.

A more integrated approach aimed at identifying hazard zones, both in terms of destruction and probability of occurrence, is shown schematically in Figure 12.4. It involves combining the following input data: (1) a map of all active and relict faults (Box 12.2); (2) a magnitude and frequency analysis of all recent and historical earthquakes within the region; (3) information about the topography and ground conditions within the area; and (4) an assessment of the susceptibility of different land parcels to secondary hazards, such as mass movements. These data are combined using some form of hazard matrix, either giving each variable equal status or providing some relative weighting. Areas of high hazard intensity should correspond to those areas close to active faults, with a high frequency of previous large earthquake events, in which ground conditions are likely to amplify shaking and which are threatened by areas, such as unstable slopes, likely to be affected by secondary hazards. In contrast, low hazard zones will be located well away from active faults, and will experience only low magnitude ground shaking very occasionally, will have stable ground conditions and will be located well away from secondary hazard sources. This type of approach is illustrated in Figure 12.5 with reference to earthquake hazards in Israel. Each of the pixels in Figure 12.5 covers an area of 25 km^2. The intensity and frequency of all historical earthquakes is used to assess the maximum intensity of damage likely to be experienced by each square over a period of 50 years, and this is then combined with information on ground conditions and surface geology to produce a hazard assessment.

Hazard maps of this sort provide a means of strategic land-use planning, but need to be combined with information on population vulnerability to produce a complete risk assessment. This is more important than with many other hazards since the quality and nature of the building stock has a major influence on the level of destruction, as shown in Table 12.1. Figure 12.6 shows the population density within Israel and when combined with the seismic hazard map in Figure 12.5 it demonstrates very clearly the areas at risk, principally the coastal belt where a high hazard potential, dense population and a concentration of industrial regions occur. At a more local scale, maps of building type coupled with hazard potential maps would identify those areas at greatest risk.

So far we have considered those areas with an identifiable seismic hazard, which most would agree are seismically active. However, in the location analysis of sensitive facilities such as nuclear power stations, nuclear waste disposal sites or chemical plants manufacturing toxic material there is a need to consider even the most remote seismic threat. For example, in the last 100 years Britain has experienced at least seven minor earthquakes and the hazard potential for most activities is probably negligible, but this is not the case when considering

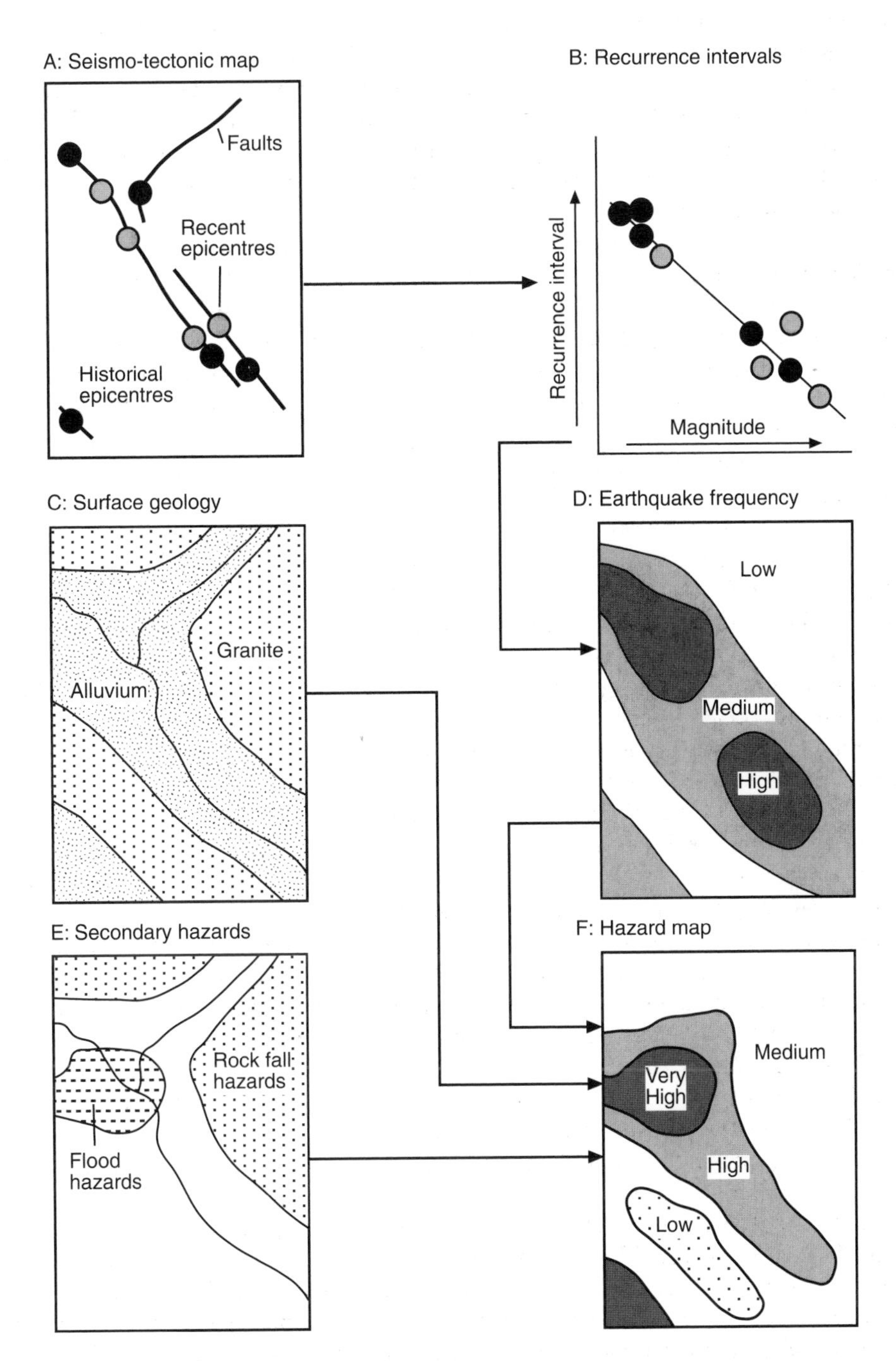

Figure 12.4 *Schematic illustration of the components within an earthquake hazard map*

BOX 12.2: THE ROLE OF GEOMORPHOLOGY IN TECTONIC STUDIES

Active faulting such as that often associated with earthquakes may leave a mark in the landscape. As Diagram A illustrates, this depends on the balance between the rate of fault displacement and the rate of erosion. If fault displacement exceeds the rate of erosion then a fault line scarp or landform feature will result. Fault line scarps can be used to map faults and are therefore of importance in reconstructing tectonic history. The degradation or degree of erosion of fault line scarps can also be used to provide a relative age

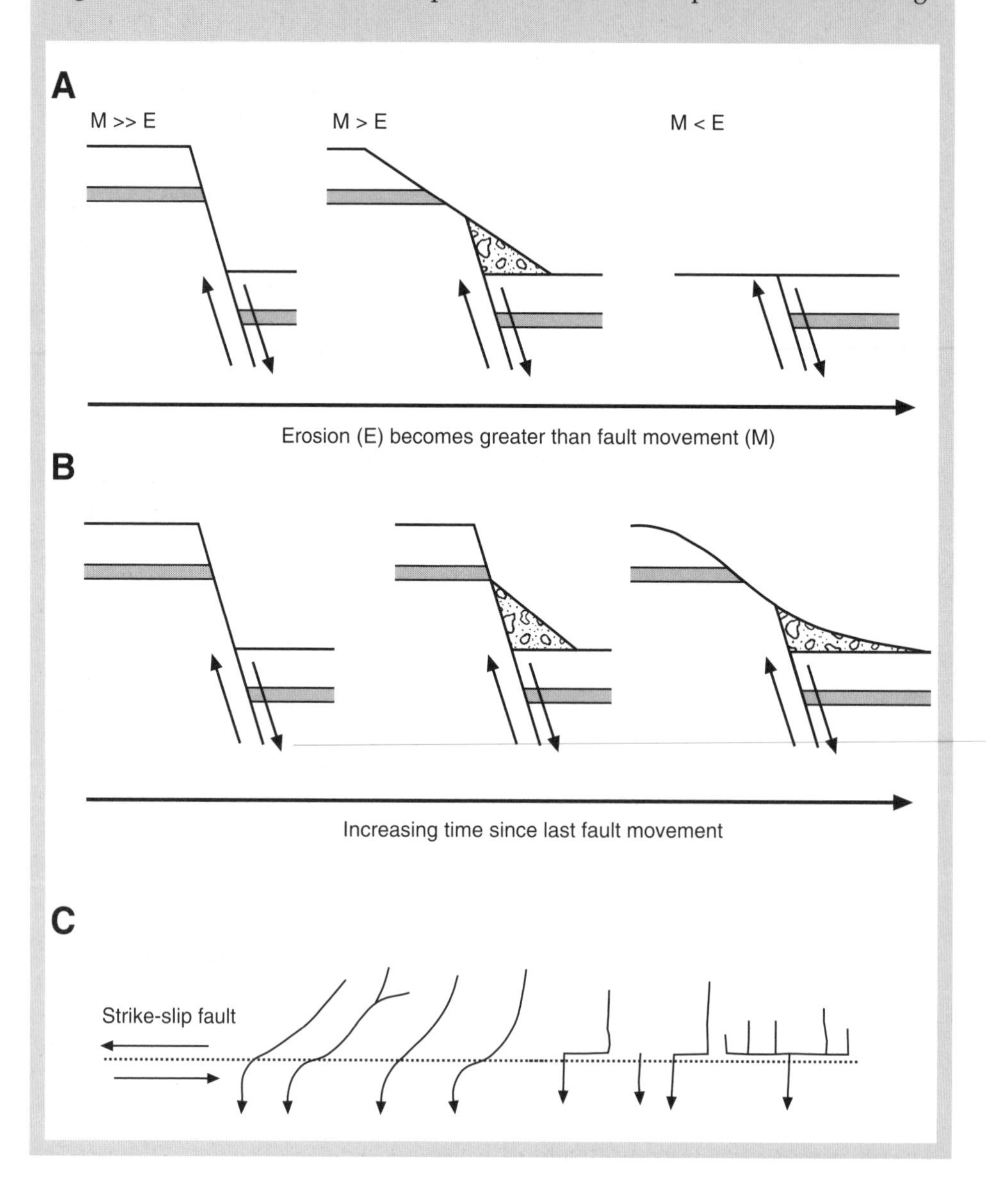

to fault movement. This is particularly important in generating a database of fault movements and therefore potential earthquake events. As Diagram B shows, the morphology of a fault line scarp becomes more degraded with time. Consequently by comparing two fault line scarps it should be possible to determine their relative age. The degree of soil development has also been used in some cases to determine the relative age of faults and, more importantly, to determine the time of last movement for some faults.

So far we have only considered vertical displacement, but in areas of lateral fault displacement, such as along the San Andreas Fault in California, landforms can also be used to identify the presence of a fault line. In this case offset rivers and streams may provide an indication of the fault line (Diagram C).

By using landscape analysis fault zones can be identified and their relative age determined as long as the rate of fault displacement exceeds erosion, which is usually the case for faulting associated with earthquakes.

Sources: Wallace, R.F. 1977. Profiles and ages of young fault scarps, north-central Nevada. *Geological Society of America Bulletin*, **88**, 1267–1281; Stewart, I.S. & Hancock, P.L. 1990. What is a fault line scarp? *Episodes*, **13**, 256–263. [Diagrams modified from: Stewart & Hancock (1990) *Episodes*, **13**, Fig. 8, p. 261]

environmentally sensitive or toxic facilities. The problem here is that there are few recorded seismic events on which to build a hazard map and an alternative approach is required. This usually involves the production of a **seismotectonic analysis** of a region consisting of a collation, or database, of all information relevant to seismic events. Typical components in such analysis include: (1) all historical reports of seismicity within the region; (2) all instrumental records of seismic events within a region; (3) a detailed map of all geological structures based on geological and geomorphological mapping (Box 12.2); (4) all geophysical data picking out subsurface structures and discontinuities; and (5) satellite image analysis to pick out all major lineaments. On the basis of the assembled data, presented in a thematic and accessible form, location decisions and hazard potential can be assessed and, if necessary, problem sites can be avoided. As the level of seismic activity increases it is possible to interpret the data further, identifying active faults and movement recurrence intervals.

12.1.2 Risk Mitigation and Management for Seismic Hazards

There are two aspects to the problems of dealing with seismic hazards. The first is dealing with the primary problem of ground shaking, and the second is dealing with the secondary hazards such as mass movements and tsunamis triggered by the earthquake. These secondary hazards have been dealt with in Chapter 11. Here we will concentrate on the problems associated with ground

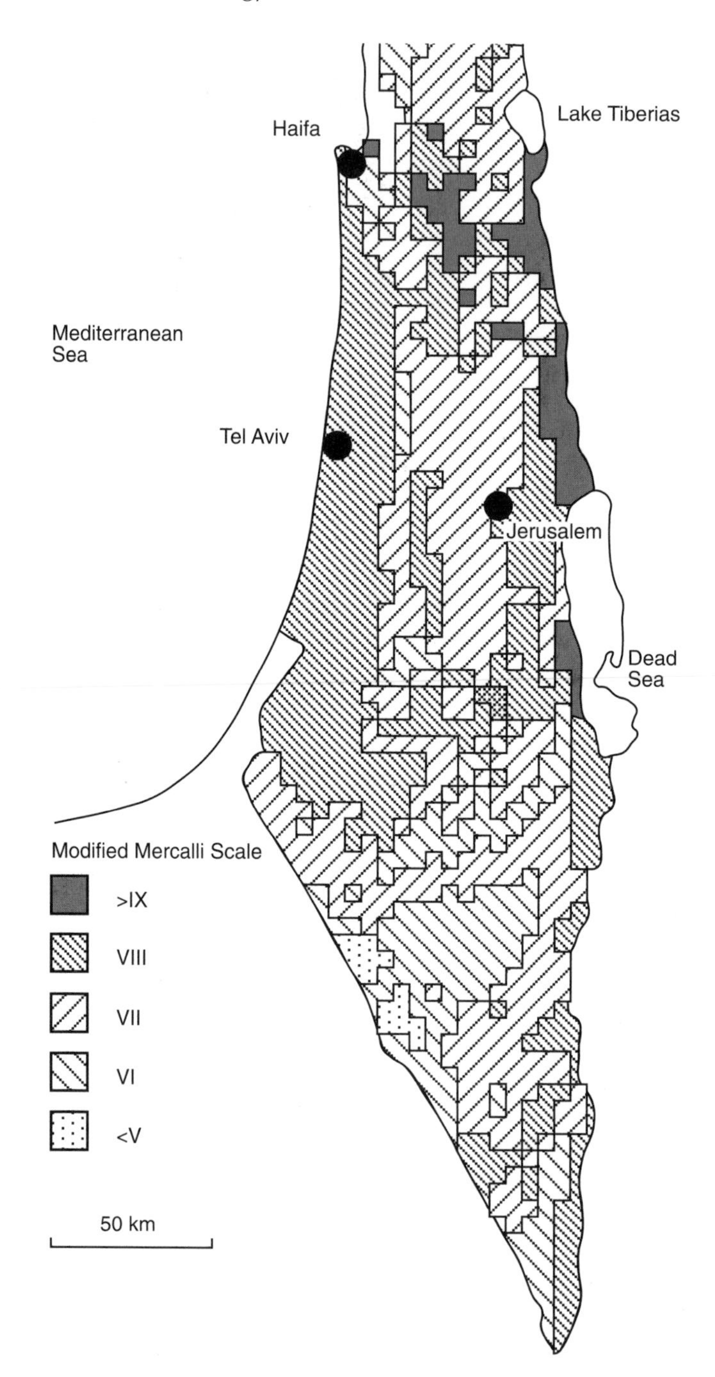

Figure 12.5 *Earthquake hazard map of Israel. Each class refers to the maximum intensity on the Modified Mercalli Scale likely to be experienced once in 50 years [Modified from: Degg (1992) In: McCall* et al. *(Eds)* Geohazards: natural and man-made. *Chapman & Hall, Fig. 4, p. 98]*

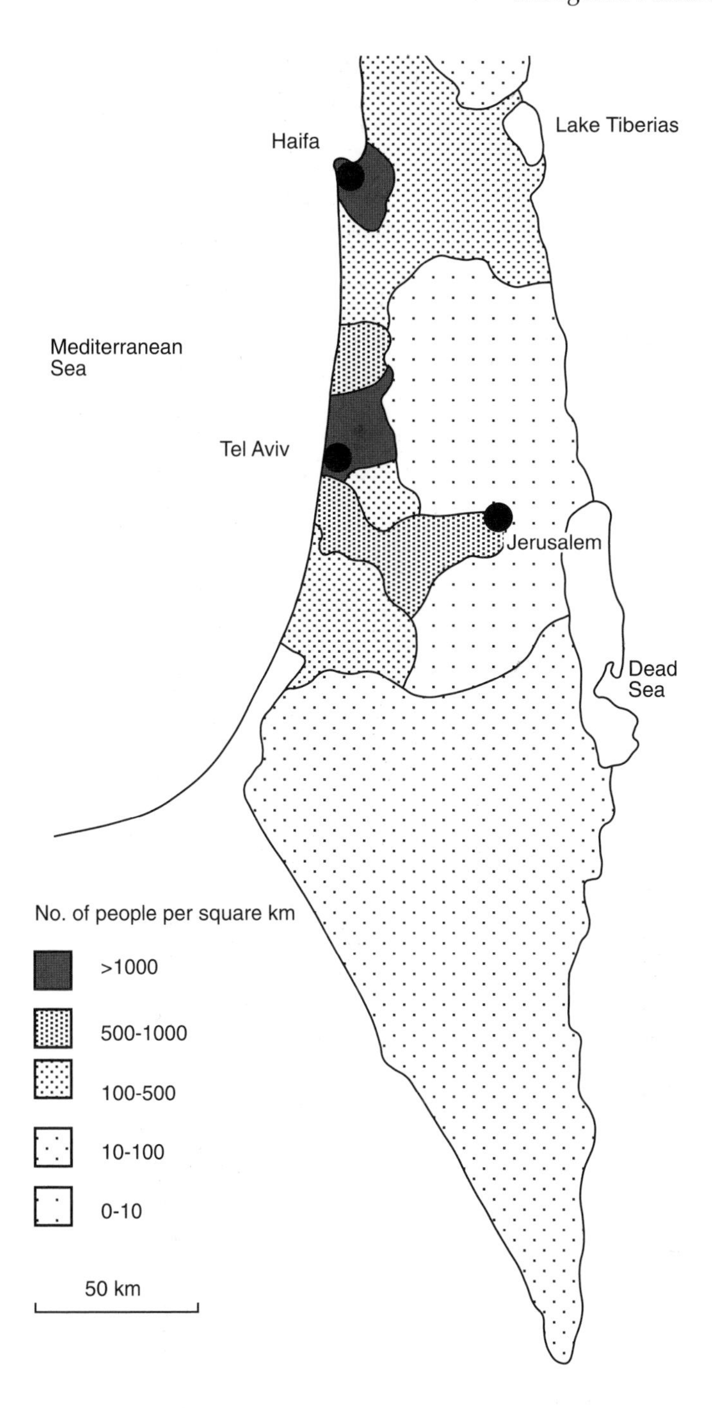

Figure 12.6 *Population density within Israel, showing how most people are located in high risk areas [Modified from: Degg (1992) In: McCall* et al. *(Eds)* Geohazards: natural and man-made. *Chapman & Hall, Fig. 5, p. 100]*

shaking. There are three approaches: (1) hazard-resistant design; (2) location control through land-use planning; and (3) forecasting and warning.

Over the last 50 years there have been major developments in the design of seismically resistant structures. As a consequence it has been suggested recently that up to 90% of the deaths and damage from earthquakes could be avoided, in that we could rebuild all the affected cities from scratch. In practice, however, much of the building stock is hazard-prone and retrofitting of safety features is both expensive, time-consuming and beleaguered with problems of identifying those liable or responsible for the cost of such work. For example, in California alone there are some 60 000 unreinforced masonry buildings which were constructed before 1933 and which are considered unsafe should there be an earthquake. The benefit of hazard-resistant design is illustrated by the Loma Prieta earthquake in California in 1989, where very few buildings collapsed, compared with the Armenian earthquake of 1988 which was of similar magnitude but killed over 25 000 people and destroyed 50 towns and cities.

Hazard-resistant design involves several basic principles: (1) the use of strong, ductile materials, such as steel, wherever possible in preference to brittle ones, like glass; (2) careful bracing of buildings to withstand lateral as well as vertical stress; (3) removal of all 'soft storeys' or weaker levels in a building such as large open spaces often introduced for functional or design reasons; (4) careful design of the linkages between different building components to provide strength and flexibility; (5) firm attachment of all fixtures and exterior panels, parapets and similar details in order to prevent them being shaken free; (6) careful choice of foundations, including deep piles, to avoid problems of surface liquefaction; (7) careful design of adjacent structures or building units orientated in different directions to possible shock waves which is important to avoid twisting or adverse interaction between them; (8) inclusion of shock-absorbing foundations, constructed from steel and rubber, for all critical buildings such as fire stations and hospitals; and (9) inclusion of safety cut-off valves and reinforced tanks for all gas, oil, petrol and water vessels to reduce fire and flooding.

Careful land-use planning in conjunction with potential hazard maps can be used to relocate buildings and guide future urban development away from problem areas. For example, this may involve the relocation of buildings or transport infrastructures which cross active faults along which displacement could occur, or in areas subject to liquefaction.

Forecasting and warning as a management strategy has attracted considerable research and funding, but with little current success. There are two scales of research: (1) frequency analysis, and (2) location-specific monitoring and prediction. **Frequency analysis** stems from attempts to find periodicity within the historical record of earthquake events and to link them to global variables such as variation in the rate of the Earth's rotation. The record does show some evidence of event clustering, although attempts to suggest that this clustering has a regular periodicity are far from unambiguous. More importantly, a global cause has not been established, although variation in the rate of the Earth's rotation can be linked to changes in ocean and atmospheric circulation which may be linked to sun spot activity.

Of more practical value is **site-specific monitoring**. This is most sophisticated in Japan and California where varying levels of success with earthquake prediction have been obtained. The first requirement for prediction is the presence of a good database of events and recurrence intervals coupled to an understanding of the structural geology of a region. Within active fault zones it is possible to identify **seismic gaps** or areas where there has been little or no movement relative to that which has been taking place elsewhere along the fault. These zones, or locked fault segments, are the likely locations for earthquakes as the stress builds up and ultimately is released as an earthquake as the fault adjusts rapidly. On the San Andreas Fault in California, careful monitoring of fault displacement using repetitive surveys has established 'locked' fault segments and 'creeping' segments. Creeping segments are characterised by continuous sliding and small vibration, while at locked segments, stress is accumulating due to the lack of movement. Monitoring is then concentrated into these locked segments. Some of these locked segments move periodically causing small earthquakes. In some cases the statistical series is good enough, owing to the regularity of the event, to be used predictively.

In most cases, predictive strategies rely on the detection of physical precursors to an event, and a considerable amount of research has been aimed at identifying these precursors in active fault zones. The problem here is understanding the linkage between these **precursor events**, fault movement and the earthquake itself and therefore their significance in making predictions and issuing warnings. Precursor events linked to imminent earthquake activity, which have been identified and are monitored, include: (1) **rock dilatancy** – as rocks dilate, they begin to deform along a fault, a phenomenon which may be detectable in the velocity of shock waves or may result in the release of radon gas into groundwater (Figure 12.7A); (2) **groundwater fluctuations**; (3) **geomagnetic** and **geoelectric variations**, which reflect groundwater variations and therefore the conductivity of the rock involved (Figure 12.7B); (4) analysis of **fore-shocks**, which appear to have foci that ring that of the main event, a phenomenon known as a **Mogi doughnut**; and (5) **ground deformation** measured using tilt meters, strain gauges and sophisticated resurveys. Attempts have been made to develop theoretical models to integrate these observations and develop a predictive theory. One of these models is the **dilatancy–diffusion model**. Following an earthquake, the strain along the locked fault segment will begin to reaccumulate. This increasing strain will cause the rocks along the fault zone to dilate after the stress on the rock exceeds half their breaking strain. **Dilatancy** involves the opening of fractures in the rock and consequently its physical expansion. It is this change in dilatancy which may be recorded by the precursor events. For example, flooding of these fissures by groundwater may reduce the electrical resistivity of the rock and release radon gas. Expansion of the rock mass may cause minor earth movements and be associated with minor foreshocks which increase in frequency until the strain exceeds the resistance along the fault zone and movement occurs causing an earthquake. This type of model is shown in Figure 12.8, and is just one of several theories about how rocks may dilate, causing precursor events, with increasing strain prior to an earthquake. Despite these

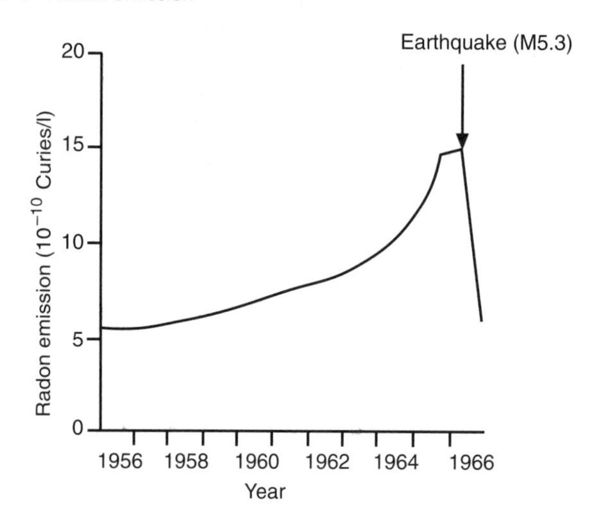

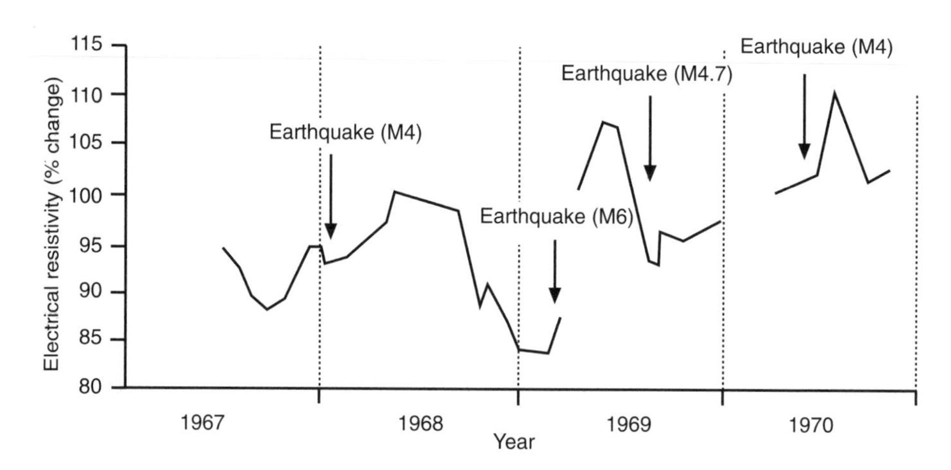

Figure 12.7 *Two examples of earthquake precursor events.* ***A.*** *Radon gas emissions.* ***B.*** *Electrical resistivity [Modified from: Keller (1992) Environmental geology. Merrill Publishing Company, Figs 8.31 & 8.32, p. 178]*

theoretical developments, the science of earthquake forecasting is still very crude. To be of use forecasts need to be specific about the magnitude, location and timing of an event; at present this cannot be achieved routinely. Moreover, many of the models which have been developed are specific to particular fault zones and dependent on detailed monitoring of the fault zone, data which are not available in many locations.

One of the problems with earthquake forecasting is that inaccurate or false predictions have an adverse effect; by 'crying wolf' once too often the credibility

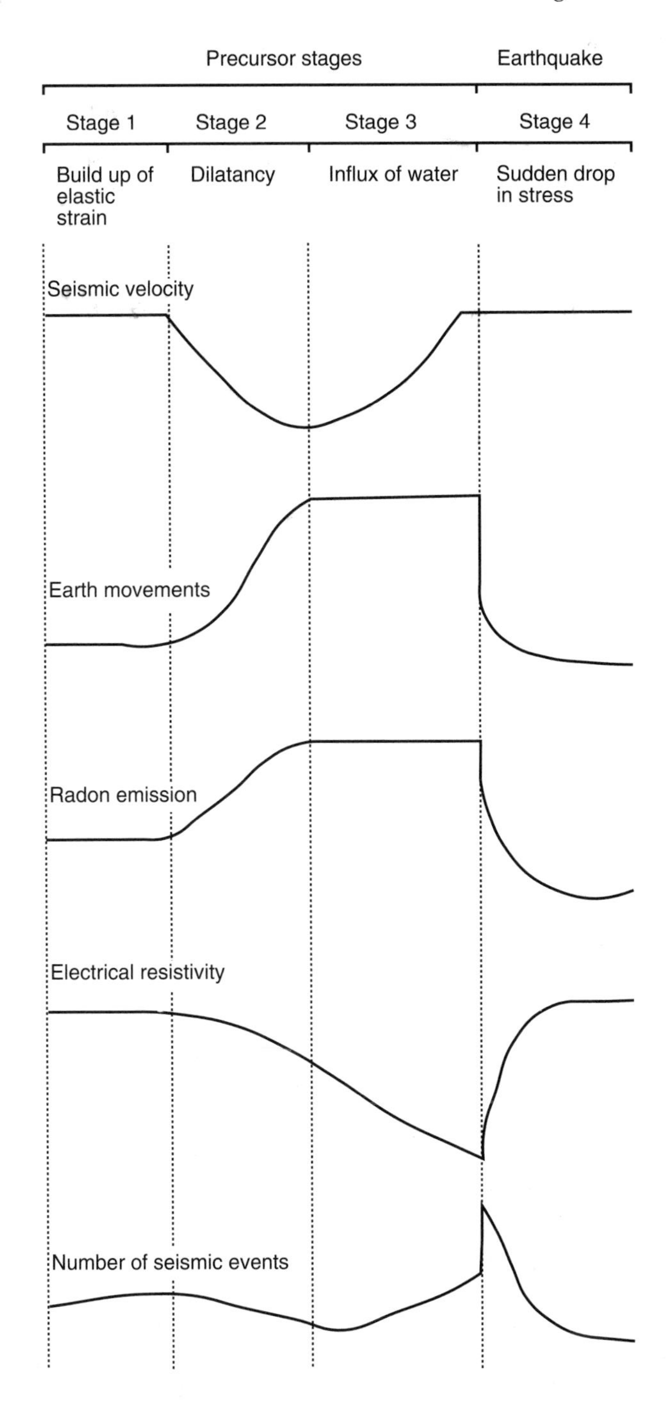

Figure 12.8 *Model of the way in which precursor events may be linked to the dilatancy along a fault zone prior to failure and earthquake initiation [Modified from: Keller (1992)* Environmental geology. *Merrill Publishing Company, Fig. 8.37, p. 183]*

of future predictions is lost. At present, predictions are simply too crude in many cases to be of practical value and have the potential to cause unnecessary socio-economic hardship if publicised widely. For example, in 1980 it was erroneously predicted that nine months later Peru would suffer its worst ever earthquake, and as a consequence of this prediction over $50 million dollars were lost in tourist revenue and in taking precautionary measures. This type of public failure gives people little faith in the warnings which are given.

The discovery in the 1960s that earthquakes could be initiated by pumping of fluids into fracture zones (Box 12.3) stimulated widespread research into the idea that earthquakes could be controlled by increasing porewater pressures within fracture zones so as to precipitate minor earthquakes and thereby prevent the build-up of potentially hazardous stress fields within a fault zone. Although several practical experiments have been conducted since to prove the basic premise, the control of earthquakes by the control of groundwater pressures is still largely a theoretical rather than a practical concept.

12.2 VOLCANIC HAZARDS

Volcanoes are found along plate boundaries and within plate interiors at locations of thermal upwelling known as hot spots (Figure 12.2). There are around 500 active volcanoes in the world, although there are many more which have erupted within the last 25 000 years and could do so again. Approximately 640 people are killed each year by volcanic activity and since 1900 around $10 billion worth of damage has been caused. More than half the deaths recorded in this period occurred in a single event in 1902 when Mount Pelée erupted on the island of Martinique in the West Indies and killed 29 000. A typical eruption releases between 10^{15} and 10^{18} joules of energy which compares favourably with the 4×10^{12} joules produced by a one kiloton atomic bomb. As with earthquakes, most volcanic incidents are best described as composite hazards, since the eruption triggers a range of secondary hazards. Primary volcanic hazards include ash fall, pyroclastic flows, lava flows, gas emissions, and volcanic collapse. Secondary hazards triggered by an eruption include mudflows and other landslides, and tsunamis, discussed in Chapter 11.

The combination of hazards produced by a volcanic event depends primarily on its explosivity, which is a function of the magma viscosity and history of the volcano. A broad continuum of increasingly explosive eruptions can be recognised, associated with increasing viscosity.

1. **Icelandic type**. Large-scale emission of mobile lava often associated with well defined linear fissures.
2. **Hawaiian type**. Emission of mobile lava usually from a single vent.
3. **Strombolian type**. A more explosive event producing less mobile lava and much more pyroclastic debris.
4. **Vulcanian type**. This involves more viscous lava which forms a crust between eruptions, thereby increasing their magnitude. This type usually involves the release of large amounts of pyroclastic debris.

BOX 12.3: ANTHROPOGENICALLY INDUCED EARTHQUAKES IN DENVER

In 1962 a deep well near Denver was first used for disposal of contaminated waste water produced as a by-product from the manufacture of chemical weapons. Waste water was pumped and injected under pressure into Precambrian rocks at over 3638 m below the Rocky Mountain Arsenal. Pressure injection of waste water began in March 1962. In April 1962 an earthquake centred on the Arsenal was recorded – the first since 1882. From April 1962 to September 1965, 710 earthquakes with epicentres in the vicinity of the Arsenal were recorded with magnitudes ranging from 0.7 to 4.3 on the Richter Scale. As the diagram below shows, these earthquake events loosely correlated with pressure injection of waste water. Public concern caused an investigation into the earthquakes and a link between the injection of water and movement along fractures generating earthquakes was identified. The significance of this story is in its potential for earthquake control. It has been widely suggested that by pumping water into fracture zones, fault movement could be induced before stress build-up causes a major earthquake when movement finally occurs naturally. Despite the promising nature of this discovery, direct pumping into fracture zones has yet to be used in the serious mitigation of earthquake hazards.

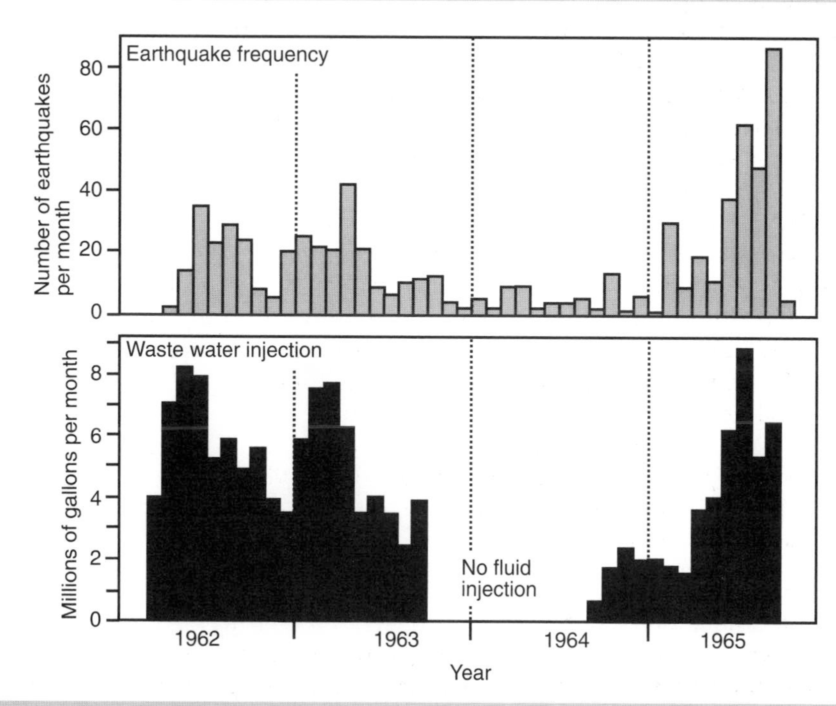

Source: Evans, D.M. 1966. Man-made earthquakes in Denver. *Geotimes*, May–June, 11–18. [Diagrams modified from: Evans (1966) *Geotimes*, May–June, Fig. 4, p. 13]

5. **Vesuvian type**. This type of eruption is even more explosive than the Vulcanian, and large clouds of gas and pyroclastic debris are produced.
6. **Plinian type**. This produces the most violent upward expulsion of gas and debris. The volcanic plume may extend for 30–55 km into the atmosphere.
7. **Peléan type**. This is potentially the most dangerous type of eruption involving violent lateral explosion of gas and debris associated with the lateral failure of a volcanic cone. The blast is directed horizontally, maximising the destructive potential.

Different types of volcanic setting are typically associated with different types of eruption. For example, divergent plate margins tend to be associated with less viscous, more mobile magma and therefore very fluid eruptions, while subduction zones are often associated with viscous or acidic magmas which typically give more explosive eruptions. In practice, however, such generalisations do not always hold since most volcanoes are polygenic, erupting several times, and may change their characteristics as different parts of the fractionated magma body beneath them is sourced.

There are five primarily volcanic hazards which may occur singly or in combination during an eruption, depending on its explosivity.

1. **Pyroclastic falls**. On average there are 60 major ash fall events each century and they have a significant impact on urban settlements and on agricultural land, as well as on commercial aviation. The amount of ash produced depends on the nature of the eruption and the prevailing winds at the time. Figure 12.9 shows the thickness of ash fall associated with the Strombolian-type eruption of the Icelandic volcano Eldjfell which threatened the town of Vestmannaeyjar. Flat-topped buildings collapsed under the weight of less than 1 m of damp ash. Ash fall may simply involve the rainout of small ash particles or alternatively may also involve the rainout of incandescent bombs which not only pose an impact hazard but may also cause fires where they crash through roofs and windows or ignite grass and brush land. Large Plinian-type eruptions in Tenerife, Canary Islands, have produced deposits of ash over 8 m thick at a distance of 20 km from the source covering an area of over 4200 km^2. Ash fall causes problems for agricultural land by destroying crops and burying grazing land for livestock; it also causes damage to the teeth and digestive tracts of livestock by the ingestion of ash. During 1963 to 1965 ash falls from the Irazu volcano, Costa Rica, destroyed the coffee crop and ruined agricultural land causing over $150 million worth of damage. When the rate of emission is very high and accumulation of hot ash rapid, it may become welded making subsequent clearance difficult. Finally, ash has the potential, when erupted into the upper atmosphere, to cause problems for aircraft engines. The ash acts as a coarse abrasive when taken into jet engines, causing serious damage and potentially affecting safety.
2. **Pyroclastic flows**. These are of two principal types: (A) hot pyroclastic flows or nuées ardentes (burning clouds); and (B) mudflows or lahars. Hot pyroclastic

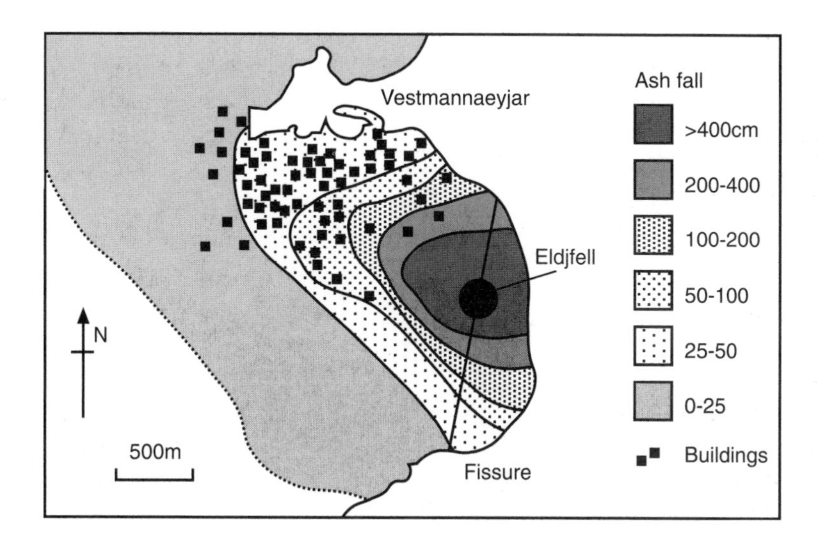

Figure 12.9 *Patterns of ash fall from Eldjfell in the first eight hours of the 1973 eruption. Much of the eastern part of the town, Vestmannaeyjar, was covered in ash and was later destroyed by advancing lava [Modified from: Booth (1979)* Journal of the Geological Society, ***136****, Fig. 1, p. 331]*

flows or nuées ardentes occur when the explosion from the volcanic vent directs a lethal mixture of hot gas, lava and ash in a downslope direction. These clouds of debris may be up to 1000°C in temperature and advance at speeds of over 100 km h^{-1}. They have caused over 70% of all volcanic deaths this century, the most famous of which was the Mount Pelée disaster when the town of St Pierre some 6 km from the vent was covered by a hot flow (700°C) travelling at 33 m s^{-1}. People exposed to such flows are killed immediately by a combination of asphyxiation and severe internal and external burns. The air blast in front of such flows may be sufficient to destroy some buildings and fell trees.

When large ash falls become mixed with either water, snow or ice, catastrophic mud flows, **lahars**, may result. One of the worst volcanic disasters of recent years involving a lahar occurred with the eruption Nevado del Ruiz volcano, Colombia, in 1985. This active volcano started to erupt in November 1984 causing melting of adjacent glaciers. A year later the volcano erupted violently causing ash to become mixed with melting glacier ice and snow. The resulting mudflow or lahar travelled over 50 km downstream and overwhelmed the town of Armero killing over 22 000 people, destroying the town and burying it under 3 to 8 m of mud.

3. **Lava flows**. Lava flows generally occur sufficiently slowly along predetermined topographic routes that they rarely threaten life, although property damage may be considerable (Figure 12.10). An exception was the 1977 eruption of Nyirangongo, Zaire which killed 72 and destroyed over 400 houses. Lava floods are the most serious of all and may convert many tens

Figure 12.10 *A house being swamped by lava in the town of Vestmannaeyjar, in Iceland, during the 1973 eruption of Eldjfell.* **A.** *During the eruption.* **B.** *The same house 10 years later [Photograph B: A.F. Bennett]*

of hectares of land into a barren stony waste. For example, Laki in Iceland erupted in 1783 along a 24 km long fissure. Over a period of five months lava flows covered an area of 560 km^2 with over 12.3 km^3 of lava.

4. **Gaseous emissions**. Volcanic eruptions are associated with the emission of a range of toxic gases including carbon monoxide and carbon dioxide. Carbon dioxide collects in hollows because it is denser than air. In 1979, 142 people were being evacuated from a village in Java when they walked into a dense pool of volcanically released carbon dioxide and were asphyxiated. Similarly, in 1984 a cloud of gas rich in carbon dioxide was released from the volcanic crater of Lake Monoun, Cameroon, and asphyxiated 37 people, while a year later 1746 lives were lost along with 8300 livestock by a gas release from the Lake Nyos crater, also in Cameroon. In the village of Nyos less than 1% of the original population survived. The Laki eruption of 1783 in Iceland was associated with the release of sulphurous gases which severely damaged crops and killed livestock, causing widespread famine and reducing the population of the whole of Iceland by 20%.
5. **Structural collapse**. The structural collapse of large volcanoes to form calderas and huge landslips is rare, occurring at most only once a century, but is a major hazard. The structural collapse of parts of Bandai-San, Japan, in 1888 resulted in over 460 deaths and 1213 km^3 of the volcano was removed. Structural collapse is not only associated with dramatic changes in topographies but also generates very large landslides which if they enter the sea may cause tsunamis. This particular scenario is of considerable concern in parts of the Canary Islands and threatens the eastern seaboard of North America. Mount St Helens provides a good example of a large-scale structural collapse. Over a two month period a bulge formed on the side of the volcano at a rate of 1.5 m per day as magma began to rise within the volcano. A month before the main eruption, the build-up was nearly 2 km in diameter and swelled by as much as 100 m and was heavily cracked. On 18 May when the bulge had become about 150 m high, an earthquake caused it to fail and initiated the eruption. The huge debris avalanche which resulted contained 2.7 km^3 of material and radically reshaped the mountain.

12.2.1 Risk Assessment for Volcanic Hazards

Risk assessment of volcanic hazards involves determining: (1) the likely extent of different volcanic hazards given an eruption; and (2) the vulnerability of settlements and human infrastructure within a region. Hazard mapping is based on a detailed understanding of the volcanic history and the topography of a cone or volcanic fissure. The relative proportion of different volcanic products erupted is based on past history and the direction of any lava or pyroclastic flows is deduced from the topography of the volcano. This approach makes the assumption that future eruptions will follow a similar pattern to historical ones. This may not always be the case since magma chemistry and therefore the style of eruption can change with the degree of fractionation and therefore time. More importantly, it may work for polygenic volcanoes but is of little use for monogenic eruptions involving new vents or fissures.

Volcanic hazard maps are produced in a variety of ways depending primarily on local circumstances. The best way to demonstrate the general principles is through an example. Mount Etna is the largest continental volcano in the world; it reaches a height of 3300 m and covers an area of 1750 km^2. It is a strato volcano with over 200 parasitic cones, and its slopes have a long history of human occupation. Lava may be generated either from the central vent or from one of the parasitic cones. The historical extent of summit lava can be mapped, but predicting flow from a parasitic cone is more difficult. In the hazard map shown in Figure 12.11 an arbitrary value of one cone per square kilometre was taken and the lava flow from the edge of this zone was predicted on the basis of topography and the average slope gradient. The majority of lava flows would need to be sourced within this zone to threaten the villages and towns on the flanks of the volcano. Lava could be generated by parasitic cones outside this zone but are unlikely to flow sufficiently far to threaten urban areas. Using this one cone per square kilometre source area it is possible to categorise the volcano flanks into areas which would be topographically protected from lava flow and those which would not. In this way a hazard map can be produced (Figure 12.11). Like all hazard maps its reliability is limited by the data and assumptions are made in its construction. In this case a major assumption is that topography is the only control on lava flow. Subterranean lava flows may complicate this picture in some regions. In addition, the hazard map in Figure 12.11 does not contain information about ash fall. Figure 12.12 shows a hazard map for Tenerife, Canary Islands, which also includes potential ash fallout values, based on those observed for previous eruptions.

12.2.2 Risk Mitigation and Management for Volcanic Hazards

There are few practical approaches to mitigation of volcanic hazards, and most management strategies are based on forecasting and the implementation of an emergency plan.

Most volcanic hazards cannot be controlled, although attempts have been made to control lava flows, with varying degrees of success (Box 12.4). Three methods exist for controlling lava flows.

1. **Bombing**. Bombing lava flows high on a volcano may cause the flow to spread and solidify. Fluid lava flows in channels, the walls of which can be bombed in an attempt to divert the flow. This method was used with some success in the diversion of fluid magma flows on Mauna Loa in 1942. An alternative is to bomb the vent itself in an attempt to promote lava flow over a wider area, thereby preventing the formation of lava streams. In practice, although some attempts have been made there is not enough practical experience to develop a well defined strategy for this approach.

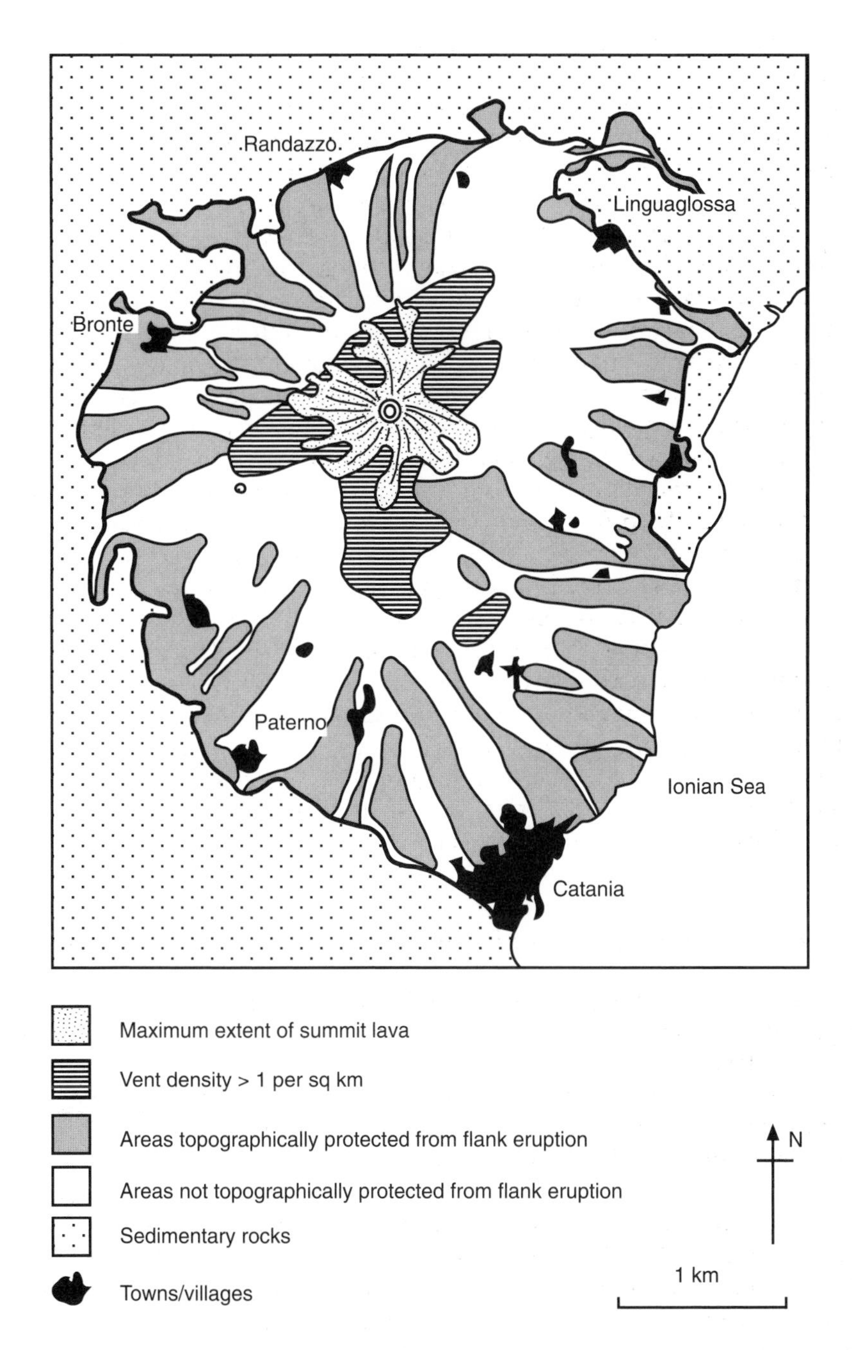

Figure 12.11 *Hazard map for Mount Etna [Modified from: Guest* et al. *(1981)* Geographical Journal, ***147****, Fig. 6, p. 176]*

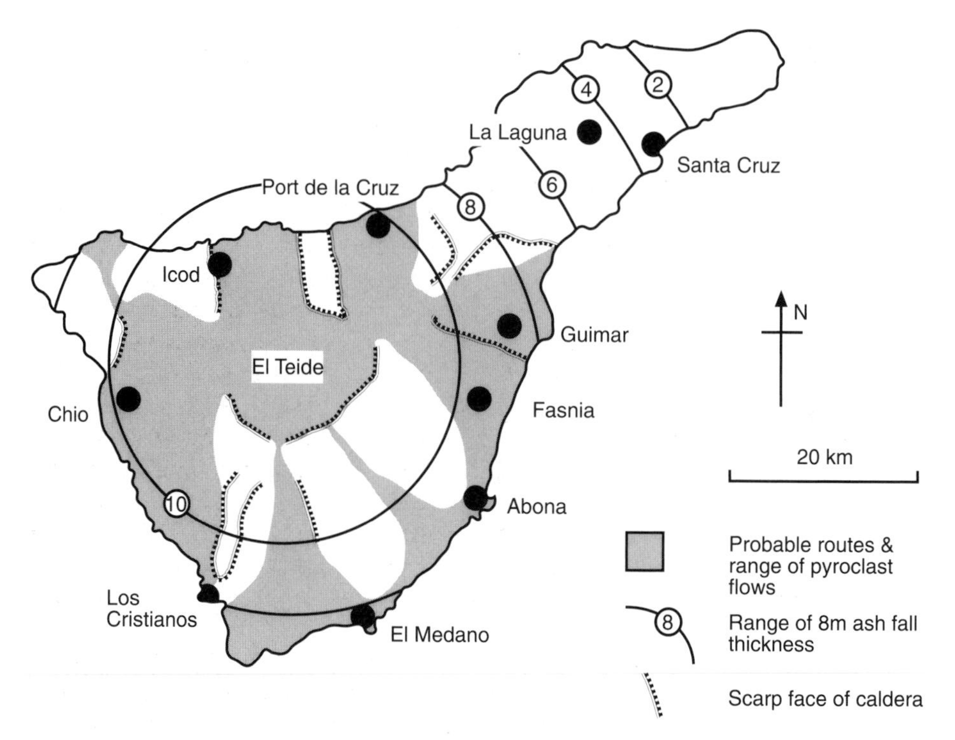

Figure 12.12 *Hazard map for the island of Tenerife, Canary Islands [Modified from: Booth (1979)* Journal of the Geological Society, ***136****, Fig. 3, p. 334]*

2. **Artificial barriers**. Given favourable topography, the idea is to divert the flow using a barrier. This is only practical for thin fluid flows, there being too much force behind thicker flows. Although barriers have been constructed from earth in a number of places there is little empirical evidence of their success to constrain future design.
3. **Water sprays**. The use of water sprays to induce lava cooling was first attempted on Hawaii during the eruption of Kilauea in 1960 and was used on a large scale during the eruption of Eldjfell in Iceland. Approximately 1 m^3 of water is required to cool 0.7 m^3 of lava from 1100°C to 100°C. Availability of large volumes of water, usually sea water, is therefore essential. At Eldjfell in Iceland, pumping rates from the adjacent sea reached 1 m^3 s^{-1} at their height, chilling 60 000 m^3 of lava. The pumping process continued for 150 days and was consequently of considerable expense, but it worked, leaving a lava wall along the advancing front up to 20 m high in places.

Mitigation works have also been attempted for some lahar hazards. Artificial mounds have been constructed in parts of Indonesia to provide a safe refuge for villagers when faced with a lahar. It is doubtful, however, whether they could attain safety quickly enough in the event of disaster since this depends on adequate warnings being issued. Direct engineering has been used on the Kelut

BOX 12.4: STOPPING THE FLOW: MOUNT ETNA 1992

An example of the use of direct intervention is provided by the eruption of Mount Etna on 14 December 1991. Two fractures opened at an elevation of 3000 m, one of which propagated down to 2700 m. Lava began to pour out of the lowest end of the fracture. Hazard maps (Figure 12.11) of the volcano suggested that the village of Zafferana Etna with a population of 7000 was at risk. Three attempts at intervention were made to safeguard the town. The first involved the construction of a 234 m long, 21 m high embankment containing 370 000 m^3 of earth. The dam contained the lava for one month. Three additional smaller earth barriers were constructed to slow the flow further.

The second attempt to control the flow involved work to stop or divert the flow from its natural and heavily tunnelled channel at an elevation of 2000 m. Initially helicopters were used to drop concrete blocks and antitank obstacles into the entrance of a tunnelled section of the channel. In addition the channel walls were blasted to help with this blocking work. This caused a diversion of the channel and temporarily halted the flow in an adjacent area for two weeks. However, the lava flow restarted along its original course after two weeks.

The final and successful intervention was carried out in May. An artificial channel was dug adjacent to the current one in order to divert the flow away from the village. The wall between the artificial channel and the natural one was thinned and then blown with 7000 kg of dynamite. Two-thirds of the flow was spontaneously diverted into the channel and complete diversion was achieved when the natural one was blocked with 230 m^3 of lava boulders. The lava flow finally stopped just 850 m away from the village of Zafferna.

This example illustrates the role of direct intervention in dealing with lava hazards and the difficulty in its control.

Source: Gasparini, P. 1993. Research on volcanic hazards in Europe. *Science*, **260**, 1759–1760.

volcano in Java. In 1919 an explosive eruption expelled large quantities of water which generated lahars killing over 5000 people. In an attempt to reduce this hazard, artificial drainage of the crater via a series of tunnels was undertaken to reduce the volume of stored water. This was successful and an eruption in 1951 failed to generate significant lahars. However, during this eruption the crater was deepened further and dangerous water levels developed. Owing to the poor design of further drainage tunnels the water level was not successfully reduced and in an eruption in 1966 significant lahars formed, claiming a number of lives. Subsequently, improved drainage tunnels have again been used to reduce the water volume and hazard potential. Building design has a role to play in the development of structures which shed ash fall, and which can either withstand blasts or provide survival spaces on collapse in which inhabitants may survive.

In practice, most hazard management schemes rely on early warning and the implementation of emergency evacuation procedures. Prediction of volcanic activity is difficult and there have been few successes. Prediction is based on the monitoring of phenomena known to be associated with the onset of an eruption, including seismic events, land deformation, geomagnetic and geoelectrical fluctuations, and variations in temperature and gas emission. However, there are few well established relationships. The problem is exacerbated because the aim in hazard management is not to predict the initiation of an eruption, but the onset of a volcanic catastrophe. No volcanic catastrophe yet observed has occurred at the very start of an eruption. The question of importance in hazard management is to determine whether a developing eruption will culminate in a dangerous climax and, if it does, when and how it will occur. The problem is illustrated by the residents of Guadeloupe in the West Indies who were told to evacuate because of an imminent eruption from the volcano of La Soufriére. They waited for 15 weeks, bringing economic ruin to many before a small and totally harmless eruption took place. Predicting the onset of a volcanic catastrophe as opposed to the start of an eruption is very difficult and in practice it can only be done on the basis of probability founded in a detailed knowledge of past events where sufficiently good records exist. Recording the extent of past volcanic events and their chronology is of vital importance in establishing a database with which likely scenarios and some assessment of their probability can be made. In practice what is required is the identification, using past experience, of a series of threshold or critical levels of activity likely to indicate an imminent eruption. These threshold or critical levels can then be used to initiate emergency plans.

Emergency plans will vary from location to location depending on the available resources and nature of the hazard; however, three broad stages can be identified.

1. **Alert phase**. Mobilise fire fighting and medical services; prepare reception centres to receive evacuees; prepare emergency food and shelter; assemble the necessary transportation to effect an evacuation; brief press and inform public of the level of risk. The time required for this phase depends on the degree of preparation and planning, but on average may involve between five and 15 days.
2. **Readiness**. Emergency services prepared for immediate action; evacuation of sick and infirm; evacuation of valuable property. This should only take two to five days.
3. **Evacuation**. Removal of all persons from the danger area. This should in most cases be achievable in one or two days.

The length of time required for each step will vary but the maximum time required based on the estimates given would be 22 days. This process must therefore be initiated at least 22 days before the event occurs. Determining the risk threshold at which each phase is initiated is based, where possible, on past experience or historical data of the behaviour of this or similar volcanoes. A threshold may be based on a single parameter or on several found by past experience to be linked to an eruption (e.g. seismic activity, crustal deformation, temperature of gaseous emission, quantity of ejecta). The reliability of these thresholds depends primarily on the length and quality of the data set available for a given or similar volcano.

Until our ability to predict the onset and evolution of a volcanic eruption improves we need to err on the side of caution in initiating emergency evacuations, irrespective of their cost and unpopularity.

12.3 RADON GAS

In recent years there has been growing concern about natural radiation causing clusters of ill health, in particular lung cancer. Attention has focused on the naturally occurring radioactive gas **radon** which is produced by the radioactive decay of ^{238}U present in a variety of rocks, the most noticeable of which is granite. The radioactive decay of ^{238}U produces the daughter nuclide radon as part of the chain of decay which leads ultimately to the stable ^{210}Pb isotope. Radon has a half-life of only 3.8 days (or 91 h 48 min) after which it converts to a highly radioactive but solid isotope, and is therefore short-lived. It is a colourless, odourless and tasteless gas and as such cannot be easily detected. Radon contributes to the natural background radiation, and is harmless when diluted in air outdoors. However, it may become concentrated within enclosed spaces such as tunnels, mines and buildings. Domestic radon exposure is estimated to be responsible for between 10 000 and 15 000 deaths from lung cancer in the USA and Canada each year. Thus one death for every 25 000 people is due to radon, which compares with one in 1800 due to smoking and one in 2.6 million due to the nuclear industry. Radon exposure is therefore one of the most important radiation hazards accounting for 55% of all radiation health risks. However, public perception and concern are poor, which is reflected in the USA in the amount of investment in prevention. In 1991 prevention costs were estimated at $200 000 per life saved from radon exposure, compared with $200 million for radioactive waste and $2 billion for reactor accidents. Radon exposure can, however, be effectively eliminated by simple domestic works.

Concentration of radon in buildings may occur in several ways: (1) radon is soluble in water and may enter from water abstracted from wells, being released as water flows from taps; (2) radon may seep from soil and rock beneath buildings and become concentrated, particularly in basements and foundation voids when poorly ventilated; and (3) it may enter buildings from construction materials such as granite blocks, some gypsum (plaster) boards, Portland cement, some cinder blocks and phosphate fertilisers. The most important source is the seepage of radon gas from foundation soils. Seepage occurs in response to pressure variations. Low pressure within a building will cause foundation soils to exhale radon, while higher pressures will drive radon into the soil. Pressure variations occur primarily in response to meteorological factors such as the passage of low pressure systems and fluctuations caused by wind shear on buildings. As a consequence, radon concentrations may fluctuate dramatically over relatively short time periods. In general, modern insulated buildings are more at risk than draughty old ones.

The main source rocks for radon include: (1) granite; (2) phosphate-rich rocks; (3) black shales and other rocks rich in organic matter; (4) uranium mine tailings; and (5) glacial tills derived from uranium-rich rocks. Defining the extent of radon hazard is based on the identification of likely geological host rocks coupled with detailed testing. Remedial work to counter the build-up of harmful radon concentrations for the most part involves simply improving ventilation with basements and foundation voids to prevent its accumulation. More effective grouting around incoming services to seal buildings may also form part of a works programme. Where groundwater is the principal source, settling pools can be introduced into the water supply system to ensure radon decay prior to its entry into domestic buildings. In severe cases, extreme measures may include the removal of radon-rich building materials.

12.4 SUMMARY OF KEY POINTS

- There are three main endogenic hazards: seismic, volcanoes and radon gas. The first two are the most important and are best described as composite hazards. Each is associated with a primary phenomenon but it is the secondary exogenic hazards that they trigger which often cause the greatest loss of life.
- The primary seismic hazard is ground shaking, while secondary hazards include fault displacement, ground subsidence, liquefaction, mass movement and fire within urban areas. Ground shaking experienced at a given location is a function of the magnitude, frequency and duration of the earthquake as well as the distance from it, in conjunction with the prevailing ground conditions which have an amplifying effect. Areas underlain by soft sediment experience greater ground shaking than those underlain by hard rock.
- The magnitude of earthquakes is measured on the Richter Scale while the degree of damage is measured on the Modified Mercalli Intensity Scale. Owing to the role of ground amplification, two earthquakes with similar magnitudes as measured on the Richter Scale may be associated with different degrees of destruction as measured by the Modified Mercalli Intensity Scale.
- A variety of methods exist for producing earthquake hazard maps, but the most effective maps integrate data on earthquake magnitude and frequency with data on ground conditions and exposure to secondary hazards.
- Earthquake hazard mitigation measures include: (1) the design of resistant buildings; (2) hazard mapping associated with land-use planning and in some cases urban relocation; and (3) prediction and warning. Earthquake forecasting is an imprecise science and relies on the recognition of suitable

precursor phenomena or events. Although integrated earthquake prediction models exist which link these precursor events to the initiation of an earthquake, they are at present far from universal and lack precision.

- Typical volcanic hazards include pyroclastic falls, pyroclastic flows, lava flows, gas emissions and structural collapse. In addition, they may trigger secondary hazards such as tsunamis.
- Volcanic hazard mapping is primarily based on the assumption that future eruptions will follow a similar pattern to historical ones and consequently hazard zones can be predicted on this basis.
- Hazard management consists of: (1) the direct control of lava and ash flows; (2) the building of blast- and ash-shedding buildings; and (3) warning and emergency evacuation. The prediction of volcanic eruptions is difficult, not least because the aim is not to predict the onset of an eruption, but to predict the escalation of an eruption into a volcanic catastrophe which could threaten lives and infrastructure.
- Radon gas is a quiet hazard but in the USA and Canada alone probably accounts for between 10 000 and 15 000 deaths each year. It is therefore as significant as the more dramatic endogenic hazards. More importantly, by identifying the problem and taking a series of simple remedial measures its impact can be removed almost completely.

12.5 SUGGESTED READING

The books by Wood (1986), Bryant (1991) and Smith (1996) provide useful overviews of both volcanic and earthquake hazards. In addition the paper by McCall (1996) provides an excellent review of the subject. The effects of the Alaskan earthquake in 1964 are reviewed in Hansen (1965) and Eckel (1970), while Degg (1992b) provides a useful review of the 1985 Mexican earthquake which emphasises the role of ground conditions in the amplification of shock waves. Borcherdt (1975) contains a series of papers on seismic hazard in the San Francisco Bay region which provide a summary of the range of problems encountered in seismically active regions. Degg (1992a) provides information about the construction of earthquake hazard maps as does the paper by Cross *et al.* (1985). Keller (1992) provides an authoritative review of the attempts to develop predictive models for earthquakes.

Information about the Mount St Helens eruption can be obtained from Hammond (1980), Rosenfeld (1980) and Swanson *et al.* (1983). The hazard posed by lahars is covered by Mothes (1992) and Verstappen (1992). Walker *et al.* (1992) discuss the impact of gas release in 1986 from Lake Nyos in Cameroon. A series of papers on the assessment of volcanic hazards can be found in the *Journal of the Geological Society*, Volume **136**(3); papers by Booth (1979), Fournier d'Albe (1979), Guest & Murray (1979) and Taxieff (1979) are of particular note. The work of Pyle (1995) discusses volcanic hazards in urban areas. The development of hazard maps for Mount Etna is covered in a paper by Duncan *et al.* (1981). Nossin & Javelosa (1996) discuss hazard zonation for Mt Pinatubo in the Philippines. Williams & Moore (1973), Mason & Foster (1983) and Gasparini (1993) provide information on attempts to control lava flow. Probably the best account of the radon gas problem is provided by Tilsley (1992).

Booth, B. 1979. Assessing volcanic risk. *Journal of the Geological Society of London*, **136**, 331–340.

Borcherdt, R.D. 1975. *Studies for seismic zonation of the San Francisco Bay region*. US Geological Survey Professional Paper, 941A.

Bryant, E.A. 1991. *Natural hazards*. Cambridge University Press, Cambridge.

Cross, M., Degg, M. & Johnson, J. 1985. The escalating earthquake risk in the Middle East. *Risk and loss management*, 16–19.

Degg, M.R. 1992a. The ROA Earthquake Hazard Atlas project: recent work from the Middle East. In: McCall, G.J.H., Laming, D.J.J.C. & Scott, S.C. (Eds) *Geohazards: natural and man-made*. Chapman & Hall, London, 93–104.

Degg, M.R. 1992b. Some implications of the 1985 Mexican earthquake for hazard assessment. In: McCall, G.J.H., Laming, D.J.J.C. & Scott, S.C. (Eds) *Geohazards: natural and man-made*. Chapman & Hall, London, 105–114.

Douglas, L.A. 1980. The use of soils in estimating the time of last movement of faults. *Soil Science*, **129**, 345–352.

Duncan, A.M., Chester, D.K. & Guest, J.E. 1981. Mount Etna volcano: environmental impact and problems of volcanic prediction. *Geographical Journal*, **147**, 164–178.

Eckel, E.B. 1970. *The Alaskan earthquake, March 27 1964: lessons and conclusions*. US Geological Survey Professional Paper, 546.

Fournier d'Albe, E.M. 1979. Objectives of volcanic monitoring and prediction. *Journal of the Geological Society of London*, **136**, 321–326.

Gasparini, P. 1993. Research on volcanic hazards in Europe. *Science*, **260**, 1759–1760.

Guest, J.E. & Murray, J.B. 1979. An analysis of hazard from Mount Etna volcano. *Journal of the Geological Society of London*, **136**, 347–354.

Hammond, P.E. 1980. Mt. St. Helens blasts 400 meters off its peak. *Geotimes*, **25**, 14–15.

Hansen, W.R. 1965. *The Alaskan earthquake, March 27 1964: effects on communities*. US Geological Survey Professional Paper, 542A.

Keller, E.A. 1992. *Environmental geology*, sixth editon. Merrill Publishing Company, Columbus.

McCall, G.J.H. 1996. Natural hazards. In: McCall, G.J.H., De Mulder, E.F.J. & Marker, B.R. (Eds) *Urban geoscience*. A.A. Balkema, Rotterdam, 81–125.

Mason, A.C. & Foster, H.L. 1953. Diversion of lava flows at Oshima, Japan. *American Journal of Science*, **251**, 249–258.

Mothes, P.A. 1992. Lahars of Cotopaxi Volcano, Ecuador: hazard and risk evaluation. In: McCall, G.J.H., Laming, D.J.J.C. & Scott, S.C. (Eds) *Geohazards: natural and man-made*. Chapman & Hall, London, 53–63.

Nossin, J.J. & Javelosa, R.S. 1996. Geomorphic risk zonation related to June 1991 eruption of Mt. Pinatubo, Luzon, Philippines. In: Slaymaker, O. (Ed.) *Geomorphic hazards*. John Wiley & Sons, Chichester, 69–94.

Pyle, D.M. 1995. Reduction of urban hazards. *Nature*, **378**, 134–135.

Rosenfeld, C.L. 1980. Observations on the Mount St. Helens eruption. *American Scientist*, **68**, 494–509.

Smith, K. 1996. *Environmental hazards. Assessing risk and reducing disaster*. Second edition. Routledge, London.

Swanson, D.A., Casadevall, T.J. & Dzurisin, D. 1983. Predicting eruptions at Mount St. Helens, June 1980 through December 1982. *Science*, **221**, 1369–1376.

Tazieff, H. 1979. What is to be forecast: outbreak of eruption or possible paroxysm? The example of the Guadeloupe Soufriére. *Journal of the Geological Society of London*, **136**, 327–329.

Tilsley, J.E. 1992. Urban geology 2. Radon: sources, hazards and control. *Geoscience Canada*, **19**, 163–166.

Verstappen, H.Th. 1992. Volcanic hazards in Colombia and Indonesia: lahars and related phenomena. In: McCall, G.J.H., Laming, D.J.J.C. & Scott, S.C. (Eds) *Geohazards: natural and man-made*. Chapman & Hall, London, 33–42.

Walker, A.B., Redmayne, D.W. & Browitt, C.W.A. 1992. Seismic monitoring of Lake Nyos, Cameroon, following the gas release disaster of August 1986. In: McCall, G.J.H., Laming, D.J.J.C. & Scott, S.C. (Eds) *Geohazards: natural and man-made*. Chapman & Hall, London, 65–79.

Williams, R.S. & Moore, J.G. 1973. Iceland chills a lava flow. *Geotimes*, **18**, 14–18.

Wood, R.M. 1986. *Earthquakes and volcanoes*. Mitchell Beazley, London.

13
Environmental Geology: an Urban Concept

By the end of the twentieth century more than half of the world's population will live in cities, at least 60 of which will have populations in excess of five million, and over 3.5 billion people will live on only 1% of the Earth's surface area. Environmental geology has been defined in this book as the interaction of humans with the geological environment, both its physical constituents and the surface processes. Given that the Earth's population will be concentrated in urban centres in the next century, the main focus for environmental geologists will be increasingly urban. Within the urban framework, the environmental geologist will have three broad tasks: (1) to manage the provision of mineral, construction, water and conservation resources; (2) to provide appropriate geological information in construction, engineering and waste management projects; and (3) to manage and mitigate against the natural and human-induced hazards which threaten an increasingly concentrated, and therefore vulnerable, population. These three tasks are the subject areas of environmental geology. The urban environment draws together all of these aspects and provides its greatest challenges. In this chapter we examine the urban environment as the modern frontier of environmental geology, in controlling both the constraints and the opportunities for continued urban growth. Two case studies, based upon the ancient city of Rome and the modern city of Hong Kong, demonstrate how the urban environment draws together, and provides a focus for, all aspects of environmental geology.

13.1 URBAN GROWTH AND THE GEOLOGICAL ENVIRONMENT

Urban growth is a feature of all of the industrialised nations of Europe, North America and Japan, where it is common for between 60 and 80% of the total population to be concentrated into the urban centres (Table 13.1). Concentration

Table 13.1 *Populations of urban areas expressed as a percentage of the total population of a country [Modified from: Legget (1973)* Cities and geology, *Table 1.1, p. 8]*

Country	Population in urban areas (%)	
	c.1920	c.1960
Europe		
Belgium	57.3	62.7
Denmark	43.2	69.9
France	46.4	55.9
W. Germany	70.5	76.8
Netherlands	45.6	60.4
UK	79.3	80.0
Former USSR	17.9	47.9
North America		
Canada	49.5	69.6
USA	51.2	69.9
Australasia		
Australia	64.0	81.9
Asia		
Japan	18.1	63.5

of people into urban centres is a continuing trend which began with the Industrial Revolution, and which has increased exponentially ever since (Figure 13.1A). For example, in 1800 there were only 50 cities in the world with populations exceeding 100 000 people, and only one city, London, with a population of over 1 000 000; by 1970 there were 1000 cities of over 100 000 people, and 100 with a population of over 1 000 000. Although the rate of concentration is now slowing to about 2% per annum for the developed nations (Figure 13.1B), average growth rates for Africa and southern Asia are increasing, typically to 5%. Growth is particularly strong in Africa, where a predominantly rural population is becoming increasingly urbanised, and the urban population is expected to grow from 23.7% to 42.4% by the end of the century. The trend, then, is towards 'mega-cities', each with attendant problems associated with the concentration of humanity into a small part of the Earth's land surface.

13.2 THE HISTORY OF URBAN DEVELOPMENT AND GEOLOGY

Historically, geology has had a major role in the siting and development of urban centres. For example, the majority of ancient cities in Europe, Asia and the Americas were sited to take advantage of some geological or geomorphological feature for defence or commerce. Typically, these include: the use of geomorphological features such as river crossings or transport corridors through mountainous areas; river mouths for the development of ports; or defensive positions

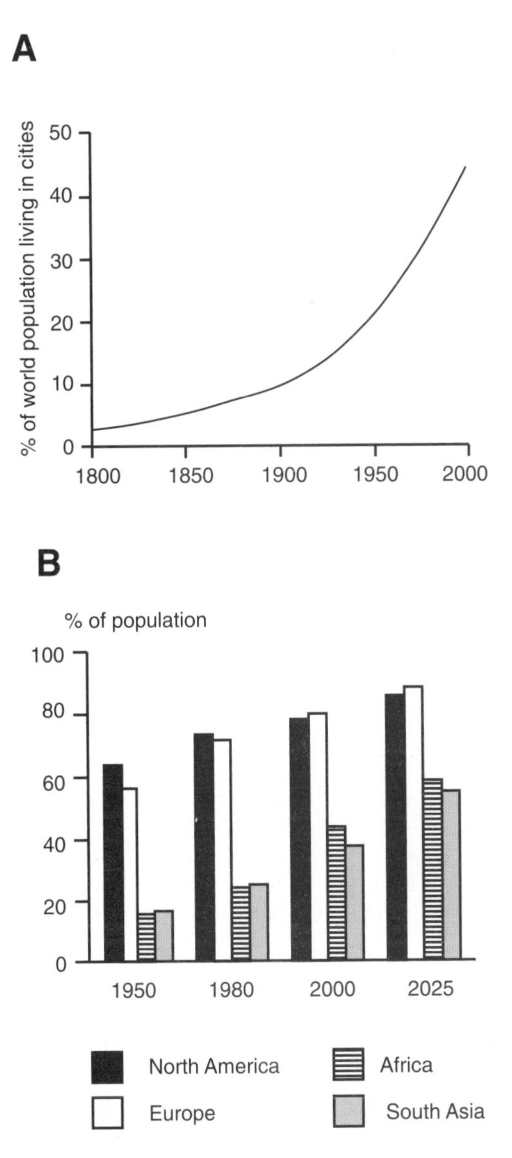

Figure 13.1 *The world's population of city dwellers.* **A.** *The increasing percentage of the world's population living in cities.* **B.** *Percentage growth of city dwelling in North America, Europe, Asia and Africa. [A Modified from: Legget (1973)* Cities and geology. McGraw Hill, *Fig. 1.2, p. 7. B Modified from: De Mulder (1996) In: McCall* et al. *(Eds)* Urban geoscience. *Balkema, Fig. 1.1, p. 2]*

in river meanders or on topographic high points. Early cities were also located to make use of abundant natural resources such as water, construction materials (Box 13.1) or economic minerals. The same is true of the more recently founded North American cities, as the European settlers of the 'New World' found suitable locations for transportation or resource provision (Box 13.2).

BOX 13.1: GEOLOGY AND VERNACULAR ARCHITECTURE IN BRITAIN

Geology had its most profound effect in influencing the location and character of early pre-industrial settlements. It determined the availability of water, building materials (Diagrams A & B), mineral resources, agricultural soils and defensive landforms. In these settlements, the building materials used and the type of building styles (vernacular architecture) were strongly influenced by the availability of construction materials. Stone construction is common wherever there is an abundant source of hard rock: the sandstones, limestones, slates and crystalline rock characteristic of Britain northwest of the line of the River Tees to the River Exe. Timber and clay, often fired as bricks, were used more in the soft rock areas of the southeast. The building style is influenced by the nature of the building material. For example, Diagram C shows the design of typical drystone walls from across the country, each reflecting the availability and natural shape of undressed stone. Blocky shapes produced by limestones and sandstones in northern England give rise to square cut walls requiring flat tie stones to bind each face of the wall. In contrast, in North Wales fissile mudrocks and cleaved slates produce abundant flat and elongate stones which naturally tie the wall together. The distribution of roofing materials shows a similar story. In the northwest, roofs are dominated by slate or fissile sandstones, while tiles made from clay tend to dominate in the southeast, and thatch or turf is common where other roofing materials are unavailable. The importance of local building materials was due to the difficulty in their transportation prior to the Industrial Revolution and is still strongly controlled, in many cases, by the cost of transportation today.

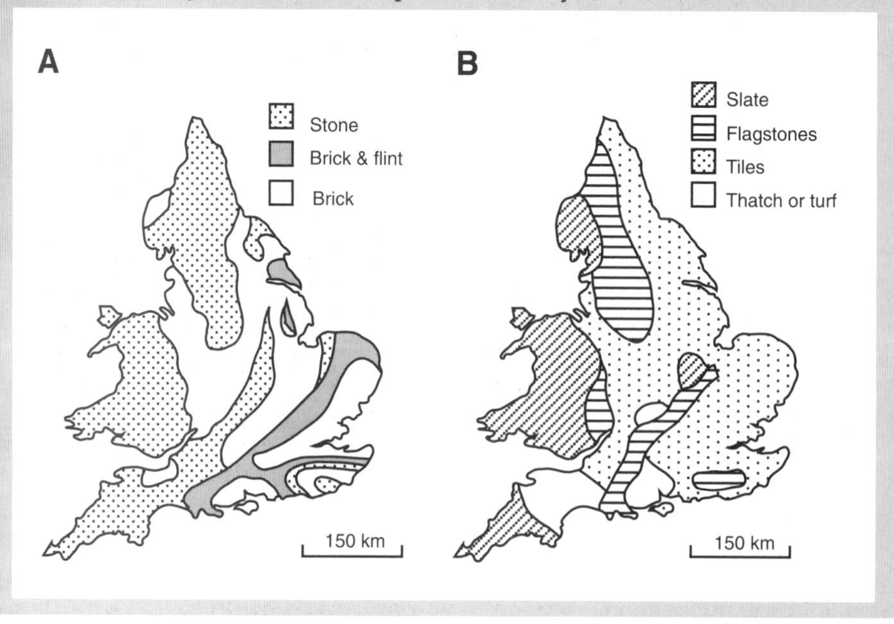

C Dry stone wall construction and variation in the shape of undressed stone

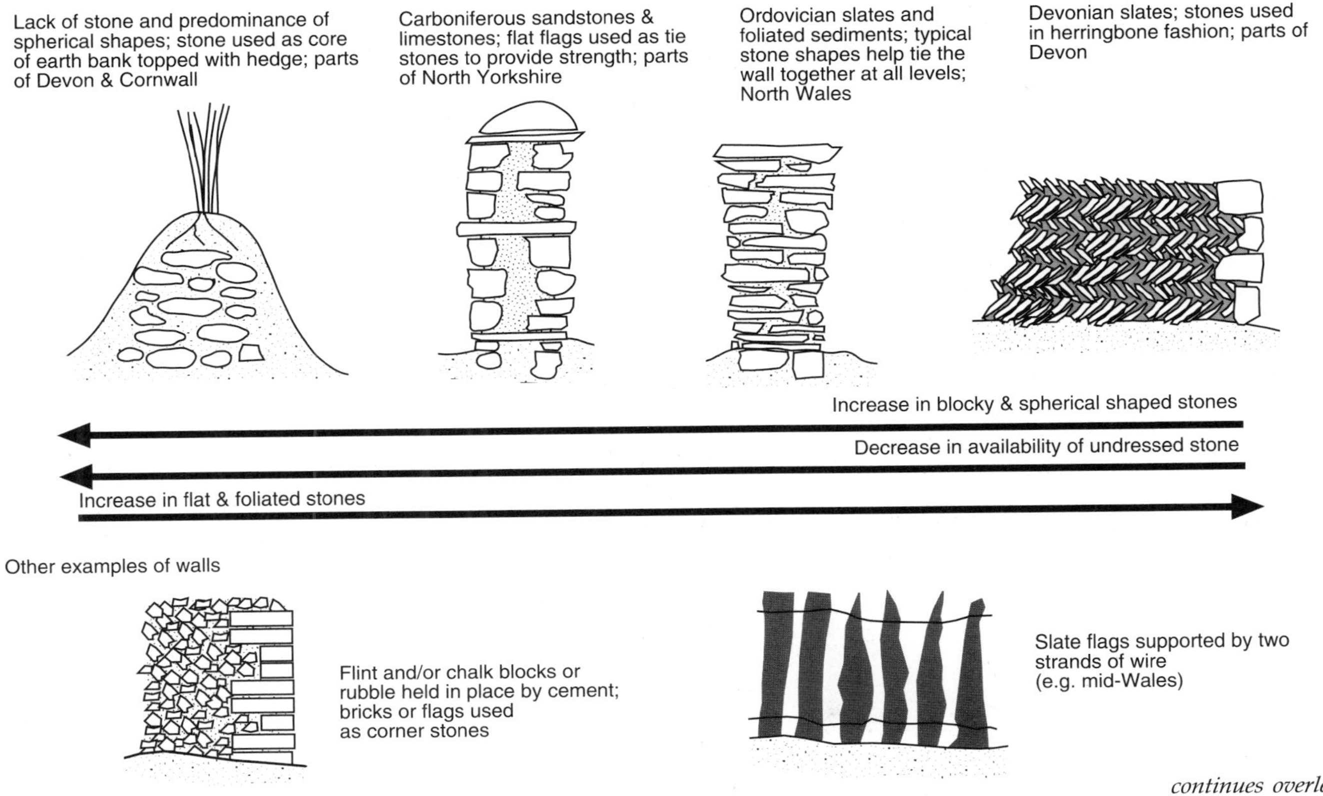

Lack of stone and predominance of spherical shapes; stone used as core of earth bank topped with hedge; parts of Devon & Cornwall

Carboniferous sandstones & limestones; flat flags used as tie stones to provide strength; parts of North Yorkshire

Ordovician slates and foliated sediments; typical stone shapes help tie the wall together at all levels; North Wales

Devonian slates; stones used in herringbone fashion; parts of Devon

Increase in blocky & spherical shaped stones

Decrease in availability of undressed stone

Increase in flat & foliated stones

Other examples of walls

Flint and/or chalk blocks or rubble held in place by cement; bricks or flags used as corner stones

Slate flags supported by two strands of wire (e.g. mid-Wales)

continues overleaf

BOX 13.1: *(continued)*

The traditional pattern of building was disturbed with the onset of the Industrial Revolution. With the development of cost-effective transport methods the towns and cities of Britain changed, severing the link between the fabric of a town and its local geology. Most are now uniform in their mixture of brick, exotic stone and concrete derived from all over Britain and, in many cases, the world. Cost was also a factor, and the huge conurbations of Manchester and Liverpool are built mostly from cheaply produced local bricks rather than the traditional red sandstones, which were less easy to mass produce. The effect was felt far and wide; for example the Welsh town of Llangollen is dominated by brick buildings, reflecting the fact that the brick could be brought cheaply to the town from northwest England via the Shropshire Union Canal, while all the surrounding settlements use local stone. Links with local geology are increasingly lost as building stones are derived from across the globe, chosen not for their local availability but for their aesthetic impact. Only the historic parts of a town often preserve this link.

Source: Clifton-Taylor, A. 1987. *The pattern of English building*, fourth edition, Faber & Faber, London. [Diagrams A & B modified from: Woodcock (1994) *Geology and environment in Britain and Ireland*. UCL Press, Fig. 3.4, p. 16]

Urban growth can be precipitated by a wide range of factors, including economic, political, historical and geological considerations. The geological environment continued to influence settlements during the Industrial Revolution. In Britain, urban growth as a consequence of the Industrial Revolution was largely controlled by the distribution of mineral wealth (Figure 13.2) with perhaps the exception of London, which grew through commerce and political influence. For example, the rapid urban expansion of towns in northern and Midland England was driven by industrial growth facilitated by the availability of mineral fuels and industrial/metal minerals in these regions. A similar story is true for most European countries, and can also explain the rapid urban growth of the eastern states of the USA.

In some cases, the original geological rationale behind the location of a town or city may be obscure, particularly if the city was founded on the basis of a natural resource which has long since been worked out. As urban development continues it is less influenced by direct geological factors and more by inertia; the very size of the urban centre ensures and maintains its growth (Figure 13.3). Links with the local geology become obscure, particularly as local building stones are replaced by others derived from across the globe. Although such local links may be lost, geology is of increasing importance in the supply of mineral and other natural resources and because of the engineering constraints placed upon urban expansion.

BOX 13.2: GEOLOGY AND AMERICAN CITIES

The ancient cities of Europe are all traditionally founded on the basis of their geology. This is usually observed as the utilisation of a geomorphological feature for defence, or as part of a transport route, or adjacent to a geological resource such as water or mineral deposits. Cities founded by the European settlers of North America have at most a 300 year history, but they too were located on the basis of physical features of the landscape. **Boston** is one of the earliest of the European settlements in North America. The city was founded on alternating glacial deposits of till and gravels which yielded several large springs caused by the local development of artesian conditions. **New York** was situated in a defensive position on an island between two bodies of water, the Hudson River and Long Island Sound, providing the important river and sea transportation links. The resistant Manhattan Schists which helped in the formation of the island have since provided sound foundations for the 'sky-scrapers' so characteristic of this city. **Albany** in New York State was sited at the head of the navigable part of the Hudson River, at the junction of the Hudson and Mowhawk rivers. Albany owes its success to its proximity to industrial minerals: iron ores from the Adirondack Mountains; iron-moulding sand from the Albany Lake plain; and clay and limestone for the local manufacture of Portland cement. **New Orleans** was founded in 1717 on a levee which formed the east bank of the Mississippi River close to the delta front, providing trade routes along the river and out to sea. This has led to some problems given its rapid expansion into a major city, particularly with regard to foundations. Finally, **San Francisco** was located on a rocky peninsula which forms a natural harbour, again allowing it to develop because of fisheries and coastal trade. This location has given San Francisco its characteristic steeply inclined streets.

Source: Legget, R.F. 1973. *Geology and cities*. McGraw-Hill, New York.

Urban growth in the late nineteenth and twentieth centuries has followed at least three distinctive patterns, mostly associated with the constraints of the geological environment: (1) lateral spreading; (2) building upwards; and (3) utilising or creating underground space. **Lateral spreading** is a common feature of most cities and can be achieved in many different forms, often coupled with the development of transport links. Geological constraints on lateral growth are usually associated with the geomorphology and engineering properties of the region, as the initial foundation of the city will usually be in the optimum area for development, with further development on floodplains, inclined slopes or coastal regions which can often increase the impact of both exogenic and endogenic hazards. Lateral growth may also be constrained by proximity to resources, as quarry or mine activity may become included within city boundaries, thereby sterilising future resource exploitation. The pressure for this type of growth may

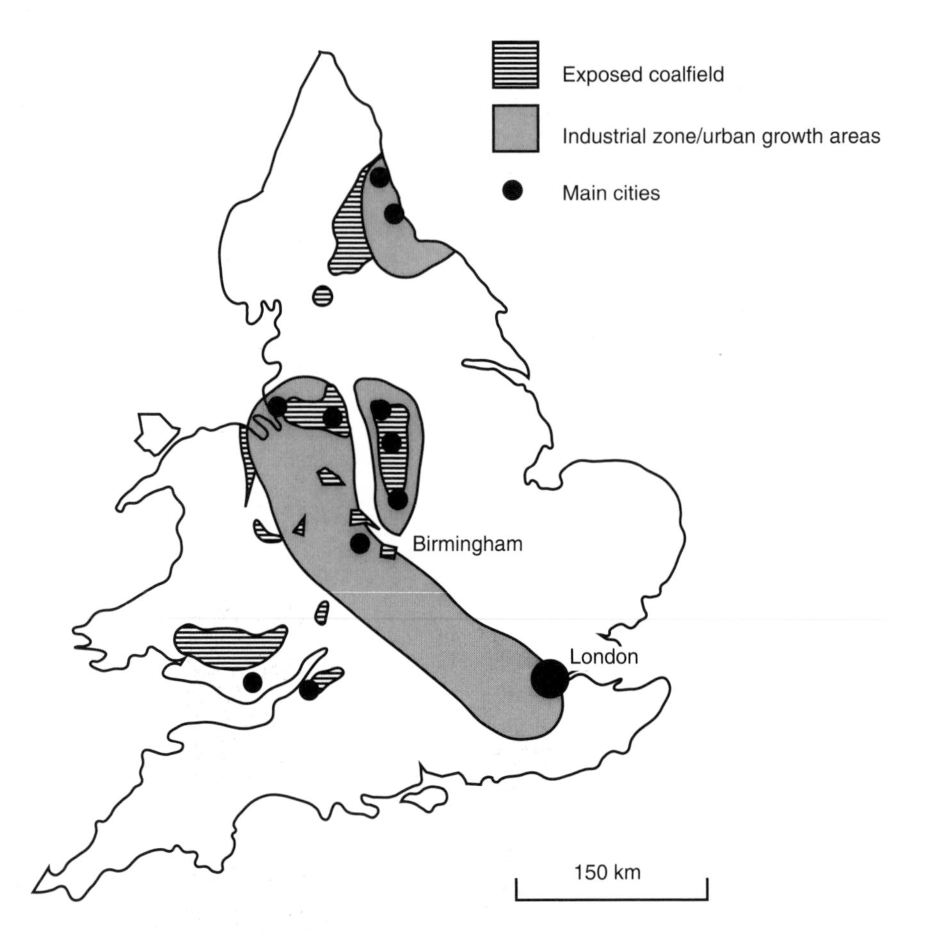

Figure 13.2 *Distribution of major urban centres in England and their relationship to major coalfields [Modified from: Woodcock (1994)* Geology and environment in Britain and Ireland. *UCL Press, Fig. 3.6, p. 18]*

threaten land belts of agricultural, recreational and conservation importance which are often maintained around cities. The use of old industrial sites and other contaminated 'brown land' areas can relieve pressure on such resources ('green land'), although there can often be a problem from such factors as mining. For example, mine subsidence exerts an important negative influence on the growth of the Scottish city of Glasgow. Here, old and inadequately documented pillar-and-stall coal mine workings have led to significant subsidence problems for new housing developments. Transport corridors for large cities are also an important part of the lateral growth of towns and cities, and the building of large road and rail systems, often with the need for adequate foundation works and with adequate hazard protection works, is a major undertaking.

Building upwards, thereby creating 'cities in the sky', was a utopian vision for many architects and town planners in the middle part of the twentieth

Figure 13.3 *A model of the role of geology in urban development. Initial urban settlement is strongly influenced by local geology, an influence which declines with time. Geological control continues, however, through the needs for resource provision, engineering geology and waste management*

century. Although the dream of high-rise dwellings is now less attractive, the use of 'sky-scrapers' by commercial and financial operations is a characteristic of the central business districts of most major cities. The creation and security of tall buildings is a function of the nature of the foundation works, themselves ultimately controlled by the bedrock and soil conditions. **Utilising underground space** has always been a consideration in urban areas, particularly in the development of underground infrastructure and transport. Many ancient cities have extensive underground tunnels which have at times been used for habitation, such as in Arras in northern France, and Rome in Italy. Recently, there has been much interest in a number of North American cities in the use of underground space for industrial purposes, often in old mining regions. For example, in Kansas City over 3000 people work in 25 locations under the city, utilising 2 km^2 of a total space of 15 km^2 left after the cessation of underground pillar-and-stall limestone mining. Typical operations include food freezer rooms, the manufacture of precision instruments requiring stable conditions, and a local college.

13.3 URBAN PLANNING AND GEOLOGY

The aim of urban planning is to identify opportunities for development, determine a particular course of action, and respond to perceived or real problems associated with it. In the development of an urban strategy it is perhaps appropriate to consider the urban environment as a kind of machine (Figure 1.1). The urban machine consumes **inputs**, usually derived from the natural environment. These inputs are largely: (1) water, derived locally from springs and aquifers, or as part of a reservoir and transfer system; (2) raw materials, in the form of mineral resources for industry and construction, traditionally derived locally, but with the greater development of transport links, now often globally obtained; (3) food resources, derived from farming of agricultural lands, but also from effective transport links with other nations; and (4) energy, derived from the use of the Earth's physical resources. The machine produces **outputs**, in the form of: (1) products of industry and commerce created from the raw materials imported into the city; (2) waste, in the form of worn-out materials, by-products from industry, and day-to-day wastes from domestic, commercial and industrial sources; and (3) pollution, caused by poor waste management overloading the ability of natural atmospheric, land and water systems to recycle and redistribute the waste gases, solids and liquids produced by industry, commerce and domestic activities. Like any machine, the urban system needs regular maintenance, particularly in the upgrading of its infrastructure of utilities and transport. Equally, the integrity of its building materials, particularly where open to enhanced attack from the acid rain produced by most industrialised cities, is another consideration and aspect in need of regular maintenance.

The efficient running of the urban machine depends on strategic planning, both in terms of recognising opportunities for development, and in identifying constraints. Geologically speaking, typical **opportunities** reflect the supply of inputs to the system and low maintenance costs, and would include favourable

Table 13.2 *Human-induced hazards in the urban environment [Based on data in: Mather, J.D.* et al. *(1996) In: McCall, G.J.H.* et al. *(Eds)* Urban geoscience. *A.A. Balkema]*

Hazard	Principal causes and effects
Surface movements	Extraction of water significantly altering the water table; subsurface extraction of minerals. Causes subsidence and damage to housing and infrastructure
Contaminated land	Heavy industry in the urban environment leaving derelict buildings and infrastructure, commonly with concentrations of chemical substances. Potentially hazardous to health; needs detailed assessment and clean-up and/or removal
Groundwater fluctuations	Particularly rises in water table caused by rebound from mining operations; a decreased use of inner city supply wells through destruction and declining water quality; leakage from sewers and water mains. Impacts include basement and tunnel floods; foundation instability
Reduced groundwater quality	Reversal in groundwater gradients introducing sea water into the aquifer; leakage of sanitation systems, particularly where they are unsewered; surface pollution and contaminated land, such as unconfined leachates from landfills and leakage from storage tanks at petrol stations
Waste disposal	Disposal of solid, liquid and gaseous wastes, usually in landfill containment or as direct release, treated or untreated, into the atmosphere or water system. Typically causes contaminated land and leads to the modification of groundwater and atmospheric conditions

locations and ground conditions, the presence of natural resources such as water and minerals, and the absence of significant hazards, whether natural or human-induced (Table 13.2). Most **constraints**, which limit the success of the city and the level and extent of its output, are actually a function of the same set of factors: unfavourable locations and ground conditions, the absence of natural resources, and the presence or growth of significant hazards (Table 8.1). Urban planning, therefore, is a recognition of the balance between these opportunities and constraints in the development of a town or city, most of which are a function of environmental geology – the human interaction with the geological environment. There are four aspects to this interaction: (1) the provision of mineral resources for industry and construction, including water; (2) the assessment of ground conditions for transport links, urban expansion, urban redevelopment or for the creation of new urban areas, particularly in areas of hostile terrain; (3) the disposal and management of the waste products generated by urban and industrial areas and their utilities, such as power stations; and (4) identification and mitigation against the threat from natural hazards.

Planning responses may be both **proactive** and **reactive**. Proactive responses are associated with forward planning in order to: (1) ensure adequate resourcing of the population, commerce and industry; (2) plan for expansion without jeopardising amenity, conservation or recreational values; (3) ensure the use of appropriate construction methods, materials and foundations; (4) consider the development of an infrastructure which is appropriate to meet the needs of the population, particularly in the provision of utilities such as water, power and transport; (5) develop an adequate waste management policy; and (6) deal with the threat of natural hazards, developing appropriate mitigation and civil defence strategies. Reactive responses are appropriate in dealing with unforeseen or infrequent factors, such as: (1) the use of emergency services in case of major disasters; (2) undertaking small-scale remedial works; and (3) monitoring change where there is no immediate threat.

Essential to effective planning is the availability of accurate, understandable and pertinent data with which decision-makers can make informed choices. The provision of geological information, as a means to identifying both the opportunities and constraints for development, is an essential part of urban planning. Unfortunately, geological information is an underused resource in planning, despite the fact that the majority of cities owe their historical development to geology. The planning process is complex, not least because its legislative framework varies from state to state, and from country to country. Despite this legislative diversity, urban planning usually focuses on the following factors: economic and financial constraints; population and social considerations; impact and enhancement of the natural environment; existing land use; the provision of appropriate buildings for their end-use; transport infrastructure; and aesthetics.

An effective urban strategy will therefore involve: (1) land-use planning to guide development and appropriate use of land areas; (2) the identification of specific uses for designated sites; and (3) the regulation and granting of permissions for development. In all these stages it is essential to be able to make informed judgements, and the provision of appropriate geological information is therefore essential. Such information can help in the prevention of costly remedial foundation works and other factors. Typical information needed includes: (1) opportunities for development in the form of the nature and extent of mineral, soil, water and undeveloped land resources; (2) the nature of constraints on development, such as the assessment of contaminated land, geotechnical information about ground conditions, and the nature of the hydrology; and (3) sources of expert guidance where necessary. This will be achieved through data collection from desk survey, field survey and monitoring, and terrain analysis, and its presentation in a form which is accessible to all, usually in a thematic set of maps which display the geology, geomorphology, resources, hazards and so on (see Section 1.5.3).

In planning new urban areas, planners must realise that a given land area will not be a 'blank piece of paper' on which they can simply 'draw' their town or city. Instead, the landscape is a function of numerous geological factors, including the nature and form of the bedrock, and the endogenic and exogenic processes which form its land surface and hydrogeology. It is the role of environmental geologists to advise on the nature of these conditions, and to consider the impacts of prevailing conditions in the future, particularly with regard to hazard mitigation and the effects of urban growth. Effectively, it is important to identify those elements in the geological environment which provide opportunities for, and constraints to, urban development.

13.4 URBAN CASE STUDIES

We have argued so far in this chapter that the future challenge within environmental geology lies in the urban environment. It is here that human interaction with the physical environment is at its most intense, although in other aspects, particularly those of resource provision, significant interaction may take place outside the urban centre. Towns and cities encompass almost all aspects of environmental geology, from the provision of resources, the construction of buildings and provision of infrastructure, and the management of wastes and pollution, to the mitigation of natural and human-induced hazards.

In order to emphasise the central role that environmental geology has within the urban environment, we present two case studies based on the ancient European city of Rome and the modern Asian city of Hong Kong. Each case provides an integrated example of many of the aspects of environmental geology discussed in this book and is for the most part derived from a single key review paper. In these case studies, details of the four main areas of urban environmental geoscience are presented: (1) resource management; (2) engineering constraints; (3) waste management; and (4) hazard mitigation.

13.4.1 Rome
(Thomas, 1989)

Rome is an ancient city with a history which extends back to at least the fourteenth century BC. Today, metropolitan Rome has an area of 312 km^2 and a population of 2.4 million. Rome is situated straddling the River Tiber, which historically formed a barrier to the movement of ancient peoples. Rome was therefore most probably founded at a crossing point, close to Tiberina Island, the only place on the Tiber with a narrow enough floodplain to allow easy crossing points.

General geology of Rome

The geology of Rome is structurally uncomplex, and comprises mostly Neogene sediments and pyroclastic deposits which have no significant folding, although some normal faults cross the region (Figure 13.4). In essence, the subsurface geology of Rome is dominated by four units (Figures 13.4 & 13.5).

1. **Plio-Pleistocene marine clays**. These overlay Mesozoic basement to a thickness of 800 m, and were deposited after a period of subsidence in the early Pliocene.

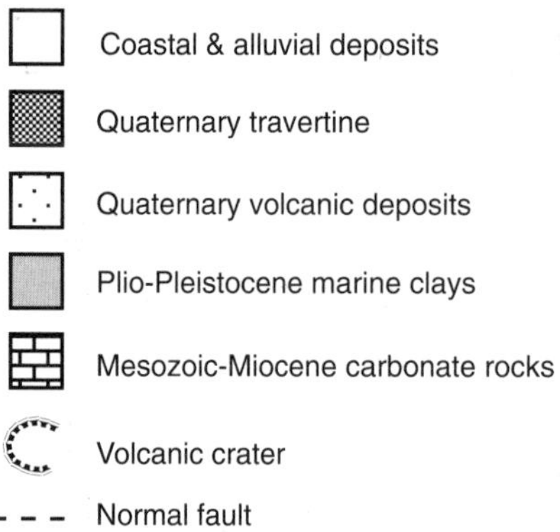

Figure 13.4 *Geological sketch map of the Rome area [Modified from Thomas (1989) Bulletin of the Association of Engineering Geologists,* ***26****, Fig. 29, p. 443]*

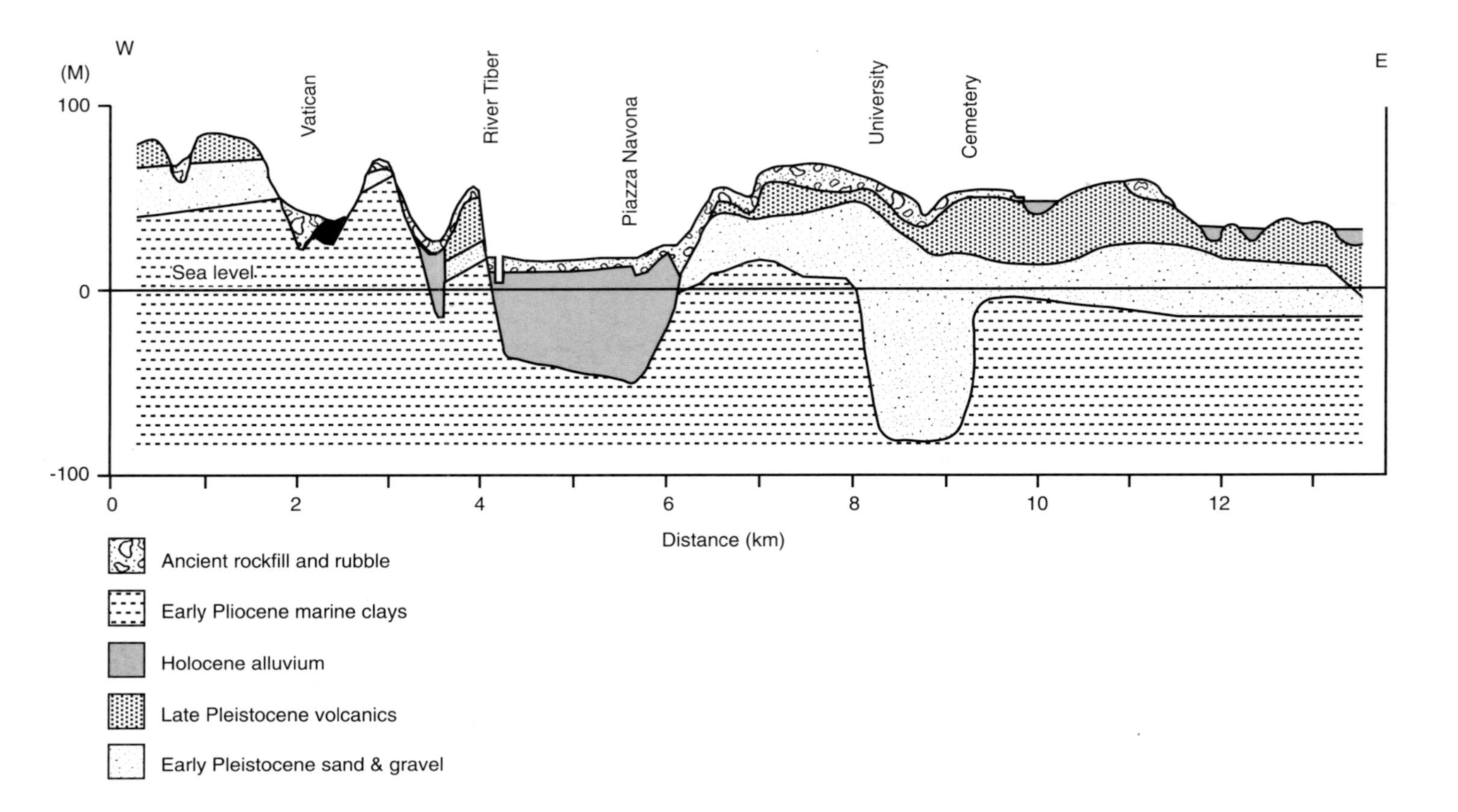

Figure 13.5 *Geological cross-section through the centre of Rome [Modified from Thomas (1989)* Bulletin of the Association of Engineering Geologists, ***26****, Fig. 31, p. 445)*

2. **Quaternary volcanic deposits**. These deposits are associated with the Alban and Sabatine volcanoes to the southeast and northwest of the city. Most activity took place during the Tyrrhenian, extending mostly from 700 000 to 30 000 years ago, depositing an extensive series of pyroclastic rocks. Commonly there are four main types of deposit: (A) hydrothermal pyroclastic rocks with clasts of variable size from 1 mm to 1 m in diameter; (B) pyroclastic flow rocks comprising massive beds with millimetre to centimetre sized clasts; (C) air-fall ash deposits, usually thinly bedded ash, lapilli and cinders, forming extensive pozzolan ash deposits used since Roman times in the manufacture of concrete; and (D) lavas. Most volcanics in the Metropolitan area are pyroclastic flow or air-fall ash deposits of no greater than 40 m in thickness.
3. **Quaternary coastal and alluvial sediments**. Late Quaternary sandy and clay-rich alluvium deposits of varying age are found, associated with a period of incision of the Tiber to at least 50 m below sea level. Post-glacial sea level rise led to the flooding of the Tiber valley, depositing mostly sands, gravels, clays and palaeosols, all of which are rich in particles of volcanic origin. The last volcanic activity was probably an ash fall 4000 years ago, and the volcanic complex, which formed a featureless volcanic ash plain, has been subject to dissection by the Tiber and associated streams to form a network of valleys separated by flat-topped ridges.
4. **Ancient rockfill and made-ground**. The long history of human habitation in this location has created unusually large amounts of rockfill (Figure 13.6A). This material is mostly of Roman age, and usually comprises heaped-up and well consolidated quantities of local and imported rock, and broken ceramics, bricks, concrete and statues.

Resource management

1. **Construction materials**. The Romans mined extensively volcanic ash units (*pozzolano*) in the metropolitan area in order to form the basic material for pozzolan concrete, providing millimetre to centimetre sized aggregates which were combined with crushed limestone derived from Tivoli, some 27 km to the east, and sand from the Tiber river bed. Extraction of the pozzolan led to the creation of the many underground pillar-and-stall excavations which characterise the subsurface of Rome (Figure 13.6B). Pozzolan is now excavated from open quarries outside the metropolitan area. Concrete was widely used by the Romans, and the Pantheon, constructed in the first century AD, is still probably the largest unsupported concrete dome in the world, without reinforcing steel or iron.

   Lithic tuffs such as the local Lionato Tuff were extensively quarried; they are easily cut, and were used in foundation works and in conjunction with concrete in the construction of major buildings, such as the Colosseum and the Theatre of Marcellus. Open quarrying using iron tools was the most common method employed, although where the overburden was particularly thick or extensive, pillar-and-stall mining was also used. Lithic tuffs are

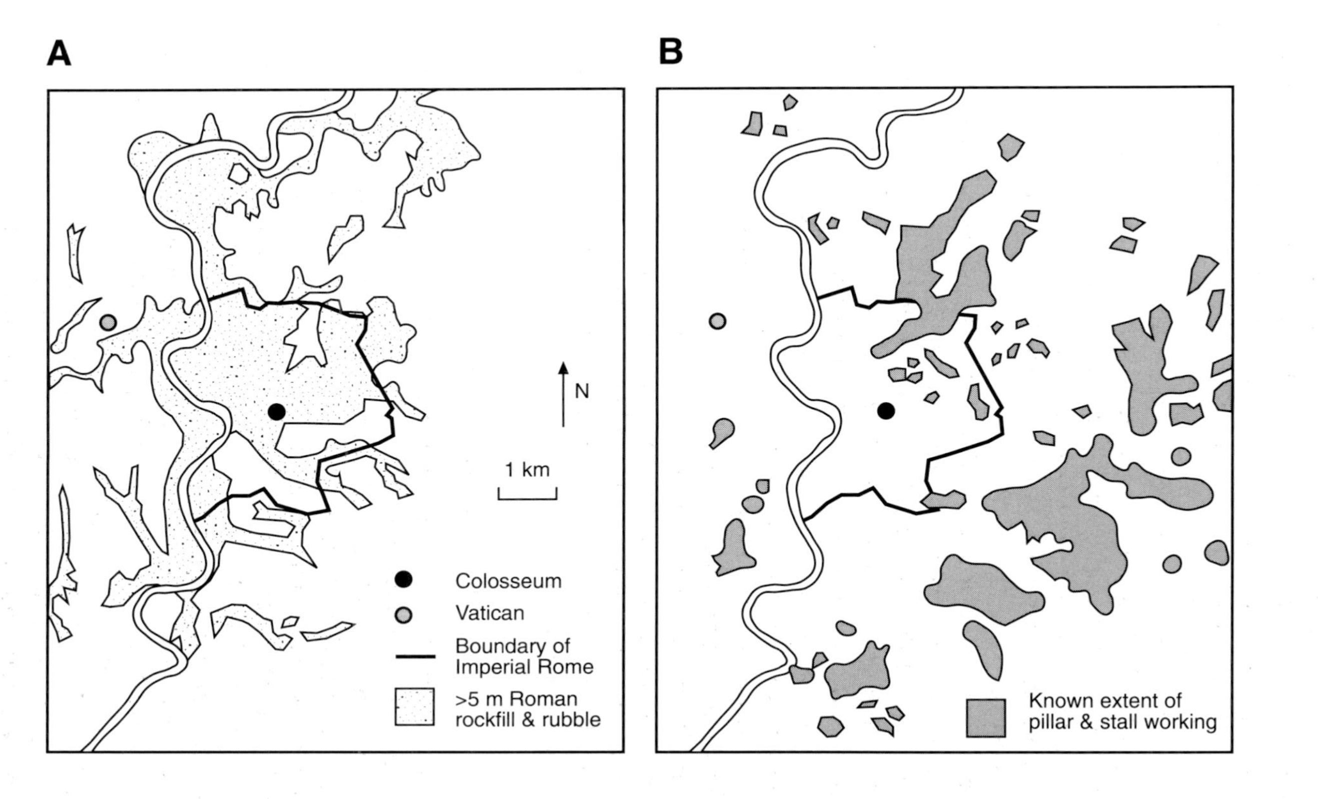

Figure 13.6 *Maps showing the extent of activities in Roman times.* **A.** *Ancient rockfill and rubble accumulation.* **B.** *Subsurface pillar-and-stall workings for pozzolan ash deposits [Modified from Thomas (1989)* Bulletin of the Association of Engineering Geologists, **26**: *A. Fig. 26, p. 440; B. Fig. 39, p. 454]*

still extensively used today in sawn blocks and as coarse aggregate for concrete.

Travertine was widely used as a decorative stone in ancient Rome, and still is today. It is derived from quarries some 15–20 km to the east of the city. Other building and sculptural stone was derived from the Roman hinterland, in particular basalt, used for paving and extracted from the Alban Hills to the south of the city, and white marble from Carrara, north of the city.

2. **Water supply**. Water supply in Roman times was mostly derived directly from the Tiber, and from the Tiber alluvium and porous pozzolan ash beds which provided groundwater from at least the sixth century BC. However, these frequently dried up during three months of the year and were supplemented using rainwater cisterns. Wells were used as a supplementary source of water up to 1986, but are now only used in industry, commerce and horticulture. The Tiber alluvium is in hydraulic continuity with the river, and floodplain groundwater is generally less than 4 to 8 m deep. At higher elevations, groundwater depth may be between 30 and 40 m, following the topography; the Pliocene clays form the base for unconfined groundwaters in the porous and permeable pozzolan ash units, Pleistocene sands, and ancient rockfill.

   By the fourth century BC, an increased need for water saw the large-scale construction of aqueducts, to bring water into the city from springs, constructed either above ground or as tunnels. These springs occur in the Mesozoic limestone karstic areas which surround the city to the northwest. There were probably 11 major aqueducts in Roman times, the water being supplied by gravity systems, and providing an estimated one million cubic metres per day. The majority of people obtained their water from free-flowing fountains supplied by the aqueducts, but some used lead piping to their homes. By the turn of the twentieth century, Rome was still supplied by the springs used by the Romans, and today a more extensive water system taps these same springs.

Engineering constraints

1. **Geotechnical aspects**. The geology of Rome causes some problems in construction. Artificial fill of Roman age does not normally constitute a problem, even where saturated; later historical fill material is more of a problem, and will not support more than 0.5 to 1 kg cm^{-2}. Recent alluvium and older fluvio-lacustrine sediments are of variable quality, the unconsolidated Holocene river alluvium presenting the greatest problems, with the older, partly cemented sediments providing more favourable ground conditions. The volcanic rocks which underlie most of the Rome area are a good foundation material, although the internal characteristics may vary rapidly over a small area. Generally, most tuffs will support loads of 5 to 6 kg cm^{-2}, although maximum loads for the porous and frequently wet pozzolan ash layers are about 2 to 3 kg cm^{-2}.

2. **Underground excavations**. Rome is riddled with an extensive series of underground passages and tunnels, mostly created from the mining of pozzolan ash deposits and tuffs (Figure 13.6B). In most cases, tunnels were cut with tuffs forming the roof, and were small in order to minimise labour. Pillar-and-stall methods were most often used in the extraction of ash deposits, and this has left a legacy of some difficulties today, as often the pillars left were small, and spalling and fracturing are now leading to their collapse, causing a significant hazard. These limitations prevent widespread use of the tunnels and underground excavations today. An underground railway system was commenced in the 1930s and was mostly constructed using the cut-and-cover principle.
3. **Foundation works**. The foundations of ancient Rome, prior to the widespread use of pozzolan concrete, were constructed mostly from blocks of lithic tuffs founded on consolidated sediments or volcanic rocks, although some piling was also used in unconsolidated sediments. Concrete became a common foundation material in the first or second century BC, with concrete foundations commonly having footing areas larger than the area of the walls they support. The Colosseum is founded on concrete in this manner. Modern foundation works have to take into account three important factors: (A) the preservation of ancient monuments and artefacts uncovered by new excavations; (B) unstable, water-saturated alluvial sands, usually handled through concrete piling; and (C) the collapse of ancient underground excavations, requiring shoring.

Waste management

1. **Solid wastes**. Waste management during Roman times took the form of open dumps outside the city walls, often building up to such an extent as to lessen the defensive capabilities of the wall. The largest dump is the Monte Testaccio which was built to a height of 30 m, and has an estimated volume of 200 000 m^3. This open dump was used for the storage of broken amphora (storage jars). Today, the majority of domestic waste is handled through tipping into a landfill site 30 km west of Rome. This is located in Pleistocene marine sands which dip gently towards the sea, and it is expected that any leachates produced will migrate away from the coastal plain. Hazardous wastes are mostly disposed of through incineration.
2. **Waste waters**. In Roman times, waste waters including both runoff and sewage were collected in cut-and-cover drains lined with tuff blocks, which opened directly into the Tiber. These drains are still in use, but only for the disposal of runoff city waters, although gates were constructed in the nineteenth century to prevent backflow during high tide states of the Tiber. Sewage collector drains were constructed in the late nineteenth century, and the first large sewage treatment plant was built in 1970. The sewage system is still under construction, and untreated sewage is still released into the Tiber downstream of Rome, causing considerable contamination with faecal

coliform bacteria and heavy metals, as well as causing a marked drop in pH and dissolved oxygen.

Hazard mitigation

1. **Human-induced hazards**. One of the most important aspects of hazard mitigation is in the prevention of ground movements through the collapse of ancient pillar-and-stall mines caused by spalling of the pillars. This involves the casting in place of concrete piles to prevent collapse. Groundwater quality is a concern for city authorities, mostly because of the probability of recharge from leaking sewers.
2. **Natural hazards**. Endogenic hazards include only some seismic activity, as the volcanic centres have been quiet for some 4000 years. Major reconstruction of the Colosseum was required following an earthquake in about AD 350 and further earthquakes in the thirteenth and fourteenth centuries caused its partial collapse. Today, no earthquake has been recorded beneath Rome in recent times, although there are some recorded in a 15 to 20 km radius. Major earthquakes occur in active tectonic areas some 30 to 50 km from Rome. Exogenic hazards include floods, which are periodically experienced by Rome, particularly in the historical and ancient past. Flood prevention measures have led to the construction of 18 m high embankments to the Tiber, 12 m above low flow levels.

13.4.2 Hong Kong
(McFeat-Smith *et al.* 1989)

Hong Kong has a total area of 1052 km^2 and comprises the island of Hong Kong, the peninsula of Kowloon to the north, and a larger piece of mainland with associated islands known as the New Territories. The main urban development is situated around Hong Kong Harbour, a 2 km anchorage between Hong Kong Island and the Kowloon Peninsula which provides protection during the typhoon season, when wind speeds of 75 m s^{-1} have been reached. The majority of the 5.5 million population live adjacent to this strait, on the north of Hong Kong Island, and to the south of the Kowloon Peninsula. Despite an otherwise inhospitable and hilly terrain, Hong Kong's population developed rapidly from a few thousand people with a mostly rural existence in the early part of the nineteenth century, to over a million in the mid-part of the twentieth century, rapidly increasing to what it is today.

General geology of Hong Kong

Hong Kong is underlain by a wide variety of geological units ranging in age from the Devonian to the Tertiary, which have been subjected to a range of tectonic and igneous activities, all of which contribute to the geological setting of the urban centre of Hong Kong (Figure 13.7). Tectonic activity is associated with

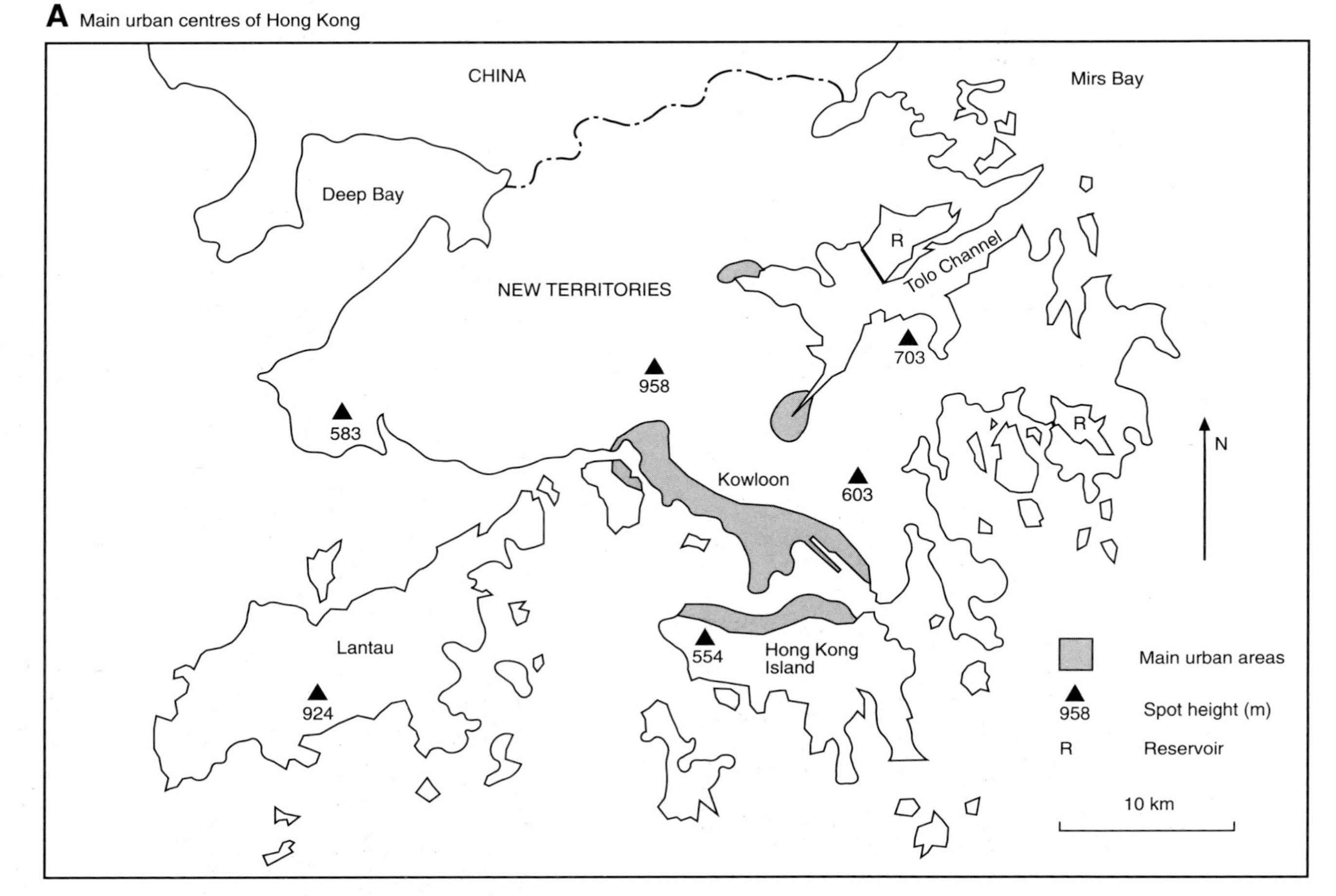

Figure 13.7 *Hong Kong.* ***A.*** *Location of the main urban areas.* ***B.*** *Geological sketch map of Hong Kong [Modified from McFeat-Smith* et al. *(1989)* Bulletin of the Association of Engineering Geologists, ***26***, *Fig. 6, p. 33]*

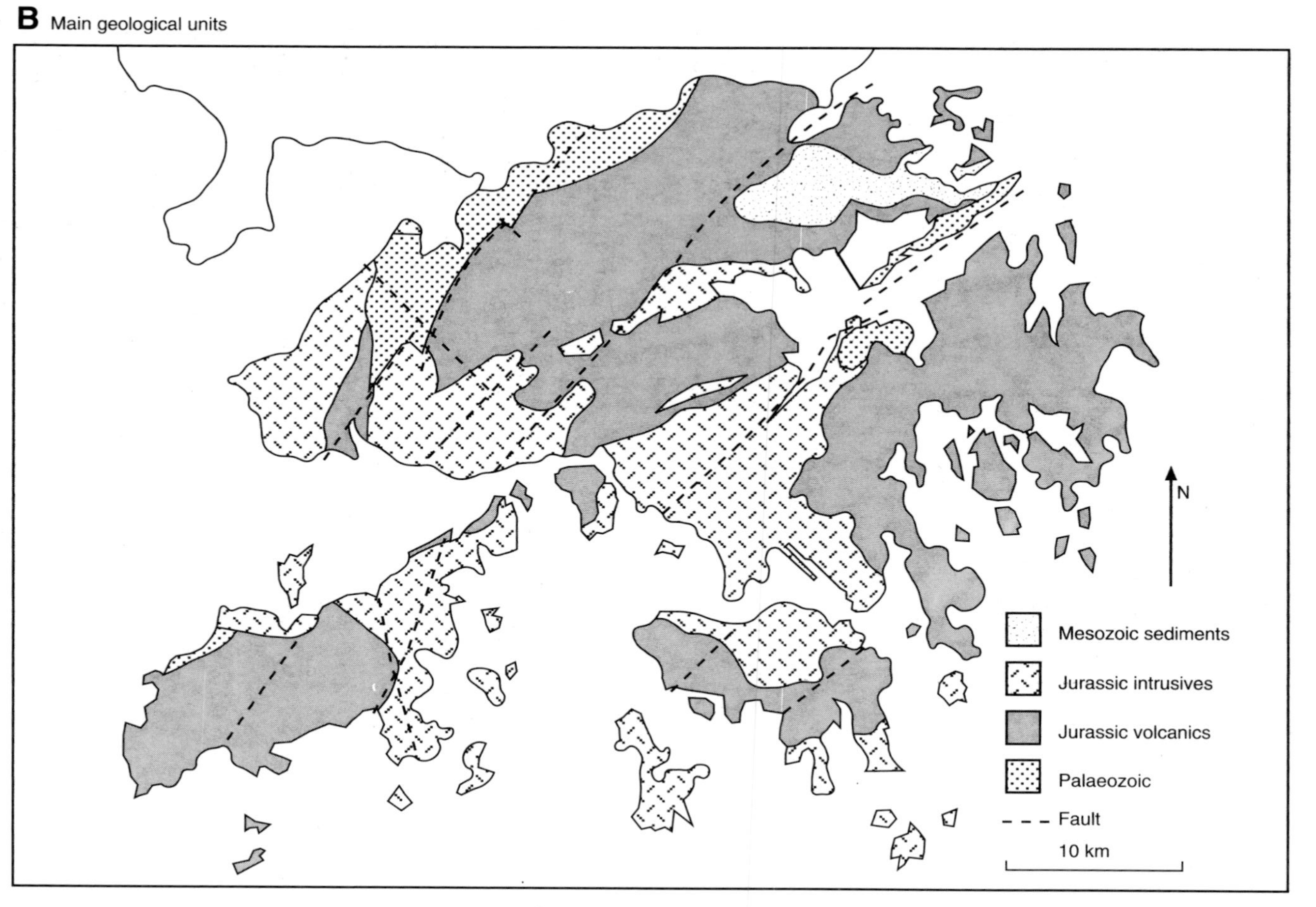

Figure 13.7 *(continued)*

Palaeozoic and Mesozoic orogenic activity, leaving folded successions of this age. Faulting is associated with a series of elongate linear faults which trend east-northeast along the south China coast. Several faults are recognised in the Hong Kong area (Figure 13.7). The main geological units are as follows.

1. **Palaeozoic sedimentary rocks**. These comprise strongly folded sedimentary or metasedimentary rocks of Devonian to Permian age and are exposed only in the north of the Hong Kong territory. These rocks were intensely folded during the Carboniferous–Permian, Triassic and Jurassic–Cretaceous intervals. The regional strike and axial races of the major folds are typically northeast–southwest although no really clear picture of the deformation is available given the level of exposure and the structural complexity of the rocks.
2. **Mesozoic igneous rocks**. During the Jurassic and Cretaceous large granitic bodies were intruded, accompanied by the eruption of silicic lavas and ash. Much of Hong Kong is underlain by the Repulse Bay Volcanic Group of Jurassic to Cretaceous age. This group consists of air-fall ash and lapilli tuffs, rhyolitic lavas and a variety of water-borne sediments reworked from the volcanics. Intrusive rocks of Upper Jurassic age, consisting of granitoid plutonic bodies, have been intruded into at least the lower part of the Repulse Bay Group. These are accompanied by microgranite dyke swarms also of Late Jurassic age. The granitic emplacement led to the warping of the volcanic rocks such that they are often folded, sometimes quite strongly.
3. **Mesozoic continental sediments**. Non-marine sediments of the Port Island Group, comprising conglomerates and sandstones, overlie the Repulse Bay Group in the northern part of the New Territories. These have been deformed, and although they have relatively low angles of dip (<20°), they have been intensely sheared, with the development of a fracture cleavage associated with thrusting activity.
4. **Late Quaternary marine and estuarine deposits**. These comprise pre-Holocene gravels, sands and muds which are found mostly offshore.
5. **Alluvial deposits**. Deposits of alluvial origin are found in most of the major river basins, and in great thickness offshore, particularly in Hong Kong Harbour.
6. **Colluvium**. The products of mass-wasting cover large parts of the area of Hong Kong, typically forming deposits of around 10 m, although they reach up to 36 m in places. Colluvium mantles many hillsides with slopes of 30° or more (Figure 13.8), and often forms fan-shaped deposits at the foot of many other slopes. Colluvium commonly comprises silty or sandy soils with around 30% clay and with a high percentage of angular to subrounded cobbles and boulders, often reaching 5 to 8 m in diameter. Voids are common, caused by piping and back-sapping of springs. The colluvium was probably deposited during three distinct intervals and reflects an environment in which sea level was very low and was backed by a mountainous topography covered by a stony regolith. Much of the colluvium is the product of mass movements: large, fluid debris flows capable of transporting

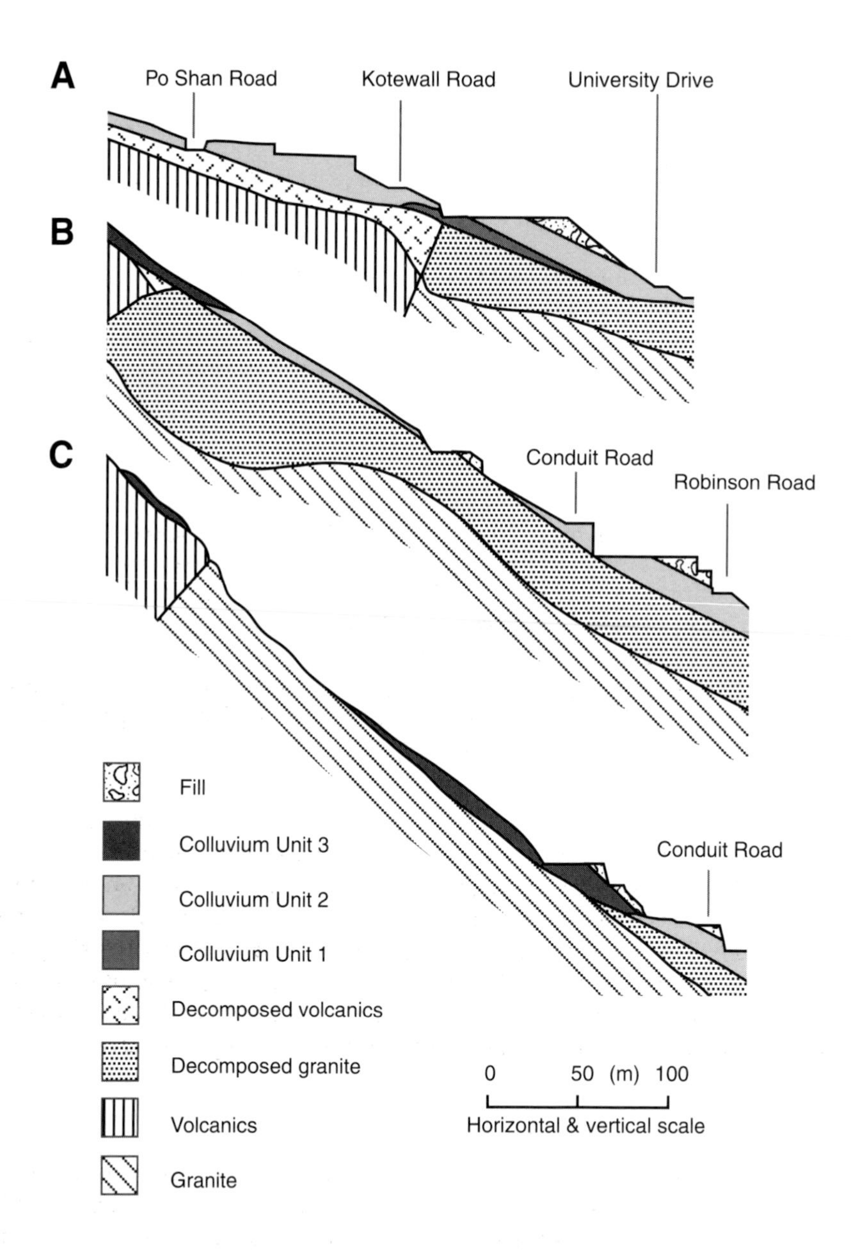

Figure 13.8 *Geological cross-sections of slopes of northern Hong Kong Island, showing the depth of weathered granite and colluvium deposits. These cross-sections also illustrate the problems of the cut and fill necessary for construction on such steep slopes [Modified from McFeat-Smith* et al. *(1989)* Bulletin of the Association of Engineering Geologists, **26**, *Fig. 7, p. 34]*

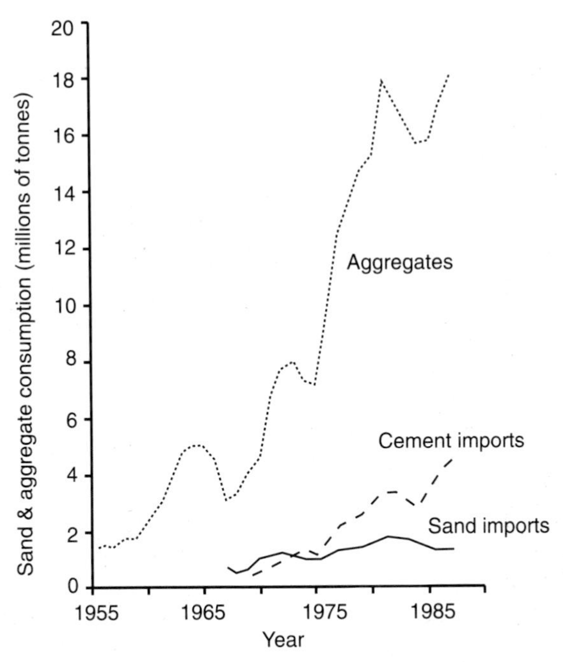

Figure 13.9 *Consumption of aggregates in Hong Kong for the period 1955–1990 [Modified from McFeat-Smith* et al. *(1989)* Bulletin of the Association of Engineering Geologists, **26**, *Fig. 15, p. 15]*

very large boulders. More recently, mass movements within this material have been generated by human activities such as deforestation.

7. **Residual soils**. Much of the regolith of Hong Kong is a residual soil derived from the weathering of granite or rhyolitic tuff. Chemical weathering is enhanced by the humid subtropical climate, and in some places the granites are weathered to a depth of 100 m or more, producing a soil of clay mineral residues and quartz (Figure 13.8). The depth and nature of the weathering is controlled by the joint spacing of the granite and rhyolitic lavas.

Resource management

1. **Economic minerals**. Some economic minerals occur in association with contact metamorphism formed by the intrusion of the granitic plutons. Hydrothermal deposits of iron, wolframite and galena were recovered up to the 1970s in the territory of Hong Kong. Feldspar from pegmatites, quartz from veins, and kaolin derived from chemical weathering of the granite bodies are still worked in small amounts.
2. **Construction materials**. The consumption of aggregates for concrete and roadstones in Hong Kong has increased dramatically since the mid-1970s (Figure 13.9). Imported sand and fine aggregates make up 40% of the total needed; the remainder is obtained from crushed stone derived from granite

and tuff quarries in the Hong Kong area, although increasingly this is now being imported from China. Material is also needed for reclamation projects, and mostly comprises locally derived residual soils, colluvium and other deposits, with rockfill derived from foundation excavations.

3. **Water supply**. Hong Kong has a history of water supply problems, and severe droughts have occurred more than once in its history. Only small quantities of water are derived from subsurface sources, as the majority of the geological units are either of low porosity (the igneous units) or low permeability (the Mesozoic sediments). Water is mostly derived by aqueducts from mainland China, or from a series of large-capacity storage reservoirs. The need for water has led to the development of innovative water storage schemes such as the damming of coastal inlets to form freshwater reservoirs (Figure 13.7A).

Engineering constraints

1. **Geotechnical considerations**. The majority of strata within the urban area of Hong Kong are weak and relatively poor for foundation works. In particular: (A) colluvium deposits are mechanically weak and prone to failure along reactivated slip surfaces, especially when wet; (B) granites are usually deeply weathered and subject to gully erosion; solid rock when reached at depth is suitable for moderate loading; and (C) tuffs are of variable quality and are known to fail on oversteepened slopes, particularly when wet and where there are closely spaced joints and other discontinuities.
2. **Foundation works**. Foundations have to withstand a wide range of conditions, particularly since much new land is created using reclamation schemes underlain by soft, saline clays and silts. Even in the granite terrain, the rockhead may be at a considerable depth owing to the intensity of the chemical weathering, and is often overlain by a great deal of clay mineral-rich material. In view of this, buildings are mostly founded on piles although rafts are also sometimes used. Piles are emplaced either through excavation and casting, or are driven.
3. **Land formation and reclamation**. Given the overall hilly nature of Hong Kong, flat land is created on slopes through the use of retaining walls and by filling. Reclamation has also had an important role in the development of Hong Kong since the late nineteenth century, with over 3000 ha being reclaimed to date. The majority of reclamation schemes are founded on marine clays and silty sands, and are formed with either precast concrete sea walls or armourstone berms, and backfilled with compacted decomposed rock and sand layers. This fill material often requires considerable time to compact and consolidate, four years being the norm. Large-scale building projects on reclaimed land usually require extensive pile foundations.
4. **Underground excavations**. Tunnelling in urban areas is complicated by the variable depth to the unweathered rockhead. Unweathered granites at depth provide the most suitable tunnelling material, as above the rockhead there are considerable complications in tunnelling through mixed terrains. Typical

problems include: (A) large water inflows which dewater the overlying soils, leading to settlement; (B) highly fractured rock needing considerable support; (C) weaker sediments needing support; (D) buried channels filled with colluvium and other unsuitable deposits; (E) pockets of deeply weathered rock located below the rockhead. All of these have been encountered in the construction of the underground mass-transit system.

Waste management

1. **Solid wastes**. Solid wastes are disposed of in several ways. Construction wastes are most commonly used as inert fill in reclamation projects. Domestic and industrial wastes are either incinerated, or dumped at seven large controlled waste sites. These are managed on the collect-and-contain principle, with controlled removal of leachate and venting of landfill gases. A classification of potential landfill operations has recently been carried out so that the main leachate migration pathways can be identified. Five main site types were recognised: (A) valley fill, over fresh rock or colluvium; (B) quarry fill; (C) land raise, tipping over reclamation fill; (D) marine fill, tipping over coastal sediments or rock; and (E) marine, dumping in the open sea. Each of these options is being actively considered.
2. **Waste waters**. Sewage is collected and disposed of separately from storm waters, and about 90% of the population has the use of public sewerage systems. All of the domestic sewage is disposed of into the sea by submarine sewage outfalls. Treatment of the sewage is by a screening process which is primarily aimed at the removal of large solid matter, and undergoes two processes: firstly, settlement; and secondly, biological processes aimed at promoting biochemical breakdown of organic substances. Large-scale disposal of sewage has led to the development of algal blooms or 'red tides', particularly where the material is disposed of into landlocked channels. Alternative strategies are currently under consideration.

Hazard mitigation

There are two important exogenic hazards which periodically affect Hong Kong: landsliding and typhoon damage. Landsliding is mostly caused through a combination of steep terrain and high rainfall together with the human activity of cutting and filling on steep slopes. The intensity of landslides has increased since the late nineteenth century in response to the expansion of Hong Kong into the mountain foothills. The majority of landslides are rain-induced, the average rainfall being 2225 mm, and most are induced by local short duration and high intensity rainfall which saturates the steep colluvium slopes. Measurement of rainfall over a 24 hour period has proven to be a reasonable indicator of the potential for landslides, as with over 100 mm for the period, there is increasing risk of failure (Figure 13.10).

Typhoons occur periodically and subject the shoreline of Hong Kong to erosion from the strong waves and tidal currents that they generate. During

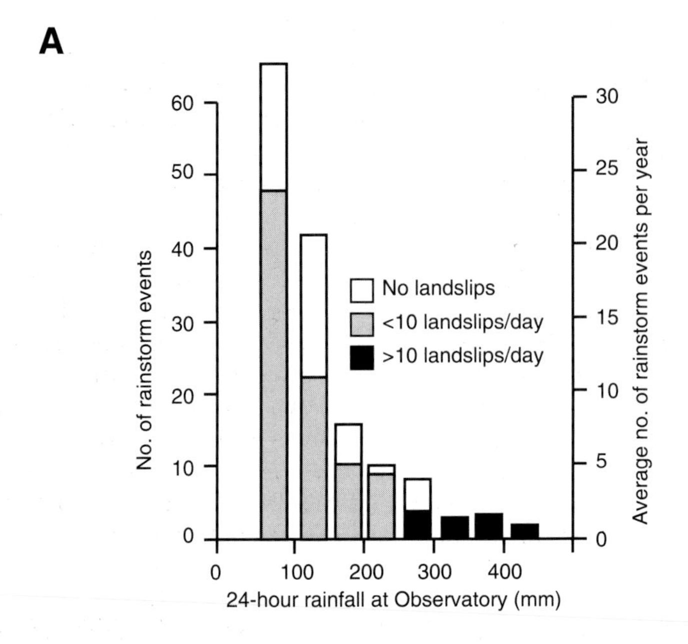

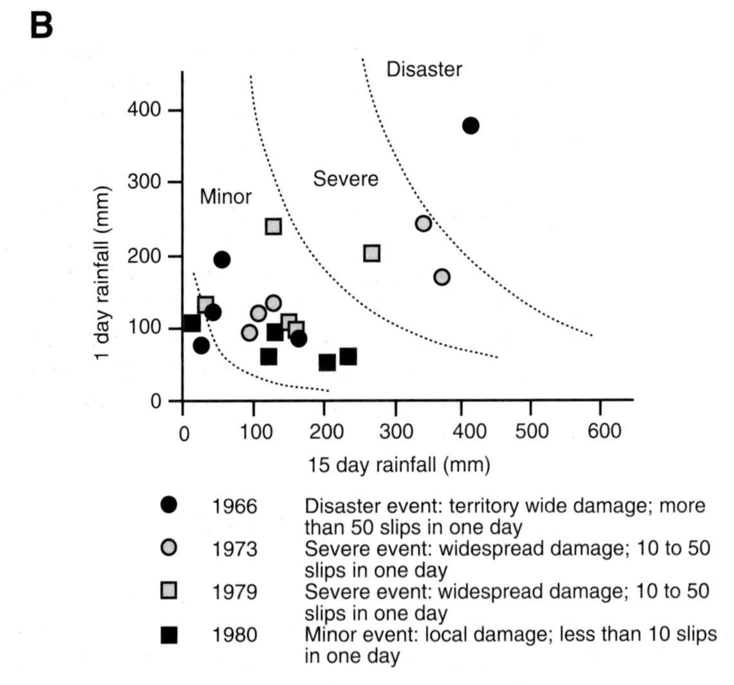

Figure 13.10 *Relationship of rainfall to mass movement events in Hong Kong.* ***A.*** *The increasing frequency of mass movement events associated with rainfall of more than 100 mm in a 24 hour period.* ***B.*** *The severity of mass movement events as a function of rainfall [Modified from McFeat-Smith* et al. *(1989)* Bulletin of the Association of Engineering Geologists, ***26****: A. Fig. 19, p. 60; B. Fig. 17, p. 58]*

typhoon conditions, over 25 000 boats may take refuge in typhoon shelters constructed from rubble mound breakwaters along the coast. All buildings constructed since 1959 have to be built to withstand intense tropical cyclones, with a requirement to withstand a horizontal ground acceleration of 0.07*g*.

The main endogenic hazard is from the seismic activity which is recorded in the area by the local observatory, although it has historically been at a low level. Only sporadic, small earthquakes have been recorded in Hong Kong, despite events of greater intensity being recorded in the nearby Chinese province of Guangdong. However, historical evidence suggests that over the last 1000 years there have been at least three earthquakes of magnitude 7 on the Richter Scale within a 500 km radius of Hong Kong. More recently, a magnitude 6 earthquake was recorded within a 250 km radius, and may have been associated with the loading of a fault during the filling of a new reservoir on the Xinfeng River in mainland China. In Hong Kong itself, all major construction works, with the exception of buildings, are now routinely designed to accommodate some seismic activity. What has yet to be addressed is the effect of a major tremor in triggering mass movement of saturated colluvium-mantled slopes.

13.5 ENVIRONMENTAL GEOLOGY: AN URBAN CONCEPT

The thesis of this chapter is that environmental geology and urban geoscience are effectively synonymous. In this book we define environmental geology as the interaction of humans with the physical environment. This interaction is at its most important where there is the greatest concentration of people: in towns and cities. The resource demands of these urban centres will, of course, determine the need for exploration, exploitation and management of natural resources in remote areas, but it is in the towns and cities where these resources are consumed, where most engineering problems are encountered, where the bulk of wastes are generated, and where the population is at its most vulnerable to natural hazards. The environmental geologist has an essential role in providing information clearly and concisely, and identifying the opportunities for and constraints to development; in many cases the safety and prosperity of our future urban world depend upon it.

13.6 SUGGESTED READING

The most influential source of information on geology and the urban environment is Legget (1973) which has influenced the majority of books on this subject that have followed in its wake. Several chapters on urban geology were included in the collected work by Tank (1973). An important new look at urban geology is provided in the book edited by McCall *et al.* (1996), with useful articles on the importance of geological information in planning provided by Marker (1996a, b). Mathewson (1982) examines some theoretical constraints on urban development, while Pyle (1995) discusses the impact of volcanic hazards in the urban environment. Aspects of geology and the planning process are discussed in the volumes edited by Bell *et al.* (1987), Culshaw *et al.* (1987) and Lumsden (1994). Case histories on urban geology are provided in the volume edited by Legget (1982), and in the papers by Boon Kong & Komoo (1990) and Fuchu *et al.* (1994). A series of urban case studies is excellently presented in the *Bulletin of the Association of Engineering Geologists* and *Environmental and Engineering Geoscience* 'Cities of the World' series of papers, of which Shata (1988), McFeat-Smith *et al.* (1989), Thomas (1989), Wyman *et al.* (1993) and Brown *et al.* (1995) are good examples. Browne *et al.* (1986) discuss the problems of mining subsidence in Glasgow, and Brose (1996) covers aspects of mine reclamation for urban development, while Gentile (1997) provides a very interesting summary of the utilisation of space beneath Kansas City. Finally, the volume edited by Bennett *et al.* (1996) demonstrates that an understanding of the geology of the built environment is of importance in an appreciation of wider geological issues.

Bell, F.G., Cripps, J.C., Culshaw, M.G. & O'Hara, M. 1987. Aspects of geology in planning. In: Culshaw, M.G., Bell, F.G., Cripps, J.C. & O'Hara, M. (Eds) *Planning and engineering geology*. Geological Society, London, 1–38.

Bennett, M.R., Doyle, P., Larwood, J.R. & Prosser, C.D. (Eds) 1996. *Geology on your doorstep: the role of urban geology in the earth heritage conservation*. Geological Society, London.

Boon Kong, T. & Komoo, I. 1990. Urban geology: case study of Kuala Lumpur, Malaysia. *Engineering Geology*, **28**, 71–94.

Brose, R.J. 1996. Subsurface mine reclamation for urban construction. *Environmental and Engineering Geoscience*, **2**, 73–83.

Brown, L.J., Beetham, R.D., Paterson, B.R. & Weeber, J.H. 1995. Geology of Christchurch, New Zealand. *Environmental and Engineering Geoscience*, **1**, 427–488.

Browne, M.A.E., McMillan, A.A. & Forsyth, I.A. 1986. Urban geology: Glasgow's hidden industrial heritage. *Geology Today*, **2**, 74–78.

Culshaw, M.G., Bell, F.G., Cripps, J.C. & O'Hara, M. (Eds) 1987. *Planning and engineering geology*. Geological Society, London.

Fuchu, D., Yuhai, L. & Sijing, W. 1994. Urban geology: a case study of Tongchuan city, Shaanxi Province, China. *Engineering Geology*, **38**, 165–175.

Gentile, R.J. 1997. Geology and utilization of underground space in metropolitan Kansas City area, USA. *Environmental Geology*, **29**, 11–16.

Legget, R.F. 1973. *Cities and geology*. McGraw-Hill, New York.

Legget, R.F. (Ed.) 1982. *Geology under cities*. Reviews in Engineering Geology, Volume 5, Geological Society of America.

Lumsden, G.I. (Ed.) 1994. *Geology and the environment in Western Europe*. Oxford University Press, Oxford.

Marker, B.R. 1996a. The role of the earth sciences in addressing urban resources and constraints. In: McCall, G.J.H., De Mulder, E.F.J. & Marker, B.R. (Eds) *Urban geoscience*. A.A. Balkema, Rotterdam, 163–179.

Marker, B.R. 1996b. Urban development: identifying opportunities and dealing with problems. In: McCall, G.J.H., De Mulder, E.F.J. & Marker, B.R. (Eds) *Urban geoscience*. A.A. Balkema, Rotterdam, 181–213.

Mathewson, C.C. 1982. Geology, foundation of the city. *Bulletin of the Association of Engineering Geologists*, **19**, 251–259.

McCall, G.J.H., De Mulder, E.F.J. & Marker, B.R. (Eds) 1996. *Urban geoscience*. A.A. Balkema, Rotterdam.

McFeat-Smith, I., Workman, D.R., Burnett, A.D. & Chau, E.P.Y. 1989. Geology of Hong Kong. *Bulletin of the Association of Engineering Geologists*, **26**, 23–107.

Pyle, D.M. 1995. Reduction of urban hazards. *Nature*, **378**, 134–135.
Shata, A.A. 1988. Geology of Cairo, Egypt. *Bulletin of the Association of Engineering Geologists*, **25**, 149–183.
Tank, R. (Ed.) 1973. *Focus on environmental geology*. Oxford University Press, New York.
Thomas, R.G. 1989. Geology of Rome, Italy. *Bulletin of the Association of Engineering Geologists*, **26**, 415–476.
Wyman, R.V., Karaouzian, M., Bax-Valentine, V., Slemmons, D.B., Peterson, L. & Palmer, S. 1993. Geology of Las Vegas, Nevada, United States of America. *Bulletin of the Association of Engineering Geologists*, **30**, 33–78.

Index

Aberfan (Wales) 74, 391
Abstraction of water 153–7
  rivers 153–4
  groundwater 154–7
Acetogenic phase 427
Acid immersion test 98
Acoustic logs 59
Activated sludge treatment 326
Active layer 246–7
Adit mining 64–6
Adobe 424
Affliction Glacier, British Columbia (Canada) 406, 409
Africa 137, 458–9
Aggregates 111–18
  roadstone 113–18
  railway ballast 118
Aggregate abrasion value (AAV) 114–18
Aggregate crushing value (ACV) 114, 117–18
Aggregate impact value (AIV) 114, 117–18
Agrostis stalonifera 76
Alaska (USA) 245, 247, 260–3, 379, 406, 414
Alaskan earthquake (1964) 427
Albert Memorial, London (England) 100–2
Alfa (Saudi Arabia) 272
Allingletscher (Switzerland) 411
Alluvial fans 265–6
Almeria (Spain) 288
Alpine mountains 212
Alston, N. Yorkshire (England) 360
Angle of internal friction 208
Antarctica 138, 317, 332–3, 335
Anthracite 40
Antiquities Act 1906 (USA) 186
Appalachian coalfields (USA) 66
Applied geomorphology 5
Applied research 10
Aqueducts 140
Aquiclude 134
Aquifer 133–4
  contamination 298, 336
Arctic Basin 245
Arctic Ocean 249
Area of Outstanding Natural Beauty (AONB) (UK) 174
Area strip mining 66–7, 286
Argentina 38, 126, 336, 373
Armenian earthquake (1988) 436
Armero (Columbia) 341, 443
Armourstone 105–11
Artesian 133–4
Asfordby (England) 74
Asia 212, 421, 458–9
Åspö (Sweden) 318
Assay 60
Atacama desert (S. America) 138
Atmospheric hazards 343
Atterberg limits 203, 206–207
Attock (India) 408
Audit 24, 62
Aufeis 248
Australia 138, 150, 168, 375, 377, 428
Avalanche hazard zones 23
Avalanches 411
Avon (England) 230
Ayers Rock (Australia) 168

Bahrain 270
Baltic coast 374
Bangladesh 334, 351, 372
Battersea Power Station, London (England) 232–3
Batura Glacier (Himalayas) 406
Bayer process 69
Beach nourishment 384–5, 389
Beachy Head, Sussex (England) 378
Bearing capacity 217
Beccium homblei 57
Belgium 129, 156, 168
Bell pit 64–5
Belvoir coalfield (England) 74
Benefication 73
Bexhill (England) 378
Big Bell gold deposit (Australia) 60
Big Thompson Canyon, Colorado (USA) 355
Bilbao (Spain) 196
Biodegradable waste 290
Biosphere Reserves 174
Birmingham (England) 285, 464
Bitumen 41
Bitumen & aggregate landfill liners 306
Bituminous coal 40
Black Butte Reservoir, California (USA) 362
Black smokers 46
Blasting 231, 235
Bleanau Ffestininog (Wales) 404
'Blue-sky' research 10
Boreholes 198–9
  light percussion drilling 199
  logs 201
  rock probing 201
  rotary coring 201
Box models 11
Brahmaputra River 334, 357
Brazil 391
Breakwaters 105, 387
Brickearth 123
Bricks 122–5
Bridges 261
Brighton, Sussex (England) 378
Britain 76, 105, 111, 118, 130, 136, 141, 145, 147, 151, 154–8, 162, 165, 179–80, 183, 187, 196, 198–9, 217, 228, 279, 281, 285, 293, 317–18, 320–1, 324, 328, 359, 372, 374, 375, 378, 382, 388, 390–2, 422, 460–2
British Geological Survey 155
British portable tester 115
'Brown land' 464
Bunds 305
Calabria (Italy) 374
Calcining 120
California (USA) 348, 362, 391, 394, 424, 428, 433, 436–7
Calliper logs 57
Callao (Peru) 138
Cambering 227, 230–1
Cameroon 445
Canada 151, 245, 247, 252, 406, 409, 451
Canadian Cordillera 412
Canary Islands 378, 442, 445, 448
Cannon's Brook, Essex (England) 353
Capillarity 95
Capillary fringe 267
Capital costs 29
Carbonate terrain (engineering geology) 217–19
  pinnacled rockhead 217–19
  sinkholes 217–19
  foundations 219
Carbon cycle 332
Carrara marble 100
Castleton, Derbyshire (England) 392
Çatak (Turkey) 391
Catch fences 81, 237, 403
Cathedral Glacier, British Columbia (Canada) 409
Caves 218
Cement 118–20
Cenozoic Ice Age 19, 50, 52, 228–9, 232–3
Chalk grassland 80
Cheshire Basin (England) 47, 64, 72
Cheyenne Mountain, Colorado (USA) 213
Chile 138
China 229, 247, 352, 369, 391, 477
Chong Kumdangle (Himalayas) 408
Chungar (Himalayas) 391
Clamp (bricks) 124
Clarifiers 326
Clay sealing (landfill) 306–7
Clays (engineering geology) 219–22
Clearwater Lake, Alaska (USA) 261
Client–consultant relationship 10
Clinical waste 281
Cluff Lake, Saskatchewan (Canada) 52
Coal 39
Coal rank 39
Coastal hazards 372–89
  erosion 374–7
  flooding 376–8
  mitigation/management 382–9
  risk assessment 378–82
  tsunamis 378–9

Coastal protection/defence 382–9
conservation 189, 190–2
hard engineering 384
soft engineering 191, 384
Coastal weathering zones 109
Co-disposal 298
Cohesion 208
Coliseum, Rome (Italy) 472, 475
Colliery waste 285–6
Colorado 186, 213, 353
Colorado River (USA) 177
Colour in mineral exploration 56
Columbia 391, 443
Columbia Glacier, Alaska (USA) 415
Concentrate and contain 299, 283
Conductivity surveys 200, 220
Cone penetration test 208
Confined aquifer 134
Conservation 161–93
areas 174
in Britain 179–80
rationale for 161–5
resource 165–72
management 172–89
voids 184, 187–8
Conservation management 182–9
integrity sites 182, 166
exposure sites 182, 170–1
threats 183–4
compromise 184–9
Conservation resource 165–72
landscapes/landforms 165–8
rock outcrops 168–9
urban landscapes 169–72
Conservation strategies 172–89
awareness 181
management 182–9
site assessment/selection 172–80
site enhancement 182
Consistency index 207
Consolidation, clays 221
Contamination 280–2
Contaminated land 282, 467
Copper production 30–1
Commercial reality/practice 8
Commodity price 28
Computer models 19
Concrete 120–2
Concrete failure 121–2
Constraint maps 14
Construction minerals (Geomaterials) 44, 87–127
Construction stone 87–111
dimension stone 88–100
decorative stone 100–5
armourstone & rip-rap 105–11
Contour strip mining 66–7
Controlled blasting 64, 236
Controlled waste 280
types 281
Conway Valley (Wales) 231
Cordillera Blanca (Peru) 412
Costa Rica 442
Cost-benefit analysis 346–7
Coulomb failure envelope 203–8
Countryside Commission for Wales 179
Craig Goch Dam (Wales) 151
Craigleath Quarry, Edinburgh (Scotland) 81
Crenulated bay model 191, 387
Crown hole 71
Crude oil 41
Crystalline terrain (engineering geology) 212–14
Crystallisation tests 97
Cudegong River (Australia) 150–1
Cumbria (England) 320
Cumsec 353
Cut-off drains 237, 403
Cut slopes 236

Dams 105, 144–6
arch dams 144
failure 145
gravity dams 144–5
Darcy's Law 132
Dartmoor (England) 103, 169
Deaf Smith County, Texas (USA) 320–3
Decorative stone 100–5
Deep cavern mining 64–6
Deformation till 222
Denmark 156–7
Dental masonry 81
Denver, Colorado (USA) 441
Deptford, London (England) 232
Derbyshire (England) 83
Derived maps 14
Derrick stone 107
Desalinisation 157
Deserts (engineering geology) 265–74
Desk survey 12
Desoto car 38
Devon (England) 302
Diesel 41
Dilatancy-diffusion & earthquake prediction 437–40
Dilute & disperse 282, 299
Dimension stone 88–100
Dinorwic (Wales) 68

Dinosaur National Monument, Colorado (USA) 186
Dogger Bank (North Sea) 19
Dolines 386
Domestic & commercial waste 288, 296
Dongziang (China) 391
Dorset (England) 168
Drainage 403
Drainage basin 131
Dressed stone 88
Drilling programme (mineral exploration) 57–61
Drystone walls 461
Dunes 386
Durham (England) 286

Earth Heritage Conservation 161–93
Earth Summit (1992) 7
Earthquakes 146, 348, 379, 394, 421–40
  anthropogenic 441
  hazard maps 431
  prediction/precursor events 437–40
Eastbourne (England) 378
East Durham coalfield (England) 78
Economic geology 5
Economics of minerals 29–30
Edinburgh (Scotland) 22, 81
Edmonton (Canada) 223–4
Effluent 281
Ekofisk oilfield (North Sea) 72
Eldjfell (Iceland) 444, 448
Electrical surveys 200, 220
Electromagnetic surveys 52, 55
Element maps 14
Endogenic hazards 343, 420–55
  radon gas 451–2
  seismic 421–40
  volcanic 440–51
Endogenic mineralisation 45–6
Engineering geology 5–6, 195–243, 244–76
  carbonate terrain 217–19
  crystalline terrain 212–14
  deserts 265–74
  maps 14, 209–11
  Hong Kong (China) 482–3
  Rome (Italy) 474–5
  site investigation 195–209
  soft rock terrain 214–22
  surficial sediments & landforms 222–31
Engineering soils 203
England 76, 79, 112, 132, 168, 175, 221, 227, 282, 302
English Nature 179, 190–2

Environmental impact 32
  flood control 366–9
  gas emissions/global warming 331–7
  groundwater abstraction 154–7
  landfill/raise 303–5
  mineral workings 70–84
  oil & gas extraction 72–3
  reservoirs/flow regulation 147–53
  river abstraction 153
  sea-level rise 336–7
  sewage disposal 329
  water transfers 157–8
Environmental geology 1, 4, 457
Environmental geology maps 14
Environmental management 2, 9
  proactive 7, 9
  reactive 7, 9
Environmental modelling 18–19
Environmental monitoring 18–20
  direct 18, 311
  indirect 18–20
Environmentally Sensitive Areas 174
Environmental Protection Act 1990 (UK) 179
Environmental science 1
Epilimnion 148
Equilibrium price 30
Equipotential surface 132
Ergodic reasoning 20–1
Erratic tracing (mineral exploration) 58
Eskers 51, 134–6, 289
Essex (England) 81, 190
Eton (England) 368
Eurasian 245
Europe 105, 156, 174, 222, 369, 374, 457–9, 462
Eustatic sea level 334
Evapotranspiration 130
Excavation methods 231–40
Exogenic mineralisation 46–7
Exogenic hazards 343, 351–419
  coastal 372–89
  fluvial 352–72
  glacial 405–15
  mass movement 389–405
Exploration models 50–2
Extreme geological environments 246
Exxon Valdez 414
Egypt 90, 112

Factor of safety 390
Fal Estuary, Cornwall (England) 79
Fault line scarp 432
Festuca rubra 76
Ffestiniog (Wales) 68, 223

Field mapping 12–14
Flash floods 265, 353
Float mapping (mineral exploration) 58
Floods 265–6, 348
  by-pass channel 368
  coastal 376–8
  embankments 367–8, 386
  frequency curves 359
  hydrograph 353–4
Flooring stone 88
Florissant Fossil Beds National Monument, Colorado (USA) 186
Flotation (mineral extraction) 68
Flow regulation, response 148–53
  channel modification 149–51
  water quality 148–53
Fluvial hazards 352–72
  bank erosion 355–7
  channel metamorphosis 355–7
  flooding 353–5
  mitigation/management 366–72
  risk assessment 357–66
  thresholds 365
Fly ash 111
Forest Park 174
Fossil Butte National Monument, Wyoming (USA) 186
Foundations
  carbonate terrain 219
  mining subsidence 73
  permafrost 252–7
France 156, 168, 286, 373–4, 466
Free market economy 29
Frost heave 247, 251
Frost wedges 228–9
Fuel oil 41

Gabions 236, 403–5
Ganges-Brahmaputra-Megha rivers/delta (Indian subcontinent) 334, 352, 372
Gangue 68
Gamma-ray logs 55, 57, 60
Garden stone 103, 106–7
Gas oil 41
Gasoline 41
Gemstones 37–8
Geobotany 57
Geochemical anomalies 57, 59
Geochemical exploration 57
Geographical Information Systems (GIS) 15
Geological Conservation Review (GCR) (UK) 175, 180
Geological hazards 5, 342
Geological maps 13–14
Geomaterials 44, 87–127
  aggregates 111–18
  cement & concrete 118–22
  construction stone 87–111
  glass sand 125–6
  gypsum 125
  structural clay 122–5
Geomembranes (geotextiles) 307
Geomechanics system (rock mass strength) 203–4
Geomorphological maps 15, 53
  morphogenetic 13
  morphological 13
Geophones 55
Geophysics 55
Geosphere 277
Geotechnical properties 203–8, 225
Geotope 172
Germany 34, 103, 156
Ghulkin Glacier (Himalayas) 406
Giant's Causeway (N. Ireland) 174
Glacial hazards 405–15
  mitigation/management 414–15
  risk assessment 414
Glacial landsystems 222
  subglacial 17, 222–5
  supraglacial 225–6
  valley 226
Glasgow (Scotland) 464
Global warming 249, 332
  sea-level rise 332–7
Goaf 64–5
Gossan 47
Grade 32
Graftham water (England) 141
Grand Canyon, Arizona (USA) 168, 177
Graphite 40
Gravestones 103
Gravity surveys 52
Greece 156
Greenhouse gas 331–3
Greenland 248, 332, 335
Green River, Utah (USA) 186
Grímsvötn (Iceland) 414
Grindelwald (Switzerland) 401
Grit chamber 325
Ground penetrating radar 199–200
Ground truthing 13
Groundwater 129–36, 130–1
  contamination 155
  management 135–6
  recharge 135–6, 154
  saltwater 267–70
Groynes 385
Guadeloupe (West Indies) 450

Guatemala City (Guatemala) 394
Gulf of Tarento (Italy) 374
Gulls 227
Gypsum 125

Hamford Water, Essex (England) 190
Hamilton City, California (USA) 362
Hamilton County, Ohio (USA) 394
Hanbury, Worcestershire (England) 285
Hanford Reservation, Washington (USA) 320–1
Hardwick Airfield, Norfolk (England) 300
Harlow, Essex (England) 353
Hastings, Sussex (England) 378
Hawaii 379, 448
Hawaiian type eruption 440
Hazard management 345–7
  coastal 382–9
  earthquakes/seismic 433–40
  fluvial 366–72
  generic responses 349
  glacial 414–15
  Hong Kong 483–5
  mass movement 401–5
  Rome 476
  volcanic 446–51
Hazard maps/zones 14–16, 346
  fluvial 370–2
  mass movement 396–401
  seismic 428–31, 434
  volcanic 445–8
Hazardous waste 281, 286
Helgoland (Germany) 374
Hells Canyon, Idaho (USA) 176
Himalayas 197–8, 212
Historical data 20, 22–3, 198
Holborn Hill, London (England) 173
Holland 126, 377
Homathko Valley, British Columbia (Canada) 413
Hong Kong (China) 137, 259, 213, 215, 290, 336, 383, 391, 457, 476–85
Hoo, Kent (England) 201
Hot springs 46
Huaraz, British Columbia (Canada) 413
Huascaran (Peru) 394, 407–8, 427
Hudson River, New York State (USA) 463
Humberside (England) 374
Hurleford (Scotland) 224
Hutton Roof, Cumbria (England) 166
Hydraulic cement 119
Hydraulic pick (pecker) 231, 234
Hydrocarbons 39–42
  production/extraction 63
Hydroelectricity 140–1
Hydrogeologist 134
Hydrogeology maps 13
Hydrological units 134
Hydrographs 353, 410
Hydrology/hydrogeology 129, 324
  systems 129–36
  landfill/raise 296–8, 303, 317
  reservoirs 142–4
  water supply 136–7
Hydrophones 55
Hydrothermal veins 46
Hydrothermal water 46
Hypolimnion 148

Ice avalanches 411
Icebergs 138, 414
Ice-cored moraine 412
Ice-dammed lakes 411
Iceland 129, 248, 408, 414, 442, 444–5
Icelandic type eruptions 440
Ice wedges 228–9
Icings 248
Idaho (USA) 176
India 372
Industrial minerals 45
  formation 45–8
Inert waste 281, 296
Istanbul (Turkey) 103
Institute of Hydrology (UK) 155
Iran 174
Irazu volcano (Costa Rica) 442
Ireland 39, 105, 156
Iron Mountain, Missouri (USA) 55
Isle of Portland, Dorset (England) 119
Israel 430, 434–5
Italy 100, 156–7, 222, 374, 391, 394, 408, 469–76

Japan 138, 379, 382, 388, 390–1, 405, 457
Järvokrog esker (Sweden) 289
Java 391, 448
Joint Nature Conservancy Council (UK) 179
Jökulhlaups 409–10

Kames 51, 134
Kame terraces 51
Kamijima (Japan) 391
Kansas City, Missouri (USA) 466
Kansu (China) 391, 394
Keppel Cove, Cumbria (England) 145
Karakorum Highway (Himalayas) 406, 408
Karst 134
Kelut (Java) 448

Kennicott Copper Mines, Alaska (USA) 406
Kent (England) 123
Kentish stock bricks 123
Kerogen 40
Kilauea (Hawaii) 448
Kilns (brick)
  chambered 124
  tunnelled 124
Kobayashi earthquake (Japan) 427
Kure (Japan) 391

Laguna San Rafael (Chile) 138
Lagunillas oilfield (Venezuela) 72
Lahars 443
Lake District (England) 76, 168
Lake Nostectucko, Britsh Columbia (Canada) 413
Lake Nyos (Cameroon) 445
Lakes (water supply) 139–53
Laki (Iceland) 445
Landes (France) 374
Land reclamation 482
Landscape perception 166–8, 176–8
Landscaping with waste 289
Landfill gas 290–8
  emissions/monitoring 311
  hazards 294, 302
  production 291, 295
Landfill/raise 80, 288–312
  area method 307
  cell method 308–10
  cells 306, 308–10
  containment types 298–301
  decomposition 291
  design 304
  development 313
  hydrology 296–7
  gas 290–8
  leachate 290–8
  operation 306–12
  site selection 301–5
  trench method 307
LANDSAT 272
Landsystems 15, 17, 197–8, 222–9
Landuse maps 13
Latino-Americana Tower, Mexico City (Mexico) 221
Lava flow 443–5
Leachate 290–8
  composition 292
  production 291, 295
Leaching 69
Leaning Tower of Pisa (Italy) 424
Lens (France) 286
Le Plata (Argentina) 38
Levees 368
Leats 141
Lignite 40
Limestone pavement 166
Limestone Pavement Orders (UK) 106
Linoplankton 153
Liquefaction 424
Liquid limit 203, 207–8
Liquidity index 203, 225
Little Ice Age 405, 411
Liverpool, Merseyside (England) 462
Llangollen (Wales) 462
Llyn Peninsula (Wales) 50
Loch Sloy (Scotland) 145
Lodgement till 17, 223, 300
Loess 123, 229, 394
Loma Prieta earthquake (1989; USA) 424
London (England) 81, 96, 100, 103, 123, 232–3, 283, 313, 353, 458, 462, 464
London Basin 154
London Clay 190–1, 221, 232–3
Long Island Sound, New York (USA) 463
Longwall mining 64–5
Longyearbyen (Svalbard) 254, 259
Los Angeles, California (USA) 394
Loscoe, Derbyshire (England) 294

Magnetic surveys 52, 55, 199–200, 220
Maldives 336
Malthusian paradigm 32–3
Managed waste 280
Manchester (England) 392, 462
Manhattan, New York (USA) 213, 216
Mantaro Valley (Peru) 391
Man Tor, Derbyshire (England) 392–3
Mar del Plata (Argentina) 373–4
Marine Nature Reserves (MNR) (UK) 174
Martinique (West Indies) 440
Masonry stone 88
Mass movement hazards 343, 348, 389–405
  causes 395
  Hong Kong (China) 484
  mitigation/management 401–5
  risk assessment 396–401
  types 397–8
Mass movement hazard maps 396–401
  inventory maps 396
  numerical scouring 399–400
  planning process 402
  qualitative slope classification 399–400
Material testing 202–9
Mattmark Dam (Switzerland) 411
Maud Glacier (New Zealand) 412–13

Mayonmarca (Peru) 391
Melbourne (Australia) 377
Mercer River, California (USA) 177
Merrivale Quarry, Dartmoor (England) 103
Metallic minerals 45
  formation 45–8
Methane clathrate 41, 249, 331
Methanogenic phase 291
Mexican earthquake 424–6
Mexico City (Mexico) 425–6
Microporosity 95
Midlands (England) 285
Middle East 32, 265–6, 424
Mine operation 77–9
Mineral colour (exploration) 56
Mineral conservation 33
Mineral exploration 48–62
Mineral extraction 62–9
  fluid removal 62–4
  impacts 70–84, 360–1
  mining 64–6
  quarrying 66–9
Mineral fuels 39–43
Mineral processing 68–9, 287
Mineral resources 37–86
  formation 37–48
  types 38
Mine restoration 79–84
  spoil tailings 75–7
  tips/dumps 285–7
Mine waste 73–7, 286–7
Mine water 79
Mining methods 62–9
Mining subsidence 70–3
  damage 71–2
  mitigation 73
Mississippi River (USA) 21, 463
Mississippi valley type mineral deposits 46
Modified Mercalli Intensity Scale 427–8
Mogi doughnut 437
Mohr's circles 208
Mono-dispersal 298
Montana (USA) 169
Moose-warmers 261
Mount Etna (Italy) 446–7, 449
Mount Fletcher (New Zealand) 409
Mount Pelée (St Martinique) 443
Mount St. Helens, Washington (USA) 445
Mount Vesuvius (Italy) 119
Moraines 51, 134, 412
Morphogenetic maps 13
Morphological maps 13
Mowhawk River, New York State (USA) 463

Naled 248
Nantle (Wales) 68
Naphtha 41
National Crushed Stone Association (NCSA) 115
National Flood Insurance Programme (USA) 370
National Geoscience Data Centre (UK) 198
National Monuments (USA) 186
National Nature Reserves (UK) 174
National Parks 162, 174–5, 186
National Parks & Access to the Countryside Act 1949 (UK) 175, 179
National Scenic Areas (UK) 174
Natural disasters 342
Natural gas 39–41
  cap-rock 40, 42
  maturation 40
  migration 41–2
  traps 41–2
Natural hazards 5, 342–455
  atmospheric 343
  endogenic 343, 421–55
  exogenic 343, 352–419
  frequency 342
  human 467
Neotectonics 432
Nepal 197, 391
Neutron logs 57, 60
Nevada (USA) 320
Nevadodel Ruiz volcano (Columbia) 443
Newhaven (England) 378
New Orleans (USA) 463
Newspaper records 22
New York (USA) 103, 213, 216, 463
New Zealand 353, 370–2, 406
Nimby 285
Nimtoo 285
Non-hazardous waste 281
NORAD, Colorado (USA) 213
Norfolk (England) 223, 300, 374
North America 165, 222, 245, 247, 262, 286, 428, 457, 459, 463
North Sea 32, 377–8, 380
Norway 140, 156, 248
Norwegian System (rock mass strength) 203, 205–6
Nottinghamshire coalfields (England) 78
Nuclear power 43
  fuel cycle 314
Nyirangongo (Zaire) 443

Ohio (USA) 394
Oil 39–41
 cap-rock 40, 42
 maturation 40
 migration 41–2
 traps 41–2
Oil & gas pipelines 245, 262–5
Okie Fenokie swamp (USA) 39
Ontario (Canada) 55
Open dumps 285–8
Operational costs 29
Orba Glacier (Italy) 408
Oreland, California (USA) 363
Ores 45
Organic Act 1916 (USA) 186
Organisation of Petroleum Exporting Countries (OPEC) 32
Outwash fans 51
Overbuden 66
Oxidation lagoons 327

Pakistan 406
Palace of Fine Arts, Mexico City (Mexico) 221
Pampas (Argentina) 168
Paris (France) 103, 125
Parametric terrain analysis 14, 399–401
Particulates 330
Pavement construction 88
Peak District (England) 83
'Pea-souper' 283
Peat 39–40
Pecker 231, 234
Peléan type eruption 442
Pennines (England) 76
Pennsylvania (USA) 394
Pennsylvanian coalfields (USA) 39
Periglacial environments 245
Periglacial landsystems 226–9
Permafrost engineering 245–65
 design options 253
 foundations 252–7
 oil & gas pipelines 262–5
 sanitation 257–60
 transport 260–2
Permeability 132
Phreatic zone 132
Peru 138, 391, 407, 412–13, 424, 440
Petrol 41
Petroleum 39–41
Philippines 379
Physiographic terrain analysis 15
Piezometric surface 132
Piles 254–6, 263
Pillar & stall working 64–5
Pingo 227, 232–3
Pinnacled rockhead 217
Placer 43, 47
Plaster 125
Plasticity index 203
Plastic limit 203, 206–7
Plate tectonics 421, 423
Playas 267
Plinian type eruption 442
Plymouth, Devon (England) 302
Point load test 202
Point Lonsdale, Melbourne (Australia) 375, 377
Polished stone value (PSV) 116, 118–19
Pollution 280, 282–3
Porosity 95
 capiliarity 95
 percentage 95
 primary 132
 saturation coefficient 95
 secondary 132
Port Cordova, Alaska (USA) 406
Portland Cement 119
Portland Stone 102, 105
Portugal 156
Porewater pressure 207
Postglacial sediments 229, 233
Potentially Damaging Operations (PDO) (UK) 175
Potential maps 14
Pozzolan 119, 472
Precious minerals/metals 45
 formation 45–8
Presplitting 236
Processed based response 7
Proton precession manetometer 220
Prudhoe Bay, Alaska (USA) 262–4
Pump storage schemes 140–1
Pyrenees 355
Pyroclastic falls 442
Pyroclastic flows 442

Quarry 67
 face stability 82
 flooding 81
 operation 77–9
 restoration 79–84
Quarrying 67
Quebec (Canada) 428
Queens, New York (USA) 216

Rabbit Lake, Saskatchewan (Canada) 52
Rodelquilar, Almeria (Spain) 287
Radioactive waste management 312–24
 disposal options 317

Radon gas 438, 451–2
Radstock, Avon (England) 230
Ramsar (Iran) 174
Ranrachirca (Peru) 391, 407
Recurrence intervals 344, 359, 365
Recyling 33, 283–4, 315
Render 118
Research 10
  applied 10
  'blue-sky' 10
Reserve 27–8
Reservoirs 139–53, 366
  costs of 146
  dam design 144–6
  environmental impacts of 147–53, 362–4
  hazards of 146
  hydrogeology of 142–4
  location/site 142
  life expectancy 146
  supply & demand 139
Resistivity logs 57
Resistivity surveys 200
Resource 27–9
Resource conservation 33
Resource management 27–35
  Hong Kong (China) 481–2
  Rome (Italy) 472–4
Restoration blasting 83–4
Revetments 386
Rhine-Ruhr region (Germany) 103
Ricardian paradigm 32–3
Richter Scale 427
Rio Aguas, Almeria (Spain) 288
Rio de Janeiro (Brazil) 391
Rio Santa (Peru) 407
Rio Shacsha (Peru) 407
Ripping 231
Rip-rap 105–11
Risk assessment 345–7
  coastal hazards 378–82
  fluvial hazards 357–66
  glacial hazards 414
  mass movement hazards 396–401
  seismic hazards 427–32
  volcanic hazards 445–6
Risk maps 14
River Axe (England) 358–9
River Bollin-Dean (England) 359, 365
River Conway (Wales) 355
River Cound (England) 359
River Crawford (N. Ireland) 359
River Dnieper (Ukraine) 139
River Elan (Wales) 151
River Exe (England) 359, 460
River Ganges (India) 334
River Indus (Pakistan) 408
River Meghna (Bangladesh) 334
River Nent (England) 361
River Rede (England) 151
River Severn (England) 359
River Svratka (Czech Republic) 152
River Tame (England) 158
River Tees (England) 152, 460
River Thames (England) 233, 368
River Tiber (Italy) 471
River Torrens (Australia) 357
River Waihou (New Zealand) 314
Roadheader 239
Road stones 113–18
Rock anchors 237, 240, 403
Rock bolts 237, 240, 403
Rock excavation 231–40
Rockhead 202, 213, 217, 220, 223, 239
Rock mass strength 202–6
  excavation 235
  Geomechanics System 203–6
  Norwegian System 203–6
  tunnelling 240
Rock probing 201
Rock quality designation (RQD) 203
Rock slope stabilisation 237–8
Rocky Mountain Arsenal, Colorado (USA) 441
Rocky Mountains (N. America) 186
Romagna (Italy) 374
Rome (Italy) 103, 457, 466, 469–76
Roofing materials 88
Rotary coring 201
Rotary milling head 239
Royal Society for the Protection of Birds (RSPB) (UK) 162
Rutland Water (England) 141
Russia 247
Rye, Sussex (England) 378

Saas Valley (Switzerland) 411
Sabkhas 267
Sachs Harbour, North West Territories (Canada) 252
Sacramento River, California (USA) 362
Saline groundwater 155
Salt efflorescence 267
Salthill Quarry, Yorkshire (England) 81
Salt water intrusion 155
Salt weathering 91, 93
San Andreas Fault, California (USA) 433, 437
San Francisco, California (USA) 362, 394, 421, 463
San Jose, California (USA) 394

San Salvador (El Salvador) 421
Sand & gravel exploration 51, 53–4
Sanitary landfill 298
Santa Clara Valley, California (USA) 156
Saturation coefficient 95
Scaling 401
Schmidt hammer 202
Scotland 39, 58, 141, 145, 175, 464
Scottish Ice Sheet 59
Screed 118
Sculptural stone 100
Sea Birds Protection Act 1869 (UK) 179
Seattle, Washington (USA) 394
Sea-level rise 332–7
  consequences 336–7
Sea walls 384, 386, 388
Secondary currents (rivers) 357–8
Seine Maritime (France) 374
Seismic gap 437
Seismic hazards 421–40
  mitigation/management 433–40
  risk assessment 427–33
Seismic surveys 52, 55, 200, 220
Seismotectonic analysis 433
Sellafield, Cumbria (England) 320–3
Sevenoaks, Kent (England) 227
Sewage effluent 281
  treatment 324–8
Siberia 248
Sieve maps 15
Sinkholes 199, 217–20
Site investigation 195–209
  geophysics 200
  material testing 202–9
  reconnaissance/regional scale 196–7
  reports/maps 209
  site documentation 197
Site of Special Scientific Interest (SSSI) (UK) 106
Skaftafell (Iceland) 414
Skeirdarajökull (Iceland) 414
Slate 68
Shale oil 41
Shear strength 207–8
Shear stress 208
Sheffield, W. Yorkshire (England) 392
Shotcrete 237, 403
Shropshire Union Canal (Britain) 462
Shyok River (Himalayas) 408
Snake River, Idaho (USA) 176–7
Slope stabilisation 236–39, 401–5
Snow avalanches 411
Snowdonia (Wales) 164
Snowfall 22
Soft rock terrain (engineering geology) 214, 222
  carbonate 214–19
  clays 219–22
Soil classification 207
Soil testing 203
Somme (France) 168, 374
Sonic logs 59
Source rocks (oil & gas) 40
South America 424
Southern Alps (New Zealand) 406
Space-time transformation (ergodic reasoning) 20–1
Spain 156, 196
Special Protected Areas 174
Spitsbergen (Svalbard) 29, 248, 254
Spoil 69, 73–7
Spread footings 257
Staffordshire (England) 83
Standard penetration test (SPT) 207–8, 224
St. Kilda (Scotland) 174
St. Lawrence (North America) 428
Stockholm (Sweden) 135, 289
Stone cladding 88
Stonehenge (England) 90
Stone restoration 99
Stony Creek, California (USA) 362
Stope 66
Storegga slide (North Sea) 380
Storm surges 377–8
St. Paul's Cathedral, London (England) 96
Strike-slip fault 432
Strombelian type eruption 440
Structural response 7
Subglacial landsystem 17
Subglacial meltout till 223
Subsidence 70
  bowl 70–1
  groundwater 154–7
  mining 70–3
  mitigation/mining 73
  oil & gas 72–3, 380–1
  wave 72
Subtropical deserts 265
  engineering 265–74
  salt weathering 267–70
  sand/dust drift 268–74
Suez (Egypt) 266
Summit Lake, British Columbia (Canada) 419, 411
Supply & demand 29–30, 139, 143
Supraglacial meltout till 223
Surge-type glacier 406
Sussex (England) 378

Sweden 135, 156, 289
Swiss Alps 23, 411
Switzerland 156, 401, 411

Tailings 69, 73–7
Talik 247
Tangshen (China) 421
Te Aroha (New Zealand) 371
Tenerife (Canary Islands) 442, 448
Terrain analysis 14–18, 197–8
  parametric 14
  physiographic 15
Teton National Park, Wyoming (USA) 177
Texas (USA) 320
Thames Barrier (England) 384, 388
Thames Estuary (England) 330, 388
Thematic maps 14, 23
Thermocline 148
Thermokarst 247, 249, 252
Thurrock, Essex (England) 81
Till 51, 57, 135, 222
Titanic 414
Toe protection 403
Tokyo (Japan) 391, 421
Tools in environmental geology 12–24
  data presentation 20–4
  desk surveys 12
  environmental monitoring 18–20
  field mapping 12–14
  terrain analysis 14–18
Total organic carbon content (TOC) 39, 292
Toxic waste 281
Trade effluent 281
Tragedy Canyon, British Columbia (Canada) 413
Trans-Alaskan pipeline 262–4, 415
Triaxial test 207–8
Trickling filter 325
Trimmingham, Norfolk (England) 223
Tsunami 348, 378–81, 389–90, 440
Tunnelling 213, 215–16, 239–40
Turbines 140
Turkey 103, 391
Ty-nant (Wales) 237

Unconfined compressive strength test 202
United Nations Earth Summit (1992) 7
Unit hydrograph 353
Uppsala (Sweden) 135–6
Uranium 43, 50, 52, 314
Urban environment 457–85
Urban machine 3, 466
Urban planning 466–9
Urban population 459
USA 39, 104, 111, 113, 115, 126, 129, 146, 154, 156–7, 168–9, 176, 222, 284, 290, 318, 320–1, 353, 355, 370, 378, 390–1, 427–9, 451, 462
Utilidors 259

Vadose zone 132
Valdez, Alaska (USA) 262, 414
Valont (Italy) 391
Valparaiso (Peru) 138
Vancouver (Canada) 413
Vatican, Rome (Italy) 471
Vatnäjökull (Iceland) 414, 408
Vernacular architecture 460–2
Vestmannaeyjar (Iceland) 442–4
Vesuvian type eruption 442
Victoria Falls (Africa) 168
Virginia (USA) 391
Volcanic hazards 348, 440–51
  lava flows 443–5
  pyroclastic falls 442
  maps 446–8
  mitigation/management 446–51
  pyroclastic flows 442–3
  risk assessment 445–6
Vulcanian type eruption 440
Vulnerability maps 14

Wales 50, 68, 74, 76, 141, 175, 234, 237, 286, 355, 404, 460
Walton-on-the-Naze, Essex (England) 190
War graves 104
Washington State (USA) 320, 394
Waste 279–81
  controlled 280
  costs 284
  disposal options 278
  dumps 285–8
  effluent 324–30
  gas & particulates 330–7
  landfill/raise 288–312
  managed 280
  minimisation 283–4
  radioactive 312–24
  reclamation saving 284
  recycled 383–4, 315
  types 281
  unmanaged 281
Waste management 277–339
  concentrate & contain 283, 299
  dilute & disperse 282, 299
  Hong Kong 483
  Rome 475–6
Waste Reclamation Always Pays (WRAPS) 284

Water injection (waste) 441
Water resources 129–59
  distribution/transfer 137, 157
  resource acquisition 129
  treatment 157–8
Watershed 130
Water shortages 130
Water supply 136–9
  Hong Kong (China) 482
  periglacial regions 257–60
  Rome (Italy) 474–5
Water table 133–4, 144
Waterloo (Belgium) 168
Watts Branch (USA) 359
West Indies 440, 450
Westminster, London (England) 232
Weathering
  chemical 93
  coastal zones 109
  engineering problems 213–16
  frost 93
  mineral formation 47
  periglacial 227
  salt crystal growth 91–6, 267–70
  thermal 93
  tills 224–5
  wetting & drying 93
Wheal Jane, Cornwall (England) 79
Wild brining 64, 72
Wildlife & Countryside Act 1981 (UK) 106
Wilmington Oilfields, California (USA) 72–3
Windermere Dam (Australia) 150–1
Windsor, London (England) 368
Wireline (well) logs 57–61
Wisloka (Poland) 359
Wolverhampton (England) 229
Woolwich, London (England) 388
Worcestershire (England) 285
World Heritage Site 174
World Trade Centre, New York (USA) 214
Wyoming (USA) 186

Yaciton (Peru) 391
Yucca Mountain, Nevada (USA) 320–1
Yellowstone River, Wyoming (USA) 177
Yellow River (China) 369
Yellowstone National Park, Wyoming (USA) 177
Yokohama (Japan) 421
Yosemite National Park, California (USA) 177
Yorkshire (England) 91, 282, 360, 374
Yungay (Peru) 391, 407–8

Zafferna (Italy) 447, 449
Zaire 443
Zambian Copper Belt 57